MW00845905

Lamb's

Questions and Answers on the Marine Diesel Engine

Eighth Edition

Stanley G Christensen

B.Sc., C. Eng., F.I. Mar. E., Extra 1st Class Engineer,
M.O.T., Silver Medallist Institute of Marine Engineers and
William Nevins Prize Winner; Member, Society of Naval
Architects and Marine Engineers, USA

Formerly Professor, Dept. of Engineering, US Merchant
Marine Academy, Kings Point, New York, USA

ELSEVIER
BUTTERWORTH
HEINEMANN

AMSTERDAM · BOSTON · HEIDELBERG · LONDON · NEW YORK · OXFORD
PARIS · SAN DIEGO · SAN FRANCISCO · SINGAPORE · SYDNEY · TOKYO

Elsevier Butterworth-Heinemann
Linacre House, Jordan Hill, Oxford OX2 8DP
200 Wheelers Road, Burlington, MA 01803

First published 1921 by Arnold
Sixth edition 1950
Seventh edition 1978
Eighth edition 1990
Reprinted 1999
Reprinted 2001, 2004
Transferred to digital printing 2004

British Library Cataloguing in Publication Data
A catalogue record for this book is available from the British Library

Library of Congress Cataloguing in Publication Data
A catalogue record for this book is available from the Library of Congress

ISBN 0 852 64307 1

For information on all Elsevier Butterworth-Heinemann
publications visit our website at www.bh.com

Typeset by Colset Private Ltd, Singapore

PREFACE

The late John Lamb wrote his first book *The Running and Maintenance of the Marine Diesel Engine* during 1919. The first edition was published by Charles Griffin and Co. Ltd in 1920.

Readers of *The Running and Maintenance of the Marine Diesel Engine* then gave many expressions of thanks to the author and made interesting enquiries regarding diesel engines.

Following these expressions of thanks for his earlier book and the interesting enquiries a need was recognised for a second book. A first edition of this book was then created in the form of a categorized series of questions and answers and was published in 1922.

In the preface to the first edition of his second book John Lamb wrote 'The Question and Answer method seemed most serviceable for the purpose as giving at once essential teaching and enabling the student to express his knowledge'. The need for a person to express himself or herself is as valid to day as when John Lamb wrote these words so many years ago. He also said at this time that the book made no claim to completeness.

Over the many years the *Questions and Answers* book has been in publication it has been used by apprentices and students, seagoing engineer officers, and shore-based technical staff. It has found use both as a book for study and for reference.

The late A.C. Hardy wrote of John Lamb in the *History of Motorshipping* with the words:

---

* The late Cornelius Zulver was technical head of the Royal Dutch Shell fleet of tankers. He was also an early innovative pioneer in the use of the diesel engine in marine propulsion and introduced the under-piston method of pressure-charging in conjunction with Werkspoor of Amsterdam during 1929. This simple method of pressure-charging was used in four stroke cycle crosshead engines up until their demise in the years following World War II.

Although only these two names are mentioned it must be remembered there were many others who pioneered this most efficient means of propelling ships and brought it to the perfection it enjoys today.

'Best known of all to technical people is, of course, John Lamb of boiler-oil-for-diesels fame. A quiet-speaking 'Geordie' with a rich practical experience of motorships and a wide diesel engine knowledge, for many years he was Zulver's* right hand man and during this period he produced two noteworthy books on marine diesel engineering which are as popular today in their up to date form as ever they were'.

What A.C. Hardy wrote in 1955 is still true today. It has been the aim of the present author to maintain a precept of John Lamb and the publishers, that is to keep this book fully up to date. Again no claim is made to completeness in its content, it may however be claimed that it is very complete as a guide for those wishing to make an in-depth study of the diesel engine irrespective of where it is used.

It must be remembered that diesel engines drive the largest and fastest of ships, the largest and smallest of tug boats, fish factories and fishing craft, the largest and smallest pleasure craft, the largest trucks and lorries, the smallest passenger vehicles, perform stand-by duty in hospitals and factories for the supply of electrical power under emergency conditions, and in most nuclear power plants world wide the diesel engine is there ready to keep a reactor cool and safe under the worst emergency conditions.

The name of John Lamb was incorporated into the present title to perpetuate the name of one of the early pioneers of the diesel engine.

Stanley G. Christensen
1989

INTRODUCTION

Many things have happened within the field of the marine diesel engine since the last edition of this book was published. Every endeavour has been made to cover these changes in as full a manner as possible within the constraint of a book that must of necessity cover all aspects of marine diesel machinery.

For a start one of the world's most well known passenger liners the 'Queen Elizabeth II' became a motorship. The change over to diesel machinery was made to keep the ship competitive and obtain the many advantages of diesel electric propulsion. Diesel electric propulsion gives a wider range of economic speeds with a much lower fuel cost than may be obtained by a conventional geared steam turbine or turbo electric drive. Another advantage of a multi engined diesel electric propulsion system is the ease with which the survey of the propulsion machinery may be carried out within the short turn round time requirement necessary for profitable ship operation. The normal turn round time for a passenger liner or cruise ship is now only a matter of a few hours instead of days. To survey an engine, disassembly may take place during the passage, an examination by a surveyor takes place while in port. The assembly of the engine is carried out after the ship is back to sea. A surveyor then sees the engine under operating conditions when next in port, that completes the survey on the engine. There is nothing new in diesel electric propulsion systems. The history of marine engineering shows many fine examples of diesel electric propulsion involving passenger liners, refrigerated fruit carriers, the largest of dredgers, fish factory ships, trawlers, etc.

The largest part of the operating or voyage costs in many ships is shown in the fuel cost when expressed as a percentage of the total operating cost. Using figures that may only be considered as approximate for many classes of ship, fuel cost has risen from something around 10% before the fuel crisis in 1973 to over 50% of the total operating cost a few years later.

This began the demand for engines with the lowest specific fuel consumption. The modern high efficiency turbocharger has enabled the designer of uniflow scavenged engines to expand the combustion gases further down the piston stroke and so increase the thermal efficiency of the engine. This has resulted in a much lower specific fuel consumption that could ever be obtained with loop scavenged engines where the necessity for efficient scavenging limits the ratio of the bore and piston stroke.

Today, all slow speed engines are uniflow scavenged two stroke engines and follow the same uniflow scavenge principle adopted more than fifty years ago by Burmeister and Wain, when they placed single and double acting two stroke cycle engines on the market. Uniflow scavenging goes back to the days of the large industrial engines built in the last century. These engines used producer gas or the gas supplied in towns for illumination purposes before electricity was available.

During the span of fifty years between the thirties and the early eighties diesel engine builders have proclaimed the merits of their respective engines; their followers were generally divided into two camps. Those who favoured the cross scavenged engine or the loop scavenged engine because the absence of exhaust valves made for simplicity, and, those who favoured the lower fuel consumption of uniflow scavenged engines in spite of their added complication. The complication being the exhaust valves irrespective of their types or the extra bearings in an opposed piston engine where the piston acts as an exhaust valve.

When the major builders of cross scavenged engines changed over to uniflow scavenging in the eighties they had to tell the world of the volte face they were making.

Burmeister and Wain had reached a position of preeminence with their four stroke cycle single and double acting engines in the late twenties. At this time they also recognised that two stroke cycle engines could be designed to develop considerably more power in the same space as their four stroke cycle engine. Out of this their two stroke cycle double acting and single acting engines came into being. The uniflow scavenging method was chosen for this range of engines. A Dr H.H. Blache, the leader of the design team, had the difficult public relations task of telling the world of the change from cross scavenging to uniflow scavenging.

Now all slow speed diesel engines are very similar.

It is pertinent to remark that over fifty years ago the first specially designed hydraulic spanners and wrenches were supplied with the new range of Burmeister and Wain two stroke engines. They were used to precisely control the tightening of threaded fastenings with the correct degree of tension. This prevented the failure of parts subjected to alternating or fluctuating stresses.

The increasing use of finite element analysis has made it possible to correctly ascertain the magnitude of stresses in both fixed and moving engine parts. This has led to reductions in the dimensions of engine parts without in any way impairing reliability. It may also be said that finite element analysis has increased reliability to a great extent. Large savings in weight and material cost has then been made possible. Finite element analysis has also been used in the study of heat transfer and this has also improved design.

The quality of residual fuel and some refined products supplied today has deteriorated since the last edition was published. The deterioration has come about from advances in oil refining techniques. These advances made it possible to increase the percentage yield of the more valuable oil products from crude oil during the refining processes. The concentration of impurities has then increased as the amount of residue has been reduced.

Centrifuge manufacturers have responded well to the challenge of dealing with these low quality fuels. The fuel purification equipment available today is capable of handling low quality very high specific gravity fuels in an efficient manner. The separator, as we knew it before, now has no place in the treatment of low quality blended fuel oil. It has been replaced by the self cleaning clarifier with sophisticated surveillance and control equipment that only operates the cleaning process as and when required. The frequency of the cleaning process then depends on the amount of waste matter in the fuel.

Lubricating oil characteristics are still being improved with better and more powerful additives. The quality and characteristics of lubricating oils have advanced beyond all belief from the early days of burning heavy fuel oil in the late forties and early fifties. Then, additives giving alkalinity and detergency were compounded in an oil emulsion. Additives are now held in solution and do not separate while in storage. When additives were held in suspension or in an emulsion separation sometimes occurred when the lubricants were held in storage on the ship. It can be said the work of the lubricants chemist has played an enormous part in the commercial success of burning heavy low quality fuel in the marine diesel engine.

Increases in injection pressure together with other advances have made it possible for medium speed engines to use fuels with lower cetane numbers than could formerly be considered. Some of these fuels have such poor ignition qualities that new methods of comparing the ignition quality of fuel have had to be devised. New generations of medium speed engines are being designed specifically with the use of these low quality fuels in mind.

Speed control governors of the electronic type are now being used increasingly and electronic control of the fuel injection process is being used more and more as time progresses.

The seventh edition of this book was brought right up to date. No material has been deleted in this new edition with the exception of a question on engine scavenging, some questions and answers have been replaced in order to update them and comply with the latest practice. The major part of the matter keeping the book up to date is in the form of additional material. The first page of the book has remained the same and so belies the changes and additional material in this new edition.

Drawings and sketches have now been included with the text to help students and others to better understand and clarify many of the answers.

ACKNOWLEDGEMENTS

The author acknowledges and affirms his thanks to the following mentioned companies for their assistance in supplying drawings used in the production of this book.

Brown Boveri Corporation, Baden, Switzerland

Lucas Bryce Limited, Gloucester, England

MAN-Burmeister and Wain, Copenhagen, Denmark and Augsburg, FDR

Sulzer Brothers Limited, Winterthur, Switzerland

The author also thanks the following for their friendship and the assistance so freely given in supplying material and data enabling this book to be kept fully up to date.

Mr Denis Eley
Mr Ernst P. Jung
Palle B. Jørgensen
Mr Gerald Losch
Mr Tom Moore
Mr Poul R. Nielsen
Mr Hans Roefler
Mr Claus Windelev.

CONTENTS

1

HEAT AND ENGINEERING SCIENCE

■ 1.1 Give a definition of the term 'matter' and state the constituents of which it is composed. Show how matter exists in its various states.

In technology, matter is sometimes referred to as material substance. It can be defined as anything known to exist and occupy space. Any material substance consists of minute particles known as molecules; these are the smallest particles of a substance which can exist and maintain all the properties of the original substance. A molecule is made up of a combination of two or more atoms of the elements. The atom consists of various parts which are held together by forces, recognized as being electrical in character. The forces of attraction come about from unlike electrical charges. The constituent parts of an atom are the central core or nucleus which has a positive charge and one or more electrons. The electron has a negative charge. The nucleus is composed of protons and neutrons (except the atom of hydrogen). Protons have positive electrical charges and the neutrons are electrically neutral. When an atom is electrically neutral it will have the same number of protons and electrons. The number of electrons contained in an atom is shown by the atomic number of the element. An atom becomes an ion when the number of electrons is more or less than the number of protons. The ion will be positive or negative according to the predominant electrical charge.

The atom of hydrogen is the simplest; it consists of one proton and one electron. If the electron is removed from the atom of hydrogen the remaining proton becomes a hydrogen ion which will be positive. In some cases two atoms of the same element may differ in the number of neutrons contained in the nucleus. The atomic weights will therefore be different and the atoms are described as being isotopes of the element. The isotopes of an element have identical chemical properties but differing physical properties. The electrons outside the nucleus control the properties of the atom, and the protons and neutrons in the nucleus determine its atomic weight. The electrons are considered to form a series of orbital envelopes or cases around the nucleus, each envelope containing a set pattern of electrons. Other particles exist but need not concern us in this study.

If a molecule can exist as one atom of an element it is referred to as being monatomic; if two atoms joined together form a molecule they are referred to as diatomic; three atoms joined to form a molecule are triatomic. A substance formed from identical atoms is known as an element. Substances formed from molecules of different atoms are called chemical compounds. If the molecules retain their identity and do not join together, the substance formed is a chemical mixture.

Material substances usually exist in one of three physical states: solid, liquid, or gaseous. Substances existing in the solid state have the ability to retain their original shape unless acted upon by externally applied forces. The ability to retain their shape is due to the relatively large cohesive forces existing between the particles forming the substance. Solids have three elastic moduli (Question 1.9). When a substance is in the liquid state the cohesive force between the molecules is much less than in solids, the molecules having the ability to move with respect to one another. Liquids possess a property which allows them to take up the shape of their container, and they offer a large resistance to change of volume when subjected to pressures. Liquids have two of the elastic moduli: bulk modulus and modulus of tension. The ability of a liquid to resist small amounts of tension is due to the cohesive force existing between the molecules. This cohesive force creates an apparent shear resistance between two adjacent planes in a liquid body and gives rise to the phenomenon which we know as viscosity. With substances in the gaseous state the cohesive force between the molecules is so small that they are almost completely free to move in any direction. This property allows gases completely to fill their container. Gases have only a bulk modulus. Some substances do not have all the properties of solids. They exist in an indeterminate state between that of a solid and a liquid. These substances are said to be amorphous; on heating they approach nearer the liquid state but do not have a precise melting point.

■ **1.2   What do you understand by the following terms: mass, force, velocity, acceleration, work, energy, potential energy, kinetic energy, momentum, power? Give units, derived units and symbols where applicable.**

Mass is the quantity of matter in a body. The unit of mass is the kilogram (kg).

Force is that which tends to cause a body at rest or in motion to alter its state of rest or motion. The unit of force is the newton (N).

$$1 N = 1 \text{ kg m/s}^2$$

Velocity is the rate of change of position of a point or body; it may be linear or angular. Linear velocity is measured in metres per second (m/s). Angular velocity is measured in radians per second (rad/s).

Acceleration is the rate of change of velocity; it may be linear or angular. Linear acceleration is measured in metres per second squared (m/s^2). Angular acceleration is measured in radians per second squared (rad/s^2).

Work is done when a body is moved against a resistance. It is the product of the force required to overcome the resistance and the distance through which the body is moved. The unit of work is the joule (J). One joule equals one newton metre.

$$1J = 1N\,m$$

Energy is the capacity for doing work. The various forms of energy are kinetic energy, potential energy, heat energy, electrical energy, chemical energy, nuclear energy and radiant energy. All forms of energy except radiant energy can exist only in the presence of matter. The unit of energy used in mechanical and marine engineering is the same as the unit of work.

Potential energy is the energy contained in a body by virtue of its position. Its value is measured by the work that could be done by the body in passing from one defined position to another defined position.

Kinetic energy is the energy contained in a body by virtue of its velocity. Its value is measured by the work done by the body during some change in velocity.

Momentum is the product of the mass and velocity of a moving body.

Power is the rate of doing work; the unit of power is the watt (W). One watt is equal to one joule per second.

$$1W = 1J/s$$
$$1hp\ (metric) = 735.5W$$
$$1hp\ (British;\ U.S.) = 746W$$

■ 1.3 Define the following terms: moment of force, couple, torque, twisting moment.

Moment of force is a measure of the tendency of a force to turn a body on which it acts about some axis. Its value is the product of the force and the perpendicular distance from its line of action to the axis.

Couple. A couple consists of two equal forces acting in opposite directions along separate parallel lines of action. The moment of a couple is the product of one of the forces and the perpendicular distance between the lines of action.

Torque and *twisting moment* are synonymous with the moment of a couple.

■ 1.4 What are centripetal and centrifugal forces?

Centripetal force. When a body moves in a circular path it can be shown that it has an acceleration acting towards the axis of its orbit. The force required to produce this acceleration is called centripetal force.

Centrifugal force is the reaction to centripetal force. Centrifugal force acts radially outwards; centripetal force acts radially inwards and constrains the body to move in a circular path.

■   **1.5   Give definitions of inertia, moment of inertia, and radius of gyration.**

Inertia is that property of a body which resists changes in its state of rest or uniform motion in a straight line.

Moment of inertia of a rotating body is the sum of the products of each particle of mass and the square of its distance from the axis of the rotating body.

$$\text{Moment of inertia} = m_1r_1^2 + m_2r_2^2 + \ldots$$

where $m_1 + m_2 + m_3 + \ldots = $ total mass of body.

Note   The term moment of inertia can have various definitions depending on its use and application.

Radius of gyration is the radius at which the whole mass of a rotating body may be considered as acting. If k is the radius of gyration, then

$mk^2 = $ moment of inertia
where $m = $ total mass of body.

■   **1.6   What are stress, strain, unital stress, unital strain?**

Stress may be defined as the load that is applied externally to a body, or in effect as the force acting between the molecules caused by the deformation or strain.

Strain is the change that occurs in the shape or dimension of a body subject to the action of stress.

Unital stress is the stress acting on unit area of material.

Load/area resisting load = unital stress

Unital strain is the ratio of the change in dimension to the original dimension of the body before stress was applied.

Note   When the terms stress and strain are used in the following text the single word stress will imply unital stress and the single word strain will imply unital strain. Should it be required to distinguish between the terms they will be written in full.

Load/area resisting load = stress
Change in dimension/original dimension = strain

■   **1.7   Some materials are referred to as elastic. What does this imply? What is an isotropic material?**

Most materials in a solid state when subject to stress, experience a change in shape. If, when removing the stress, the material returns to its former shape the material is said to be *elastic*. In studies of strength of materials it is often assumed that a body is isotropic. An *isotropic material* is one which has identical properties in all directions from any point within the body.

Note   In practice most metals used in engine construction are non-isotropic due to the grain structure which exists within the metal.

■ 1.8 What are tensile stress, compressive stress, shear stress?

Tensile stress. A body is subject to tensile stress when it is acted on by a load which causes an increase in its length.

Compressive stress. A body is subject to compressive stress when it is acted on by a load which causes a decrease in its length. In each case the change in length takes place in the line of action of the applied force. Tensile and compressive stresses are sometimes referred to as linear or direct stresses.

Shear stress. If the opposite faces of a cube are subjected to a couple acting tangentially to the faces, the sectional planes of the cube parallel to the applied force are under the action of a shear stress. The strain will be such that the cube will take up the shape of a prism with the section forming a rhombus. The shear strain is measured from the angle formed by the sloping side of the rhombus and the side of the cube before it was stressed. If diagonals are taken across the corners of the rhombus, one of the diagonals will be longer than it was originally and the other shorter. From this it may be deduced that some load is set up along the diagonals which has caused the change in their length. Where the diagonal has increased in length a tensile stress has been set up which is acting on the plane of the shorter diagonal; where the diagonal is shorter a compressive stress is set up which is acting on the plane of the longer diagonal. In a somewhat similar manner it can be shown that when a piece of material is subjected to a direct stress, a shear stress exists on any plane taken at 45° to the line of action of the force producing the direct stress.

■ 1.9 Define Hooke's Law, elastic limit, Young's Modulus, shear modulus, bulk modulus, and Poisson's Ratio.

Hooke's Law states that stress is proportional to strain within the elastic limit.

Elastic limit. If a body is subjected to increasing stress a point will be reached where the material will behave as only partially elastic. When this point is reached and the stress is removed some of the strain will remain as a permanent deformation. The elastic limit is the point where the behaviour of the material changes to being partially elastic; up to this point strain completely disappears when stress is removed.

Young's Modulus, shear modulus and *bulk modulus* are the three moduli of elasticity.

Young's Modulus (E) is the ratio of direct stress and the resulting strain.

$$E = \text{stress/strain}$$

Shear modulus, also known as the modulus of rigidity or modulus of transverse elasticity (G), is the ratio of shear stress and the resulting shear strain.

$$G = \text{shear stress/shear strain (measured in radians)}$$

Bulk modulus (K). If a cube of material is immersed in a liquid and subjected to hydrostatic pressure it will be seen that the cube is acted on by three equal forces acting mutually perpendicular to each other. The cube will suffer a loss in volume as the hydrostatic pressure is increased. The change in volume is the volumetric strain and the intensity of the hydrostatic pressure will be the equivalent compressive stress.

K = equivalent compressive stress/volumetric strain

Poisson's Ratio. If a body is subjected to a direct stress it will suffer from linear strain in the direction of the line of action of the applied force producing this stress. It will also suffer some lateral strain in a plane at 90° to the line of action to the force. Lateral strain is proportional to linear strain within the elastic limit.

$$\frac{\text{lateral strain}}{\text{linear strain}} = \text{a constant } (\sigma) \text{ depending on the material}$$

This constant (σ) is known as Poisson's ratio.

Note   The relationship between lateral strain and linear strain is of great importance in calculating the dimensions of coupling bolts made with an interference fit.

■ **1.10   What is resilience?**

Resilience or elastic strain energy is a term used to denote the storage of the work done in producing strain within a material which is strained. If a piece of material is subjected to an increasing stress the strain will also increase. As work done is the product of force and distance moved by force, the energy stored in a material subject to stress will equal the average force producing the stress multiplied by the distance it has moved through, which will be the total strain. Then

$$\begin{aligned}\text{resilience} &= \text{mean total stress} \times \text{total strain} \\ &= \tfrac{1}{2}\text{total stress} \times \text{total strain}\end{aligned}$$

If the material is subjected to a stress beyond the elastic limit some of the work done is lost in the form of heat which is generated as the material yields.

■ **1.11   What are fluctuating stress, alternating stress, and cyclic stress? How do they differ from simple stress? Why are they important?**

Simple stress comes about from some static form of loading, and the value of the stress does not change.

Fluctuating stress. Diesel engines when operating are not subject to static forms of loading. Due to cylinder pressure variations and dynamic effects of the moving parts, the forces acting on any part of an engine are always changing. As the forces change so the stresses in the various parts change. The changing values of stress experienced on parts of a machine are referred to as fluctuating stress. At any instant in time the value of a stress can be related to a simple stress.

Alternating stress is said to occur when the value of a stress changes from some value of tensile stress to a similar value of compressive stress. An example is the overhung flywheel where a particle in the shaft surface will change from tensile loading at its uppermost position of rotation to compressive loading at its lowest point. The overhung flywheel may be considered as a cantilever with a concentrated load.

Cyclic stress. When a certain pattern of stress change repeats itself at equal time intervals (for example, each revolution of an engine or shaft) the pattern of stress is referred to as cyclic stress.

Fluctuating, alternating, or cyclic stresses are of great importance as they are very closely associated with a form of failure of machine parts known as fatigue failure. Alternating or cyclic stresses are sometimes referred to as fluctuating stresses as a general term to distinguish them from simple stresses.

■ 1.12 What do you understand by the term 'stress raiser'? How can stress raisers be obviated or reduced?

Stress raisers occur at abrupt sectional changes of machine parts or members. They are sometimes referred to as notches. Fillets are made in way of abrupt sectional changes to reduce the abruptness of the change in section.

The effect of abrupt changes on the stress pattern across a section of material in way of the section change is such that where the change occurs, the stress is not uniform across the section. It is higher at the corner or shoulder made by the change. Material at the surface will yield earlier than material remote from the shoulder and under conditions of simple or static stress some redistribution of stress occurs. This does not have time to take place when a machine part is subjected to fluctuating stresses. It is therefore of the utmost importance to design properly and remove all stress raisers. This is done by making fillets between shoulders or having easy tapers on section changes with the end of any taper rounded in at its small end. This reduces to a minimum the chances of fatigue failure.

Stress raisers cannot usually be obviated but the effects are reduced by the use of proper fillets which give a better stress distribution and reduce stress concentrations and variations in way of the change of section. An example of a fillet which we have all seen is the radius formed between the coupling flange and the parallel portion of an intermediate shaft. It can be seen then that a close relationship exists between the ability of an engine part or member to resist fatigue failure and the profile of the fillet in way of section changes. As a section change becomes more abrupt by reduction of fillet radius, the risk of fatigue failure is greatly increased.

Stress raisers may also occur in welded joints due to bad design, undercut (see Questions 6.34 and 6.37), or discontinuities in the way of the joint due to lack of penetration between the filler and the parent material.

■ **1.13   How would you define failure of an engine part or component? State what is done to correct a failure?**

When an engine part or component is no longer fit to carry out the duty for which it was designed and manufactured it is said to have failed. Failure may come about due to wear and tear or because of a failure in the material.

If a part has failed due to normal wear and tear it may be scrapped or withdrawn from use. The part scrapped or withdrawn may then be replaced by a new part or a similar part previously withdrawn and overhauled or reconditioned. Economic factors must be considered; if a part costs more to recondition than a newly manufactured part it is better to renew rather than recondition.

In some circumstances the wear and tear may have occurred over an unexpectedly short space of time. In such cases the accelerated wear and tear should be investigated to ascertain the cause and rectify it.

In other cases failure may come about due to a breakdown in the material of the part or component. If this occurs an investigation must be made to find the cause or causes leading to the breakdown of the material.

■ **1.14   How may diesel engine component materials fail?**

Failure of the materials from which engine parts are made may occur in various ways – either from a single cause or from a combination of causes.

Material failure may occur due to the material stretching or compressing under load so that its shape is permanently altered. This occurs when a material is loaded beyond its yield point. The shape of the component is then altered so that it does not fit correctly or causes misalignment. Fracture of the material may follow on from excessive deformation if the loading on the material is further increased.

Another form of fracture may occur when there is no permanent deformation. This form of fracture is known as brittle fracture. It is not likely to be found in parts from diesel engines, but we should be cognizant of this cause of failure, particularly so in the case of parts manufactured by welding processes and not subject to rigid inspection of material and welds.

Failure may occur due to a part fracturing after a relatively small number of cycles of stress variation. This form of fracture is referred to as low cycle fatigue failure.

When fracture of a part occurs after a large number of stress variation cycles it is commonly referred to as fatigue failure. In written reports and documents it is more correct to describe it as high cycle fatigue failure.

When unsuitable materials are subjected to high temperatures excessive deformation and ultimate failure may occur due to creep (see Question 6.48).

Note 1   Plastic materials may fail due to a variety of other causes.

Note 2   Brittle fracture occurred in some welded ships built during the Second World War and the immediate post-war years. The causes of this form of failure were investigated by various government bodies and ship classification societies. Rules were drawn up with a view to its prevention and it is virtually unknown in ship construction today.

■ 1.15 What is fatigue failure? How does it occur and how would you recognize it?

Fatigue failure comes about usually when some engine part is improperly designed or made from unsuitable material, when a correctly chosen material is given incorrect heat treatment, or when parts are badly machined or badly adjusted. The cause may be a combination of those mentioned. The term for fatigue is really a misnomer as metals do not get tired. Fatigue failure is most common in materials subject to fluctuating tensile stress.

Fatigue failure occurs when a machine part is subjected to fluctuating stress and some form of slip occurs between the grain boundaries of the material, usually at some point of stress concentration. Once slip occurs a crack is initiated which gradually extends across the section of the stressed material. Due to the stress changes which occur under the action of fluctuating stress the strain also follows the stress pattern, and movement in the form of chattering takes place across the opposite surfaces forming the crack. The chattering movement smooths the rough surfaces in way of the crack. The speed of crack propagation increases across the material section until it reaches a point at which yield – and therefore sudden failure – occurs in the remaining material. In ferrous materials, failure that is due to fatigue can be recognized by the fact that there will be two forms of failure in the fracture: the relatively smooth portion where initial failure and cracking progressed across the section, and the portion where final failure took place, which would exhibit the normal appearance of failure in tension. If the material is ductile a cup and cone form of final failure may be seen. Less ductile materials may only have a rough cyrstalline appearance, but the two distinct phases can be easily seen. If the material is working in a corrosive medium, fatigue failure may come about much more quickly.

Note The study of fatigue is quite complex and a knowledge of metallurgy is necessary for full a understanding. The foregoing, however, describes some of its mechanics and how it may be recognized. Later questions and answers will show how the risk of fatigue failure can be reduced.

■ 1.16 How does the engineer guard against fatigue failure when designing important parts of a diesel engine?

When designing important parts of a diesel engine the engineer responsible for the design will use various factors in the stress calculations. These factors make the stresses coming on to the sections in way of the discontinuities acceptable.

Common examples of discontinuities are the oil holes bored or drilled in a crankshaft, re-entrant or negative fillets at the junction between crankwebs and adjacent crankpins or journals, and counterbores in shafting flanges made so that the coupling bolt heads and coupling nuts may sit flat on the flange surfaces.

The factors referred to are known as stress concentration factors (SCF). The value of the factor will depend on the geometry of the part. Stress concentration factor values can be found from charts showing various types and proportions of discontinuities in graphical form.

The allowable stress on the net sectional area in way of the discontinuity in a part will then be equal to the yield stress (YS) or the ultimate tensile strength (UTS), whichever is applicable, divided by the product of the stress concentration factor and the factor of safety (FS). Then

allowable stress = (YS or UTS) / (SCF × FS)

In practice, difficulties often arise in obtaining the stress concentration factor. When complications arise the designer must resort to other methods of stress analysis (see Question 6.51).

In computer aided design (CAD) the design engineer can use a mathematical technique known as finite element analysis. This technique utilizes the power of the computer to find the final results of complicated equations in an iterative manner.

Computer programs are available for building up the node points and connecting networks required for the analysis and then solving the equations arising out of the network. The answers obtained will indicate the location and value of the maximum stresses.

Finite element analysis is also a valuable mathematical technique when working in the fields of heat transfer and fluid mechanics.

The subject is advanced in nature and involves the work of specialists. An engineer should, however, be aware of its availability and have some knowledge of the fields of its use.

■ **1.17   What is an electron microscope? Where can it be used and for what purpose?**

The electron microscope comes in two forms: the scanning electron microscope (SEM), and the transmission electron microscope (TEM). The microscopes consist of an electron gun, a form of magnetic lens or a video amplifier, photographic plates or a fluorescent screen, or a video monitor.

The transmission electron microscope uses very thin specimens and the electrons pass through the specimen. Faults in the structure of the material are in effect opaque to the passage of electrons and show up on the photographic plate or on the fluorescent screen.

The scanning electron microscope bombards with electrons the surface of the specimen under examination. At the point of impact secondary electrons are generated; these are detected and measured. The electron beam is made to scan the surface of the specimen in synchronism with the scanning of a video monitor. The picture obtained from the secondary electrons is shown in a three-dimensional form on the monitor.

The magnifications obtained with electron microscopes far exceed those of the optical microscope. Scales can be used to obtain the dimensions of faults.

Electron microscopes are used in laboratory studies of materials and in failure analysis studies to ascertain the causes of fractures. A good example of their use is in the examination of a fatigue failure. With a scanning electron microscope it is possible to find the actual microscopic point or nucleation site where the initial slippage occurred in a fracture and indicate the cause of the slippage.

Scanning electron microscopes are also used in the analysis of fuel and lubricating oils.

■ 1.18 Define the terms 'temperature' and 'heat.'

Temperature is a measure that compares the degree of 'hotness' of various bodies or masses of material. The difference in temperature between different bodies also determines the direction in which heat will be transmitted from one body to another. Heat is transmitted from a body at higher temperature to a body at lower temperature and transmission of heat continues until both bodies are at the same temperature.

Heat is a form of energy that is possessed by matter in the form of kinetic energy of the atoms or molecules of which the matter is composed. The kinetic energy is obtained from the movement of the atoms or molecules. In gaseous substances the movement is quite complex and involves translatory, rotary and vibratory motion. Translatory motion refers to linear movement of molecules, which may occur in any plane. Rotary motion involves rotation of molecules about some axis, and vibratory motion includes both internal vibration of molecules and external vibration involving relative cyclic movements between two molecules.

■ 1.19 What are the known effects of heat on matter? What are the latent heat of fusion and the latent heat of vaporization?

When the heat content of matter in the solid state is increased, vibratory movement of the atoms and molecules increases and their kinetic energy increases. This is shown by a rise in temperature and some change – usually an increase – in dimensions (thermal expansion). Increase in heat content without change of temperature occurs during the change of state from that of solid to that of liquid. The heat required to effect this change of state without change of temperature is known as the *latent heat of fusion*.

When the liquid state is reached the cohesive forces between the molecules of the liquid are much reduced. Continued application of heat to matter in the liquid state increases the kinetic energy in the molecular movement which is again shown as a temperature rise and usually as an increase in volume. When the boiling point of the liquid is reached large numbers of the molecules gain enough kinetic energy to overcome the cohesive forces between them and break away from the surface of the liquid. As heat is applied to the liquid, vaporization or change of state continues without change of temperature. The heat required to effect this change of state is the *latent heat of vaporization*.

Note Some molecules of liquids will have enough kinetic energy to overcome the cohesive forces acting between them before the boiling point is reached, and some evaporation will occur at a temperature below the boiling point. Evaporation and vaporization are accompanied by large increases in volume. Continued application of heat to the vapour raises it to some critical temperature above which it behaves as a gas.

■ 1.20   What are sublimation and dissociation?

Sublimation. Some chemical compounds have the ability to change directly from a solid state to a vapour by application of heat. This change is known as sublimation.

Dissociation. Thermal dissociation occurs when heat breaks down a portion of the molecules of a chemical compound in the gaseous state, to their constituent molecules or molecules of other compounds. As the temperature falls the decomposed portion recombines.

Note   Thermal dissociation occurs in the combustion space of a diesel engine during combustion of the fuel charge. Other forms of dissociation are electrolytic, when the molecules are split into ions.

■ 1.21   Define the term specific heat.

The specific heat of a substance is the amount of heat required to raise the temperature of unit mass of substance through one degree Celsius (one kelvin). Specific heat is now often referred to as specific heat capacity.

Note
$$1 \text{ Btu} = 1.055 \text{ kJ}$$
$$1 \text{ kcal} = 4.19 \text{ kJ}$$

■ 1.22   What are endothermic and exothermic reactions?

An *endothermic* reaction or process takes place with accompanying absorption of heat.

An *exothermic* reaction or process takes place with the release of heat.

■ 1.23   How does the transmission of heat occur between hot and cold bodies?

A temperature difference is necessary for heat to flow or be transmitted. Thermal conduction is the passage of heat through matter caused by the inter-action of atoms and molecules possessing greater kinetic energy with those possessing less. Normally when we speak of conduction of heat we are referring to transmission of heat through solids. Transmission of heat through liquids and gases is almost entirely by convection. Convection of heat is the trans-ference of heat in liquids and gases (fluids) by actual movement of the fluid caused by density differences at higher and lower temperatures. The less dense (hotter) portion of the fluid is displaced by the denser portion and forced to rise. A convection current or circulation is set up (provided the source of heat is well below the upper parts of the fluid) and will continue until all particles or portions of the fluid are at the same temperature.

Radiation or, more correctly, thermal radiation comes about when the vibration of atoms and molecules in the hot body sets up waves which are transmitted to the cold body. This in turn increases the kinetic energy of the atoms and molecules in the cold body which is manifested by a rise in temperature of the cold body. The rise in temperature continues until the cold body is at the same temperature as the hot body. When the cold body has a constant temperature it will be radiating as much energy as it is receiving.

Radiant heat waves are known to be electromagnetic, as are other radiations such as visible light, ultra-violet rays, cosmic rays, gamma rays, etc. They are specified by a wavelength or a frequency; their velocity in space is the same as that of light. The wavelength of heat rays or infra-red radiation falls between that of visible light and radio waves. At one time it was thought that emission was a continuous process but this is now known to be incorrect. Polished surfaces reflect thermal radiation; they are good reflectors but poor absorbers. Dark and dull surfaces are poor reflectors but good absorbers. Radiant energy is the only form of energy that can exist in the absence of matter and be transferred without the aid of some form of matter.

■ **1.24 What is the relationship between the pressure, specific volume, and temperature of a perfect gas?**

Boyle's Law states that the volume of a given amount of any gas varies inversely as the pressure acting on it while the temperature of the gas remains constant.

$$P V = \text{a constant}$$

Charles' Law states that the change in volume of a given amount of gas is directly proportional to its absolute temperature while the pressure of the gas remains constant.

When these laws are embodied the characteristic equation of a gas is formed. This gives the relationship between the pressure, specific volume, and temperature of a perfect gas and is

$$P V = R T$$

where R is the gas constant.

Note Real gases only approach the behaviour of perfect gases at low pressures. Other equations are used to give a more true relationship.

■ **1.25 Name the energy transformation processes that take place in the theoretical air cycles of internal combustion engines.**

Constant pressure process is one in which the pressure of the air remains constant throughout the change.

Constant volume process is one in which the volume of the air remains constant throughout the change.

Adiabatic process is one in which no transfer of heat to or from the air takes place during the change.

Isothermal process is one in which the temperature of the air remains constant during the change.

■ **1.26 What effects occur when air is heated at (a) constant pressure and (b) constant volume?**

Air being a gas has two specific heats. When air is heated at constant volume the pressure and temperature of the air rises. As there is no change in volume no work is done. When air is heated at constant pressure the volume and temperature of the air increase and work is done as the volume increases. The specific heat is therefore higher. A relation between the two specific heats is used in thermodynamic calculations.

> Specific heat at constant pressure $= c_p$
> Specific heat at constant volume $= c_v$
> For air $c_p/c_v = 0.24/0.17 = 1.4 = \gamma$

Since the values of specific heats increase with increase of temperature, mean values are used for the ratio of specific heats.

■ **1.27 Describe the energy transformation processes that make up the Carnot Cycle. Give a statement showing the efficiency of this cycle. What does the efficiency statement show?**

The Carnot Cycle is a theoretical air cycle used in the study of heat engines. The compression stroke of the cycle begins with an isothermal compression process and finishes with an adiabatic compression process. The expansion part of the cycle commences with an isothermal expansion process and is completed with an adiabatic expansion process.

The efficiency of the cycle can be obtained from the basic statement of efficiency:

'Thermal efficiency = heat equivalent of useful work/heat input; where the heat equivalent of useful work = heat added during the cycle – heat rejected during the cycle.'

From this statement it can be shown that the thermal efficiency of the Carnot Cycle is

$$\text{efficiency} = (T_1 - T_2)/T_1$$

where T_1 is the maximum absolute temperature during the cycle and T_2 is the minimum absolute temperature during the cycle.

The thermal efficiency statement shows us that if the difference between T_1 and T_2 is increased the efficiency is increased. It can also be used to show that efficiency is dependent on the ratio of expansion of the air during the cycle and increasing the expansion ratio increases the efficiency. The ratio of expansion is also related to the ratio of compression.

The Carnot Cycle has the highest efficiency of the standard air cycles.

■ 1.28 Which theoretical air cycle does the modern compression ignition engine follow?

Modern compression ignition engines, or diesel engines as they are commonly known, operate on the dual combustion cycle. The theoretical dual or mixed combustion cycle is a combination of the constant-volume (Otto) cycle and the constant-pressure (Diesel) cycle.

In the Otto cycle the theoretical pressure-volume diagram is formed from two constant-volume and two adiabatic processes. The air in the cylinder is compressed adiabatically. Heat is added to the air at constant volume. Work is done during the adiabatic expansion and then heat is rejected at constant volume.

In the Diesel cycle the theoretical pressure-volume diagram is formed from two adiabatic operations, one constant-pressure and one constant-volume operation. Air is compressed adiabatically and then heat is added at constant pressure. Adiabatic expansion takes place and then heat is rejected at constant volume.

In the dual cycle, air is compressed adiabatically, then heat is added, part in a constant volume process and the remainder in a constant pressure process. Expansion takes place adiabatically and then heat is rejected at constant volume.

Note The theoretical air cycle can take place only in an engine based on theoretical assumptions. It is assumed that the piston is frictionless, the cylinder walls and piston consist of non-heat-conducting material, and that the cylinder head behaves sometimes as a perfect heat conductor and sometimes as a perfect heat insulator.

We must then imagine that the cycle starts with a cylinder and compression space full of pure air at some temperature T. The piston is forced in and work is done on the air in compressing it and raising its temperature. During the compression stroke the cylinder head is behaving as a perfect insulator as are the piston and cylinder walls. Under these conditions no heat is lost during the compression stroke. At the end of the compression stroke the cylinder head is assumed to become a perfect heat conductor and heat is added to the compressed air from some external source applied to the cylinder head. After the addition of heat to the air the cylinder head is assumed to become a perfect insulator again and the air at high pressure and temperature forces out the piston against some imaginary resistance, and work is done at the expense of the heat in the air. As no heat has gone into the piston, cylinder head or walls, no heat can be given to the air and the expansion will be adiabatic as was the compression. When the piston is at the end of the stroke the cylinder head is imagined to become a perfect conductor again. A cold body is then put against the head and some of the heat in the air goes into the cold body and continues until the temperature is back to T again. The process is repeated without changing the air.

The heat added in the theoretical cycle is related to the heat content of the fuel injected into the cylinder in practice. The heat rejected is related to the heat lost in the exhaust gases.

■ **1.29   State Avogadro's Law. Where may the law be used in practice?**

Avogadro's Law or hypothesis states that equal volumes of any gas contain the same number of molecules, provided the temperature and pressure conditions are the same in each gas.

It is common for the equipment used in analysing the contents of exhaust gases to give the results in terms of volumetric ratios. By using the molecular weights of the constituents found in a sample of exhaust gas, and Avogadro's Law, it is easy to convert the figures of a volumetric analysis to an analysis based on weight.

Note Exhaust gas analysis is used mainly in laboratories during engine development and for the testing of fuels.

■ **1.30   What is a mole?**

In the SI system the mole (abbreviation mol) is used as a measure of the amount of substance within a system which contains as many units as there are carbon atoms in a specified amount of a particular form of carbon.

The engineer uses a slightly different concept of this and defines the mole as the mass of a substance (in some weight units) equal to its molecular weight. For example, the molecular weight of carbon is approximately 12. A kilo mol of carbon would therefore be 12 kilos in weight. In gases the volume occupied by a mol of gas is termed the molal volume.

These units are very useful when dealing with fuel combustion and exhaust gas analysis problems.

■ **1.31   What do you understand by the kinetic theory of gases?**

The kinetic theory of gases deals with the movements of molecules and their effect on the pressure, temperature, and heat in a gas. The theory assumes that the molecules are moving with very high velocity and during their movement they collide with one another and with the walls of the vessel containing the gas. As the molecules are considered to be perfectly elastic no velocity is lost during impact. The pressure which the gas exerts on the walls of the vessel is considered to be due to molecular impact. The velocity of the molecules is considered to be related to the temperature of the gas, and the heat in the gas is considered to be due to the kinetic energy of the molecules. If the temperature is increased the velocity of the molecules increases with consequent increase in their kinetic energy and increase in heat. The theory also deals with the molecular formation of monatomic, diatomic and triatomic gases and the various degrees of movement in each group. The theory can be explained mathematically and can be used to verify Avogadro's Law.

■ **1.32   How are theoretical air cycles used and what are the deviations that occur in practice?**

Theoretical air cycles are used in thermodynamic calculations to ascertain the theoretical efficiencies of the various heat cycles. These calculations give as a result the *air standard efficiency* of the cycle. This relates the efficiency in terms of temperatures, ratios of pressures, or a combination of these factors.

In making the calculations of theoretical efficiency the following assumptions are made: the air contained in the cylinder is pure air behaving as a perfect gas; no heat is transferred or lost during the adiabatic changes; the temperature and pressure of the air in the cylinder is the same in every part at any instant during the cycle; no heat loss occurs during the heat-addition part of the cycle.

In practice, air does not behave as a perfect gas, due to the increase in value of the specific heats with increase of temperature; also, during part of the cycle the air is contaminated with products of the combustion of fuel, causing further changes in the specific heat values; and heat transfer and losses occur during adiabatic changes. The pressure and temperature of gas in the cylinder are not always the same throughout the cylinder at any instant in the cycle.

During the cycle heat losses occur to the engine coolant, also losses from friction in the moving parts, and the losses which come about in getting the air into the cylinder and the exhaust gases out (pumping losses).

In the development of a new type of engine, the air standard efficiency of the cycle would be studied. Any changes in factors affecting efficiency will be carefully calculated; the calculations will be developed, making corrections covering earlier assumptions, so that a very close approximation of what can be expected in practice will be finally obtained. The design department of an engine builder compares the results obtained from the prototype on the test bed with their earlier calculations and uses the information gained in later design studies.

2

INTERNAL COMBUSTION ENGINES

■ **2.1   What is an internal combustion engine? Name the various types.**

An internal combustion engine is one in which the fuel is burnt within the engine. It is usually of the reciprocating type. Combustion of the fuel and the conversion of the heat energy from combustion to mechanical energy takes place within the cylinders. Internal combustion engines can also be of the rotary type, such as the gas turbine and the rotary engine developed by Dr Felix Wankel.

Reciprocating internal combustion engines may be of the spark-ignition or compression-ignition type. Spark-ignition engines use gaseous or volatile distillate fuels and work on a modified Otto cycle. They operate on the two- or four-stroke cycle. Compression-ignition engines may also be of either two- or four-stroke cycle type. They use distillate liquid fuels or, where conditions allow, a blend of distillate and residual fuels. This type of engine is usually designed to operate on the dual-combustion cycle or a modification of it. In some cases the cycle is such that the whole of combustion takes place at constant volume.

Some engines are designed for dual-fuel operation and may use either liquid or gaseous fuel. When gaseous fuel is used a small amount of liquid fuel is injected to initiate combustion.

Note   Different names are used for compression-ignition engines. Nomenclature was discussed by a committee of distinguished engineers in 1922 and is still a matter of discussion and argument today. The name Diesel is in common use and has reached the point where it is often spelt with a lowercase 'd'. The modern oil engine bears little resemblance to the engine developed by Dr R. Diesel, but more closely resembles the engine developed by H. Akroyd Stuart at Bletchley, near London, in about 1890 – some few years before Dr Diesel took out patents for the engine he developed at Augsburg in Germany. In using the name Diesel we must not forget the work done by Akroyd Stuart.

■ 2.2   Describe the events which take place in the cylinders of four-stroke cycle and two-stroke cycle diesel engines.

The fundamental requirements for the operation of a diesel engine are a supply of fuel, the necessary air for combustion of the fuel, and some means to get the air and fuel into the cylinders and the products of combustion out.

The stages in the operation of a diesel engine are as follows:

1   Supply of air.
2   Compression of the air to raise its temperature high enough to initiate combustion of the fuel.
3   Supply of fuel.
4   Expansion of the hot high-pressure gas which forces out the piston against the resistance of the load on the crankshaft.
5   Removal of the products of combustion.

These stages may be performed in two or four strokes of the piston (one or two revolutions of the engine crank).

Consider first the four-stroke cycle engine, which has air inlet and exhaust valves. By the opening and closing of these valves in proper sequence, the piston can be made to perform not only its main function of transmitting power to the crank, but also the subsidiary functions of drawing air into the cylinder, compressing the air, and subsequent expulsion of the exhaust gases.

Starting with the piston at top centre, with the air inlet valve open, downward movement of the piston lowers the pressure in the cylinder, and air flows in. During the period when air is flowing into the cylinder, the air in the inlet passages to the inlet valve will gain a high velocity and, in turn, kinetic energy. Use is made of this effect to keep the air inlet valve open until the piston is past bottom centre. The air then continues to flow into the cylinder until its kinetic energy is lost and air flow ceases. The air inlet valve completely closes after the crank has moved 20° to 40° past bottom centre. The gain of kinetic energy of the air moving in the air inlet passages, and the use made of it, is known as the ram effect.

With upward movement of the piston the air is compressed to a pressure which may be between 24 and 63 bars depending on the engine design and speed. Injection of the fuel commences when the crank is between 25° and 10° before top centre position. After fuel injection begins, a short delay occurs before the fuel begins to burn. Combustion continues until the piston and crank pass over the top centre position. Injection of fuel usually finishes shortly after the top centre position, depending on engine speed, load, and original design.

The high-pressure gases in the cylinder, which may be between 54 and 108 bars, force the piston downwards, so rotating the engine shaft and doing work in the process. Movement of the piston continues downwards as the combustion gases expand. The exhaust valve begins to open before the piston reaches the end of its stroke. This allows a large part of the exhaust gas to be blown out of the cylinder during the period in which the cylinder pressure equalizes with the pressure in the exhaust line. This is referred to as the blow-down period. The pressure in the cylinder will be approximately 3 to 4 bars when the exhaust valve begins to open, and the crank angle will be from 50° to 40° before bottom centre. By the time the piston reaches bottom centre, the exhaust valve will be at, or nearly at, its fully open position.

When the piston moves upwards the exhaust gases are expelled by the piston movement. As the exhaust valve is fully open, the resistance to gas flow is at a minimum and any pressure build-up during the exhaust period is also minimal. Continued upward movement of the piston expels the remaining exhaust gas. Before the piston reaches the top position the air inlet valve will begin to open in sufficient time to be fully open soon after the piston passes over the upper position. The operations are then continued as a new cycle.

Note 1   Work against an external load, i.e. propeller or generator, is only done on the expansion stroke. During the air inlet stroke and exhaust stroke work must be obtained from that stored in the flywheel or from other cylinders, which is a loss. The loss is referred to as pumping loss because the piston is, in effect, working as a pump.

Note 2   There is a significant time lapse between the commencement of opening of a valve and its arrival at the fully open position; dependent upon the acceleration imparted to it by its operating cam.

In two-stroke cycle engines the events described above as taking place in four strokes of the piston are contrived to take place in only two strokes of the piston. In two-stroke engines the exhaust gases are expelled from the cylinder, and the cylinder is charged with air, during the period that the crank is passing from approximately 45° to 40° before bottom centre position until 40° to 45° after bottom centre position. The remaining part of the cycle is identical with the compression, combustion and expansion phases in the four-stroke engine.

To accomplish expulsion of the exhaust gases and the supply of air charge within 90° of crank rotation requires the assistance of a low-pressure air supply. This is referred to as the scavenge air. In simple two-stroke engines, where the exhaust and scavenge ports are situated in the lower parts of the liner, the scavenge air pressure will be 0.06 to 0.25 bars. Movement of the piston covers and uncovers the scavenge and exhaust ports. Following expansion of the gases in the cylinder the piston uncovers the exhaust ports when the crank is approximately 45° to 40° before bottom centre and blow-down of the gases into the exhaust manifold occurs. The speed of opening of the exhaust ports is very rapid and the pressure of the gas falls quickly.

By the time the pressure of the gases in the cylinder has fallen slightly below the scavenge air pressure, the piston uncovers the scavenge ports and scavenge air blows into the cylinder forcing out the remaining exhaust gases. The scavenge ports begin to be uncovered by piston movement when the crank is approximately 35° before bottom centre. After the piston has passed bottom centre the scavenge air supply is stopped when the crank is 35° past bottom centre. A small amount of the air in the cylinder escapes through the exhaust ports before they are closed by further upward movement of the piston. When the exhaust ports are covered by the piston the compression phase of the cycle commences and all events are similar to the four-stroke cycle until the exhaust phase begins again. The maximum pressure in simple two-stroke engines is lower than in four-stroke engines.

Note Variation in the heights of the scavenge ports alters the timing at which the events occur. In order to reduce scavenge air wastage between the closure of the scavenge and exhaust ports some engines are fitted with non-return valves, which are located between the inlet to the scavenge ports and the scavenge ducting. This allows the tops of the scavenge ports to be made higher than the tops of the exhaust ports. The non-return valves prevent the exhaust gases from blowing back into the scavenge ducting. When the gas pressure within the cylinder falls to a lesser value than the scavenge air pressure the non-return (scavenge) valves open and scavenge air flows into the cylinder. When the piston rises the exhaust ports are first closed, and scavenge air continues to flow into the cylinder until the scavenge ports are closed by the rising piston. The use of non-return valves in the manner described makes it possible to increase the power output of an engine by 8% to 20%.

The non-return valves are often referred to as scavenge valves.

■ **2.3 Explain the difference between crosshead and trunk-piston type engines. What is the function of the crosshead and piston trunk?**

The main difference between crosshead and trunk-piston type engines is the manner in which the transverse thrust from the piston and connecting-rod is taken up and the nature of the bearing assembly at the upper part of the connecting-rod. Crosshead engines have a piston-rod and trunk-piston engines do not.

The working parts of a crosshead engine consist of a piston head and rod, connected together. The crosshead block, pins and slippers form an assembly which is attached to the lower part of the piston-rod. The slippers slide up and down with the crosshead assembly in the engine guides. The crosshead assembly is connected to the crankshaft through the crosshead bearings (top-end bearings) and the connecting-rod bearing (big or bottom-end bearing). When the crank moves away from the top- and bottom-dead-centre positions the connecting rod is at an angle to the line of piston stroke and, consequently, there is angularity. The downward force exerted by the piston together with the upward reaction from the connecting-rod cause a transverse thrust to be set up (this can be shown with a triangle of forces). This transverse thrust is transmitted by the guide slippers on to the engine or cylinder guides. The transverse thrust is referred to as guide load.

There are fewer parts in trunk-piston engines. The working parts consist of the piston, piston trunk, gudgeon bearing assembly and connecting-rod. The transverse thrust or guide load is transmitted by the piston trunk or skirt on to the cylinder. The function of the crosshead and piston trunk is to play a part in the conversion of the reciprocating movement of the piston to the rotary motion of the crankshaft. They also transmit the transverse loads on to the fixed parts of the engine designed to take these loads (see Fig. 7.11).

Note The guide load comprises the resultant of the piston-rod and connecting-rod loads caused by the cylinder pressures (static load) and the dynamic loads caused by inertia of the moving parts.

■ **2.4   What are the relative advantages of crosshead and trunk-piston type engines?**

Crosshead type engines are able to develop much higher power at lower rotational speeds than trunk-piston type engines, because the space available for the crosshead bearings is greater than the space within the piston for the gudgeon bearing assembly. Trunk-piston engines have the advantage of requiring less head room than crosshead engines. Their working parts are fewer in number and much less costly to produce because their design lends itself to mass production methods. The gudgeon bearing assembly is not particularly suited for highly rated two-stroke engines unless special arrangements are made for its lubrication. Cheaper quality fuels may be used in crosshead engines as it is possible to isolate the cylinder space from the crankcase, thus preventing acidic residues entering the crankcase. The total cost for lubricants is less with crosshead engines than with trunk-piston engines of equivalent power.

■ **2.5   What is an opposed-piston engine and how are the cranks arranged? What advantages and disadvantages do these engines have?**

An opposed-piston engine has two pistons working in the same cylinder, which is much longer than normal. The cranks are arranged so that movement of the pistons towards each other takes place at the same time, as does movement away from each other. The opposed-piston engine always works on the two-stroke cycle with the uniflow method of scavenging. The combustion chamber is formed in the space between the heads of the pistons and the small exposed section or belt of the cylinder left between the pistons. The fuel injection valves, air starting valve, cylinder pressure relief valve and pressure-indicating cock are fitted to the cylinder in way of the belt left between the two pistons when they are at their inner-dead-centre position.

Opposed-piston engines may have two crankshafts, one at the top of the engine for the upper pistons and one in the conventional place for the lower pistons. Engines with two crankshafts are arranged as trunk-piston engines for both upper and lower pistons. The two crankshafts are connected through a train of gears.

Another form of opposed-piston engine has one crankshaft. For each cylinder there are three cranks: the centre crank is connected to the lower piston through a connecting-rod and crosshead, and the two outside cranks, which are in the same line and opposite to the centre crank, are connected to the upper piston through connecting-rods, crossheads and tie or side rods. Movement of the pistons uncovers and covers the exhaust ports which are in the top of the cylinder and the scavenge ports which are at the bottom of the cylinder.

A third variation of the opposed-piston engine uses eccentrics for the upper piston instead of the two side cranks.

The advantage of the opposed-piston engine over other types of engine is that no firing loads are transmitted from the cylinders to the bedplates holding the crankshaft bearings. In consequence of this they may be constructed to lighter scantlings and therefore have a good power to weight ratio. Another advantage is that a high degree of balance may be more easily achieved with opposed-piston engines than with conventional types.

Their disadvantage is the amount of headroom they require in comparison with other engines of equivalent power and rotational speed.

■ **2.6 What do you understand by the following terms: swept volume, clearance volume, compression ratio, volumetric efficiency, scavenge efficiency, air charge ratio, natural aspiration, supercharging? What other names are used for the supercharging process?**

Swept volume. This term refers to the volume swept by the piston during one stroke and is the product of the piston area and stroke.

Clearance volume is the volume remaining in the cylinder when the piston is in the top-centre position. The difference between the total cylinder volume and the swept volume is equal to the clearance volume. The clearance volume space forms the combustion chamber.

Compression ratio. This is the value obtained from dividing the total cylinder volume by the clearance volume and will be from 12 to 18, depending on the engine design. If the compression ratio is below 12 the engine may be difficult to start. High speed engines with small cylinders usually have high compression ratios. Slow speed direct-propulsion engines have compression ratios of around 14.

Volumetric efficiency. This is the ratio of the volume of air drawn into the cylinder (at normal temperature and pressure) to the swept volume. In naturally aspirated four-stroke engines the volumetric efficiency will be from 0.85 to 0.95.

Scavenge efficiency. This is the ratio of the volume of air (at normal temperature and pressure) contained in the cylinder at the start of compression to the volume swept by the piston from the top edge of the ports to the top of its stroke.

Air charge ratio. This is the ratio of the volume of air (at normal temperature and pressure) contained in the cylinder at the start of compression to the swept volume of the piston. This term has now more or less replaced the previous two terms. It is sometimes referred to as air mass ratio or air supply ratio. In four-stroke engines the value will vary from 0.85 for naturally aspirated types up to 4 or more in highly supercharged engines. In two-stroke engines the value will be from 0.85 for simple engines with ported scavenge and exhaust, up to 2.5 for supercharged engines.

Natural aspiration is a term applied to four-stroke engines where the air charge is brought into the cylinder only by the downward movement of the piston without other aids.

Supercharging is a term used to indicate that the weight of air supplied to the engine has been considerably increased. This allows more fuel to be used per stroke with a consequent increase in engine output power. More power is developed by a supercharged engine than by a non-supercharged engine of the same bore, stroke and speed. Supercharging has had the effect of lowering the specific weight of diesel engines, i.e. more horsepower is obtained per ton of engine weight. The term pressure-charging is now used generally instead of supercharging. Where use is made of an exhaust-gas turbo-driven compressor, the term turbocharging is often used.

■ **2.7   Name the factors an engine designer considers in the selection of the compression ratio for a compression ignition engine. Give some examples of compression ratio values.**

The ratio between the total surface area of the cylinder space and the volume of the space is such that as the cylinder dimensions increase the ratio between the values decrease.

In a small engine this means more heat is lost to the cylinder space surface during compression than in a larger engine. For this reason smaller engines require a higher compression ratio than larger engines.

An engine started in low ambient temperatures without preheating requires a higher compression ratio than an engine started in higher ambient temperatures.

The factors considered by the designer are therefore the cylinder dimensions and the ambient starting temperature of the engine's operating environment.

Common values of diesel engine compression ratios are as follows.

Slow-speed two-stroke cycle engine used for ship propulsion 12:1.
Medium-speed turbocharged four-stroke cycle engine used for propulsion purposes 12:1.
Emergency electrical generator set 14:1 to 16:1.
Small, high-speed, naturally aspirated four-stroke cycle automotive engine fitted with glow plugs up to 23:1.

Note   Large engines are usually preheated by raising the temperature of the cooling water. This aids starting and reduces cylinder liner wear. The lubricating oil is also preheated to reduce its viscosity and to assist starting by reducing the friction in bearings.

For cylinders with identical proportions, the total area of the cylinder surfaces varies as the square of the linear dimensions, and the volumes vary as the cube of the linear dimensions (see Question 2.14 and questions on fuel atomization in Chapter 4).

■ **2.8   Describe with the aid of sketches the loop-scavenge, cross-scavenge, and uniflow-scavenge methods of scavenging two-stroke cycle engines currently in use.**

Loop- and cross-scavenged engines are relatively simple in design because the cycle of operations is carried out without requiring an exhaust valve or valves. Instead the air–exhaust gas exchange process (gas exchange process) is carried out by using ports cut or cast in the lower part of the cylinder liner.

Some of the ports (called scavenge ports) carry the scavenge air between the scavenge trunk or scavenge manifold into the cylinder space. In a similar manner the exhaust ports carry the exhaust gases into the exhaust pipes or the exhaust manifold.

The opening of the ports occurs when the piston moves downwards to a position near the bottom centre point and the ports become uncovered giving access into the cylinder space. The ports are closed by upward movement of the piston blanking them off.

In cross-scavenged engines the scavenge and exhaust ports are arranged diametrically opposite one another and the scavenge air flows from the scavenge air ports to the exhaust ports and crosses from one side of the cylinder to the other. The scavenge ports are usually sloped in an upwards direction to scavenge the upper part of the cylinder space. (Fig. 2.1)

In loop-scavenged engines the exhaust ports are placed above the scavenge ports on the same side of the cylinder liner. The pattern of air flow takes place across the diameter of the cylinder, then upwards into the upper part of the cylinder space, then down the opposite side of the cylinder into the exhaust ports. The air flow forms a loop pattern inside the cylinder space. (Fig. 2.2)

Note In later generations of slow-speed engines the cross-scavenged engine was superseded by the loop-scavenged engine. The loop-scavenged engine was later superseded by the uniflow-scavenged engine. Slow-speed cross-scavenged and loop-scavenged engines are no longer built.

Uniflow-scavenged engines have a row of scavenge ports arranged around

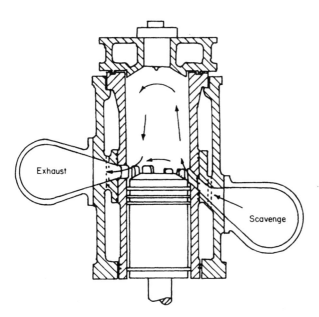

Fig. 2.1 Section through a cylinder of a cross-scavenged engine showing direction of air flow during scavenging.

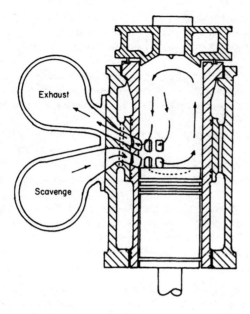

Fig. 2.2 Section through a cylinder of a loop-scavenged engine showing direction of air flow during scavenging.

the circumference of the lower part of the cylinder liner. The ports connect directly with the scavenge space formed around the outside of the lower section of the cylinder liner.

An exhaust valve is fitted in the cylinder cover or cylinder head and connects directly with the exhaust pipes in older engines or the exhaust manifold in engines of later generations. (See section on methods of pressure-charging in Chapter 5.)

As the piston approaches the bottom centre position the exhaust valve is made to open allowing the relatively high-pressure exhaust gases to blow out of the cylinder. The pressure in the cylinder rapidly falls, the scavenge ports are then opened by the downward movement of the piston and scavenge air passes upwards in one direction through the cylinder space. The remaining exhaust gas is then expelled and the cylinder is left with a new air charge ready to commence another cycle. (Fig. 2.3)

■ **2.9   What effect does an increase in the stroke/bore ratio have on cross- and loop-scavenged engines?**

The stroke/bore ratio of an engine is the number obtained by dividing the length of the stroke by the diameter of the cylinder.

The value of the stroke/bore ratio in loop-scavenged engines is about 1.75; in cross-scavenged engines it may go higher reaching a value between 2.00 and 2.20. Some limiting value will be very little higher than this.

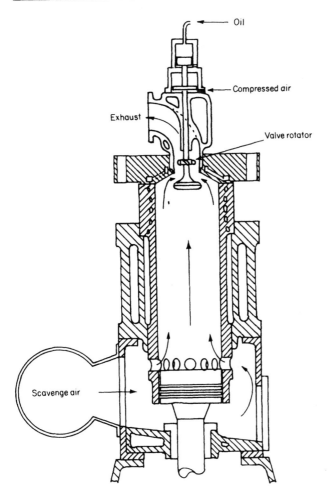

Fig. 2.3 Section through a uniflow-scavenged engine with an exhaust valve in the cylinder head. Some swirl can be imparted to the air by giving the scavenge ports a near tangential entry instead of a radial entry. Some eddies occur in the air flow below the exhaust valve.

If the stroke/bore ratio of cross- and loop-scavenged engines is increased beyond the values given, it becomes increasingly difficult to get the upper part of the cylinder space properly scavenged of exhaust gas. Mixing of the remaining exhaust gas and the incoming scavenge air takes place. The oxygen content in the cylinder at the start of compression is then reduced. In order to correct this it will become necessary to reduce the quantity of fuel injected during each cycle; the output of the engine will then be reduced. (Fig. 2.4)

Note The final design of any engine is a finely balanced compromise between

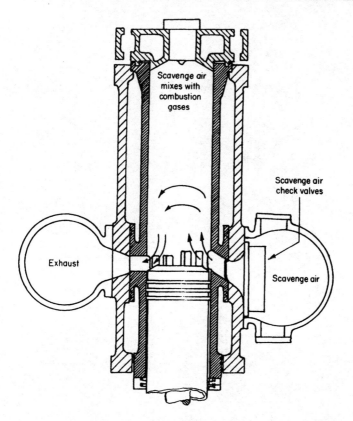

Fig. 2.4 Section through a cross-scavenged engine with a high bore/stroke ratio. **Note.** Here, the scavenge ports are shown higher than the exhaust ports, backflow of the combustion gases into the scavenge air manifold being prevented by the scavenge valves. The scavenge valves allow air flow only into the cylinder and act as check valves or non-return valves and so prevent a backflow.

the extremes of various desirable features. For example, if the stroke/bore ratio is increased it may be possible to obtain an increase in engine efficiency. If this increase in efficiency requires a larger size cylinder to maintain the same power output on the same or a very slightly lower specific fuel consumption, any commercial advantage gained is badly offset by the required increase in the size of the engine.

■ **2.10   What are the advantages and disadvantages of cross-scavenged and uniflow-scavenged engines?**

Cross-scavenged engines do not require exhaust valves or scavenge valves so some simplicity is obtained over other engine types.

The cylinder liners of cross-scavenged engines require a complicated pattern of scavenge and exhaust ports in the lower part of the cylinder. The surfaces left by a core in the casting process of the liner are inadequate in their profile and surface finish. In order to be acceptable the ports must be milled out to give a correct shape and a smooth surface finish. The height of the ports extends relatively high in the cylinder liner and the effective stroke for expansion of the gases is reduced. The cross-sectional area of the ports is relatively large compared with the area of the port bars. This often leads to an excess of liner wear in way of the port bars.

Piston ring breakage is more common in cross-scavenged engines than in uniflow-scavenged engines.

Because they are so complicated the cost of a cylinder liner for a cross-scavenged engine is considerably more than for a uniflow-scavenged engine of similar dimensions.

Uniflow-scavenged engines require an exhaust valve or valves, the number depending on engine speed and cylinder size.

In slow-speed engines only one exhaust valve is required. When one exhaust valve is required two or more fuel injection valves must be fitted whereas in the cross-scavenged engine only one centrally located fuel injection valve is required.

The cylinder liner for a uniflow-scavenged engine has the scavenge ports fitted around the whole of the circumference of the liner. The full circumferential space available allows the ports to be made circular. This arrangement of ports does not extend as far up the cylinder liner so the effective length of the piston stroke is considerably more in a uniflow-scavenged engine than in a cross-scavenged engine of similar dimensions.

Cylinder liner wear in way of scavenge port bars in uniflow-scavenged engines shows no increase over those parts above and below the ports.

The cylinder liners of uniflow-scavenged engines cost considerably less than those for equivalent cross-scavenged engines.

The arrangements for sealing the bottom of the cooling water space are much simpler in uniflow-scavenged engines.

■ **2.11 Why has the cross-scavenged engine been superseded by the uniflow-scavenged engine?**

The cross-scavenged engine cannot take advantage of an increase in thermal efficiency by increasing the stroke-bore ratio. The stroke-bore ratio of modern uniflow-scavenged engines may be between 2.4 and 2.95. This allows for a greater ratio of expansion; the increase in thermal efficiency reduces the specific fuel consumption and so reduces fuel costs. As fuel costs make up a large part of the daily running cost of a ship, engines, if they are to be commercially attractive, must have the lowest possible specific fuel consumption.

Note The ratio of expansion is governed by the compression ratio, the bore-stroke ratio and the timing of the opening of the exhaust valve. The opening point of the exhaust valve is related to the power demand of the turbocharger. An increase in the efficiency of the turbocharger allows the exhaust valve to be opened later. Opening the exhaust valve later increases the thermal efficiency of the engine and lowers the specific fuel consumption.

Note By 1981, only one of the three principal slow-speed engine builders was still building cross-scavenged engines. The other two builders had always built uniflow-scavenged engines. Today all slow-speed engine builders and their licensees build uniflow-scavenged engines only, but large numbers of loop- and cross-scavenged engines will remain in service for some years to come.

■ **2.12   Why is it necessary to cool the cylinder heads or covers, cylinder liners and pistons of diesel engines? What is used as the cooling medium?**

The temperature inside the cylinders of diesel engines rises to approximately 2000°C during combustion of the fuel and drops to approximately 600°C at the end of expansion. With temperatures in this range the metal of the cylinder covers, cylinder liners and pistons would quickly heat up to the point where its strength would be insufficient to withstand the cylinder pressures; also, no oil film would be able to exist on the cylinder walls, and lubrication of the cylinder and piston rings would break down. Cooling is necessary to maintain sufficient strength in the parts and to preserve the oil film on the cylinder.

The cooling medium for cylinder liners and covers is a flow of distilled or fresh water: the medium for cooling pistons is also distilled or fresh water, or oil from the crankcase system. The amount of heat extracted from the various parts must be such that they operate at temperatures well within the strength limits of the materials used. The coolant flow patterns must also be arranged so that the surfaces of all parts are as near uniform temperature as possible to prevent large thermal stresses being set up.

With modern highly rated engines the temperatures of the parts subjected to combustion temperature are much lower than in earlier engines. This has been made possible by the availability of better temperature measuring devices and the research carried out by engine builders. The temperature of the combustion chamber surfaces of cylinder covers, piston crowns and cylinder liners varies between 200°C and 350°C in modern highly rated engines. The variation in temperature of the different parts of the surface of cylinder covers will be within about 50°C to 100°C, and for piston crowns the temperature variation will be 75°C to 100°C. Cylinder liners show greater temperature variation throughout their length, but in the highly critical area at the top of the liner the variation is kept to within approximately 100°C.

Small diesel engines with pistons less than about 150 mm (6 in) diameter have only the cylinders and covers cooled by water. The piston crown will be cooled by excess lubricant from the gudgeon bearing and by the heat transfer to the walls of the piston which are then cooled by the cylinder liner. Small high-speed diesel engines may also be cooled by forced air flow passing over fins fitted on the outside of cylinders and cover. It should be noted that air-cooled diesel engines have very low cylinder wear.

Note With pressure-charged engines the air flow during the scavenge period (in two-stroke and four-stroke engines) over the hot internal surfaces of the cylinders, covers and piston crowns helps to maintain low surface temperatures. It also reduces the temperature gradient across the material section and in turn lowers the thermal stresses.

■ 2.13 How is the combustion chamber formed in diesel engines? What governs its shape?

In normal engines the combustion chamber is formed in the space between the cylinder cover and the piston crown. The upper part of the cylinder liner usually forms the periphery to the space.

The shape of a combustion chamber may vary between that of a spheroid which will be formed from a concave piston crown and cylinder cover, to that of an inverted saucer, formed from a concave cylinder cover and a slightly convex piston crown. In opposed-piston engines the combustion chamber will be spheroidal. The piston crowns on the upper and lower pistons are usually identical in form. Combustion chambers of the shapes mentioned are referred to as open types.

The shape of a combustion chamber must be such that all parts of the space are accessible to the fuel sprays. If any part is not accessible, the space is wasted and combustion has to take place in a reduced space, which causes further difficulties due to less air being available in the region of the fuel spray. The wasted space is sometimes referred to as parasitic volume. The shape of the various parts must also be satisfactory in respect of their strength as they must be able to withstand the pressures in the cylinder without flexing.

With high-speed engines, open combustion chambers can create problems with very high rates of pressure rise due to the shortness of time available for injection and combustion. To overcome this problem the fuel is injected into a separate chamber which is connected to the main combustion chamber by a restricted passage. The restricted passage is at a high temperature, the fuel spray is long and narrow. Following injection the fuel commences to burn in the separate chamber and issues from the restricted passage at a high velocity due to the pressure rise in the chamber. The fuel enters the main combustion chamber as burning vaporized particles and combustion is then completed. The small chamber is about one-third of the clearance volume and is called a pre-combustion chamber or antechamber. Its use allows high-speed engines to operate over wide speed ranges without combustion difficulties, and is a necessity in automotive engines. It is met in the marine field when automotive engines are used for electrical generation or other auxiliary purposes.

■ 2.14 Why is it necessary to atomize the fuel when it is injected into a diesel engine cylinder?

When combustion of liquid fuel takes place the fuel must go through various changes before it can begin to burn. These changes require absorption of heat and temperature rise. After the changes have taken place oxygen is required to complete combustion and this is present in the compressed air charge.

The rate at which the fuel can receive heat to raise its temperature will be dependent on the surface area of the fuel particles in contact with the hot compressed air in the combustion space. The speed with which the changed fuel particles can burn will be dependent on the supply and availability of air.

Let us assume that one cubic centimetre of fuel is used per cycle. If the fuel were to enter the cylinder as a single cube it would have a surface area of six square centimetres. The heat absorption by this cube would be relatively slow and the required changes would take place only on the surface, which would then start burning. The volume of air closely surrounding the cube would soon become oxygen-starved, and combustion would slow up and could proceed only as oxygen made itself available.

If the cubic centimetre of fuel is broken down to one thousand cubes, each a cubic millimetre, the total surface area will be six thousand square millimetres. With these smaller particles the temperature rise necessary to prepare them for burning will occur much more quickly. Assuming the cubic millimetre particles are distributed evenly throughout the combustion chamber, each will be surrounded by air and the speed of combustion of each small particle will be faster. The time for combustion of the total fuel mass will then be the time required for one particle to burn.

If we proceed to break down the particles further, with each decrease in particle size a larger increase in the total surface area of the fuel droplets comes about. This allows a still greater rate of heat transfer to the fuel and a quicker temperature rise, so the fuel is prepared for burning in a much shorter time. If the particles are dispersed evenly throughout the combustion chamber space, each particle will have an adequate air volume around it and oxygen starvation will not take place. Completion of combustion can then come about in a shorter time period.

From the progressive steps we have taken it is readily seen that to enable a fuel charge to go through the various stages of combustion in a very small time interval, it is absolutely essential for it to be broken down to minutely small droplets, i.e. become atomized.

Note   The time span (for a two-stroke engine) for injection to take place will be T seconds when

$$T = \frac{\text{period of fuel-valve opening in degrees}}{360} \times \frac{60}{\text{rev/min}}$$

For an engine operating at 100 rev/min with a fuel-valve opening period of 20°

$$T = \frac{20}{360} \times \frac{60}{100} = \frac{1}{30} \text{ second} = 33 \text{ milliseconds}$$

The time required for the changes to take place in the fuel before it can burn is 1.5 to 3.0 milliseconds.

■ **2.15   How are the bearings of diesel engines lubricated? What do you understand by the terms fluid-film lubrication and boundary lubrication?**

Diesel engine bearings are lubricated by fluid films. The journal is always smaller than its surrounding bearing. When the shaft is static it will make contact with the bearing and this contact will be a line. On each side of this line the normal distance between the shaft and the bearing will increase gradually and will in effect be a curved wedge. When the shaft revolves in the presence of an adequate liquid supply (lubricating oil), the oil is pulled into the wedge and pressure is set up. If these liquid pressures were to be plotted on lines drawn radially from the centre of the bearing it would be seen that the plot of these pressures would form a bulge something like a cam profile. The pressure of liquid in the wedge-shaped space sets the shaft over to one side and lifts the shaft away from the bearing so that it is supported on an oil film. The position where the oil film thickness is least will be a small distance away from the static contact line in the direction of shaft rotation. For pressures to be built up to a value high enough to separate the shaft from the bearing, the oil must have sufficient viscosity and the speed of the shaft must be above a certain value. This form of lubrication is referred to as *fluid film* or *hydrodynamic*.

Boundary lubrication occurs when the rotational shaft speed falls and the oil wedge is lost. Metal to metal contact then occurs. To prevent metallic contact under boundary conditions greases may be used or additives may be added to the oils. The bearings of a diesel engine do not work under boundary conditions. Very highly loaded crosshead bearings in two-stroke engines may approach boundary conditions.

Diesel engine bearings are lubricated by oil films built up under the conditions described. The bearings are supplied with large amounts of oil which are used to maintain the oil film and remove the heat generated. Removal of the heat generated keeps the working parts at temperatures that will not reduce the oil viscosity to values low enough to allow breakdown of the oil film.

■ **2.16   How are the air inlet and exhaust valves of a diesel engine opened and closed? What forces must be applied to open exhaust valves and where is the force obtained from?**

The air inlet and exhaust valves of four-stroke engines and the exhaust valves of two-stroke engines are opened by cams, and closed by springs. In four-stroke engines the camshaft runs at half the crankshaft speed; in two-stroke engines the speed of camshaft and crankshaft are the same.

When a valve is opened the coil spring is compressed and loaded. When the cam roller rides off the cam the resilience in the spring closes the valve. During the closing period the spring may set up a reverse torque on the camshaft by driving the cam. The force required to open an air inlet valve or an exhaust valve will be the sum of the following forces: the product of the valve lid area and pressure difference on the valve, the acceleration forces during the opening period, the force to overcome the spring and the force to overcome friction of the moving parts.

The torque on an engine camshaft may have wide variations, even to the extreme condition in which, during valve opening the crankshaft drives the camshaft, but during valve closing the camshaft feeds back work into the crankshaft.

The mechanism consists of a cam engaging with a cam roller. The roller may be fitted between the forked end of a valve lever which receives its motion from the action of the cam. Upward movement of the cam end of the lever causes downward movement at the end connected to the valve and the valve is opened. In other cases the cam roller may be connected to a push rod which is connected at its upper end to the valve lever. Where push rods are fitted in large engines a hydraulic loading device is fitted at the foot of the push rod; this permits smaller tappet clearance without fear of the valve being kept off the seat during the closed period. The camshaft is connected to the crankshaft through gearing or roller chains.

■ **2.17 Basically two different areas of maintenance work are involved in keeping diesel engines in good operational order. Name the two areas and list the maintenance requirements.**

The two areas requiring maintenance are those associated with (a) combustion, and (b) bearing adjustment, and maintenance of correct alignment in *all* running parts. There is some overlap between the two areas of activity.

Maintenance work associated with combustion involves scavenge port and valve cleaning, piston ring replacement, air inlet and exhaust valve changes and overhaul, cleaning turboblower blading, compressor air inlet filters, scavenge and charge air cooler, and attending to instrumentation associated with combustion. The items mentioned cover all types of engines.

The other type of maintenance work covers all the moving and static parts of the engine, and includes bearing examination and adjustments, lubrication and cooling services, examination of bedplates, frames, cover, safety devices, etc.

■ **2.18 List the characteristics by which diesel engines can be classified and compared. Using these characteristics, briefly specify the various types of diesel engine found in marine practice.**

Diesel engines can be classified as follows.

  1 Four-stroke   (a) naturally aspirated   (b) pressure charged
  2 Two-stroke   (a) low-pressure scavenge   (b) pressure charged
  3 Trunk-piston type
  4 Crosshead type
  5 Vertical cylinder in line
  6 V cylinder arrangement
  7 Slow-speed   65/70–15 rpm
  8 Medium-speed 300–1200 rpm
  9 High-speed   120, up to about 3500 rpm
10 Fuel viscosity (a) 370 CST 1500 sec Redwood No. 1
               (b) 85 CST 350 sec Redwood No. 1
               (c) diesel oil
               (d) gas oil
11 Crankshaft supported on bearings in bedplate
12 Crankshaft underslung from engine frame
13 Bedplate continuously supported on short spaced solid chocks
14 Bedplate continuously supported on resilient chocks
15 Bedplate point support   (a) solid
                     (b) resilient

Slow-speed diesel engines directly coupled to the propeller shaft will have characteristics **2b, 4, 5, 7, 10a, 11** and **13**.

Normally a designer aims to keep propeller speeds low to obtain greater efficiency from the propeller. In medium- and lower-power main engine installations the speed may be allowed to go higher to keep machinery weight low at the expense of propeller efficiency.

Medium-speed diesel engines directly connected to the propeller, as found in smaller vessels such as coasters, trawlers and service vessels, may have engine speeds up to 750 rev/min. Their characteristics will be within the group 1b, or 2b, 3, 5 or 6 (depending on power requirements and space available), 8, 10b or 10c, 11 or 12, and 13.

Propulsion machinery installations using medium-speed engines and gearing will have characteristics 1b or 2b, 3, 5 or 6, 8 (engine speed will be between 400 and 600 rev/min), 10b, 11 or 12, and 13.

Engines used for electrical generation purposes commonly have characteristics 1a or 1b, 3, 5 or 6, 8 (speed will be 800 to 1200 rev/min) 10c, 11 or 12, 13 or 14.

Where high-speed engines are used for electrical generation the engine characteristics most likely are to be 1b, 3, 5, 9, 10d, 12, 15a or 15b. Engines of this type follow the standards of automotive diesel engine practice. In some cases they may be pressure-charged two-stroke uniflow engines with two or four exhaust valves in the cylinder-head.

■ **2.19 What is the value of the maximum load that a diesel engine cylinder cover and piston are subjected to? Give an example of the magnitude of this load in a slow-speed two-stroke diesel engine. Show how these loads are transmitted through the engine structure.**

The value of the maximum load on a cylinder cover and piston will be approximately the same, and will be the product of the area of the piston and the maximum gas pressure. In the case of the cover area it will be the projected area of the cylinder cover measured to the outer edge of the joint spigot. In some engines this may be considerably more than the piston area.

Example
   Cylinder diameter = 980 mm
   Maximum pressure = 78.5 bars (80 kg/cm²)
   Maximum load = $0.7854 \times 98^2 \times 80/1000$ tonnes
              = 600 tonnes approx

The load on the cylinder cover is transmitted into the cylinder beam through the cover studs. The load on the cylinder beam is passed down to the bedplate through the tie-bolts and transverses supporting the crankshaft main bearings. The upward direction of the forces on the cylinder cover is balanced by the downward forces on the piston, which are transmitted through the piston-rod, crosshead block, bearings, connecting-rod, crankshaft and crankshaft main bearings.

The system of parts may be likened to a square flat plate tied to a square frame below it by four long tie-bolts at the corners. If a round shaft is placed across the frame, bearing on the two sides, and a jack is inserted between the shaft and the plate, and loaded, we may say that the system of engine parts is simulated. The load on the jack, which is simulating the firing load on the piston and transmission of the load to the shaft, produces tension in the tie-bolts, a bending moment on the shaft and a bending moment on the two sides of the frame supporting the shaft. The flat plate at the top will also be subjected to a bending moment. The actual parts of an engine will be subjected to the same loads as the simulation rig. The engine tie-rods will be in tension and the transverses supporting the main bearings will be subjected to a bending moment. The cylinder beam will also be subjected to bending moments.

Note   The gas load coming on to the cylinder cover studs will be calculated on the area to the outer edge of the spigot. The tension on the tie-bolts from gas load will be calculated on the area of the piston.

The stresses coming about from the bending moment on the shaft in the simulated rig will be additive to the other stresses set up during engine operation.

3

FUELS, LUBRICANTS – TREATMENT AND STORAGE

■ **3.1** What are fossil fuels and how do they differ from other types of fuel? Which fossil fuels are used in diesel engines?

Fossil fuels are the remains of prehistoric animals and plants and are found below the surface of the earth; they may be solid, liquid or gaseous.

Solid fuels. Coal is the most important solid fuel used commercially.

Liquid fuels of a wide variety are obtained from distillation and other processes carried out on crude oil. The products obtained are essentially engine fuels, boiler burner fuels, and lubricants.

Note The oil industry is also a large supplier of chemicals used in other industries such as plastics, paints and compositions, synthetic rubbers and the like.

Gaseous fuels may exist naturally in the ground or be produced from coal or crude oil. Liquefied petroleum gases (LPG) are increasingly used.

The fossil fuels are essentially carbon-hydrogen compounds. The energy is derived from them by the exothermic action of converting the carbon to carbon dioxide and the hydrogen to water, which will be in the form of steam at the end of the combustion process.

The other types of fuel used are nuclear, which are fissile materials used in a reactor. One of the isotopes of uranium is commonly used.

The fuels used in diesel engines are the gas oils and diesel oils which boil off from crude at temperatures between approximately 200°C and 400°C, or blends of diesel oil and residual fuel which have higher boiling points.

Note Liquefied petroleum gases must be stored under pressure or in refrigerated conditions, since their boiling points are low.

3.2   Name the various types of crude oil and briefly describe the refining processes by which petroleum fuels and lubricants are produced.

There are no universally accepted standards for classifying crude oil. For our purposes crude oil can be classified as paraffinic, as found in Pennsylvania, naphthenic as found in the Caucasus, and asphaltic as found in Texas. Many types of crude oil are found throughout the world but the majority will be within the groups paraffin base, naphthenic base, or some intermediate base.

Crude oil is a mixture of hydrocarbons, and, although there are considerable differences in the physical properties of the various hydrocarbons, the variation in chemical analyses is small. The carbon content varies from 83% to 87% and the hydrogen content from 14% to 11%. The balance is made up of sulphur, sodium, vanadium, water, etc., which may be classed as impurities.

The molecular structure of the fuel determines its physical properties. This structure can have numerous forms and may be such that the carbon atoms form either chains or chains with side chains, or have a ring structure. The hydrocarbons with the ring structure are more stable chemically.

The oil refinery processes are generally devised with a view to obtaining the highest yield of fuels in the range from the the liquefied petroleum gases through to the paraffins (kerosine) and gas oil. The crude oil is first stored in a settling tank to separate water, sand and earthy matter. After separation of the heavier impurities the crude oil is pumped into an oil- or gas-fired heat exchanger (pipe still) and heated to approximately 350°C, which brings a large part of the crude oil to above its boiling point. The heated crude oil is then passed to fractionating towers. The fractionating towers are in effect vertical condensers with horizontal partitions. The heated crude oil is passed in near the bottom and the major part flashes off and passes up through the fractionating tower. The rising vapour condenses at the various levels of the horizontal partitions and is piped off from them as various grades according to their boiling points. The vapour leaving the top consists mainly of petroleum gases, part of which is condensed, while the remainder may be used as fuel for heating processes within the refinery. The condensed portion may be recirculated through the first tower. The bottoms from the first fractionating tower are passed through a second tower from which petrol is produced. The bottoms from the second tower are passed to a third tower from which a range of other products are obtained. The residuum from the third tower may be treated in various plants, in which the molecular formation of the residuum is reformed to increase the yield of the light constituents.

If the base of the crude is satisfactory the residuum may become the stock for production of lubricants. Lubricants can be produced from most types of crudes and their properties will vary according to the crudes from which they are produced. The residuum contains waxes, resins, asphalts and unstable hydrocarbons.

Lubricants are produced by solvent refining and acid refining. In solvent refining the solvent is pumped into the top of a tower and the stock is pumped in at the side. The solvent takes out the unwanted constituents in the stock, which pass to the bottom of the tower. The refined oil is passed out from the top of the tower. It is further treated to remove waxes, impurities and discoloration.

Modern oil refineries are highly automated and much of the equipment used has now found its place in ships' engine rooms and other industrial plant.

■ **3.3 What are the reasons for the deterioration in the quality of the fuel supplied for use in marine diesel engines?**

The sale of energy in any form of the three types of fossil fuel is a highly competitive business. When the cost of crude oil rose sharply during 1973 the suppliers of refined crude oil products were forced to compete at a considerable disadvantage with the suppliers of fuels such as coal and natural gas. Furthermore, at about this time some countries were bringing in legislation to reduce and eventually stop the supply of leaded fuels. This was done to reduce the very harmful atmospheric pollution resulting from increased use of the automobile and the resulting increase in gaseous pollutants containing compounds of lead.

The customary forms of refining crude oil are very briefly set out in the answer to Question 3.2.

Oil refining techniques were updated to meet the increasing demand for unleaded petrol or lead-free gasoline having an acceptable octane rating, and to increase the yield of the more valuable fuels from the crude oil stock. This modification and updating gave a greater yield of the more valuable distillation products and reduced the amount of the remaining less valuable residual products.

Increased yields are obtained by subjecting the residue from the atmospheric distillation process to a vacuum distillation process. This increases the amount of distillate from that part of the residue having a higher boiling point. While under a reduced pressure the boiling point of the liquid is lowered and distillation then takes place without subjecting the residue to such high temperatures.

The distillate from the vacuum distillation process may then be reheated and treated in a catalytic cracking reactor.

Note There are many different forms of the catalytic cracking process.

The fluidized solid catalytic cracking process uses silicon oxide (silica) and aluminium oxide (alumina) as the catalyst. It is used in a powdered form so that it behaves like a fluid when in a stream of air or vapour. Some of the particles break up and catalyst dust is formed. The dust is referred to as catalyst fines or CC fines.

The cracked oil vapours or light hydrocarbons from the reactor create gases, petrol or gasoline, and light fuel oils. The residue left from the process often contains some of the catalyst carried over from the reactor.

The other cracking process used is known as thermal cracking. This may be used for altering the molecular structure of distillates and residues from the atmospheric distilling process. The thermal cracking process uses distillate to increase the yield of high octane petrol or gasoline, and the residue to increase the yield of light fuel oils.

A form of the thermal cracking process may also be used to reduce the viscosity of residual products. This is known as 'Visbreaking'.

These modifications in oil refinery practice result in a reduced amount of residuum. The impurities such as sulphur, vanadium, sodium, barium, calcium, and ash, etc., while remaining the same in a unit amount of crude oil, become much more concentrated in the lesser amounts of residue.

Similarly carry-over of silica and alumina from the fluid catalytic cracking process also shows a greater concentration in the lesser residue amount.

The fuels supplied to diesel-propelled ships are obtained by blending a residual fuel having a relatively high viscosity with a distillate fuel having a lower viscosity. The resultant blend then has a viscosity complying with the viscosity stated in the order for the fuel. When the residual component of the blend has a viscosity lowered by the visbreaking process, the amount of distillate fuel (the 'cutter stock') required to bring the blend to the required viscosity is again reduced in amount. This leads to a further increase in the concentration of the impurities.

Another complication arising and leading to more problems with blended fuels is that in many cases the cracking processes increase the amount of the aromatic constituent. The increased aromatic constituent may then lead to problems with combustion and the cleaning of the fuel with centrifugal separators and clarifiers.

The following changes in quality may be apparent. In some cases most of the mentioned changes may be present while in other cases only one or a combination of two or more may be present.

Increase in aromatics giving a high density
Increase in ash content
Increase in asphaltic material content
Increase in carbon residue content
Increase in catalytic cracking fines content
Increase in sodium content
Increase in density
Increase in sulphur content
Increase in vanadium content

Note The cracking or molecular reforming of liquid hydrocarbon fuels are not recent advances in oil refining techniques. The first forms of the thermal cracking process were begun at about the time of the First World War; catalytic cracking processes were started during the mid 1930s.

■ **3.4 Give a list of the properties or tests by which distillate and blended fuels may be specified or decisions made on their fitness for use.**

Density
API Gravity (API: American Petroleum Institute)
*Colour
Viscosity (kinematic)
*Cloud point
Pour point
Flash point
Fire point (open cup flash-point)
Ignition point (self-ignition point or auto-ignition point)

*Distillation range
Calorific value (thermal value, heating value)
*Cetane number and cetane index
*Aniline point
*Diesel index
Carbon residue
Alumina content
Asphalt content
Silica content
Sodium content
Sulphur content
Vanadium content
Compatibility (blended fuels)
*Copper strip corrosion

Notes   The terms marked with the asterisk are generally applicable to distillate fuels.

The tests used to find the cetane number or the cetane index do not generally yield reliable results with many of the high-viscosity fuels marketed today.

The use in some modern engines of very high pressures for the injection of fuel into the cylinder has had a beneficial effect on engine operation when using very high-viscosity blended fuels.

■ **3.5   Give a list of the properties or tests by which a lubricating oil may be specified or a decision made on its fitness for further use.**

Density
Colour
Viscosity (kinematic)
Viscosity SAE number (SAE: Society of Automotive Engineers USA)
Viscosity index
Cloud point
Pour point
Flash point
Total base number (TBN)
Total acid number (TAN)
Ash content

An analysis can also be made of the strong acid content of oil samples removed from the crankcase. Spectrographic analysis can be carried out on samples of crankcase system oils to determine the content of different metals that may be present.

■ **3.6   What is the flash-point of an oil and what dies it indicate?**

The flash-point is the lowest temperature at which an oil will give off sufficient inflammable vapour to produce a flash when a small flame is brought to the surface of the oil. The flash-point may be measured as an open or closed flash-point figure. Fuels for use aboard ships are tested in a Pensky–Martens instrument which measures the closed flash-point. The Department of Trade & Industry sets the lower limit of 65°C for the flash-point of fuels used aboard merchant vessels and also stipulates that fuel in storage tanks must be kept at temperatures at least 14°C lower than its flash-point. The flash-point of an oil gives no indication of its suitability for use in a diesel engine. It only serves as a guide to the temperature below which it can be stored and handled with reasonable safety. A knowledge of the flash-point of the lubricating oil used in the crankcase of a diesel engine is useful, since lowering of the flash-point inducates that the lubricant may be contaminated with fuel.

■ **3.7 Define the following terms: viscosity, viscosity index, cloud point, pour point.**

Viscosity can be defined as that property of a fluid (liquid, vapour, or gas) which is related to its ability to flow. We are all aware of what we commonly refer to as thick and thin liquids, and know that thin liquids flow more easily than thick liquids.

The resistance to flow is caused by the shear resistance between adjacent layers in the moving fluid. On this basis we can define viscosity as that property of a fluid which tends to prevent relative movement between adjacent parts within itself.

Viscosity index. The viscosity index (VI) of a lubricating oil indicates the change of viscosity that occurs with a change of temperature. A lubricating oil with a high-viscosity index has only a small change in viscosity over a large temperature change. Additives are used to improve the viscosity index of lubricating oils. The viscosity index is a dimensionless number.

Cloud point. The cloud point of an oil indicates the temperature at which waxes begin to form. The figure is important because as the wax crystallizes there is always the possiblity that filters will become clogged with the wax crystals.

Pour point. The pour point of an oil is the temperature at which it ceases to flow. The pour point of some oils may be lowered by the addition of additives known as pour-point depressants.

■ **3.8 How is the viscosity of an oil measured and what are its units?**

Work done by Newton established what is now known as Newton's law of viscosity. This law shows that the shear force set up in a slow-moving liquid is equal to the product of its absolute viscosity (now commonly known as dynamic viscosity) and the extremely small distance between adjacent layers within the liquid divided by the extremely small change in velocity between the adjacent layers. (Fig. 3.1) This can be written as an equation.

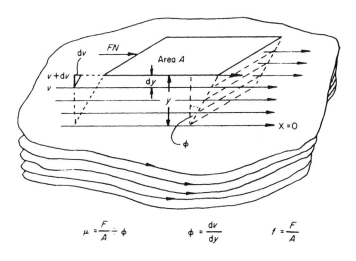

$$\mu = \frac{F}{A} \div \phi \qquad \phi = \frac{dv}{dy} \qquad f = \frac{F}{A}$$

Fig. 3.1 Diagrams showing the derivation of dynamic viscosity.

Let f = shear stress
  μ = dynamic viscosity
  v = fluid velocity
  y = depth from surface

By definition

$$\mu = \frac{f}{\phi} = \frac{f}{dv/dy}$$

or $\mu = \dfrac{\text{force} \times \text{time}}{\text{length}^2} = \text{N s/m}^2$

Viscometers based on Newton's law of viscosity have been devised to obtain values for the dynamic viscosity of a liquid. (Fig. 3.2)

One such instrument uses two hollow cylinders. The inside diameter of the larger cylinder is slightly greater than the outside diameter of the smaller cylinder. When the smaller cylinder is fitted in the larger there is a clearance between the two cylinders equal to dR. The inner cylinder is fixed to a slender vertical shaft. When the outside cylinder is rotated the shear stress set up in the liquid by movement of the outside cylinder is shown as an angular deflection on

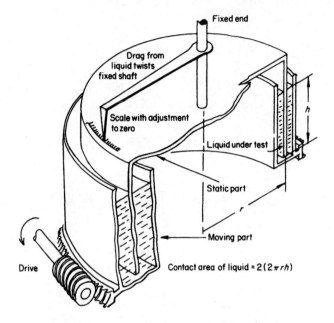

Fig. 3.2   Viscosity meter used for measuring dynamic viscosity.

the shaft. The deflection can be used to find the torque, and the shear stress on the liquid can then be calculated.

The SI unit for dynamic viscosity is the newton second per metre squared. In symbolic form it is written

N s/m², or Pa s

since 1 Pa = 1 N/m². This unit has the cumbersome name of pascal second, in France it is known as the the *poiseuille* (P1)

A unit commonly used in England and other countries is the poise (P) named after the French scientist Poiseuille.

1 poise (P)   = 0.1 N s/m²
1 centipoise (cP)   = 0.001 N s/m²

During the middle of the 19th century Poiseuille investigated the viscous friction encountered by various liquids when moving in capillary tubes. Poiseuille's work and that of others led to a set of equations making it possible to calculate the kinematic viscosity and the dynamic viscosity of liquids.

This led to the construction of viscometers such as the Saybolt viscometer, the Redwood viscometer, the Ubbelohde viscometer and other similar instruments. Viscometers of this type allow a defined volume of the liquid being tested to pass through a small orifice. The instrument can be calibrated by timing the passage of liquids of known viscosity. From this it is possible to a obtain a constant *C* for the instrument.

The set of equations mentioned above show that the product of the constant

C and the time taken for the liquid to pass through the orifice is equal to the dynamic viscosity of the liquid divided by its density. The dynamic viscosity of a liquid divided by its density is known as the kinematic viscosity. The kinematic viscosity of the liquid is obtained by multiplying the constant for the instrument by the time taken for the flow.

It is sometimes necessary to use the kinematic viscosity of a liquid in calculations. This is particularly so in the subjects of fluid mechanics and naval architecture and therefore it is useful to have values for the kinematic viscosity of liquids.

The SI unit for kinematic viscosity is the metre squared per second, m^2/s. An older unit has been named the stokes (St) after Sir George G. Stokes, a British scientist and mathematician.

$$
\begin{aligned}
1 \text{ Stokes (St)} \quad &= 0.0001 \text{ m}^2/\text{s} \\
&= 10^{-4} \text{ m}^2/\text{s} \\
1 \text{ centistokes (cSt)} &= 0.000\,001 \text{ m}^2/\text{s} \\
&= 10^{-6} \text{ m}^2/\text{s} \\
&= 1 \text{ mm}^2/\text{s}
\end{aligned}
$$

A third method of determining the viscosity of a liquid is based on the work of Stokes. His research led to what is known as Stoke's Law. This research investigated the time taken for steel balls to fall in a column of a viscous liquid.

Viscometers have been devised in which a steel ball is allowed to fall in a tube filled with the liquid under test. Some simple types of this viscometer are available for use in the field. These work by measuring the time taken for a steel ball to fall in a liquid under test, or by comparing the time taken for a ball to fall in a liquid of known viscosity with the time taken for an identical ball to fall in a liquid being tested.

For the mathematical background to Newton's law of viscosity, Poiseuille's work, Stoke's Law and the relationship of these and other laws, the student should refer to books dealing with fluid mechanics.

The viscosity of fuel and lubricating oils is usually quoted giving the viscosity value in cSt (centistokes) at some temperature.

In some cases the viscosity obtained at two separate temperatures is given.

Example

Kinematic viscosity (cSt) = 152.0 at 40°C
Kinematic viscosity (cSt) = 14.9 at 100°C

The viscosity of distillate fuel is quoted in cSt at a temperature of 40°C and the viscosity of residual fuel is quoted in cSt at a temperature of 80°C or 100°C.

Note Viscometers giving a digital readout of viscosity based on electronic circuitry are now available.

■ **3.9 State how pure mineral lubricating oil may be improved for use in diesel engines.**

Oils produced from various types of crude-oil stocks possess differing characteristics. Oil from paraffinic base stocks has generally a lower s.g. and a higher viscosity index than oils produced from naphthenic base stocks. The carbons formed from paraffinic base lubricants are generally harder than carbons formed from the naphthenic base lubricants. Naphthenic base lubricants have some natural detergency properties. Blending of oils from different base stocks makes for some improvement, but to obtain improvements necessary for the commercial success of the newer generation of high-output engines, additives must be used in the lubricants. The use of additives in internal combustion engine lubricating oils commenced in the 1920s and gained some impetus in the early 1940s from research carried out earlier. With the gain of knowledge from the field of organic chemistry the results from the use of additives have been spectacular in the last few years.

Crankcase oils in both crosshead and trunk-piston engines deteriorate in use through changes brought about by oxidation. When this occurs the viscosity of the oil increases, corrosive material in the form of organic acids is created, and sludges may form. Additives in the form of complex compounds of sulphur, phosphorus or both may be used. Another class of additives obtained from modified compounds of ammonia or derivatives of aromatic organic compounds is also used. These additives are referred to as antioxidants. Oils used in the crankcases of slow-speed engines where large amounts are in circulation do not suffer from oxidation at the same rate as high-rated high-speed auxiliary engines. In these engines the temperatures are higher, and much smaller amounts of oil are in circulation. The use of antioxidants reduces or prevents the breakdown of crankcase lubricating oil due to oxidation. Different additives act in different ways.

Cylinder oils used in crosshead engines and the crankcase oils used in trunk-piston engines, if untreated, allow carbon to build up in the piston-ring grooves. This is caused by the high temperatures involved. Combustion residues also form sludges which mix with the crankcase oil in trunk-piston engines. A group of additives referred to as detergent–dispersants are used to prevent carbon build-up in piston rings and to prevent the formation of deposits from sludges. The detergent–dispersants are generally complicated compounds of barium or calcium. Some types of engine develop lacquers on the pistons and cylinder liners from untreated lubricants; the use of detergent–dispersants prevents lacquer formation.

Certain engine parts and gearing may at times work very near boundary-lubrication conditions. To reduce wear rates under these conditions, fatty acids, or compounds of sulphur, phosphorus or chlorine may be used. The additive selected will depend on the severity of operating conditions and possible temperatures. Additives to improve the resistance of oils to very heavy loads are called extreme pressure (E.P.) additives. Additives may also be used to improve the viscosity index of an oil, i.e. to give an oil a higher viscosity at higher temperatures. They are usually polymers of high molecular weight.

High-molecular-weight polymers may also be used as additives in paraffinic base lubricants to lower the pour point. These additives are referred to as pour point depressants.

Oil used for lubrication of reduction gearing and in the crankcase of high-speed engines becomes aerated and may allow large volumes of foam to build up. Foam formation is prevented by using small quantities of an organic silicon compound in the form of a polymer.

Most of the treated oils in common use today are fairly stable in normal storage conditions.

■ **3.10 Name the properties or constituents that may be found in a blended fuel having a high viscosity and a high carbon content. Explain how they may cause problems in engine operation.**

Density. The ability of a centrifugal separator to function correctly and remove water and other foreign matter from fuel oil is dependent on the differences between the density of the oil, the water, and the foreign matter.

As the density of the oil increases the difference in the separating forces between the oil, the foreign matter, and the water is reduced; the ability of the centrifugal purifier to function correctly is then impaired.

Limits on the density of the fuel oil are fixed by the density it will have at the operating temperature of the centrifugal separator. The operating temperature of the separator must be less than the boiling point of water due to the problem of losing the water seal.

Viscosity. High viscosity values have a similar effect as high fuel densities on the action of a centrifugal separator.

If the viscosity of the fuel is such that it cannot be reduced by heating to the requirements of the fuel injection system, problems may arise with combustion and failure of parts in the injection system due to the high pressures that are created.

Pour point. Fuels having a pour point higher than the expected ambient storage temperatures must be maintained at some safe temperature above the pour point to prevent waxes coming out of suspension, or the oil congealing.

If some form of solidification occurs the action is often irreversible and causes serious technical problems and a heavy financial commitment to remove the solidified fuel.

As the pour point increases above the ambient temperature the demand for heating steam increases and may reach a point where the exhaust gas boiler cannot meet the total steam demand. In such cases it may be necessary to supplement the heat from the exhaust gases by firing the auxiliary boiler or shutting down the steam turbine generating set if one is fitted and running an auxiliary diesel set for electrical power.

Whatever is done, some of the saving from utilizing lower-cost heavier fuel is lost when the pour point is higher than the ambient storage temperatures.

Carbon residue. Fuels with a high carbon residue value often run into problems with combustion and the build up of carbon and other materials in the combustion chamber and exhaust system. This may affect exhaust valves, the exhaust gas section of turbochargers, the heating surfaces within exhaust boilers, and spaces in silencers or mufflers.

Asphalt content. When a fuel is high in asphalt or asphaltenes and low on certain aromatics some of the asphaltenes will not be held in solution and trouble may be experienced with fouling of filters and separators.

Fuels with a high asphaltene content burn relatively slowly when compared with fuels from a paraffinic base crude.

Generally the effect of high asphaltene content is similar to the effects of high carbon residue content.

Note   The paraffinic series of hydrocarbons have a chain-type molecular structure while the napthenic series of hydrocarbons have a ring structure. Asphaltic base crudes and napthenic base crudes are synonymous terms.

Sulphur is known to cause corrosive wear in cylinder liners but the problem of high sulphur content fuels has been overcome with alkaline cylinder oils. When burnt, sulphur forms gases having various combinations of sulphur and oxygen. Hydrogen when burnt creates H_2O in the form of steam vapour. If at any point in the exhaust system the exhaust gases fall in temperature below their dew point, corrosive acids are formed. These acids cause corrosion damage at the place where condensation of the acid vapours occurs.

Trouble, manifested by high rates of cylinder liner wear, has been experienced with very low sulphur fuels when used with some of the high-alkalinity cylinder lubricants available.

Silica and alumina. These fuel contaminants are very abrasive. If they are not removed when the fuel is cleaned in the separator and clarifier they may cause extensive wear of the fuel injection equipment in a very short space of time.

Sodium and vanadium. These contaminants are chemically combined with the fuel and cannot be removed by centrifuging. In conjunction with each other after combustion they are highly corrosive in the liquid state. If the exhaust valves cannot be operated at a sufficiently low temperature the corrosive products in the liquid state stick to the valve seating surfaces and lead to early problems with gas leakage. This results in burnt valves and low compression pressures leading to a loss of efficiency.

3.11   How can lubricating oils be tested for alkalinity or acidity?

Lubricating oils can be tested for alkalinity or acidity by a method known as titration in much the same way as the water taken from a boiler is tested for alkalinity. Titration is an old experimental method of volumetric analysis used by chemists. It is known that if an increasing amount of an acidic solution is added to an alkaline solution the resulting solution will eventually be made neutral. The measured amount of acidic solution required to neutralize the alkaline solution can be then used as a measure of the strength of the alkali in the alkaline solution.

In a similar manner acidic solutions can be tested for the strength of the acid content by carefully measuring the amount of alkaline solution to make the resulting solution neutral.

Acidic reagents such as standard solutions of hydrochloric acid can be titrated into alkaline oil solutions to measure their alkalinity; alkaline reagents such as standard solutions of potassium hydroxide can be titrated into acidic oil solutions to measure their acidity.

Colour indicators (e.g., methyl orange, litmus, phenolphthalein, etc.) are used to find the end point during titration as in boiler water testing. These indicators are not, however, suitable for finding the end point when testing solutions of lubricating oils.

Potentiometry is then used to find the end point of the titration. This involves placing electrodes or half cells in a solution of the oil under test and measuring any potential difference across them with a millivoltmeter. When the reagent is being added to the oil solution the potential difference measured on the millivoltmeter is recorded together with the amount of reagent added to the oil solution. The results are plotted on a graph with the end point shown by the midpoint where the plot changes from concave to convex curvature. It takes considerable skill to operate this type of analytical apparatus and interpret the results.

The final results are calculated and then reported as the weight of reagent required to neutralize some standard mass of the oil sample. Common units are milligrams of reagent to 1 gram of sample. The numbers corresponding to these units are then expressed as Total Acid Number (TAN) and Total Base Number (TBN).

Notes The specifications for these testing procedures covering equipment, the standard solutions, solvents, etc., are given in various standards adopted by different nations throughout the world.

Examples of the standards most commonly used by engineers internationally are ANSI/ASTM Standards D 664 and D 2896. These correspond respectively to the PI Standards 177 and 276.

ANSI: American National Standards Institute
ASTM: American Society for Testing and Materials
PI: Petroleum Institute

■ **3.12   How often should crankcase system oils be tested?**

The period allowed between testing of crankcase system oils will be dependent on whether the engine is of the crosshead type or the trunk-piston type.

For an engine of the crosshead type an acceptable period between laboratory testing is three months provided there is no bad case history of oil contamination for the engine.

For an engine of the trunk-piston type a satisfactory time interval for laboratory testing of crankcase oil will be dependent on a number of factors. These factors relate to engine design, engine rating, risk of fuel contamination, daily amount of oil make up relative to the quantity of oil being circulated, type of fuel being used, filtration of the oil, type of centrifuge treatment of fuel, type of bearings (e.g., white metal-Babbitt lined bearings or hard alloy thin wall bearings), period of time between oil changes if changed on the basis of number of engine operating hours and the degree and amount of on board oil testing carried out.

■ **3.13   What are the benefits of subjecting crankcase system lubricating oil samples to spectrographic or spectrochemical analysis at regular intervals of time?**

If an oil sample is analysed spectrographically the presence and the amount of any foreign metallic substances in the oil can be obtained.

The amount of the various metals shown in the analysis is then compared with the metal contents previously shown and recorded. An abnormal increase of one or more of the metal contents indicates abnormal wear and enables the investigating engineer to ascertain the region in which the increased wear rate is occurring. For example, if a rapid increase in the iron content and the aluminium content is being shown it could indicate an abnormal increase in the amount of wear in cylinder liners and on piston skirts.

An investigation should then be carried out to establish the causes of the abnormal wear and eventually remove them.

■ **3.14   What do you understand by the term ferrographic analysis? How does it differ from spectrographic analysis?**

Ferrographic analysis can be carried out on lubricating oil samples to achieve the same objectives as spectrographic analysis. In ferrographic analysis the sample of lubricating oil is first thinned with some solvents and then allowed to pass slowly down a slide set within a powerful magnetic field. The particles of ferrous and other metallic materials are then graded according to size along the slide.

The slides are examined in a special microscope using both reflected light and transmitted light together with red and green filters. The shape of the particles is used to identify the source of the wear debris.

In spectrographic analysis the wear debris in the oil sample is found by an examination of its spectra. Each individual component in the wear debris can then be identified by the wavelength of its electromagnetic radiation.

The testing of lubricating oil for use as a wear monitor is mainly applicable to high-speed engines and highly rated medium-speed engines. There is, however, nothing to preclude their use for slow-speed engines.

■ **3.15   The term 'microbial degradation' of oils is sometimes seen. To what does it refer?**

The term microbial degradation is sometimes referred to as biodegradation. It is the name given to the process whereby micro-organisms increase in number and decompose a hydrocarbon fuel or lubricant and eventually render it unfit for its duty.

This form of decomposition requires the presence of water together with other favourable environmental conditions including temperature, acidic conditions (pH value), and nutrients. With favourable environmental conditions the increase in microbial count may take place very quickly and cause rapid breakdown of the fuel or lubricant.

In the case of lubricating oils the additives in the oil may function as the nutrients.

■ **3.16   How will lubricating oil degradation (due to microbial growth) be noticed and what will be its effects?**

The indications of attack may be seen as follows.

Creation of sulphurous gases having a smell similar to bad eggs.
Build up of yellowish-coloured film on the inside of crankcases and the polished steel surfaces at the sides or unworn parts of bearings.
The colour of the oil darkening.
The oil tending to become opaque with a milky appearance.
Inability of the lubricating oil centrifuge to separate water from the oil due to the creation of stable emulsions.
Plugging of lubricating oil filters due to thick sludges.

The effect of degradation usually shows up on bearings and bearing journals as a corrosive attack in the form of pitting in both the journal and the bearing, or a breakdown of the bearing surface. This may show itself as staining and in extreme cases as a breakdown of the bearing lining alloy.

■ **3.17   How does microbial degradation of fuel oil manifest itself?**

Degradation of high-viscosity blended fuels does not usually occur due to the high temperature to which they are subjected in land-based storage, in bunker and settling tanks, when being pumped, and during cleaning treatment when passed through centrifugal separators and clarifiers. The high temperature sterilizes the fuel by killing off the organisms causing the degradation.

Degradation from various organisms only occurs with distillate fuels when water and the appropriate environmental conditions are present.

The degradation shows itself by smell, problems with filters becoming clogged at frequent intervals, the development of a slimy build-up in pipes and on storage tank surfaces, and by the creation of sludge.

■ **3.18   How can microbial degradation of distillate fuels and lubricating oil be prevented?**

Biocides and fungicides can be used to kill and prevent the spread of organisms within a distillate fuel oil. They can also be used in lubricating oils provided their use is approved by the oil supplier. Most of the treatments available cause some deterioration in the lubricating properties of the oil and their use should be carefully followed and observed.

The known organisms causing degradation are killed by preheating the lubricating oil to a temperature of 82.5°C during continuous separation treatment and when preparing to centrifuge the whole of the system oil charge.

Note   The temperature at which lubricating oils are heated prior to centrifuging should never exceed the supplier's recommendations.

Care should be exercised in preventing leakage of cooling water into the system oils in both crosshead and trunk-piston engines. Modern non-toxic anti-corrosion additives may act as a nutrient to the organisms causing degradation.

Some of the older additives are toxic to the organisms, but their use is banned in cooling systems used for heating low-pressure distilling plants producing potable water.

■ **3.19   What do you understand by SAE numbers?**

The SAE number of an oil is an indication of its viscosity based on a classification involving two temperatures. The authority responsible for the classification is the Society of Automotive Engineers.

Lubricating oils are marketed throughout the world on SAE number specifications. Crankcase lubrication oil or system oil usually has an SAE 30 number. Cylinder lubricants may be SAE 30, 40, or 50 depending on the service and type of engine.

■ **3.20   What are 'Supplement' and 'Series' lubricating oils? How do they differ from treated cylinder oils?**

Supplement 1, Series 2, and Series 3 lubricating oils are specially developed oils for use in highly rated trunk-piston engines. They are all high detergent–dispersant lubricants. The testing of these lubricants is performed in a test engine using a low-quality fuel. Test procedures are governed by rules drawn up by various authorities.

Series 3 oils under test are found to have twenty times the detergency level of untreated mineral oils; Series 2 and Supplement 1 lubricating oils have lower detergency levels.

Treated cylinder oils for use in crosshead engines remain in the cylinder until they drain down to the diaphragm space, and are used once only. The cylinders of trunk-piston engines are lubricated by mist and splash from the crankcase so the oil lubricating the cylinder is in a process of regular change with oil from the crankcase. On the basis of equal fuel quality, the cylinder lubricant of a crosshead engine, owing to once-only use, is subject to more rigorous conditions than the oil lubricating the cylinder of a trunk-piston engine. The use, in crosshead engines, of fuels containing larger amounts of impurities makes the conditions of crosshead engine cylinder lubrication even more severe.

Treated cylinder oils for use with lower-grade fuels have higher detergency and are more alkaline than the Series and Supplement oils. This is reflected in the total base number of the oil. The TBN of a cylinder lubricant is not, however, a direct guide to its effectiveness in holding a piston and piston rings clean and keeping ring and liner wear rates low.

The additives in cylinder oils are now usually oil soluble. This type of treated oil is more stable than the dispersion-type additive oils and the emulsion-type cylinder lubricants, both of which gave problems in storage. Dispersant additives tended to come out of suspension and form a gel on the bottom of the storage tank; the difficulty with emulsion-type lubricants was that water tended to separate out of the emulsion.

■ 3.21 What do Stokes's laws tell us about cleaning fuel oils?

One of the many studies made by the the scientist Stokes was that dealing with the behaviour of small spheres falling in viscous liquids. By mathematical analysis he came up with the following statement.

'The resistance to a sphere falling slowly in a viscous liquid is equal to six times the product of π, the fluid viscosity, the radius of the sphere, and its velocity.'

If

η = the coefficient of viscosity
r = the radius of the sphere
v = velocity of the sphere

then

$$\text{resistance to motion} = 6\pi\eta r v$$

The resistance to motion of the falling sphere is equal to the downward acting force causing motion. This is equal to the difference between the mass of the sphere and the buoyancy it receives from the fluid. The force acting downward is equal to

$$\frac{4\pi r^3 \delta_2}{3} - \frac{4\pi r^3 \delta_1}{3} = \frac{4\pi r^3}{3}(\delta_2 - \delta_1)$$

where δ_1 = density of fluid and δ_2 = density of sphere. Then

$$\frac{4\pi r^3}{3}(\delta_2 - \delta_1) = 6\pi\eta r v$$

from which

$$v = \frac{2r^2(\delta_2 - \delta_1)}{9\eta} \qquad \text{(Eq. 1)}$$

and

$$\eta = \frac{2r^2(\delta_2 - \delta_1)}{9v} \qquad \text{(Eq. 2)}$$

Examination of Eq. 1 shows that if the viscosity increases the velocity of the falling sphere decreases, and as the viscosity decreases the velocity of the falling sphere increases. For this reason it can be seen that when a viscous fuel is heated and its viscosity is decreased, any particles of water contained in the fuel will separate out more rapidly because of their higher velocity.

If Eq. 1 is further examined it can be shown that as the density of the fluid is reduced the velocity of the falling sphere is increased.

Water will separate more easily in a fuel which has a low density. Heating the fuel lowers its density and therefore makes separation of the water occur more easily. If the density of the the fuel is equal to the density of water the difference in the densities will be zero and the value of the velocity will become zero. Heating high-viscosity fuels is necessary to reduce their viscosity and their density in order to make separation of water occur more easily.

■ **3.22 What do you understand by the term separating force when applied to water particles contained in a fuel oil as applicable to water separating out in a settling tank? How does the separating force differ in a centrifugal purifier, or a centrifugal clarifier?**

It can be shown that the separating force between a very small volume of water and the oil surrounding it is equal to the difference between the mass of the water particle and the buoyancy it receives from an equal volume of oil.

If the particle or small volume of water is spherical this statement can be written in mathematical form with the separating force f_s equal to

$$f_s = \frac{4\pi r^3 g}{3}(\delta_2 - \delta_1)$$

$$= \frac{\pi d^3 g}{6}(\delta_2 - \delta_1)$$

since r $= \dfrac{d}{2}$

where f_s = SEPARATING FORCE newton
  r = radius of sphere of oil and equivalent sphere of water
  d = diameter ditto
  g = 9.81 m/s^2
  δ_1 = density of oil kg/m^3
  δ_2 = density of water kg/m^3

Note δ is the density of the oil and water at separation temperature.

If water is separated from the fuel in a centrifugal separator, the separating force experienced by the water particle in the settling tank will be multiplied by the centrifugal force acting on the particle when passing through the centrifugal separator. It should be noted that there is a limiting size to the particles of liquid or solid matter that can be separated, and particles of matter below a certain size will not be separated. The size is governed by throughput, viscosity, densities, rotational speed, and the internal construction of the bowl. The separating force in the centrifugal separator f_{sc} is equal to

$$\frac{\pi d^3 g}{6}(\delta_2 - \delta_1)\frac{v^2}{rg} = \frac{\pi d^3 v^2}{6r}(\delta_2 - \delta_1)$$

where

  d = diameter of water particle
  v = linear velocity of water particle
  r = radius of rotation of water particle

With separator bowl speeds of 6000 to 7000 rpm the separating force is increased many thousands of times.

■ 3.23 What is coalescer? Describe where it is used and how it operates.

A coalescer is a filtering device used for the separation of water and solid impurities from pure distillate fuels.

It has the advantage of being a static device and so requires little attention. However, it cannot cope with high-viscosity blends of distillate and residual fuels; also, if large amounts of water are present the cost of filter cartridge renewal becomes prohibitive. The space requirement for a coalescer is much greater than for normal separation equipment of the same capacity.

The coalescer is often in three parts, the first being a cascade-type filter through which the fuel is passed to remove the water and solids that separate out more easily. This stage may be heated and can be used in place of a settling tank.

The second stage consists of metallic filter elements of chosen dimensions that will filter out all solids above a certain size. Phenol-impregnated paper cartridges are incorporated with these elements. The paper elements cause the fine water particles to collect together or coalesce, and form water globules. The larger globules will separate from the oil straight away and collect in the bottom of the second stage. The smaller globules will pass with the fuel to the third stage.

The third stage consists of pleated paper elements. The paper in these elements is treated with a material to render it preferential to the oil and opposed to the water. The fuel is able to pass the element but the water globules (coalesced water particles) remain on the outside of the vertical element and drain down to a sump where they collect. The water-free fuel is then led out from the top of the third stage of the filter.

The material used to impregnate the pleated paper elements in the third stage must be such that it has the capability of reducing the surface tension of the oil in contact with the impregnated paper, so allowing the oil to pass through the capillaries formed by the porosity in the paper, while at the same time it must not affect the surface tension of the water globules, which will be too great to allow them to pass through the paper.

Note The surface tension of a liquid is dependent on inter-molecular forces which cause the molecules of a liquid to be attracted to one another. The surface may be the open surface or the interface or boundary surface between two different liquids as in oil and water. Materials that decrease the surface tension of liquids are referred to as wetting agents. Materials that decrease the surface tension of liquids other than water are referred to as hydrophobic wetting agents.

■ 3.24 What is a homogenizer? Describe where it might be used and how it operates.

A homogenizer is a piece of equipment used to create a material with a stable uniform structure (homogeneous structure) from a mixture of two or more finely divided solid materials or a mixture of immiscible liquids. It can be used to break down relatively large water particles within a heavy fuel into a homogeneous structure or emulsion consisting of water particles of the minutest size uniformly distributed throughout the resulting liquid.

A homogenizer can also be used to reconstitute an emulsion that has separated out from some heavy fuel, in order to give it some stability.

A homogenizer works by severely agitating the mixture being homogenized. The agitation can be carried out by mechanical means such as pumping the mixture through very fine orifices, or by acoustic means such as pumping the mixture in a thin layer over a surface being agitated at an ultrasonic frequency (above twenty kilohertz). The agitation can be created with any electronic device that will create ultrasonic pressure waves.

■ **3.25   What do you understand by the terms 'batch treatment' and 'continuous treatment' when applied to lubricating oil and fuel oil centrifugal separators?**

Originally the terms continuous treatment and batch treatment applied only to the purification of main engine crankcase lubricating oil.

In continuous treatment the separator was operated for the whole of the time the main engine was in operation. Lubricating oil was supplied to the separator by its own pump taking oil from the main engine drain tank, or was bled from the pressure supply to the main engine. The separator was never shut down except for short periods when it was being cleaned.

Batch treatment referred to the system whereby the engine was shut down and the whole of the sump lubricating oil charge was pumped up to the dirty lubricating oil tank in the upper part of the engine room. The lubricating oil was heated in the tank and left as long as possible to settle out solids, sludge and any water. It was then slowly purified in one batch, hence the name.

Today the terms have different meanings. Batch treatment centrifugal separators and clarifiers must be shut down for cleaning after treating a batch of oil.

Continuous treatment machines are capable of being cleaned or sludged without being shut down. These machines are also termed automatic self-cleaning separators and clarifiers or automatic separators and clarifiers.

■ **3.26   What are the essential differences between a separator and a clarifier? What are dam rings and sealing rings?**

A purifier is fitted with a dam ring which controls the position of the separation line or interface between the water and oil when the bowl is rotating.

If the inside of the dam ring is too large in diameter the separation line or interface moves outward towards the outer periphery of the bowl and some oil will be discharged with the water from the water outlet. If the inside diameter of the dam ring is too small the interface moves inwards and some water will be discharged with the oil. The diameter of the dam ring is governed by the density of the oil being treated.

If the diameter of the hole in the dam ring is increased, the interface between the oil and water contained in the bowl moves outwards. If the diameter is increased excessively, oil globules will be discharged with the sludge and water. If the hole diameter is reduced unduly, the interface moves inwards and particles of water will be discharged with the clean oil.

Holes are placed in the conical plates making up the plate stack to allow the oil to feed upwards into the clearance spaces between the conical discs. Water and heavy stable emulsions are discharged through the dam ring and spin off the bowl or are removed from the rotating bowl by a paring disc. Clean oil is discharged from the bowl and spins off or may be removed by another paring disc. Heavy solid matter is held in the bowl. (Fig. 3.3(a))

Clarifiers do not have a dam ring, a plain ring (also called a sealing ring) is fitted in its place, and the water outlet is sealed off. Solid matter and water that has passed through the separator are retained within the clarifier bowl until it is opened and the water and sludge are discharged. On modern machines surveillance and control devices watch out for the build-up of water within the bowl or when a very small amount of water is discharged with the clean oil. The bowl is then caused to open. Modern centrifuges are operated as clarifiers and controlled in this manner.

The conical disc stack may not be fitted with feed holes, but if they are fitted a blank conical disc without feed holes will be fitted at the bottom of the conical disc stack.

When a centrifugal separator is started up it must be filled with water to establish a seal which prevents oil leaving the separator at the water outlet. Clarifier bowls do not have to be filled with water after start up.

The main purpose of the clarifier is to remove traces of foreign material not removed when the oil passed through the separators, and to act as a second line of defence against the accidental passage of contaminant material into the clean side of the oil storage system. Clarifiers are not normally used to clean lubricating oil unless the oil is almost free of any water content. (See Fig. 3.3(b))

■ **3.27   What are the differences between the various types of self-cleaning separators and clarifiers?**

Self-cleaning separators and clarifiers found their place on board ship in the very early 1930s. The differences between these machines and others were in the walls of the bowl. On self-cleaning machines the bowl side or wall sloped at an angle instead of being vertical. A cross-section across the diameter of the bowl resembles the letter 'W'. The internal parts of the bowl are fastened in place and locked in position.

In order to clean the bowl it is only necessary to stop the machine, open the hinged cover and remove the bowl cover. The hinged cover on the separator casing is then closed and the bowl is run up to speed. The solid material, sludge, and water in the bowl is moved up the sloping sides by centrifugal force and then spun clear of the bowl which is left clean. Occasionally, however, the disc stack must be removed for cleaning the spaces between the conical discs.

At first this type of machine was used mainly for lubricating oil treatment. It was a good step forward at this time, particularly with engines that had

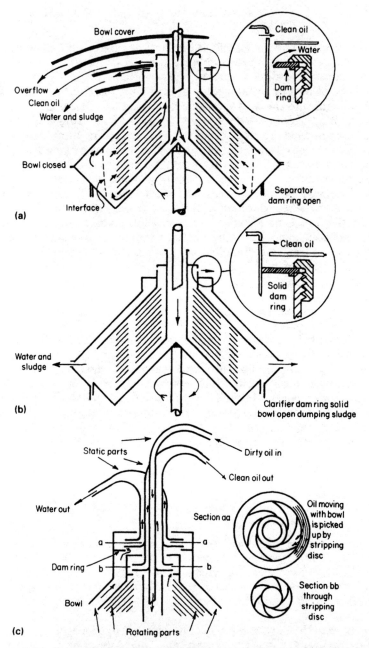

Fig. 3.3 (a) Diagrammatic arrangement showing a centrifuge arranged as a purifier. **Note**. The specific gravity ring is sometimes called a dam ring.
(b) Diagrammatic arrangement showing a centrifuge arranged as a clarifier. Note that the bowl has a solid dam ring. (c) Dirty oil feed through centrally located tube and discharge of oil and water with stripping discs as fitted in modern centrifuges. The stripping discs are static and the oil and water moving with the bowl are picked up by the outer periphery of the stripping disc.

oil-cooled pistons. The shut down time for cleaning the separator was reduced and the cleaning was made much easier.

Automatic self-cleaning separators were available to shore-based industries at about the same time. They were mainly used for the production of animal fats and vegetable oils. These automatic self-cleaning machines were not fitted into ships until some time later.

Automatic self-cleaning purifiers have the lower part of the bowl fitted into a lower fixed casing. The inner part can be made to slide up and down by hydraulic pressure. In the lower position (the cleaning position) the outer periphery of the bowl is open and allows solids, sludge and water to be spun outwards leaving the bowl clean. In the upper position the bowl is closed for normal operation.

After the bowl is closed sealing water is admitted and the separator is ready to receive dirty oil. Cleaning is completed in a matter of seconds without stopping the machine.

Clarifiers can be cleaned in a similar manner.

In older ships the cleaning cycle or dump cycle was carried out by manually operated valves controlling the water flow to the bowl's opening and closing mechanism. Later separators had timing devices to control the frequency of cleaning, and solenoid, pneumatic, or hydraulic actuated valves to control the water supply. The automation and increased frequency of the cleaning operation made the equipment suitable for unmanned engine rooms.

In modern ships, properly fitted out for handling high-viscosity, high carbon content fuels, the automatic cleaning operation is controlled by electronic timing equipment through a computer. Sensing devices in the dirty water outlet can be used to trigger bowl cleaning if this is necessary within the normal cleaning frequency, or a water-sensitive device can be fitted in the clean oil outlet to monitor the clean fuel for traces of water. If water is present the bowl-cleaning mechanism is triggered.

This type of electronic equipment is used on both separators and clarifiers. (See Question 3.30).

■ **3.28 Describe a centrifugal separator used for the purification of fuel and lubricating oils**

The main part of a centrifugal separator is the bowl, which is mounted on the top of vertical spindle supported in two bearings. On the spindle between the bearings is a multiple-tooth helical worm. This worm is driven by a helical-toothed gear wheel connected to a driving motor through a centrifugally operated clutch.

The bowl is cylindrical and closed at the bottom by a concave conical end. The spindle fits within the concave space formed by the bottom. On the outside of the upper end of the cylindrical bowl there is a coarse thread or a breech-block type thread. The bowl cover nut screws on to this thread. A section across the diameter of the bowl resembles the letter 'W'.

The fuel or lubricating oil is led into the separator bowl by a distributor similar to an inverted filling funnel. Radial ribs on the inside of the conical end give a uniform clearance between it and the end of the bowl. On the outside of the distribution pipe are four or six longitudinal fins. The conical separator plates fit over the fins and are driven by them. Radial fins on the conical separator plates hold them equidistant.

Over.the conical separator plates a conical inner cover is fitted. This inner cover keeps the separator plates in position. The radial clearance between the outer edge of the conical separator plates and the cylindrical bowl sides is approximately 30 mm in small separators and increases with increase in bowl size. The radial clearance between the inner cover and the sides of the bowl will be approximately 5 mm. A sleeve on the top of this inner cover and integral with it surrounds the distributor pipe and extends upwards to some point lower than the top of the distributor pipe. The space between the sleeve and the distributor pipe is the purified-oil outlet.

The top cover has a conical profile similar to the inner cover. Near its periphery is a groove which houses an oil- and heat-resistant rubber O-ring. This makes a seal between the cover and the vertical or sloping sides of the bowl. The top cover is held in place by a ring-nut which screws on the thread on the side of the bowl. On the top of the outer cover a thread is machined to take another ring-nut which holds the specific gravity ring or dam ring in place. A set of these rings having different internal diameters is supplied with every purifier. They are identified by a number or a specific gravity figure.

Some separators are fitted with specific gravity plugs instead of dam rings and function in a similar manner. The impurities separated from the oil are discharged over the dam ring or through the specific gravity plug.

As good balance is essential, the components of the bowl are designed so that they can be assembled in one position only, in relation to one another.

The fixed parts of the separator are the frame and the upper and lower spindle bearings housed in it. The upper bearing bush is held in a flexible housing. The lower bearing is arranged to support the weight of the bowl and thrust from the gearing. The frame also houses the bearings for the worm wheel and has a flange to support the flange-mounted driving motor.

The gearing is lubricated by splash from an oil bath and the bearings by forced lubrication from an oil pump driven by the motor. The frame also supports a flange-mounted tachometer and a hand priming pump for bearing lubrication prior to start-up.

On the upper part of the frame is fitted a hinged cover with shallow conical horizontal partitions inside. The diameter of the hole and height of these partitions is arranged to suit the various parts of the bowl assembly. The top partition takes the liquid spun off from the distributor pipe if an excess amount causes overflow from the bowl. The second partition takes the purified oil which spins off from the upper end of the inner cover. The lower space accepts the water, sludge and solid particles discharged over the dam ring or through the specific gravity plug. A centrally located nozzle in the cover supplies liquid to the purifier and extends downwards into the distributor when the cover is closed.

In use, the separator is first run up to its operating speed and then water is fed into the bowl until it discharges from the dam ring. This water provides a seal and prevents discharge of oil from the dam ring. After providing the initial water seal the mixture of oil, water and impurities is fed into the bowl through the distributor pipe.

After entry into the bowl the mixture is subjected to a large angular acceleration and then moves with the bowl. The speed of rotation subjects the liquid to
centrifugal force which breaks up the mixture into its constituents. The heavier
parts of the mixture, which include water and solids, move to the outer part of
the bowl and displace an equivalent volume of water from the dam ring. The
medium weight sludge and oil pass into the spaces between the conical separator
plates; the sludge breaks away from the oil and moves outwards, while the
purified oil continues to move towards the centre of the bowl and then upwards
to the purified-oil outlet.

A stable state develops with the heavier material at the outer part of the bowl,
the lighter, purified oil surrounding the distributor pipe, and the mixture within
the distributor pipe. The process of separation is continuous as liquid is fed into
the bowl. The efficiency of purification will depend on the amount of foreign
matter in the oil and the rate of flow through the separator. (Fig. 3.3(a))

■ **3.29 How are high-viscosity fuels prepared for use in the older generations of diesel-propelled ships?**

Fuel is prepared for use in an engine by settling and purification, the objective
being to remove the maximum amount of impurities.

After receipt on board, fuel is stored in double-bottom tanks, wing bunker
spaces, thwartship bunker spaces or forward deep tanks. The fuel is transferred
from the storage space with a transfer pump of the positive displacement type.
It may be a steam-driven simplex or duplex pump, or a motor-driven gear or
helix type pump. The fuel is heated to reduce its viscosity and so facilitate the
pumping operation. The transfer pump discharges the fuel to a settling tank.

The fuel may be further heated in the settling tank and retained there for as
long as possible, which may be from twelve to twenty-four hours. While the fuel
is in the settling tank some of the solids, emulsions, and water separates out and
collects at the bottom of the tank. The water can be drained off through the
drain connections which must be of the self-closing type. Separation takes place
owing to the difference in the specific gravities of fuel oil and the foreign
matter.

The equipment used for cleaning the fuel will consist of two centrifugal
separators associated with a stand-by machine, or two separators and one
clarifier.

The machines are connected up with pumps, heaters and associated piping
together with piping used to connect the cleaning system with the settling tanks
and clean oil or daily use tanks. The pumps in the cleaning system are usually of
the gear type driven through gearing by the separators or clarifiers. The fuel
heaters are heated by steam and the correct outlet temperature of the fuel is
maintained by the thermostatic control.

The separators can be used either singly or connected in parallel as in single-
stage purification. In other cases a separator and clarifier may be used together.
The fuel is first treated in the separator to remove the water, sludge and the bulk
of the solid matter. The separated fuel is then treated in the clarifier to remove
as much as possible of the remaining solid matter. This is referred to as two-
stage cleaning, the separator and the clarifier being connected in series.

The fuel in the settling tank runs down by gravity flow through a filter to a

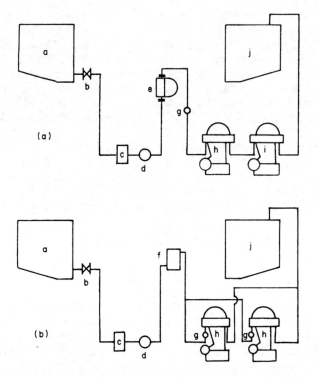

Fig. 3.4 (a) Arrangement of two centrifuges in series.
(b) Arrangement of two centrifuges in parallel.
a Dirty-oil tank or settling tank.
b Stop valve with remote closing control.
c Strainer – filter.
d Pump supplying oil to heaters and centrifuges.
e Heater (see fig. 15.1(c)).
f Plate-type heater.
g Flow control valve.
h Centrifuge arranged as a purifier.
i Centrifuge arranged as a clarifier.
j Clean-oil tank or day tank.

pump on the first centrifugal separator. The pump forces the fuel through the heater; it then flows to the first centrifugal separator, which is fitted with a gravity ring or gravity plug that allows water and other materials separated from the fuel to be discharged out of the separator. When a centrifugal separator is fitted with a gravity ring or gravity plug it is often referred to as a purifier. While passing through the purifier the untreated fuel is subjected to centrifugal action by rotation within the purifier bowl. This action considerably increases the forces causing separation of the foreign matter and water from the fuel.

The purified fuel leaves the purifier bowl and flows down to the purifier discharge pump which pumps it to the second heater (if fitted) and then to the second centrifugal separator. This separator is fitted with a closed gravity ring or plug and further impurities separated from the fuel are retained within the bowl. A centrifugal separator fitted with a blank gravity ring or solid plug is usually called a clarifier.

After the fuel has been treated in the clarifier, the clarifier discharge pump discharges the treated fuel to the clean-oil tank, where the purified and clarified fuel is stored until it is required by the engine. The fuel will then flow by gravity through a strainer to the fuel oil surcharge pump. This pump may be of the centrifugal type. The surcharge pump discharges the oil through a steam heater which will heat the fuel and give it the required injection viscosity. The viscosity of the fuel leaving the heater. will be controlled by a viscometer which regulates the flow of steam to the heater. After being heated to the correct viscosity the fuel is passed through a fine filter to the main engine.

Note Settling tanks and purifiers can only remove water and other impurities that exist as mixtures with the fuel. Impurities that are soluble in the oil, such as the vanadium and sodium compounds, will pass through the purifiers and not be separated out.

The water, sludge, and solid matter removed during cleaning is collected in a sludge tank and held there for disposal. The sludge tank is fitted with heating coils and connections for pumping out the sludge. The tank is often used to form the foundation for the separators and clarifiers. (Fig. 3.4)

■ 3.30 Describe how the techniques for handling and cleaning fuel should be changed to accommodate the high-viscosity, high-density, high carbon content fuels now being supplied and expected in the future.

The equipment required for handling and cleaning fuels having a high viscosity, a high density, and a high carbon content is similar to that fitted in older motorships (See Question 3.29) except in the areas mentioned below.

The fuel system must be fully automated, monitored, and carefully designed to cover all the control functions, cleaning functions and fail safe in the event of failure or shut down of any part.

This reduces the risk of any disastrous consequences that may arise if the fuel cleaning system goes out of adjustment and allows improperly prepared fuel to find its way into the main and auxiliary engines. Such problems can easily occur in modern practice with reduced engine room staff and more particularly if their time is fully taken up in dealing with some other crisis.

Insulation to heated fuel storage spaces and piping must be increased. Steam tracer lines must be fitted or their capacity increased on connecting pipe lines.

The capacity of heaters and their fouling factors should be increased. The means to clean oil heaters easily and rapidly should also be provided.

Separately driven positive-displacement pumps should be provided for handling fuel taken from the settling tanks for passage through the cleaning system and each machine should have its own pump. The capacity of the pumps should be carefully sized to suit the fuel requirements of the engine in conjunction with the capacity of the individual pieces of cleaning equipment.

If two or more separators are required to handle the maximum fuel requirement of the engine, the discharge from each pump must not be brought into a common line to feed the separators. Each separator should have its own heater or the fuel should be heated prior to being handled by the separator's own supply pumps.

The lowest-cost fuels supplied today have a high viscosity, density, and carbon content. These kinds of fuel require a high temperature when being treated in a fuel cleaning system consisting of separators and/or clarifiers.

The maximum temperature for heating the fuel prior to cleaning treatment is governed by the boiling point of water, which limits the preheating temperature to something a little less than 100°C. A figure given by separator manufacturers is 98°C with a tolerance of two more degrees.

The manufacturers of centrifugal-type fuel cleaning equipment are well able to offer a wide variety of machines, well suited to the fuels supplied today and expected in the future.

These machines do not always come within the previously accepted definition of separator or clarifier as they are not always fitted with a dam ring in the case of a separator or a sealing ring in the case of a clarifier.

The main advantage of these modern machines is that they can be operated without an internal water seal. In this respect they are similar to clarifiers.

Control of the cleaning function is governed by monitoring the build up of solids, sludge, and water contained within the bowl. When solids, sludge and water build up within the bowl the interface between the water and the partially cleaned fuel moves inwards and reaches a point where the cleaned fuel will contain traces of water.

One manufacturer uses a computer programmed to control the automatic cleaning or dump function of the purifier bowl in conjunction with a device extremely sensitive to the smallest trace of water. This device is fitted in the clean oil discharge from the purifier. During normal operation of the purifier the sludge and water discharge coming from the dirty side paring disc is shut off by a valve fitted in the discharge piping from the purifier.

The computer program covers fuels having a wide range of water content. When cleaning fuel containing limited amounts of water the dump cycle is programmed to act at regular fixed intervals. When the fixed interval expires the bowl-cleaning action is triggered by the computer and the solids, sludge and water are dumped out of the bowl when it opens.

If the clean oil outlet shows traces of water before the normal time interval has expired, the water-sensitive monitoring device relays a signal to the computer and the closed valve in the water outlet is opened. Water is then discharged from the bowl through the water paring disc and dirty water line. If the device monitoring the water content in the clean fuel shows a sharp drop in water content over a short time interval, the valve in the dirty water line remains open for some given period and then closes. The normal dump cycle is then repeated at the set interval following the previous dump cycle.

If the water content in the dirty fuel increases, the cycle of operations resulting in the discharge of water through the dirty water line is repeated and will continue if any water content is shown in the clean fuel. The dump cycle is again activated after the fixed interval measured from the previous dumping cycle. When this cycle is completed the bowl is clean and a new cycle begins. The period between the bowl dumping or cleaning operation remains the same irrespective of what may occur during the time between cleaning cycles.

If water content in the dirty fuel is more than the purifier can handle, the computer measures the rate of change of water content in the clean fuel with respect to time, and if this is above some accepted value the control system will activate alarms and cause the water contaminated fuel to be bypassed back to the settling tank.

One manufacturer's equipment monitors the fluid taken from the lower paring disc. Again, the time interval between clearing the solids, sludge and water out of the bowl by opening it is fixed in the computer program and never alters in amount.

When a centrifugal purifier of this type is started the bowl is completely filled with oil. The oil is discharged from the lower paring disc outlet line and monitored for conductivity before going through a two-way control valve and passing back to dirty oil feed to the purifier.

As water is removed from the fuel it builds up in the periphery of the bowl and will be discharged during the cleaning or dump cycle. If the water contained in the dirty fuel is above a certain amount, the water will build up in the bowl and cause the oil water interface to move inwards to some point where water will be discharged from the paring disc outlet. When this occurs the change in conductivity will be relayed by the sensor to the computer. The computer control will activate the two-way control valve and divert the water flowing through it away from the dirty oil supply and into the alternative route to the water outlet from the purifier. When the time for the dump cycle is reached, the solids, sludge and water, are discharged out of the bowl, in the normal manner and a new cycle commences when the purifier is filled with dirty oil again.

The remaining parts of the fuel oil system are similar to those mentioned in the answer to Question 3.28.

Normal separators and clarifiers can be used for cleaning very high-density fuel oils, but the sealing water in the purifier must have a higher density than the fuel when both are at the operating temperature of the separator. If the density of the fuel is equal to or higher than the density of the sealing water, the separator will not function correctly. A liquid having a higher density than the fuel must be selected to create an interface. Liquids having a density higher than the density of the fuel oils now available can be obtained by dissolving one of a variety of salts in water for use as the sealing medium. This can provide a seal with a density and boiling point higher than that of water.

The material selected will have to be carefully chosen because of the possibilities of setting up corrosion within the fuel cleaning equipment, sludge tanks and system pipework. Corrosion inhibitors will have to be used to give adequate corrosion protection if it is shown to be necessary.

The sealing liquid and water removed from the separator can be recycled after testing for density and inhibitor content.

■ **3.31   Why are fuel and lubricating oils heated prior to treatment in a centrifugal separator? Show in a simple manner how the forces causing separation vary in a settling tank and a centrifugal separator.**

It is necessary to heat fuel and lubricating oil prior to treatment in a centrifugal separator to reduce the viscosity of the oil so that it flows easily into and out of the separator and does not cause high pumping loads.

Heating the fuel or lubricating oil, which is a mixture of oil, water and solids, lowers the specific gravity of the constituent parts. The specific gravity of the oil is reduced at a greater rate than the specific gravity of the water and solids; thus the difference in the specific gravities of the constituent parts is greater when the mixture is heated.

	At 15°C		At 70°C	
	s.g.	Viscosity in centipoise	s.g.	Viscosity in centipoise
Water	1.000	1.42	0.982	0.43
Oil	0.898	100.00	0.848	10.00
Difference	0.102		0.134	

The oil in this example is SAE 30 lubricating oil of 100 VI.

Separation of the constituents in a mixture of oil, water and solids takes place in a settling tank due to the difference in the specific gravity of the various parts.

From the table it can be seen that heating the oil and water from 15°C to 70°C increases the difference in the specific gravity from 0.102 to 0.134, which is an increase of more than 30%. The viscosity of the oil has also been reduced by 90 centipoise. There will be considerable differences in the change in viscosity related to temperature for different kinds of oil. The change of the specific gravity in a wide range of oils will be approximately similar to the values given in the table for the same range of temperature.

The forces causing separation between a mixture of oil, water, and solid material will be many times higher in a centrifugal separator than in a settling tank. But to get this advantage the mixture must be heated.

The forces causing separation will also be related to the rotational speed of the separator, the bowl radius and the specific gravity of the constituents in the mixture. Separation of the constituents becomes increasingly easier as the differences between their specific gravities increase. For this reason it becomes necessary to preheat contaminated fuel or lubricating oils.

A fuel with a density of 1.005 g/ml at 15°C will have a density similar to that of water at 50°C. When this grade of oil is at 98°C it will have a density just below that of water at the same temperature.

■ **3.32   What is the difference between a strainer and a filter? Where are strainers and filters used and how does a filter differ from a centrifugal separator?**

Strainers and filters are similar in that they both serve to filter out contaminants contained in a fluid. Generally strainers are of the full-flow type and are fitted on the suction sides of pumps to prevent foreign bodies entering the pump and causing damage. Filters may be of the full-flow or by-pass type and are fitted in discharge lines. The filtering media are usually such that contaminants down to minute size are prevented from going into circulation. The pressure-drop across a clean filter will depend on the viscosity of the liquid being filtered, the size of the filter, the size of particles being removed and the flow rate through the filter.

A centrifugal separator separates foreign material from a liquid by virtue of the difference in the specific gravities of the parts to be separated and the liquid from which it is separated.

■ **3.33   What are simplex, duplex, by-pass and full-flow filters?**

A *simplex filter* is a single filter fitted in a line to remove contaminants from a liquid by a filtration process. *Duplex filters* are arranged in pairs so that one filter can be in use and the other ready for use or being cleaned. An arrangement of valves or cocks is usually fitted to duplex filters so that as one filter is opened to flow the other is closed. *By pass* filters take a part of the flow from a pump and return the filtered liquid back to the suction side of the pump. *Full-flow* filters take the full amount of discharge from a pump for filtration.

Note   Care must be taken with the change-over interlocking arrangements on duplex filters to ensure that they are always in good working order. Extreme care must be exercised in removing the cover of a shut-down filter. The air leak-off and drain cocks must be opened to check that the filter case is completely isolated before slackening the cover fastenings.

■ **3.34   Fuel with some definite viscosity value can be supplied. How is fuel of this type obtained and what difficulties may be experienced in using it?**

Fuel supplied to motor vessels is usually of the high-viscosity type. This kind of fuel is a blend of low-viscosity distillate and high-viscosity residual oils. The bunker supplier has tables which give the proportional amounts of each kind of oil required to obtain some specified viscosity. After the proportions have been established the blend is produced by using two pumps arranged to discharge into a common pipe. The size of the pipe is such that turbulent flow takes place and the two kinds of oil become well mixed. The speed of the two pumps is set so that the proportion of each kind of oil passing into the common discharge is maintained correctly.

Distillates from one crude stock type do not always blend well with residuals from another. This occurs when the smaller part is not soluble in the larger part. If incompatible oils are used to produce a blended fuel, precipitation will occur. This shows itself in the operation of the centrifugal purifiers which quickly fill with asphaltic material and extreme difficulty may be experienced in maintaining an adequate throughput of fuel through the purifier for the requirements of the engine. This problem is well known to fuel oil suppliers and they take every care to blend compatible types of fuel.

Normally fuels from different sources should not be mixed aboard ship. A blend that is compatible in itself may not be compatible with a blend from another origin, and precipitation may occur.

■ **3.35   What are oil-water emulsions? Where may they occur?**

Oil–water emulsions are liquid mixtures of small amounts of water finely dispersed through the oil. Although heavier than oil the water remains in suspension.

Oil–water emulsions may occur in fuel tanks; for instance when bunker fuel is taken into a double-bottom tank which has contained water ballast and not been properly drained prior to receiving the oil. Emulsions may also occur in the crankcase of diesel engines when the system oil becomes contaminated with residues of treated cylinder oils from diaphragm or scraper box leakage and small amounts of water.

If the emulsions, due to ship movement and pumping, become stable, considerable difficulties may arise.

Additives can be used to break up unpumpable fuel–water emulsions. This may be useful when these emulsions occur in places awkward and costly to clean such as double bottom tanks and the like. These additives are sold under various trade names and are often based on mixtures of the high boiling point fractions produced in coal tar distillation. One such mixture is known as cresylic acid.

■ **3.36   What is grease and where is it used?**

Grease is essentially a material used for lubrication of moving parts of machinery. It is generally made from a lubricating oil and a metallic soap. The soap acts as a thickening agent and is dispersed in the oil; the resultant grease may be semi-fluid or solid. Additives may be used in conjunction with the lubricant to give the grease more desirable qualities.

Sometimes materials other than metallic soaps and lubricating oils are used to produce a grease having characteristics to suit special requirements. Molybdenum disulphide is used in many applications where loads are high and speeds slow.

Greases are used in electric motor bearings, roller and ball bearings and in applications not easily accessible such as rudder bearings. Grease may also be used to exclude foreign matter from bearings in circumstances in which seals would be costly or impracticable.

■ **3.37   What is the probability of a ship receiving bad fuel oil leading to problems with the main propulsion and auxiliary engines?**

Studies made by various organisations show that there are limited chances of receiving a badly contaminated fuel, high in solids and other material, leading to problems with combustion and erosion of parts. But this does not mean that it will not happen. If contaminated fuel does come aboard unnoticed, the financial costs of repairs and lost time could be disastrous. Use should, therefore, be made of the means available for testing the fuel received on board.

Ship classification societies such as the American Bureau of Shipping, Lloyd's Register of Shipping, Det Norske Veritas, and others, together with some commercial concerns, offer a fuel testing service. The ship's engineers take samples of the newly received fuel. These are sent through accredited agents by one of the available fast courier services to a laboratory for analytical tests. The results are notified to the owner within hours of the samples having been taken. The ship's staff are notified of the result and advised of any requirements that may be necessary to avoid problems and risk of damage.

Fuel tests can also be made on board by the ship's engineers when the fuel is received. These tests will give an early warning of likely problems. Testing equipment sets for shipboard use are sold by various organisations.

Problems with poor-quality fuel have sometimes begun when the department purchasing fuel opted for what appeared to be the lowest-cost fuel without consultation with the technical department (See Question 4.40).

■ **3.38 Are any specifications available covering quality of fuel oil?**

Specifications covering fuel oil standards and quality have been drawn up by various organisations. Some organisations involved in this work are the British Standards Institution (BSI); Conseil International des Machines à Combustion, commonly known by its acronym CIMAC; the International Standards Organisation (ISO); the American Society for Testing and Materials (ASTM); the American National Standards Institute (ANSI), in conjunction with many others including international and national ship-owning organisations and major oil suppliers.

Standards, including the ISO standards for intermediate marine fuels (blended fuels), are now available, but they have not been universally adopted.

4

COMBUSTION AND
FUEL-INJECTION SYSTEMS

■ **4.1   Briefly describe the principle of the fuel-injection pump and give
reasons for the important points of design.**

There is one pump for each cylinder. Fuel is pumped by a ram working in a
sleeve, the ram being operated on the pumping stroke by a cam, and on the
return stroke by a spring. It works at constant stroke, and the amount of fuel
delivered is varied by varying the point in the stroke at which the pressure side of
the ram is put into communication with the suction side. When this point is
reached the pressure drops suddenly, fuel ceases to be delivered to the engine
cylinder, and for the remainder of the stroke of the ram oil is merely pumped
back to the suction side of the pump. The means of putting the pressure side of
the ram into communication with the suction side is a helix-shaped groove on
the side of the ram, registering with a hole in the sleeve; according to the
rotational position of the ram the helix-shaped groove will register with the hole
early or late in the stroke of the ram.

The requirements that lead to fuel-injection pumps being of this design are as
follows.

1   The need to build up to full pressure very rapidly at the beginning of
injection so as to ensure full atomization immediately. Cams lend themselves to design for very rapid initial movement of the ram.
2   The need to cut-off ram pressure as suddenly as possible to prevent dribble.
The opening of a port does this more effectively than the ending of a
(variable) stroke.
3   The need to deliver extremely small quantities of fuel when the engine is
running light, with the same requirements of rapid build-up, rapid cut-off,
and full atomization. Cam drive and port cut-off is the most effective way
of achieving this.

When the cut-off point occurs and fuel discharge stops, the rapid drop of
pressure causes a shock on the inlet line to the pump. This may be damped out
by a shock absorber fitted on the side of the fuel pump and connected with the

oil supply line to the pump. Guard plates are fitted round the bottom of the ram to prevent fuel oil leakages going into the engine lubrication system.

Some fuel pumps are fitted with delivery valves. Where delivery valves are fitted the wings on the valve are not cut through to the valve mitre; this part of the valve acts as a piston. When the delivery valve opens, a large lift is required to give it opening area; when delivery stops, the volume of the space between the pump and the injection valve is increased rapidly as the pump delivery valve closes. This sharp increase in volume allows the injection valve to seat quickly and prevents dribble.

The parts of a fuel pump are very robust to withstand the discharge pressures built up in the fuel system. These pressures may be up to 450 bars (approx. 6500 lb/in^2 or 450 kg/cm^2).

The rotation of the fuel ram to meter the volume of fuel delivered by the fuel pump may be controlled by a governor, in generators and alternators, or by the fuel lever, in propulsion engines.

■ 4.2 Describe how the fuel quantity delivered by a fuel-injection pump can be controlled through the pump suction valve.

The amount of fuel delivered is controlled by the clearance or lost motion of a tappet positioned under the fuel pump suction valve. The tappet lever is connected into a collar or slot on the fuel pump ram. The lever fulcrum is a section of a small shaft which is made eccentric to the shaft bearings; rotation of the shaft alters the eccentric position and in turn varies the tappet clearance. The shaft may be connected to a governor or to the engine fuel lever.

When the fuel pump ram is in the bottom position, the suction valve is held open by the tappet. Upward movement of the ram moves one end of the lever (connected to the ram) upwards and the other end (connected to the tappet) downwards, and the suction valve comes on its seat. Further upward movement of the ram builds up pressure in the fuel line; the fuel valve opens and fuel is injected into the cylinder until the pump comes to the end of its stroke. This type of fuel pump is fitted with a delivery valve on the top of the pump chamber. Downward movement of the pump ram allows fuel to flow into the pump through the suction valve as in a normal pump.

■ 4.3 How can fuel be taken off a cylinder without stopping the engine?

If it becomes necessary to take fuel off a cylinder without stopping the engine, the fuel pump crosshead is lifted until the roller is clear of the cam peak. The by-pass on the fuel valve must be opened before lifting the crosshead.

A lever is supplied by the engine builder which connects to the pump crosshead through a bolt and finger-plate. Lifting the bolt with the lever connects the finger-plate to the underside of the pump crosshead, and lifts it as the lever is pushed downwards or upwards depending on the linkage arrangement. After lifting, the crosshead can be fastened in the lifted position.

In some cases the pump crosshead is lifted by a small pin on an axle. The pin is eccentric to the axle axis, and movement of the axle connects the pin with the crosshead, lifting the roller clear of the cam.

■ **4.4 What do you understand by the terms 'atomization', 'penetration' and 'turbulence'?**

Atomization is the break-up of the fuel charge into very small particles when it is injected into the cylinder.

Penetration refers to the distance that the fuel particles travel or penetrate into the combustion chamber.

Turbulence or swirl refers to the air movement pattern within the combustion chamber at the end of compression.

■ **4.5 Why are atomization, penetration and turbulence most important?**

Correct combustion of the fuel and economic engine operation are dependent upon proper atomization, penetration and turbulence.

If the atomization is excessive, particles of fuel will be smaller and will have insufficient kinetic energy to carry them through the combustion chamber space. The compressed air, being very dense, has a high resistance to the motion of the particles. They will tend to cluster around the fuel injector tip and then become oxygen starved during combustion. The rate of combustion will be reduced, which may lead to after-burning.

If the atomization is insufficient the particles will be larger and will have more kinetic energy when entering the combustion chamber. They will therefore travel further into the combustion chamber space and some may come to rest on the cylinder wall. This will lead to a lower rate of combustion and the possibility of after-burning. A carbon build-up will occur around the top of the cylinder liner and the side of the piston crown.

The degree of penetration is dependent on the amount of atomization. Atomization is obtained only at the expense of penetration, provided sprayer hole size, injection pressure and injection viscosity do not change.

Atomization, penetration and turbulence all contribute to obtaining the best conditions for the combustion of the fuel. The aim is to get the particles small enough to burn in the short time available, and evenly distributed throughout the combustion chamber so that adjacent particles do not become oxygen starved during combustion. The penetration and turbulence contribute to evenly spaced particle distribution.

■ **4.6 Describe a fuel-injection valve. What size are the holes in the sprayer tip?**

The body of a fuel-injection valve is normally flanged at its upper end. The lower end is threaded to accommodate the sprayer nozzle and nozzle capnut.

The body of the valve contains four holes. One is for the fuel inlet and another for the fuel priming valve; these two holes are connected through a common space within the fuel nozzle or by an annular space. The remaining pair of holes accommodate the fuel valve cooling service inlet and outlet connections.

The centre part of the body is drilled through and along its axis. This hole is tapped at the end for the spring adjusting screw and may also house an adjustment screw to control the fuel-valve needle lift. The lower end of the centre hole is counterbored to house the spring which holds the needle valve on its seat.

The end of the fuel valve or nozzle may be made in one, two, or three pieces depending on the design.

The valve needle and spindle are in one piece, and are lapped into a very accurately machined guide. Valves produced for operation on diesel fuel or gas oil have a very small clearance while valves made for operation on high-viscosity fuel have a slightly larger clearance to allow for temperature changes when working with heated fuel.

The lower end of the valve spindle is of smaller diameter and has a conical end which forms the valve. The conical end comes on to the valve seat, below which is an open space. The sprayer holes are drilled through the tip and connect into the space below the needle.

The end of the valve spindle is of reduced diameter and forms an annulus within the guide end, into which both fuel inlet drilling and priming valve drilling lead.

On the upper end of the valve spindle is a disc on which the valve spring bears. A similar disc on the top of the spring bears on the spring adjustment screw.

The nozzle tip is cooled by the circulation of coolant through the space around the bottom of the nozzle, which is connected to the cooling services through holes drilled to the top of the valve body.

The nozzle end is fastened to the body by the capnut. The faces between the parts carrying high-pressure fuel are lapped together to make them oil-tight.

The valve is held in the cylinder cover by studs passing through the flanged end of the valve body. A rubber ring may be fitted around the top of the valve to prevent fluids and dirt entering the valve pocket. Gas-tightness is made by a lapped joint between the end of the valve and the cylinder cover.

The valve is opened by a shock wave and by oil pressure within the annulus formed at the end of the valve spindle, where it is machined to a smaller diameter. Once the needle valve lifts off its seat the effective area subject to oil pressure increases from the annular area to the full cross-sectional area of the spindle. This gives the valve a good effective lift and prevents any bleeding action occuring across the valve seat.

The number of sprayer holes in a fuel-injection valve tip varies between three and eight, and they may have a diameter of 0.075 mm to 1.0 mm. They will be located to give the most effective spray pattern. Some valve atomizer tips have holes of two different sizes.

Some high-speed engines fitted with an antechamber have a fuel-injection valve with a centrally located hole. There is a small tip on the end of the spindle below the valve seat. This fits into the centrally located hole with a small clearance around it. The small piece on the end of the spindle is called a pintle and the injector is known as a pintle injector. It is less costly to manufacture than a valve with a multiple number of spray holes of very small diameter (Fig. 4.1). The fuel valves fitted on small engines do not have cooled atomizer tips. The fuel valves fitted on modern, long-stroke, slow-speed engines are not cooled in a conventional manner; a continuous circulation of fuel through the

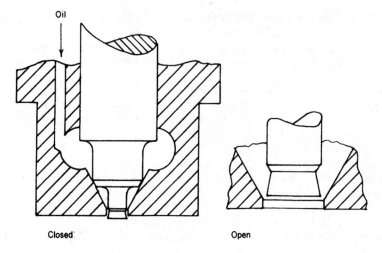

Oil

Closed: Open

Fig. 4.1   Pintle-type fuel injector used in high-speed automotive engine where fuel is injected into an ante chamber.

valve (except when injection occurs) keeps the nozzle tip cool (see Question 4.37).

■ **4.7   When a fuel valve is being tested the spray coverage is much greater than the cylinder diameter. If the injection test pressure and valve lift are correct, what occurs in the cylinder to prevent the fuel particles coming into contact with the piston or cylinder?**

When a fuel valve is tested the spray emerges into air at normal temperature and pressure, which is less dense than the air in a cylinder at the end of compression. Resistance to movement of the fuel particles in dense air is much greater than in air at atmospheric pressure; hence the spray coverage will be much less when it emerges into air at high pressure and temperature, as in the combustion chamber.

■ **4.8   What is the shape of the spray extending from a hole in a fuel atomizer during injection of fuel?**

The spray pattern of the fuel is cone-shaped. The density of the fuel particles is greater towards the centre or core of the cone.

■ **4.9   Describe the action that takes place in the fuel pump, piping and injection valves just prior to and during injection of fuel.**

Before fuel can be injected into a cylinder the pressure must rise to the point at which the fuel valve lifts. The pressure required will depend on various factors, but will be between 245 and 445 bar. In some of the latest generation of engines the fuel-injection pressures may go up to about 1000 bar; in some medium-speed engines using low-cost fuels, higher injection pressures going up to about 1250 bar may be used.

While the pressure is rising the parts of the system including the fuel pump, piping and valve will be subjected to an increasing total strain. The increase in strain causes an increase in the volume of the space containing the fuel and requires some time period for the effect to take place; the time period is very small.

The cam and fuel-pump design is such that the initial upward movement of the fuel ram continues with increasing acceleration. During this initial acceleration fuel is discharged back into the suction side of the fuel pump, and by the time the inlet port or suction valve is closed the fuel-pump ram is moving at high velocity.

The rate of pressure rise is very high and transmits a shock wave through the oil contained in the fuel-pressure piping: this shock wave causes the injection valve to lift, and injection commences. The fuel pressure continues to rise after injection commences at a rate dependent on the rate of fuel injection. The time taken for valve opening during the rising pressure period is referred to as injection lag. It may be given as a measurement of time or as a crank displacement angle.

When the fuel pressure falls at the end of injection the material of the parts of the fuel system reverts to its normal size.

■ **4.10 Fuel can be metered by controlling the fuel-pump suction valve or by helix and port control. What important difference in effect does each system have on injection timing with change of load?**

Fuel pumps fitted with only one suction valve may use this valve to control either the beginning or the ending of the injection period. This will be determined by the location of the fulcrum point on the lever controlling the valve tappet.

The fuel quantity is metered by placing the fulcrum point on an eccentric. Turning the eccentric alters the clearance of the tappet controlling the suction valve and holds it open for a longer or shorter duration during the pump stroke.

If the fulcrum point is such that the valve tappet moves upwards when the pump ram is moving upwards, the valve will control the ending of injection. When the valve is opened by upward movement of the tappet, fuel is spilled into the suction line until the ram reaches the top of its stroke. The beginning of injection will always take place at the same crank position.

If the fulcrum point is such that the valve tappet moves downwards when the pump ram is moving upwards, the valve will control the beginning of the fuel injection period and the ending will always occur at the same crank position. As the pump ram is moving upwards, fuel is spilled back into the suction line until the valve closes and injection begins. As the load on the engine increases the suction valve will close earlier and the commencement of injection will be advanced relative to the crank position, but the ending of injection will not be changed.

It is possible to control both the beginning and the ending of the fuel-injection period when two valves are fitted with each having its own connection to the fuel inlet side of the injection pump, and lever fulcrum positions located so that one valve moves upwards when the other moves downwards. Each tappet lever will require the fulcrum to be fitted on an eccentric.

With fuel pumps constructed in this manner the valve controlling the beginning of injection is the suction valve. The valve controlling the end of injection becomes the spill valve. As the fuel quantity is increased to suit an increasing load the suction valve closes earlier and injection commences earlier. By arranging controls to cover both valves it is possible to vary the injection timing. Fuel cannot be discharged while either the suction or the spill valve remains open. By arranging external controls for the eccentrics it is possible to make small changes in the injection timing without altering the angular position of the fuel pump cam on the camshaft. (Fig. 4.2)

Note   In fuel-injection pumps having a single suction valve, it is common practice to arrange for the suction valve tappet to move in the opposite direction to the pump ram. The ram then gains some velocity before the suction valve closes. This gives the high rate of pressure rise necessary in the fuel system to obtain a sharp opening of the fuel-injection valve.

In the Bosch type of fuel pump, commonly referred to as a 'jerk' pump, the angular position of the helix relative to the suction and spill ports controls the quantity of fuel injected into the cylinder during each working cycle.

The angular position of the ram relative to the position of the suction and spill ports varies the effective length of the discharge stroke to suit the fuel demand of the load on the engine.

Injection commences when the ram is moving upwards and closes off the suction and spill ports; injection ends when the continuing upward movement again opens the ports in the pump barrel.

The direction in which the helical-edged slot is cut in the top of the ram determines whether the commencement of injection or the ending of injection varies with the position of the crank.

When a left-handed helical slot is cut, injection always commences at the same crank position. The top edge of the ram closes off the ports and injection commences. The helical slot opens the ports, spill commences and injection ends. The end of injection occurs later as the engine load increases and varies with the circumferential position of the pump ram and with the crank angle.

If a right-handed helical slot is cut the commencent of the injection period begins when the edge of the helical slot cuts off the spill and suction ports, the ending of injection occurs when the circumferential groove at the lower part of the helix opens the ports and spill begins. The ending of injection always occurs at the same crank position. Commencement of injection begins earlier with increasing engine load. (Figs 4.3 and 4.4)

■   **4.11   Why is it necessary to atomize the fuel charge when it is injected into the cylinder?**

The time available for combustion of fuel is very small and by atomizing the fuel it is possible to complete combustion within this time.

The rate of combustion of fuel in compressed air increases when the particle

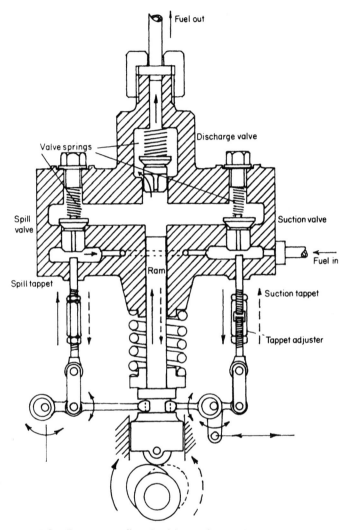

Fuel out

Discharge valve

Valve springs

Spill valve

Suction valve

Fuel in

Ram

Spill tappet

Suction tappet

Tappet adjuster

Fig. 4.2 Fuel pump fitted with suction and spill valves.

size is reduced. It is obvious that a sphere of fuel having a volume of one cubic millimetre will burn quicker than a sphere of fuel having a volume of one cubic centimetre provided neither are starved of air or oxygen. Similarly one may say, one thousand spheres of fuel each having a volume of one cubic millimetre will burn in the same time-span as one sphere having a volume of one cubic millimetre provided that sufficient air is available.

As the number of particles in a fuel charge is increased by making them smaller the sum of the surface areas of the particles increases. The smaller particles have a greater amount of contact with air and this enables the fuel charge to burn faster (see Question 4.34).

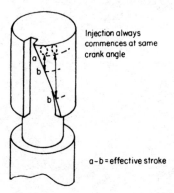

Injection always
commences at same
crank angle

a–b = effective stroke

Fig. 4.3   Fuel pump ram with left-hand helix.

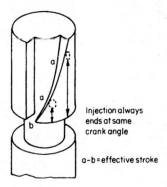

Injection always
ends at same
crank angle

a–b = effective stroke

Fig. 4.4   Fuel pump ram with right-hand helix.

■ **4.12   Describe how the combustion of fuel commences and proceeds in a diesel engine cylinder.**

Combustion of the fuel charge takes place in three distinct phases, which are shown on a pressure-crank displacement diagram or an out-of-phase indicator card.

The fuel emerges into the cylinder as small liquid particles which are surrounded by hot compressed air. They receive heat from the air, and the more volatile constituents of the fuel vaporize. The vapour formed surrounds each particle and leaves a trail as the particle moves into the combustion chamber.

The shape of the spray from each nozzle hole is conical and the particles are at a greater density at the central core of the cone; the fuel particles on the outside of the cone therefore receive heat faster than those towards the core.

After the start of injection, self-ignition of the vapour occurs and combustion commences. The time-span between commencement of injection and the start of ignition is referred to as the ignition delay period. During the ignition delay period a large part of the fuel charge is prepared for combustion. After ignition commences flame propagation proceeds very quickly in the fuel particle/fuel vapour/air mixture, accompanied by rapid temperature and pressure increase. The pressure increase is accentuated by upward movement of the piston. This second phase of combustion is shown on the pressure-crank displacement diagram as a turning-point in the compression curve, after which the pressure rise is very rapid; this point marks the commencement of the second phase of combustion.

Towards the end of the rapid pressure rise a point will be reached where the rate of pressure rise falls away quickly, and the curve flattens out towards the maximum-pressure point. The point where the rate of pressure rise changes, near and approaching the maximum-pressure point, marks the end of the second phase of combustion. This point usually occurs just prior to the end of injection.

The third phase of combustion shows only a small pressure rise, as the rate is decreased due to downward movement of the piston. The end of injection occurs approximately at or slightly beyond the maximum-pressure point.

Combustion phase	Crank rotation	% fuel charge injected	% total heat liberated	Remarks
One	3°–20°	40–80	Nil	Ignition delay period. Fuel injected undergoes physical and chemical changes. More volatile constituents vaporize.
Two	5°–10°	10–40	35–70	Ignition started. Volatile and gaseous constituents burn. Pressure rise very rapid.
Three	5° or more	10–20	30–65	Completion of combustion of remaining fuel constituents.

Note The individual values for each engine type will vary, and further variations will occur in the figures with load and fuel changes.

■ **4.13 What is after-burning, and how is it caused?**

After-burning is said to occur when the third phase of combustion extends over a long period. It may be caused by incorrect fuel grade, bad atomization, poor or excess penetration, incorrect fuel temperature, incorrect injection timing, insufficient air supply, or any combination of these.

Slow-burning, high-viscosity, high-density, high carbon content fuels may also cause after-burning of a serious nature leading to engine damage.

■ **4.14 What effects does after-burning have on an engine?**

After-burning creates high exhaust temperatures and may cause over-heating of the engine in severe cases. It is usually accompanied by some drop in the maximum firing pressure. There is a loss of thermal efficiency when after-burning occurs, due to a greater loss of heat to exhaust gases and the transfer of larger amounts of heat to the cooling water. There is a risk of damage to exhaust valves and of scavenge fires.

■ **4.15   How does the viscosity of a fuel change with increase of pressure?**

As the pressure rises in the fuel system during injection there is a considerable rise in the viscosity of the oil under pressure. The table shows the increase in viscosity of a fuel oil having the following properties:

viscosity at 100°C   =   25 cSt
density at 15°C   =   0.991 g/ml
density at 100°C   =   0.941 g/ml

P bar	0	150	300	450	600
P psi	0	2175	4350	6525	8700
Vis. cP	23.5	29.8	37.8	47.9	60.7
Vis. cSt	25.0	31.7	40.2	50.9	64.5

The values given for the increase of viscosity are approximate calculated values for a fuel obtained from a paraffinic base crude.

■ **4.16   What is the fuel viscosity range within which fuel-injector valves are designed to work?**

Sprayer hole sizes are interrelated with injection pressure requirements and penetration requirements, rate of injection and the type of fuel burned in the engine.

For engines operating on marine diesel oil the injectors are manufactured to operate with a fuel having a viscosity between 5.0 cSt and 15 cSt at injection temperature.

For engines burning heavy fuel oils the viscosity of the oil is controlled by the temperature of the fuel leaving the oil heater. The viscosity of the fuel leaving the heater is maintained between 10.0 cSt and 15.0 cSt. Engine manufacturers sometimes quote a maximum viscosity that should not be exceeded and this figure is about the same as the maximum viscosity of diesel oil, that is 20 cSt. In all cases the engine builder's recommended figure for fuel viscosity should be complied with.

It must also be remembered that overheating the fuel may lead to fouling or coking of the fuel oil heaters.

■ **4.17   What is the consequence of operating an engine on high-viscosity fuel at too low a temperature?**

If an engine is operated with a fuel of the high-viscosity type at too low a temperature the injection viscosity will be too high. This will affect the degree of atomization and penetration.

With a small decrease below the correct fuel temperature poor atomization and penetration will cause after-burning to occur. Further decrease of the fuel temperature increases the amount of after-burning. Eventually the point will be reached when fuel will come into contact with the piston crown, the sides of the piston, and the cylinder walls, and will burn on the surfaces of these parts.

The fuel coming into contact with the cylinder walls destroys the lubricant and causes increased wear. Fuel on the piston sides enters the ring grooves, forms carbon, and eventually seizes the rings in the grooves so that blow-past occurs; finally the sides of the piston burn away in localized areas.

While an engine will continue to operate with fuel at quite low temperatures, the consequences are disastrous and result is very costly repair work to correct the damage.

■ **4.18 Briefly describe the various causes of dark-coloured smoke in the exhaust gas of a diesel engine.**

Dark-coloured smoke is caused by fine unburnt carbon particles being present in the exhaust gases owing to poor or incomplete combustion of the fuel. The shade darkens as the amount of carbon particles in the gas increases. Poor combustion is usually due to faults in the fuel-injection equipment or to insufficient air supply.

Possible faults in injection equipment are dirty injector tips, incorrect spring setting pressures, incorrect valve lift, or incorrect timing of fuel-injection pumps. Leakage of fuel from individual fuel pumps, pipes and valves can cause imbalance of power and overloading of the other cylinders.

Air supply may be insufficient because of dirty air filters, faults in turbo-blowers, worn-out inlet valve cams, or dirty scavenge ports.

■ **4.19 How would you investigate and find the causes of an engine smoking?**

When starting to investigate the cause of smoke there may be some difficulty in locating which engine is smoking, since wind may drift the smoke from one exhaust pipe over another. When the offending engine has been identified it must be investigated to find if one cylinder is causing the trouble or whether the fault is in all cylinders.

The plugs or cocks that are usually found on each exhaust elbow or branch leading into the manifold should be removed one at a time, and the exhaust gases examined for smoke as they blow out. It is sometimes helpful to use a white surface with good light-reflecting qualities and extra lighting to assist in the observations. If all the cylinders are found to be smoking, the following points should be checked.

Some engines are not fitted with plugs or cocks on the exhaust branches. In such cases the fuel may be shut off each cylinder in turn while the exhaust outlet on the funnel or stack is kept under observation. The cylinder having the combustion fault is identified when the smoke clears after shutting off the fuel to a particular cylinder.

Engines operating on heavy fuel should have the fuel temperatures checked and corrections made if necessary.

If the fuel temperature is correct and the engine is pressure-charged, the air-inlet filters and downstream air pressure should be checked and related to engine load and turbo-blower revolutions. A common cause of trouble is dirty air filters, particularly if the atmosphere is dusty as can occur in certain parts of the world or when handling dusty cargoes in port. Dirty air filters usually increase the exhaust temperature on all cylinders, but the rise may not be the same on each.

Most naturally aspirated engines are fitted with an air-intake filter to the inlet valve manifold, but it is not usual to find instrumentation on the manifold. Dirty air filters on these engines will show up on the induction pressure line on a light-spring indicator card. If an indicator and spring suitable for the engine speed are available a light-spring card should be taken. If no indicator is available the air filter should be temporarily removed and a check made to see if the engines still smokes. In some cases the air filter can be removed without stopping the engine, but if there is any danger the engine should be shut down to remove the filter and then re-started to make the check, which must be made with the engine on the same load as before.

Two-stroke port-scavenged engines may smoke when the scavenge ports become fouled, but this fault will be indicated by a rise in scavenge pressure or turbo-blower discharge pressure.

Smoke from an individual cylinder is usually found to be due to a dirty fuel valve or valves. If, after the fuel valve has been changed, the cylinder unit is still found to smoke, checks will have to be made on the fuel-pump delivery pressure, fuel-pump delivery valves (if fitted), maximum firing pressures, and exhaust temperatures. If the engine speed is such that an out-of-phase indicator card can be taken, the combustion pressure rise and compression pressure should be checked with the indicator by taking an out-of-phase card and a compression card, which should be checked for abnormalities.

When a fuel valve is changed the old valve should be checked for leakage, setting pressure, and lift. The condition of the sprayer holes, hole edges and diameter of the hole should also be carefully checked. If any condition is found that might cause smoke it can be corrected when the valve is overhauled.

■ **4.20   What is the cause of patterned carbon formations building up on fuel-valve nozzles?**

This phenomenon may be found on any fuel-valve nozzle but is usually seen on fuel valves using high-viscosity fuel. The carbon builds up into 'petal' or 'trumpet' formations which interfere with the spray pattern and cause poor combustion resulting in smoke, high exhaust temperatures, and increased fuel consumption.

Between the fuel-valve seat and the spray holes is a small space, sometimes called the sac. After fuel injection the sac contains fuel which can become overheated.

This, in mild cases, causes some fuel to issue from the spray holes, which burns or cokes and forms carbon around the edge of the holes. The carbon formation gradually builds up and interferes with the spray pattern, affecting atomization and penetration and consequently causing after-burning. In severe cases the lighter constituents of the fuel may boil and burn within the sac. The trouble is usually caused by operating the fuel-valve cooling service at too high a temperature, in which case lowering the temperature of the coolant discharge will solve the problem.

A secondary cause is poor closing of the fuel valve, e.g. sluggish shut-off, which allows oil to bleed slowly at and towards the end of injection. If this secondary cause is present with the first, serious combustion problems may arise.

If the fuel-valve cooling service is kept at too low a temperature, corrosion may occur on the parts of the fuel injector having contact with combustion gases. If it is kept at too high a temperature carbon trumpets may form on the tip around the sprayer holes.

■ **4.21 What is the shape and angle of fuel-valve needles and seats?**

Fuel-valve needles and seats are now always conical, but in some older, slow-speed, two-stroke cycle engines flat-ended needles and seats may still be found. In some cases modern needle-type fuel valves have been retrofitted where flat-ended seats were used.

There is no set angle for the needle end of the valve, as manufacturers decide on the angle which they feel gives the best service. Generally the included angle of the needle used by the various makers is between 40° and 70°.

When conical ends are used the outer end of the cone is ground to a slightly larger angle than the angle between the contact surface of the valve and seat. This difference in angle provides an increasing linear clearance from the outer contact line of the valve and seat to the end of the needle seat. In practice it is found that this clearance gives sharper closing of the valve and reduces any tendency to dribble during the cut-off period. The clearance also allows for longer periods between fuel valve changes or overhauls.

■ **4.22 How does atomization and penetration occur during injection?**

The fuel is at high pressure when it passes through the fuel-injection valve. The lift should be adequate so that the pressure drop across the needle and needle seat is minimal, and should also be such that needle bounce does not occur.

When the pressure wave moves up the fuel piping between the pump and the valve, the wave and its reflection cause the needle to lift by first acting on the annulus formed by the difference in the area of the needle and the valve spindle. As soon as the needle lifts off its seat the pressure acts on the whole of the sectional area of the spindle and gives a very rapid needle opening.

Fuel passes into the fuel sac and first displaces the hot fuel remaining in the sac from the previously injected charge, forcing it into the combustion chamber space. This is followed by the new charge and discharge of fuel continues.

The fuel leaves the sprayer holes in the sac with a very high velocity and as the fuel issues from a sprayer hole it meets the dense air in the combustion chamber. The fuel at the outer surface of the jet emerging from the hole is in a turbulent state from the eddies created within the sprayer hole. The friction from the dense air causes these fuel particles to separate from the outer layers of the jet. The fuel from the centre of the hole has much less turbulence and penetrates further into the combustion chamber. As the outer layers of fuel break off into separate particles the inner parts of the jet become subject to air friction.

By the time this happens the pressure within the cylinder has risen owing to the heat and pressure rise from combustion of the outer layers, and the friction effects increase as the gas pressure in the cylinder rises.

Since the fuel in the centre of the jet has less turbulence, the density of the particles in the central core is greater than that of the particles at the outer layers of the spray pattern.

The degree of atomization and penetration is dependent on the viscosity of the fuel. If the viscosity is too high atomization is reduced and penetration is increased; if the viscosity is too low atomization increases and penetration decreases.

■ **4.23 What is the direction of fuel spray relative to air movement or turbulence within an engine with an open combustion chamber?**

Usually the spray direction is approximately 90° to the direction of air motion. Some engines with two or more fuel valves inject the fuel tangentially to the cylinder bore and in the same direction as the rotation of the air. There are no clear formulae or rules on fuel-spray direction relative to direction of air movement. Generally a small amount of turbulence is preferred to large amounts. Perfecting of combustion is conducted on the test bed during development of new engines, and sometimes very small changes have large overall effects on combustion and fuel consumption.

■ **4.24 What is the shape of the hole and the material section surrounding a fuel-valve sprayer?**

The spray holes are drilled with small-diameter twist drills from the outer part of the tip into the sac space. The transverse section of the hole is circular.

The thickness of the metal of the tip is arranged to give a certain ratio of length of hole, to hole diameter, which is often of the order of 3 to 1, although wider variations may be found. An increased length of hole in relation to hole diameter increases the amount of fuel penetration. There is also a relationship between cylinder diameter and hole diameter.

The ends of the hole are usually left clean and square as any further work on them greatly increases production costs.

Some large-size fuel valves have the sac end of the sprayer hole machined with a radius to improve the coefficient of discharge. This allows the sprayer hole to be made with a smaller diameter and aids penetration and atomization. The back of the small hole may be machined out by a process known as electron discharge machining.

■ **4.25 How would you decide whether a fuel-valve nozzle end was fit for further service?**

The decision whether to scrap a fuel-valve nozzle or use it is sometimes difficult to make. Engine instruction books should be consulted for information applicable to the engine concerned.

The diameter of the hole should be checked to see that it is not over-size to the engine builder's limits. This check must be made carefully. First, the hole must be cleaned, either by soaking in solvent or by using the cleaning drills and hand chuck provided by the engine builder.

After cleaning with solvent the hole diameter should be checked with go/no-go wire gauges or, if the valves are large, with the largest cleaning drill supplied in the cleaning kit. In checking hole sizes care must be taken not to damage the holes, or to break the gauge or drill in the hole. If the hole is oversize to the maker's recommendation the nozzle should be scrapped.

The condition of the outer surface of the hole is also an indication of fitness or otherwise. If it is rounded-off and has lost its square edges it should be scrapped. Comparison of the old nozzle end with a new unused nozzle by visual examination under a good magnifying glass gives a good indication of whether it is fit for further use, especially if that particular valve has been changed previously because of a tendency to cause smoking when the engine was on load.

Fuel leakage across the seat is not usually a cause for scrapping a nozzle end, as leakage can nearly always be rectified.

Sometimes fuel valves are returned to the makers for overhaul, and when this is the case, any valve giving doubts as to condition should be suitably tagged and the facts reported to the superintendent.

Note   Most large, slow-speed engine fuel valves have the tip or nozzle portion of the valve forming the needle seat and the sac made so that an old fuel-valve tip may be removed and replaced by a new one.

■ **4.26   What is the difference between a fuel priming pump and a fuel surcharge pump? For how long and when should each be used?**

A fuel priming pump is used to prime the fuel pressure pumps, pipe lines and fuel valves on engines fitted with a common-rail fuel system as found on some opposed-piston engines. The priming pump should only be used for a few seconds, and if no response is seen on the fuel-pressure gauge the cause should be investigated and found before the priming pump is used again.

The common-rail system should be checked for open drains and open bypass valves. The fuel-injector valves should also be checked with the hand fork lever to see that they are properly closed. If any are found open, or some doubt exists as to their being properly closed, the scavenge trunk doors should be opened and the engine turned so that all the lower pistons can be examined for the presence of fuel oil. If any is present it should be cleaned out.

When the scavenge doors are off, the priming pump may be used again to check the fuel valves for leakage, since any leakage can be seen through the scavenge doors and ports, as the leaking fuel usually runs down the cylinder liner.

The cause of any lack of pressure response on the fuel-pressure gauges, when using the priming pump, must be found before any attempt is made to turn the engine on starting air. Before the engine is put on starting air it must be turned a complete revolution with the turning gear, with the indicator cocks open for observation and the scavenge drains being watched for the presence of fuel oil.

Fuel surcharge pumps are used with high-viscosity fuel to keep the suction side of fuel-injection pumps under pressure, thus preventing the hot oil from gassing and forming vapour locks within the fuel pump.

In most installations the surcharge pump is kept in service from preparing the engine for standby, during manoeuvring and passage, and until after the 'finished with engines' order is received.

The fuel surcharge pump is sometimes used only during standby and manoeuvring periods if the head from the daily service tank is sufficient to prevent gassing of the fuel in the suction lines and fuel pumps.

Use of the fuel surcharge pump will be different if the engine is manoeuvred on diesel fuel. After 'full away' the surcharge pump will be brought into use for the change-over to high-viscosity fuel and then kept in use or closed down depending on the head of the daily service tank above the main engine fuel-injection pumps.

If no surcharge pump is used during 'full away' operation the temperature of the fuel in the daily service tank must be kept sufficiently high to prevent pipeline friction becoming too much for the head available. If the temperature of the fuel in the tank drops, the fuel viscosity increases and causes line drag. Gassing may than occur on the downstream side of the fuel heater and allow a vapour lock to form in the fuel pumps.

Note   Surcharge pumps are also known as service pumps, booster pumps, fuel-feed pumps and primary pumps.

■ 4.27   Describe the removal, overhauling procedures and requirements, and replacement of large fuel-injection valves.

Removal of fuel-injection valves. The local area around the fuel valve should first be cleaned to prevent any dirt or other material entering or draining into the cylinder. The cooling services must be shut off. After all connections have been removed the valve should be lifted out from the cylinder cover with a direct lead so that the pull is in line with the axis of the valve.

When the valve is removed from the cylinder the nozzle end must be capped off to protect it during transport into racks or stores area. The opening into the cylinder should also be protected with a plug or blank to prevent ingress of liquids, dirt and solid objects.

Overhaul of fuel valves. The external parts of the fuel valve must be cleaned, and the fuel valve tested first for lift and injection pressure, and then for leakage or bleed at a pressure just below the lifting pressure. If the valve is found to be fuel-tight it will be a guide to the amount of work required on the valve.

After testing, the valve must be stripped down, the internal parts cleaned with paraffin or solvent and a lint-free rag, and then carefully examined. Any edge-type filters on the valve should also be carefully cleaned. The spray holes and sac should be cleaned either by soaking in solvent or cleaning with the kit supplied by the engine builder. After cleaning, the sac and holes should be washed out with clean paraffin or solvent using a syringe.

The valve seating surface requires very careful examination. If the valve has a conical end, it should be given the *lightest* smear of blue marking and then put into the cleaned guide and seat and carefully rotated. After removal the conical end should be checked for indication of seating width. If the seating width is narrow and the valve was found to be leaking, it may be lapped with any one of the variety of special lapping compounds available or with jeweller's rouge. After lapping, a check should again be made for seat width and if in order the valve can be reassembled and tested. The condition of the seat should also be checked.

Flat-seated valves should be treated in a similar manner, with the exception of the blue marking check for seat width. The construction of a flat-seated valve is such that the width remains very nearly constant. If the end of the flat-seated valve is not counterbored it may be lightly lapped to prevent any small ridges forming, but lapping should be kept to a minimum.

If the valve was fuel-tight under pressure it should not be lapped but must be retested for fuel-tightness after assembly.

When reassembling fuel valves the threads of the capnut should be lubricated with the lubricant recommended by the engine builder. In the absence of any instructions molybdenum disulphide grease is often used, but with this grease care must be taken to avoid overtightening parts. If the parts are assembled and tightened to rated torque figures the correct lubricant must be used; if the correct lubricant is replaced by another giving much lower coefficient of friction the parts will be overstrained during assembly and tightening.

Some fuel valves are constructed in such a manner that the spray comes from one side of the valve or must point in some special direction. These fuel-injection valves have locating pins within the various parts to ensure that the spray direction is maintained correctly. Care must be taken during assembly to see that the locating pins are fitted and that they are not damaged or bent.

Testing of fuel valves requires that the valve be set to the correct injection pressure and tested for tightness at a pressure just below the injection pressure.

The lift of the fuel valve should also be correctly set. Some fuel valves have internal spacer rings which are used to limit lift; if the valve is so fitted the lift need usually only be checked and regulated when fitting new parts to the valve.

The angle of the spray must also be checked. Some engine builders supply a gauge or shroud to carry out this test. This is fitted over the nozzle end of the valve and the jets from the nozzle must pass through the holes or slots without wetting the surfaces around the hole or slots.

The cooling space of the tip is sealed from the pressure space by lapped surfaces or composition O-rings. If the valves are fitted with O-rings they should be renewed at each valve overhaul.

When the seal between the pressure space and the cooling space is made by lapped surfaces, difficulties sometimes arise through leakage of fuel into the cooling space, thus contaminating the coolant. Testing for leakage into the cooling space may be carried out by filling the cooling space with coolant and plugging one of the coolant connections. The other side is connected to a pressure gauge which will register low pressures. Air must be removed from the cooling space before fitting the pressure gauge. The pressure side of the fuel valve is then put under pressure with the test pump and the pressure held at a value approaching the injection pressure. Leakage will show by a lowering of pressure on the fuel side and a rise in pressure on the coolant side.

Testing for fuel leakage into the cooling spaces requires the test pump to be in first-class condition, and the fuel bypass valve on the fuel valve must be absolutely tight.

The working space where fuel valves are overhauled and tested must be kept perfectly clean because of the precision required during fitting and assembly. Spanners and wrenches used to dismantle and assemble valves must be a good fit on the part to be loosened or tightened, as any slackness causes damage and raises burrs. This creates difficulties when subsequently working on the valves. Fuel valves held in a vice should be protected with copper vice clamps.

Fuel test pumps should only be filled with clean filtered oil from clean vessels.

Tins of lapping compound should have the lids replaced so that the compound does not become contaminated with any coarse material which might damage the valve.

When fuel valves are stored in racks after testing, the nozzle protecting caps should be fitted in place and the fuel and cooling connections must be capped off to prevent entry of dirt or foreign matter.

Replacement of fuel valves. After removal of a fuel valve the pocket should be carefully cleaned out and the landing seat in the pocket lapped with the lapping jig. Afterwards the seat must be cleaned and the pocket examined for condition. Prior to the fitting of the fuel valve the landing seat on the fuel valve must be cleaned, and lapped if necessary. The seal rings used to prevent ingress of dirt into the pocket must be fitted on the valve.

Examination of the cylinder space prior to fitting a valve must be done with the piston on bottom centre as this obviates blind spots and prevents any object being hidden. After making certain nothing has fallen into the cylinder the fuel valve should be immediately placed in the cylinder cover. The tightening and hardening up of the fuel-valve securing studs must be done carefully to ensure that each stud is taking the same load. This prevents distortion of the valve, which may cause difficulties in operation, and reduces the chances of gas leakage across the landing face of the valve and cover.

After opening the plugged fuel and cooling-water connections, the service pipes to the fuel valves must be connected and any tests on them carried out as soon as possible.

After the ship is under way the area around the fuel valve should be checked for any noise caused by leakage and any leakage rectified. This is sometimes overcome by tightening the valve securing nuts a very small amount. If the valve is put down on a clean, properly lapped landing no leakage problems should be experienced. If leakage is present in only small amounts it can lead to the danger of repeated false fire alarms being sounded. The landing faces on the fuel valve and cylinder head may be cut badly if leakage is allowed to persist.

The engine builder's instructions regarding allowable torque on fuel-valve studs and nuts should always be complied with. Many modern engines have a set of Belleville washers fitted between the fuel-valve flange and the nuts.

In unmanned engine rooms, when the high-pressure fuel pipe connections have been checked following engine operation, the screens and high-pressure pipe coverings must be replaced before the engine room is left unmanned. In some cases the coverings must be fitted at the time the fuel pipes are connected; in this case the drain connections from the covering must be broken and the drain checked for oil outflow from any high-pressure connections.

■ **4.28** Describe the methods of checking maximum cylinder pressures. What effects are likely to be experienced following operation with too high and too low maximum cylinder pressures? How can the maximum cylinder pressure be increased or decreased?

Maximum cylinder pressures can be checked with a normal indicator on a slow-speed engine; on medium-speed engines they can also be checked with special indicators which have low inertia, or by a piezoelectric transducer instrument built into electronic circuits arranged for measuring and recording pressure.

Note An indicator should not be used on an engine operated at a greater speed than recommended for the instrument. The inertia effects will cause the measured pressures to be higher than the actual cylinder pressure.

On high-speed engines the maximum cylinder pressures can be checked with a maximum pressure indicator which gives a reading of the cylinder pressure. When an indicator is used to check maximum pressures, a paper should be fitted on the drum, the atmospheric line should be marked and the cord slowly pulled when making the pressure recordings. This will give a wave-formed line, the height of the wave peaks giving the maximum pressures when related to the spring number. If there are variations in the maximum pressures a mean value must be taken.

When taking maximum cylinder pressures the compression pressures should also be recorded. The difference between the maximum and compression pressures gives the pressure rise from combustion of fuel.

If an engine is operated with maximum combustion pressures in excess of the designed maximum, the crankshaft will be subjected to greater stresses than the design allows for. This may initiate fatigue cracks in the crankshaft or engine structure. With modern two-stroke engines, trouble may be experienced with the crosshead bearings or gudgeon pin bearings owing to the excess load. Normally cylinder relief valves should lift and give warning of excess pressure, but if there is any malfunction of the valve serious damage may occur.

Operating an engine at a low maximum combustion pressure increases fuel consumption progressively as the maximum combustion pressure falls.

The maximum combustion pressure may be increased or decreased by advancing or retarding the fuel-pump cam relative to the crank position. Helix and port-control type fuel pumps can also be lowered or raised on their foundation to increase or decrease the maximum cylinder pressure.

In fuel pumps controlled by suction or spill valves, the cam can be advanced or retarded by means of a radial spline with very fine teeth cut on the side of a collar and the side of the cam, which is bolted to the collar. The bolts must be released and the cam eased away from the spline in order to advance or retard the cam as required; after adjustment the bolts must be tightened and locked.

Alteration of the cam position by advancing or retarding does not alter the amount of fuel injected.

Note   Before removing a cam from its spline, reference lines should always be scribed on the cam and some fixed point on the shaft or collar so that the original position of the cam can be verified, and also to check the amount of advance or retard.

With helix and port-control fuel pumps one method of advancing or retarding commencement of fuel injection is to raise or lower the fuel pump. Pumps designed for adjustment in this way have a machined plate fitted between the fuel-pump base and the fuel-pump foundation or bracket. As wear takes place on cams, bearings, and shafting, it is sometimes necessary to lift off the fuel pump, remove the plate and reduce its thickness by accurate machining. This compensates for the wear which has taken place and restores the maximum pressure to its original value. This method of adjustment is only used for making small pressure increases.

Lowering the fuel-pump body causes injection to commence earlier and also to finish earlier. The actual stroke of the fuel-pump ram remains the same. The effective stroke of the fuel-pump ram also remains unchanged.

One type of engine using a helix and port-control type fuel pump has an arrangement consisting of a thread on a circular projection of the fuel-pump cover which fits into the pump casing. A cage is screwed on to the threaded projection. The outer portion of the cage has a gear cut on it which is matched to a pinion on a spindle passing through the cover. Turning the spindle causes the cage to turn and move up or down on the thread. The cage is cut with ports which connect through to a suction valve fitted circumferentially within the pump barrel. The pump barrel fits against the lower edge of the movable cage. An axial slot is cut in the outer edge of the barrel and this engages with a dowel screw.

By slacking off the pump cover nuts it is possible to turn the sleeve or cage and move it up or down. Up-and-down movement of the sleeve forces the pump barrel up or down and the dowel screw prevents rotation of the pump barrel. A true axial movement of the pump barrel is consequently obtained without altering the circumferential relationship of the helix and port. Easy control of the maximum pressure is obtained as movement downwards causes injection to take place earlier without altering the period of injection, since the effective stroke and the actual stroke of the fuel pump have not been altered (see Fig. 4.7 and Question 18.51).

■ **4.29   Describe how you would determine whether the maximum firing or combustion pressure was correct for propulsion engines and auxiliary engines.**

Modern turbo-charged engines show a considerable rise in compression pressure and firing pressure as the load on the engine is increased. In order to check that the firing pressure is correct, it is desirable to put the engine on to full load conditions; this is easy to arrange with diesel auxiliary engines used for electrical generating purposes. When the engine is on full load the maximum pressure and the compression pressure should be checked out for each cylinder, and the figures tabulated.

The maximum firing pressures obtained during the test should be compared with the test-bed data for the engine in question. If they are low, the compression pressures can be checked to ascertain whether the low maximum firing pressure is caused by low compression pressure.

With high-powered, slow-speed, direct-coupled engines and multi-engined geared installations, it is not always possible to operate the main engines at full power (except in navigational emergency) because of Company's orders or for other reasons. In such circumstances it is necessary to obtain the engine characteristics, which are a graphical record of the test-bed results plotted against a base line of brake horsepower.

If the propeller shafting is fitted with a torsion meter, the torsion meter should be used to obtain a brake horsepower. At the time the brake horsepower is taken the maximum firing pressure and compression pressure should be taken for each cylinder and the results tabulated.

The brake horsepower obtained from the torsion meter should be marked on the base line against the corresponding brake horsepower figure, and a vertical line projected upwards to intersect the compression line and maximum-pressure line. Horizontal lines drawn through the intersection of the vertical line and the curve of compression pressure and maximum pressure will give the values for these on the scale at the side corresponding to the power at which the pressures were taken.

The values obtained from the engine may then be checked against those obtained from the engine characteristics.

If there is no torsion meter, indicator cards must be taken and the indicated horsepower of the engine obtained.

From the characteristics of the engine it will be possible to obtain a value for mechanical efficiency corresponding to the engine speed (rev/min). The product of the mechanical efficiency and the indicated horsepower gives the brake horsepower. The characteristics of the engine can then be used as described earlier.

It should be noted that large ships in ballast condition will have a higher engine speed, for the same power output, than in loaded condition. While the curves can be used as a check they will not be absolutely true.

With multi-engined, geared-propulsion installations the compression pressures and maximum pressures can be checked as described for slow-speed, direct-coupled engines, but all engines must be working and connected to the propeller.

If the engine is connected to a controllable-pitch propeller, there is every likelihood that the engine governor control progressively reduces engine speed to some idling speed at zero pitch. If the characteristic curves for the engine are drawn up according to the cube law as for solid propellers (power varies as the speed cubed), the propeller pitch control should be brought to hand control and the pitch adjusted to give a power and speed (rev/min) corresponding to the engine characteristics. This will usually bring the propeller to approximately the maximum pitch position.

Note   Some ships can manoeuvre with bridge controls by controlling propeller pitch. In such cases the propeller pitch control is from the bridge while the vessel is at sea, and the pitch control must not, therefore, be switched from the bridge to the engine room or local control without arrangement and agreement from the bridge officer, and the master and chief engineer.

If maximum firing pressures are found to be low due to low compression pressure, the condition will usually correct itself when pistons are lifted and new piston rings are fitted.

If all the maximum firing pressures are low it is usually due to wear and tear of working parts such as tooth wear on geared fuel-pump drive, chain link wear and stretch, and chain wheel sprocket wear on chain drives.

Before any change is made to fuel pump settings the causes must be fully investigated and the reasons established. The tests should also be conducted with fuel obtained at different bunkering locations.

A check of indicator card records will also show if the change has been slowly progressive.

■ **4.30   What do you understand by the term 'variable injection timing'?**

Variable injection timing is a form of fuel pump control enabling an engine to operate with the designed maximum cylinder firing or combustion pressure from approximately 75% power output to maximum power. This improves thermal efficiency and lowers fuel consumption.

The fuel consumption for an engine at any load will be related to the expansion ratio of the combustion gases from their maximum pressure to the pressure at the commencement of exhaust blowdown.

The maximum cylinder pressure is a factor used in the design of the crankshaft and other important engine parts. In a normal engine the maximum cylinder pressure is reached only at full power operation, whereas with variable injection timing the maximum cylinder pressure is reached at about 75% of the full load. The expansion ratio is therefore increased when the engine is operating under light loads right up to full load.

In a normal engine the cylinder pressure plotted against the engine load is nearly a straight line sloping upwards and reaching its maximum value at maximum load.

If the cylinder pressure is plotted against the engine load in an engine with variable injection timing, the maximum allowable pressure is seen to be reached at approximately 75% of the full load and then remains at a constant value for the remaining part of the graph. This increase in efficiency flattens the specific fuel consumption (sfc) curve and reduces the fuel consumption at part loads on the engine.

It might also be said that the strength designed into the crankshaft and other important parts is better utilised.

The diagram in Fig. 4.5 illustrates this, showing how variable injection timing allows the engine to operate with the maximum designed firing pressure. This

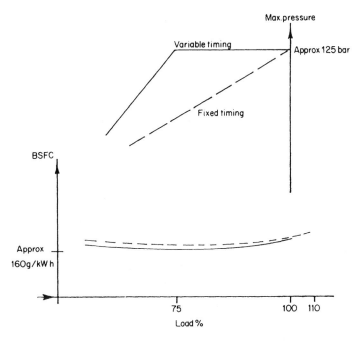

Fig. 4.5 Cylinder pressure plotted against engine load.

gives a higher ratio of expansion over a wider load range, thus increasing thermal efficiency.

■ **4.31   Describe how variable injection timing is effected.** **.**

If a fuel pump fitted with a suction valve and a spill valve is considered, it can be seen that the injection timing can be altered by changing the positions on the eccentrics controlling the suction valve and the spill valve.

If the position of the tappet controlling the suction valve is lowered, injection is commenced earlier but the fuel quantity will be increased. If the tappet controlling the spill valve is raised, the end of injection is made earlier and the increased quantity of fuel delivered is reduced and may be brought back to its original level. The timing of fuel injection is now advanced without any change in the quantity of fuel delivered.

The timing of the injection can be retarded by reversing the direction of eccentric movements.

Control of the eccentrics can be effected by mechanical means, but in modern engines it is done through the computer controlling the operation, speed and manoeuvring of the engine.

Fuel pumps with helix and ports controlling the fuel quantity can have the timing of injection advanced by lowering the pump cylinder or barrel relative to the highest and lowest points of the ram movement. The top position and the bottom position of the ram and the ram stroke do not change. The movement of the ram is controlled by the fuel-pump cam and return spring. Commencement of injection occurs when the ram moves upwards and closes off the inlet and spill ports.

If the pump cylinder is lowered the inlet and spill ports are cut off earlier and injection commences earlier. In a similar manner the inlet and spill ports are opened earlier due to their lowered position, and the end of injection takes place earlier with no change in the quantity of fuel delivered, provided the ram is not moved circumferentially.

When it is required to retard the injection period the pump cylinder is raised.

Raising and lowering the pump cylinder is brought about by putting a thread on the lower part of the cylinder and engaging the thread in a nut located between two thrust faces in the pump body. On the outer circumference of the nut a toothed gear is cut. The toothed gear engages with a rack held in a guide located in the pump body. Movement of the rack forward or backward advances or retards the injection period relative to the position of the crank by raising or lowering the pump cylinder.

The fuel cam is used to adjust the timing of the commencement of injection in a normal manner so that the maximum allowable firing pressure is reached at about 75% of full load. The timing of the commencement of injection will then be retarded as the quantity of fuel injected is increased. The amount of retardation necessary to hold the cylinder pressure constant at the desired maximum value will be designed into the control system.

The checking procedures and timing adjustment instructions will be found in the engine builder's instruction books. These must always be referred to before checking or starting any adjustments.

Regular checking of the maximum cylinder pressures should also be carried out to check the actual pressures and to check the returns given by the cylinder pressure monitoring equipment. This is necessary to safeguard the integrity of the crankshaft. (Figs 4.2 and 4.6)

■ 4.32   How is turbulence created within the engine cylinder?

It is known that a considerable amount of air movement takes place at the end of the compression stroke due to piston movement. If more movement of air or turbulence is required in some particular engine design, it is obtained by sloping the scavenge ports away from a radial direction so that the air moves into the cylinder tangentially, causing the air to rotate within the cylinder.

The same effect is achieved in engines with air-inlet valves by means of a shield or deflector fitted on one side of the valve. Where shields or deflectors are fitted, care must be exercised when assembling the valve to ensure that the deflector is properly located and that the means to prevent the valve turning is correctly fitted.

Some high-speed engines with open combustion chambers are designed with a very small clearance between the flat circumferential surface surrounding the pocket on the top of the piston and the flat bottom face of the cylinder cover or cylinder head.

When the piston is nearing the upper part of its travel, air is forced out from

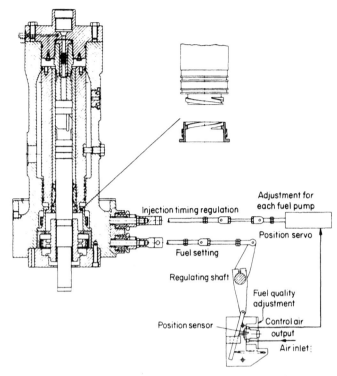

Fig. 4.6 Fuel injection pump showing variable injection timing (VIT) controls. (MAN B&W L-MC VIT System).

between the two flat surfaces. The air assumes a rotary motion as it is compressed into the piston crown pocket, and a very high degree of turbulence known as squish is obtained. Squish is necessary with high-speed engines having direct injection, as it gives the necessary mixing of air and fuel to complete combustion in the very limited time available.

■ **4.33** A propulsion engine is capable of developing 28 000 brake horsepower at 100 rev/min. The engine is of the two-stroke type and has eight cylinders. Calculate the volume of fuel injected per stroke at full power and at 32 rev/min. The specific fuel consumption is 155 and 200 g/bhp h at 100 and 32 rev/min respectively; the specific gravity of the fuel is 0.96 at 15°C.

At 100 rev/min

weight of fuel/hour	$= 28\,000 \times 155$ g
weight of fuel/min	$= (28\,000 \times 155)/60$ g
weight of fuel/engine revolution	$= (28\,000 \times 155)/(60 \times 100)$ g

weight of fuel/stroke $= (28\,000 \times 155)/(60 \times 100 \times 8)$ g
                         $= 90.41$ g
specific gravity $=$ weight of fuel/weight equivalent volume
                         water
volume of fuel/stroke $=$ weight of fuel/specific gravity
                         $= 90.41/0.96$ ml
                         $= 94.18$ ml

This is the volume of fuel at 15°C; at injection temperature there will be an increase in volume.

It is now necessary to calculate the brake horsepower at 32 rev/min; this can be obtained by using the cube law for the propeller, i.e. the power P varies as $(\text{rev/min})^3$.

$$\frac{P_1}{(\text{rev/min}_1)^3} = \frac{P_2}{(\text{rev/min}_2)^3}$$

$$P_2 = \frac{P_1(\text{rev/min}_2)^3}{(\text{rev/min}_1)^3}$$

Power at 32 rev/min $= (28\,000 \times 32^3)/100^3$
                         $= 917.5$ brake horsepower

At 32 rev/min
weight of fuel/stroke $= (917.5 \times 200)/(60 \times 32 \times 8)$ g
                         $= 11.95$ g
volume of fuel/stroke $= 11.95/0.96$ ml
                         $= 12.45$ ml at 15°C

It will be seen that the fuel-injection pumps has to meter fuel over a wide volume range between full-power and slow-speed operation.

The method of calculating the brake horsepower at slow speed should only be considered as approximate. In practice, factors such as wind from forward or from aft can cause wide variations in the brake horsepower value.

4.34   Find by what factor the surface area of a spherical volume of oil of 100 ml is increased when it is atomized into spherical particles having a diameter of 20 micron.

To find the surface area of the large sphere first find the diameter (D) of the sphere using the formula for the volume (V) of a sphere

$$V = \frac{\pi D^3}{6}$$

which is equivalent to

$$D = \left(\frac{6V}{\pi}\right)^{\frac{1}{3}}$$

V is equal to 100 cm³. Therefore

$$D = \left(\frac{6 \times 100}{\pi}\right)^{\frac{1}{3}} = 5.759 \text{ cm or } 57.59 \text{ mm}$$

The diameter can now be used to find the surface area (S) using the formula

$$S = \pi D^2$$

D is equal to 57.59 mm. Therefore

$$S = \pi \times 57.59^2 = 10\,419\text{mm}^2$$

To find the surface area of the same volume of 20 microns spherical particles first find the number of particles. The volume of the particle (v) is given by

$$v = \frac{\pi \times 0.02^3}{6} = 4.189 \times 10^{-6}\text{mm}^3 \text{ or } 4.2 \times 10^{-6}\text{mm}$$

The number of particles is given by V (100 cm³) divided by v i.e.

$$\frac{100 \times 1000}{4.2 \times 10^{-6}} = \frac{10^5 \times 10^6}{4.2}$$

Next find the surface area (S) of one spherical particle which is

$$S = \pi D^2 = \pi \times 0.02^2$$

The total surface area S_T is

$$S_T = \left(\frac{10^5 \times 10^6}{4.2}\right) \times \pi \times 0.02^2 = 30\,000\,000 \text{ mm}^2 \text{ or } 3^6 \text{ m}^2$$

The surface area is therefore increased by a factor of

$$\frac{30\,000\,000 \text{ mm}^2}{10\,419 \text{ mm}^2} = 2\,879.0$$

The diameter of a sphere having a volume of 100 cm³ is

$$\text{Vol} = \frac{\pi d^3}{6} \therefore d^3 = \frac{6 \times \text{Vol}}{\pi} \, d = \left(\frac{6 \times \text{Vol}}{\pi}\right)^{1/3}$$

$$d = \left(\frac{6 \times 100}{3.142}\right)^{1/3} = (190.9612)^{1/3} = 5.7586 \text{ cm}$$

$$= 57.586 \text{ mm} = 0.0576 \text{ m}$$

Surface area of a sphere having a diameter of 0.0576 m

$$\text{Surface area} = \pi d^2$$
$$= 3.1416 \times 0.0576^2$$
$$= 0.0104 \text{ m}^2$$

Volume of particle

$$\text{Vol} = \frac{\pi d^3}{6} = \frac{\pi \times (0.000002)^3}{6}$$

$$\text{Number of particles} = \frac{100}{1\,000\,000} \Big/ \frac{\pi(0.000002)^3}{6}$$

$$= \frac{100}{1\,000\,000} \times \frac{6}{\pi(0.000002)^3}$$

Surface area of particle $= \pi(0.000002)^2$

$$\text{Total surface area of particles} = \frac{100}{1\,000\,000} \times \frac{6}{(0.000002)} \times (0.000002)$$

Increase $= \underline{\underline{288.46}}$ times.

It can now be seen that atomizing the fuel into small particles gives an enormous increase in the total surface area of the fuel particles. A much greater surface is then in contact with air so the rate of combustion is increased.

■ **4.35** Find the theoretical velocity of the fuel passing through the injector sprayer holes if 100 cm³ of fuel is injected over a period equal to 25 degrees of crank rotation when the engine is turning at 80 rpm. Assume there are eight sprayer holes of 1.0 mm in diameter and the coefficient of discharge is 0.75. After finding the velocity find the theoretical injection pressure.

The quantity of fuel injected is 100 cm³ $(= 100 \times 10^{-6} = 0.0001$ m³). The time for injection (T) is

$$T = \frac{80}{60} \times \frac{25}{360} = 0.0926 \text{ sec}$$

so the rate of injection (R) is

$$R = \frac{0.0001}{0.0926} = 0.0011 \text{ m}^3\text{s}^{-1}$$

The effective area (A_E) of the sprayer holes is

$$A_E = \frac{\pi}{4}\left(\frac{1.0}{1000}\right)^2 \times 8 \times 0.75 = 4.7124 \times 10^{-6} \text{m}^2$$

The velocity (V) is given by

$$V = \frac{R}{A_E} = \frac{0.0011}{4.7124 \times 10^{-6}} = 229.183 \text{ m/s or } 751.95 \text{ ft/s}$$

The head is given by

$$\frac{V^2}{2g} = \frac{229.183^2}{2 \times 9.81} = 2677 \text{ m or } 8784 \text{ ft}$$

Head pressure (P_h) is 267.7 kgf/cm² or 263 bar. Injection pressure (P_i) is given by the addition of the compression pressure (P_c) and the equivalent head pressure (P_h) which is assumed to be 70 bar (or 1015 psi). Injection pressure P_i is therefore

$P_i = P_h + P_c = 263 + 70 = 333$ bar or 4830 psi

■ **4.36   What are the advantages of manoeuvring engines on a low-cost blended fuel?**

The advantages are of an economic nature and are related to the cost differential between different fuels and the number of hours spent under manoeuvring conditions. For purposes of comparison a period of operation covering a financial year is usually used. The amounts saved in a year can be quite large and make a very valuable reduction in the ship's operating cost.

There are no real technical difficulties associated with manoeuvring engines on low-cost fuels provided that the injection viscosity is kept at the required value at the point where the fuel enters the cylinders. This requires the use of steam tracer lines and the continuous circulation of heated fuel during the time the engine is stopped.

■ **4.37   Show with the aid of a sketch how fuel can be continuously circulated through fuel-injection valves on large engines while the engine is operating or under standby conditions.**

Modern fuel valves are manufactured with a sliding sleeve fitted around a part of the needle push rod and held down in place by a lightly loaded spring. The push rod is hollow and fuel passes down from the top of the fuel valve through the hollow push rod.

When the sleeve is in the lower position, holes drilled in the bottom of the push rod are uncovered and fuel passes out through these holes, up around the outside of the push rod, through a port in the valve body and back to the low-pressure part of the fuel system.

When the injection pump commences to deliver fuel the load on the spring holding the sleeve down is overcome by the rapid pressure rise. The sleeve is lifted, blocking off the holes in the lower part of the push rod and allowing fuel to pass into the annular space around the fuel-valve needle. The pressure rise overcomes the force exerted on the needle by the push rod and heavy valve spring. Fuel then passes into the cylinder. When the end of injection is reached the pressure drop allows the heavy spring to reseat the needle, then the light spring pushing on the sleeve forces it downwards and the holes in the bottom of the push rod are opened.

Fuel again passes up the valve body and returns into the low-pressure part of the fuel system. When the engine is in operation and fuel is not being discharged by the fuel-injection pump, the pump ram is in its lowest position and the roller is riding on the base circle of the cam. In this position the surcharge pump pressure causes the fuel to pass through the inlet and spill ports and pass over the top of the pump ram into the fuel piping and through the fuel injection valve.

When the engine is stopped, circumferential movement of the pump ram brings the vertical slots in the top of the ram in line with the inlet and spill ports. The fuel now circulates through the inlet and spill ports, through the vertical slots and up to the fuel valve. Fuel will circulate only when the surcharge pump is in operation and maintaining the pressure required to overcome friction through the pipe lines, pumps, valves, etc.

The circulation of fuel when injection is not taking place maintains the fuel at the correct temperature to suit injection viscosity and allows the engine to be manoeuvred on heavy fuel. The circulating fuel also prevents overheating of the valve. (Fig. 4.7)

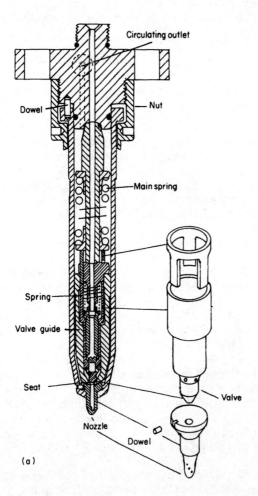

(a)

Fig. 4.7 (a) Fuel injection valve. (MAN B&W)

■ **4.38 How does combustion change between a fuel with good combustion characteristics and a fuel with high viscosity, high density and high carbon content?**

A fuel with good combustion characteristics will have an ignition delay period that is compatible with the rate of fuel injection and the rate of combustion of both the faster-burning cutter stock and the residual constituent. The rate of pressure rise within the cylinder from the beginning of ignition through the early part of combustion will also be at an acceptable value and the later stages of combustion will be completed without the problems associated with after-burning.

The first problem likely with high-viscosity, high carbon content fuel is the

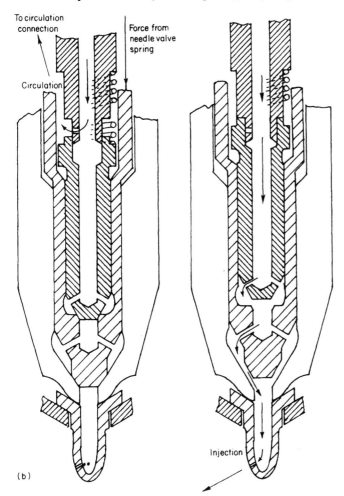

(b)

Fig. 4.7 (b) Internal parts of fuel injection valve showing circulating connection when open for circulation and when closed for fuel injection.

effect of the high boiling point of the cutter stock. Fuels with a high boiling point have a longer ignition delay period than those with a lower boiling point.

When the ignition delay period is extended and the rate of injection is normal, the quantity of fuel injected into the cylinder during the extended delay period is greater. When ignition occurs the rate of pressure rise with a large amount of fuel present in the cylinder may be increased and lead to very high maximum pressures.

Much of the residual stock is slower-burning so the problem of excessive rate of pressure rise does not always occur, but it is a possibility and will to some extent be related to engine speed.

When the residual portion of the fuel is high in asphaltenes or other slow-burning material, the rate of combustion of the slow-burning material will most likely extend into the region where after-burning is experienced.

Problems associated with after-burning are a rise in exhaust temperatures, increase in fuel consumption, increased cylinder wear, increase of cooling water temperatures, and higher thermal stresses. In extreme cases the slow progress of combustion on the relatively high temperatures involved have depleted the oil film and the cylinder liner surfaces and caused liners with limited wear to be ready for scrapping in a matter of two or three hundred hours of operation.

Fouling of the turbo-blower and the exhaust system, including the heating surfaces of exhaust gas boilers and economizers, may also occur, and more particularly so if the fuel has a high Conradson carbon value.

■ **4.39 What are the likely consequences of carry over of fluid catalytic cracking fines from the separator and clarifier into the clean oil system?**

The effects of any catalyst and catalytic cracking fines contained in the fuel oil passing through to the engine will to some extent be dependent on the clearances of the parts in the fuel pumps and fuel valves. The clearances between the fuel-pump ram and pump barrel, and the fuel-valve needle and guide, will be dependent on the viscosity of the fuel used, the sizes of the parts, and the operating temperature.

The dimensions of the catalyst and fines are known to be generally in the order of 3 micron up to about 35 micron. For propulsion machinery the clearances between pump ram and barrel and valve needles and guides start at approximately 5 micron for the smaller needles and guides and go up to about 20 or more micron for the largest-diameter fuel-pump rams.

Any catalyst and fines that can enter into the clearance spaces of parts of fuel-injection equipment can lead to accelerated wear of those parts. The wear may be very extensive in a few days of operation with contaminated fuel. Some engines normally manoeuvred on diesel oil have been unable to start after changing over to diesel fuel when entering port.

Excessive leakage through the enlarged clearances was the cause. This lowered the pump discharge pressure to below that required for the lifting pressure of the fuel-injection valves. It was then necessary for the engines to change back to heavy fuel operation to manoeuvre and enter port.

In the cases investigated no clear pattern seems to emerge from the data collected. In some cases extensive wear of fuel-injection equipment has occurred in a few hours of operation with an apparently low number in parts per million of contaminant. It must be remembered, however, that a small amount of contaminant can equate to a large amount of abrasive material passing through a fuel pump and injectors in a 24-hour period.

Consider the case of an engine developing 3500 horse power per cylinder, having a specific fuel consumption of 140 g/hp hr, and an abrasive content of 50 mg per kg of fuel. Fuel used per day equals.

$$3500 \times 140 \times 24/1000 = 11\,760 \text{ kg/day}$$

The abrasive matter passing through pump in kg per day equals

$$50 \times 11\,760/1\,000\,000 = 0.58 \text{ kg/day or } 1.28 \text{ lb/day}$$

It has also been noticed that this abrasive material causes rapid wear of the sprayer holes in fuel valves and accelerated wear in cylinder liners and piston rings.

In some cases rapid wear of cylinder liners and piston rings has occurred with fuels that have had an acceptable analysis except in the matter of alumina and silica content.

■ **4.40 What are the possible ways to reduce the risk of damage from contaminants and deleterious matter contained in fuel supplied to a ship?**

The way to avoid a problem is to be aware of it, then take steps to prevent the problem arising or to lessen its consequences.

Subscription to one of the fuel analysis services by the shipowner is the best way to get accurate information to the engineer officers on the ship, regarding the possibility of problems arising out of the constituents in newly supplied fuel.

Early warning of possible problems can be given by onboard testing of fuel at the time of delivery (see Question 3.37).

Fresh stocks of fuel should not be mixed with fuel already on board. Mixing cannot always be avoided, but every effort, within the possible limits of segregation, should be made to prevent it.

Fuel cleaning by separators and clarifiers should be carried out at the highest possible temperature and the slowest possible throughput, with the correct size gravity ring installed in the separator.

Some separation systems have been made to operate with gear-driven, fixed-capacity, positive-displacement pumps driven by the separator. A degree of automation is given by fitting the overflow from the daily use tank into the dirty oil tank. In such cases the capacity of the pumps driven by the separator matches the throughput for the best grades of fuel. This throughput is too great when cleaning lower grade fuels requiring a reduced throughput. Systems such as this will need retrofitting of flow controllers to reduce fuel throughput when cleaning the lowest grades of fuel the separator is designed to handle. The flow controller should be of a type that does not throttle the flow and cause eddies and foaming in the separator supply.

■ **4.41   How does an electronic fuel injection system operate?**

Electronic fuel injection systems are controlled by electronic means. The pressure required to inject fuel into an engine cylinder is obtained by surcharge pumps, cams, ram-type injection pumps, etc., in a normal manner.

Control is effected by inputs into a computer. These inputs are from the manual fuel control lever position, the governor, crank position from sensors, maximum cylinder pressures, exhaust temperature and the like. The computer is programmed to suit the engine operation characteristics and give output to the fuel pump for the control of injection. Monitoring functions can also be incorporated in the program to cover maximum and minimum allowable temperatures and pressures and to give an alarm call or shut down the engine if any deviation goes beyond an allowable norm.

Control of injection at the fuel pump can be effected through suction and spill valves controlled by hydraulic, pneumatic, electrical, or magnetic means. There are problems in the design of the mechanical or electrical parts of the equipment controlling the fuel pump. The problems in this area are associated with obtaining satisfactory response times. (Fig. 4.8)

■ **4.42   What proportion of oxygen is required to cause perfect combustion of (a) carbon and (b) hydrogen?**

One kilogram of carbon requires 2.6 kilograms of oxygen, and one kilogram of hydrogen required 8 kilograms of oxygen to bring about perfect combustion, i.e., the liberation of the whole of the heat in the fuel.

Note   The amount of air necessary for the complete combustion of one kilogram of liquid fuel is, theoretically, about 14 kg (air required = oxygen required/0.235). In practice, however, it is not possible to work with only the amount of air theoretically necessary, and a certain excess of air must be available to secure perfect combustion under varying conditions. About twice the amount of air theoretically required is allowed in marine oil engines.

■ **4.43   When carbon and hydrogen are burnt perfectly, what are the resulting temperatures, and how many heat units are liberated?**

When carbon is burnt to form CO_2 the temperature of combustion is 2749°C (4980°F), and the heat liberated is 34 069 kilojoules per kilogram of carbon (14 647 Btu/lb). When hydrogen is burnt to form H_2O the temperature of combustion is 2482°C (4500°F) and the heat liberated per kilogram of hydrogen is 144 444 kilojoules (62 100 Btu/lb).

Note   When carbon is burnt to give carbon monoxide, the heat produced per kilogram of carbon is only 10 269 kilojoules (4415 Btu/lb). The importance of a sufficient amount of air to bring about complete combustion to CO_2, will,

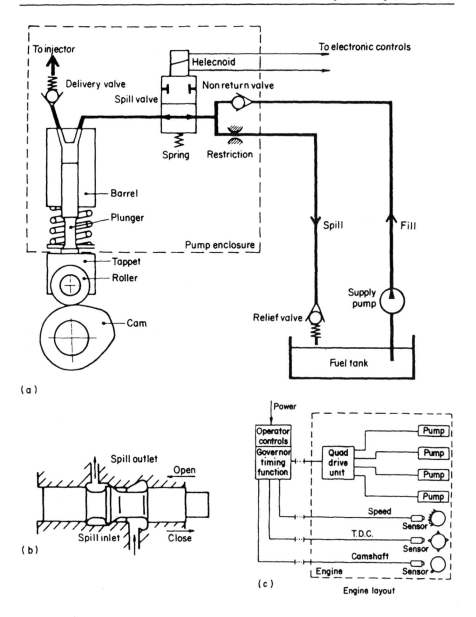

Fig. 4.8 (a) Schematic arrangement of electronic fuel injection system, using ISO/R1219 symbols. (b) Control system for electronic fuel injection – spill valve detail. (Lucas Bryce Ltd, Gloucester, UK)

therefore, be appreciated. When it is stated that the heat produced by one kilogram of hydrogen is 144 444 kilojoules (62 100 Btu/lb) when combined with a sufficient amount of oxygen, it is assumed that the product of combustion (H_2O) is in the liquid state. When hydrogen is burnt in oil-engine cylinders or under boilers, the product passes away at a temperature above the boiling point of water and, therefore, is in a gaseous state. Consequently the latent heat of vaporization is lost and the available heat produced by the combustion of one kilogram of hydrogen is 121 626 kilojoules (52 290 Btu/lb).

■ **4.44 What is the average composition of the products of combustion of ordinary diesel fuel?**

The composition is approximately:

Carbon dioxide (CO_2)	11.0%
Water vapour (H_2O)	4.0%
Oxygen (O_2)	11.0%
Nitrogen (N_2)	74.0%
Sulphur dioxide (SO_2)	trace

Note The presence of so much oxgen is due to the necessity of burning the fuel in a large amount of excess air in order to complete combustion in the limited time available. In modern engines with combustion taking place at higher temperatures, some of the nitrogen joins with the oxygen to form nitrogen oxides (NOx). These are considered to be harmful emissions and in some places a limitation is put on the allowable amount of the various nitrogen oxides that may be discharged into the atmosphere.

■ **4.45 Is it possible to burn fuel with a high water content without causing trouble with engine operation?**

If the water has a salt content similar to the content of sea water, trouble can be expected to occur. One of the problem areas will be deposits in the exhaust system, e.g., exhaust valves and turbo-blowers. If there is a high vanadium content in the fuel the sodium in the salt water will exacerbate hot corrosion problems.

A diesel engine can handle a large amount of fresh water contained in fuel, but it is necessary to homogenize the water so that it is evenly distributed throughout the fuel.

There is some evidence to suggest that small amounts of water in a homogenized mixture of fuel and water may lower fuel consumption. This comes about due to water particles in the fuel flashing off to superheated steam when injected into the high-temperature air in the combustion chamber.

It is thought that the flashing off of the water gives better atomization of the fuel. There is also the possibility that some of the superheated steam may be converted to oxygen and hydrogen by dissociation, and aid in combustion before the dissociation process is reversed.

The addition of water to the fuel lowers the temperature of combustion and helps in reducing the NOx content in the exhaust gases. Homogenizers are used in those localities where there is a limitation on exhaust gas emissions.

5

SCAVENGE, EXHAUST, PRESSURE-CHARGING SYSTEMS

■ **5.1** Describe briefly the type of pump or blower used to supply scavenge air to older, non-supercharged two-stroke engines?

Pumps used to supply air to non-supercharged two-stroke engines may be normal reciprocating pumps of the double-acting type, or Roots blowers; they are driven by the engine.

Reciprocating scavenge pumps. The pump consists of a light-weight piston working in a cylinder. Each end of the cylinder has covers fitted with suction and delivery valves, with a baffle or division plate between the suction and delivery valve groups, making two separate spaces. The suction valve space is open to the atmosphere, and the delivery valve space is connected either directly or through air trunking to the scavenge trunk. The valve plates are made from thin steel sheet, and bend or deflect off the seat to open. For standardization, the suction and discharge valve plates are usually identical. They are bolted on their seats with a covering guard which is used to control the opening lift of the valve. The valves have large opening areas, so that only a small valve lift is necessary to accommodate the air flow.

The pump piston and piston rod are attached to a crosshead which works in a set of guides. The motion of the crosshead is obtained from rocking levers linked to an engine crosshead, or from a separate crank and connecting-rod. When the scavenge pump is driven by rocking levers, the pump may be fitted on the side of the engine frames, and in this position does not increase the engine length. The width of the engine is increased, but space is usually available in the thwartships direction. If a separate crank and connecting-rod are used to drive the scavenge pump, the crank is directly connected to the engine crankshaft; the length of the scavenge pump thus increases the overall length of the engine. Reciprocating scavenge pumps, when driven by a separate crank, are sometimes arranged with two scavenge pumps, one above the other in tandem formation. Each pump has its own suction and delivery valves and the two pumps work in parallel. The stroke of a scavenge pump is usually less than that of the engine in order to keep down the piston speed of the scavenge pump; this increases pumping efficiency.

Roots blowers. Blowers of this type are also driven by the engine, either through gearing or by roller chains. They can be driven at much higher speeds than the engine and they take up less space than reciprocating scavenge pumps, but are noisier.

The moving parts of the blower consist of two rotors, each mounted on its own shaft, and connected to each other through a gear. The rotors may have two, three or four lobes. The lobes mesh during rotation but never make contact with one another. The clearance between lobes is always very small, usually just greater than the gear-tooth clearance or backlash. The rotor shafts are parallel, and supported by bearings in the end covers – ball bearings at one end, to locate the shafts axially, and roller at the other end, to allow for expansion.

As there is no contact of the lobes, the mechanical efficiency of Roots blowers is high.

The rotors work within a casing, the design of which depends on the number of lobes on the rotors. Suction is on one side of the casing and discharge on the other.

When the lobes are rotated the volume of the space on the suction side between adjacent lobes and the casing increases, drawing air into the casing. Continued rotation traps the air between the casing and the lobes; this air is passed around with the rotor until it reaches the discharge side of the casing. As the lobes mesh on the discharge side the volume of the space between adjacent lobes is reduced, and the air is expelled from the casing.

If the direction of rotation of the rotors is reversed, the air flow through the blower is also reversed. When this type of blower is used on reversible engines, a reversing arrangement must be fitted on the blower so that air is always discharged to the scavenge trunk. This arrangement consists of a large rectangular, box-shaped casing, inside which a butterfly-type flap is fitted, extending from corner to corner. The flap is mounted vertically, and keyed to a vertical shaft located centrally in bearings at the top and bottom of the casing.

Two triangular ports are cut in the base of the casing. Each triangle has its base along the opposite sides of the casing and its apex at the flap shaft. One end of the casing is connected to the air inlet filter and silencer, the other through an air trunk to the scavenge trunk. When the engine is running ahead, the air is drawn in through the silencer and filter, and passes up to the flap in the reversing casing. It enters through the triangular port on one side of the blower, and is discharged on the other side of the blower through the other triangular port into the air trunk connected to the scavenge trunk. When the engine operates astern the reversing gear swings the flap over. The blower operates in the opposite direction and the air flow through the blower is reversed.

On large slow-speed engines the mass of the rotating parts of the blower is large. In order to reduce shock caused by the inertia of the parts when starting, a spring-loaded coupling is introduced into the drive to the blower, and a damper coupling, so that cyclic variations in the engine torque are damped out and not transmitted to the blower.

Reciprocating scavenge pumps and Roots blowers are of the positive-displacement type.

■ 5.2 What effects do dirty scavenge ports have on positive-displacement scavenge pumps, and on the engine as a whole?

As the scavenge ports become dirty the resistance to air flow through the port increases, causing a pressure rise in the scavenge air system. This pressure rise increases the discharge pressure of the scavenge pump, so that more work is done in compressing the air. Since the scavenge pump requires more work to drive it, the mechanical efficiency of the engine is reduced and the fuel consumption per brake horsepower is increased.

■ 5.3 Are reciprocating scavenge-pump pistons fitted with piston rings? How is air leakage past the piston prevented or controlled? What effects do piston rings have, if fitted?

Piston rings are found in the scavenge-pump pistons of some older engines. In modern practice they are not used.

When piston rings are not fitted, air leakage past the piston cannot be prevented. It is reduced to a minimum by allowing only a very small clearance between piston and cylinder wall. Labyrinth grooves may also be machined on the piston sides. As the piston is made of light-weight soft alloy, no serious damage is done should wear on the guides allow contact between piston and cylinder; even so, guides should be adjusted, as their clearances increase, to prevent piston contact.

If piston rings are fitted, they must be lubricated. Even the small amount of oil used increases the rate of scavenge-trunk fouling. The friction between piston rings and cylinder wall is considerable, the scavenge pump therefore requires more power to drive it, and consequently the mechanical efficiency of the engine is reduced. Scavenge pumps with plain pistons, with or without labyrinth grooves, do not require as much power to drive the pump, and consequently the engine has a higher mechanical efficiency.

■ 5.4 How are scavenge pumps lubricated? What attention must be paid to the lubrication system?

The parts driving the scavenge pump, such as rocking-lever bearings, link bearings, connecting-rod bearings, and guides, are lubricated from the engine pressure-lubricating system. The scavenge-pump rod packing is lubricated by a small pump of the type used for engine cylinder lubrication. If piston rings are fitted they are lubricated by a similar pump. To prevent fouling, the lubricating oil supply to the packing and piston rings must be kept to the minimum necessary for the parts.

When an engine has been stopped in port and it is safe to open the crankcase doors, the oil flow from the parts of the scavenge-pump driving gear should be inspected to see that each bearing is supplied with oil. After the engine lubricating-oil pumps have been shut down, the driving gear, oil pipes, pipe clips, pipe unions, and any other associated fittings must be checked to see that nothing is loose. Oil pipes from lubricators to rod packing glands and cylinders must also be inspected for leakage, and loose fittings or clips.

■ **5.5   What are the factors governing the amount of air or exhaust gas flow through scavenge and exhaust ports?**

These factors are as follows.

1   The pressure differences across the ports, which in turn govern the flow velocity. (The pressure difference across the scavenge port is the difference between the scavenge pressure and the pressure in the cylinder; the pressure difference across the exhaust port is the difference between the cylinder pressure and the exhaust manifold pressure.) As the pressure difference increases, the velocity of flow increases.
2   The area of the port. An increase in area allows more gas to pass.
3   The shape of the port entry. Rounded entry edges allow air or gas to pass through without turbulence.
4   The degree of surface roughness in the port. Smooth, polished ports improve air-flow.
5   The period of time that the ports are open.

■ **5.6   In simple two-stroke cross-scavenged engines the top edge of the exhaust port is higher than the top edge of the scavenge port. Why is this necessary, and what are the objections to engines of this type?**

The top edge of the exhaust port is higher than the scavenge port, so that the exhaust port is opened first. This allows the pressure in the cylinder to fall slightly below the scavenge pressure before the piston uncovers the scavenge ports.

Engines of the simple two-stroke type are used in places where there are limited service facilities. The objection to their use is that some scavenge air is lost into the exhaust system during the period between the closure of the scavenge ports and closure of the exhaust ports. This loss of air limits the amount of fuel that can be burnt per stroke, and thus limits the mean effective pressure at which these engines can operate and still maintain a clear exhaust.

■ **5.7   What design improvements were made to older, cross-scavenged two-stroke cycle engines to increase the power developed?**

In cross scavenged two-stroke engines the relationship between the height of the scavenge ports and exhaust ports is such that during the period between the closure of the scavenge port and exhaust port some air is lost into the exhaust system. The design improvement was to increase the height of the scavenge ports to above the exhaust ports, and to fit non-return valves (referred to as scavenge valves) on the inlet side of the scavenge ports in the scavenge trunk.

When the engine is operating, downward movement of the piston first uncovers the scavenge ports; at this point the gases in the cylinder are at a pressure higher than the scavenge pressure, and they fill the port as far as the non-return valves, which prevents further backflow into the scavenge trunk.

Further movement of the piston uncovers the exhaust ports, and blowdown occurs; when the pressure in the cylinder has fallen just below the scavenge air pressure, the air in the scavenge trunk forces open the non-return valves, air flows into the cylinder, and scavenging commences. The piston continues its downward movement, passes bottom centre, comes up and covers the exhaust ports. At this point scavenge ends, but air continues to flow into the cylinder until the upward movement of the piston covers the scavenge ports.

By arranging the exhaust ports and scavenge ports in this manner more air is retained in the cylinder when compression commences than in cross-scavenged two-stroke engines. This increased amount of air enables more fuel to be burnt per working stroke, thus increasing the power developed. The power increase obtained when scavenge valves are used is approximately 20%.

Note Some cross-scavenged engines have two horizontal rows of scavenge ports in the cylinder, only the upper row of ports being fitted with non-return valves. Some cross-scavenged engines have only one row of ports, and all the ports are fitted with non-return valves.

■ 5.8 Describe the construction of scavenge valves and state where this type of valve is used. State the precautions that must be taken when assembling and fitting this type of valve.

Scavenge valves are assembled from a number of identical valve plates, made from thin flexible steel, and identical light-weight alloy castings, which function both as a valve seat on one side of the casting and as a lift guard on the other side. The parts are held together by two or four through-bolts to make the valve assembly.

The valve plates have slots at each side, through which the through-bolts pass. The material between the slots forms the valve, and is slightly deformed so that it fits tightly on its seat, which is formed from the periphery of the flat side of the light-alloy casting. A division is formed in the casting which runs diagonally from the front upper part to the back lower part. Air enters through the back lower part and pressure lifts the flexible steel plate off its seat. Air leaves the front of the valve to enter the scavenge port. This type of valve is also used as suction and delivery valves on many scavenge pumps.

When assembling the valves, care must be taken to see that the parts are all correctly positioned. If care is not taken during assembly, some parts could easily be fitted the wrong way round, because of their near-symmetry.

When fitting the scavenge valves to the trunk on the side of the cylinder, care must be taken to see that the correct side of the assembly is fitted against the cylinder. Mistakes can be prevented by marking the side of the valve assembly with an arrow in felt-tip pen, chalk or crayon showing the air-flow direction through the valve. The arrow must point into the cylinder.

The valve assemblies should be marked in a similar manner when they are being fitted in scavenge pumps. The arrow should point into the scavenge pump on the suction side and out of the scavenge pump on the discharge side.

This type of valve is used in all the loop-scavenged engines including the last generation of engines built in the early eighties.

■ **5.9 What difficulties arise in two-stroke engines with valve-controlled scavenge ports?**

The main difficulty is fouling of the scavenge ports and valves.

When the piston uncovers the scavenge ports, the gas pressure in the cylinder is higher than that within the port, and gases flow into the scavenge port as far back as the scavenge valve. These gases carry in small amounts of oil which settle on the port surfaces and on the insides of the scavenge valves, eventually fouling the valves. Some of the oil deposited on the surfaces of the scavenge ports forms hard carbon, which reduces the port area, consequently increasing the scavenge pressure. Detergent cylinder lubricants help to lengthen the periods between cleaning, but regular cleaning of scavenge valves and ports is necessary to hold an engine at peak operating efficiency.

■ **5.10 How is it possible to pressure-charge a loop-scavenged type two-stroke cycle engine where the exhaust ports are located higher in the cylinder liner than the scavenge ports? How is loss of scavenge and charge air prevented?**

The scavenge and charge air pressure is related to the exhaust gas pressure in the exhaust gas manifold. The exhaust gas pressure drops in its passage through the turbo-blower. The pressure of the exhaust gases in the exhaust gas manifold is controlled by the passages leading from the manifold into the turbo-blowers. The passages are so designed that the flow of gases to atmosphere is restricted at this point. This builds up the pressure in the manifold, and the pressure in the manifold prevents the free flow of scavenge and charge air without a large mass of the air going to waste. The power loss from the cylinders due to restricting the air flow is shown as a gain of energy in the exhaust gases. This energy is not lost, but is put back into the system by the turbo-blowers in the form of higher pressures in the scavenge and charge air system.

Note   Some older loop- and cross-scavenged engines had rotary valves fitted in the exhaust ports between the cylinder and the exhaust gas manifold. The rotary valve reduced air losses to a minimum.

■ **5.11 What do you understand by a tuned exhaust system?**

When the exhaust valve of a diesel engine opens, the gases in the cylinder rapidly expand, and gain velocity and kinetic energy as they pass into the exhaust pipe. The kinetic energy of the mass of exhaust gas carries it along the exhaust pipe, and causes a pressure build-up ahead of the mass of gas and a partial vacuum behind it.

The action may be likened to putting a close-fitting cylindrical plug in a pipe and capping one end of the pipe. If the capped end of the pipe is raised the plug will slide down the pipe. If the speed of the plug is high enough, the air in the pipe will be compressed just ahead of the plug, while a partial vacuum will be created behind the plug. If the cap is removed after the vacuum is formed, air will flow into the pipe behind the moving plug.

Similarly, if two cylinders are connected to a common exhaust pipe, it can be seen that if the exhaust valve of the second cylinder is opened when the exhaust from the first cylinder has created a vacuum in the pipe, the exhaust from the second cylinder will be discharged more easily.

This principle is used in a tuned exhaust system. By making the exhaust pipes a suitable length and arranging for two or three cylinders with suitable exhaust-valve timing to exhaust into the same pipe, the partial vacuum created by the exhaust from one cylinder is used to help exhaust expulsion from the following cylinder.

■ **5.12 What limits the power that can be developed in a diesel engine cylinder of a given size? What would happen if attempts were made to obtain greater power output by increasing the amount of fuel injected at each cycle?**

The power output of a diesel engine is limited by the weight of air in the cylinders when fuel is injected. Injecting more fuel in order to raise the power output without a corresponding increase in the weight of air would result in some of the fuel being incompletely burnt owing to lack of air. Incomplete combustion would cause the mean temperature of the cycle to be increased and lead to serious mechanical troubles, such as fouling of piston rings, inefficient lubrication of the cylinder, burnt exhaust valves, and temperature stresses so great that cylinder heads, pistons and other parts subjected to the higher gas temperatures might be fractured.

■ **5.13 How can the power of a diesel engine be increased?**

The power of a diesel engine is dependent on the amount of fuel used, and this is limited by the amount of air available to burn the fuel. If more air can be introduced into the cylinder, more fuel can be used and the power of the engine will be increased.

When contemplating the possibilities of power increase, care must be exercised because of the attendant problems of thermal stress which usually lead to early failure of engine parts subjected to high temperatures.

■ **5.14 How can more air be introduced into the cylinders of a diesel engine with a view to increasing its power?**

In four-stroke engines, more air can be introduced into the cylinder during the suction stroke by connecting the air inlet valve to an air supply above atmospheric pressure. In two-stroke engines, the same result will be obtained by increasing the scavenge pressure. In order to get the requisite air into the cylinder the timing is altered to increase the length of the air inlet period.

If the pressurized air is obtained from a pump or blower driven by the engine, some of the extra power gained will be lost in driving the pump or blower. The mechanical efficiency of the engine is reduced but considerable net gain in the power output can be achieved.

In four-stroke cycle engines the air entering the cylinders under pressure pushes the piston downwards; this reverses the effects of pumping losses experienced in naturally aspirated engines where power is lost during the inlet and exhaust strokes.

The strength of the parts and the cooling must be adequate for the increased mechanical and thermal loads.

■ **5.15   What is meant by the term 'supercharging' or 'pressure-charging'?**

When the piston of a normal oil engine is beginning the compression stroke, the cylinder should be full of air at atmospheric pressure. If means are adopted to cause the pressure at this point of the cycle to be greater than that of the atmosphere the engine is said to be supercharged or pressure-charged. The two terms are synonymous.

■ **5.16   How are modern two-stroke and four-stroke engines pressure-charged?**

Modern diesel engines are pressure-charged by utilizing the energy in the exhaust gases to drive a gas turbine connected to a rotary blower. The blower compresses air so that it is delivered under pressure to the engine cylinders. Because the air is under pressure, a greater mass can be contained in the cylinder and so more fuel can be burnt per stroke, which increases the power developed. Engines pressure-charged in this way with exhaust-gas turbo-driven blowers are often referred to as turbo-charged engines.

■ **5.17   What do you understand by the terms 'boost pressure ratio' and 'air density ratio'?**

The term boost pressure ratio (often known as pressure ratio) refers to the pressure rise across the air section of the turbo-blower.

The term air density ratio refers to the change in density which occurs when the temperature of the air rises when undergoing compression.

When graphs of boost pressure ratio are plotted against the mass air flow, the ambient conditions under which the compressor is working should be stated.

In other cases, when performance curves are plotted, two ordinates may be used; one gives the boost pressure ratio while the other gives the air density ratio. The abscissa on which the ordinates are plotted is the blower rpm or percentage of maximum rpm.

■ **5.18   Describe an exhaust-gas turbo-blower or turbo-charger.**

A modern exhaust-gas turbo-blower is essentially a single-stage impulse turbine connected through a common shaft to a centrifugal-type air blower. The turbine and blower are housed in a circular casing divided into two separate spaces by a circular division plate, which may be water-cooled, or protected by heat insulation on the exhaust-gas side. The section of the casing which houses the turbines is fitted with one or more flanged exhaust-gas inlets, which lead to the nozzle-blade ring assembly. The exhaust gases pass through this ring and are directed on to the turbine rotor blading. The gases enter the moving blades of the turbine rotor at high velocity. The passage of the gas through the rotor blades causes a change of direction in the gas flow, resulting in a change of momentum, which exerts a force on the turbine blades. This force causes the rotor to revolve at high speed. The exhaust passes from the rotor into a circular space connected to the exhaust-gas outlet branch.

The air-blower casing is fitted with filters at the air inlet to the casing. The entry passages after these filters are usually fitted with splitters to guide the air through the passages and reduce the draught losses caused by a change in the air-flow direction. Sound-absorbent material is used to cover the inside of the air passages and the splitters, to reduce wind and blower noise.

At the end of the inlet passage just before the blower impeller, curved air-guide vanes are fitted, so that the air enters the impeller without shock. The impeller takes in the air axially and discharges it radially.

The air is discharged from the blower impeller into the diffuser at high velocity. During its passage through the diffuser the air gives up much of its velocity and in so doing builds up pressure. From the diffuser the air passes into the discharge casing, which is volute-shaped and fitted with a flanged outlet connection.

The casings for the turbine and blower are built up from full circular sections held together with circumferential joints. Building the casings in this manner obviates the need to split the casing longitudinally, and reduces design problems and manufacturing costs. The turbine casings are usually water cooled, and connected with the engine cooling system.

The rotor is usually made up of a hollow shaft on which the turbine rotor and air impeller are mounted. The impeller is often made in two sections to make production easier. Three labyrinth gland sections are built into the rotor: one, at the turbine end to seal the shaft and prevent exhaust gas leakage; another, at the blower end to prevent oil being drawn from the bearing. A third is built up between the turbine rotor and impeller, forming a labyrinth gland in the casing division plate between the turbine and blower spaces. The labyrinth glands are supplied with sealing air from the blower discharge.

A thrust bearing is fitted at one end of the rotor to balance the thrust set up by the difference in gas and air pressures within the casings.

In marine practice sleeve-type bearings or ball and roller bearings may be used. In practice today the use of each type of bearing appears to be equally divided. Some manufacturers prefer to use ball and roller bearings because the cooling and lubrication of these bearings is simpler and enables the turbo-charger to be independent of an external oil supply. Ball-type bearings take up the rotor thrust and roller bearings allow for rotor expansion. In some smaller turbo-charger units ball bearings may be used at each end of the rotor shaft or in the centre of the shaft between the air and exhaust gas sections. This arrangement allows for easy access to the rotor when dismantling the turbo-charger. When ball bearings are used one bearing must be free to slide in its housing in order to accommodate the differences between rotor and casing expansion. The bearings are often fitted into spring-mounted supports.

There are three generally accepted methods of lubricating the turbo-blower bearings. One method utilizes the lubricating oil from the engine lubrication system. The oil is supplied under pressure to the bearings, from which it drains back into the engine system. The second method consists of a complete lubricating-oil system used only for the turbo-blower bearings. The system comprises a pump, which takes lubricating oil from a drain tank and pumps it through an oil cooler up to a gravity supply tank. From the gravity tank the oil passes through filters to the turbo-blower bearings, and then runs down to the drain tank. The third method used with ball and roller bearings is to make the lower part of each end cover into an oil reservoir and attach a gear oil pump to each end of the rotor shaft. Lubricant and coolant is then pumped by each gear pump from its reservoir into its adjacent bearing. The oil after passing through the bearing drains back to the reservoir. Cooling is effected by air passing into the compressor over the reservoir at the compressor end and from a cooling water jacket at the exhaust gas end. It must be remembered that ball and roller bearings will operate satisfactorily at much higher temperatures than sleeve bearings.

The rotor revolves at high speed, and it must therefore be dynamically balanced after all the parts of the rotor are assembled. The balancing will normally be carried out on a balancing machine. (Fig. 5.1)

■ **5.19 How are the exhaust pipes arranged in a constant-pressure exhaust-gas turbo-charging system?**

In constant-pressure exhaust-gas turbo-charging systems the exhaust branch or pipe from each individual cylinder is led into a common manifold. As the exhaust gas blows into the manifold, eddies are set up which help damp out any pressure wave caused by the influx of the exhaust gas. The volume of the manifold must be large enough to accommodate the gas flow from individual cylinders without causing any localized pressure rise in the manifold as exhaust gas leaves the cylinder. The exhaust gas is led from the manifold into the exhaust-gas turbo-blower at a constant pressure.

If the exhaust manifold is made too large the response of the turbine to engine load change is slowed up due to the manifold volume taking longer to bring to the higher exhaust pressure. The discharge pressure from the turbine must be higher than the exhaust manifold pressure. When the engine is operated with a low load the air discharged from the compressor may be insufficient; a separate electrically driven blower is then brought into operation.

■ **5.20 What do you understand by the term 'pulse system' of exhaust turbo-charging?**

In the pulse system maximum use is made of the energy in the exhaust gas during the blowdown period, by designing the exhaust valves so that they have the largest possible port area. The exhaust cams are designed to give the most rapid possible opening of the exhaust valve. When the exhaust valve opens, the blowdown of the exhaust into the exhaust pipe causes a pressure wave or impulse to pass down the exhaust pipe to the nozzle plate in the exhaust turbine. The static pressure of the exhaust gas during the impulse period is greater than the air charging pressure; therefore no cylinder should be exhausting at the time

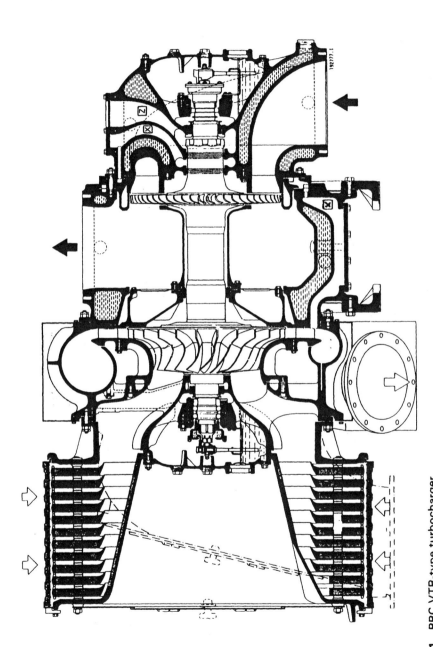

Fig. 5.1 BBC VTR-type turbocharger.
The connection marked 'x' supplies air to the labyrinth sealing rings. The connection marked 'Z' is the leak off. (Brown Boveri and Co, Baden, Switzerland).

the impulse passes in the exhaust pipe, otherwise exhaust gas may blow back into the other cylinder.

Engines turbo-charged on the impulse system must have the exhaust pipes carefully grouped according to the engine exhaust-valve timing. Each group of exhaust pipes has a separate entry into the exhaust turbine, and each entry leads to its own nozzle group (see also Question 5.11.).

■ **5.21   Describe any turbo-charging system other than the pulse or constant pressure systems.**

Other turbo-charging systems are sometimes known as quasi-pulse systems. These systems utilize a system of exhaust pipes similar to the pulse system. The turbine nozzles are made larger and two groups of pipes will be joined together, the common pipe from the two groups being led to a group of turbine nozzles. A variant of this is to lead all the common pipes from each group to a single entry into the turbine and feed all the nozzles from the common entry point. The object of these arrangements is to increase the efficiency of the turbo-charging system and reduce the size of the turbo-charger.

■ **5.22   How are the turbine blades of an exhaust turbo-blower attached to the rotor disc? Are the blade fastenings a loose or tight fit in the rotor?**

The roots of the turbine blades, where they are attached to the rotor, are machined to what is referred to as 'fir-tree' shape – because it bears some resemblance to the shape of a fir tree. On the rotor are correspondingly shaped slots into which the fir-tree roots of the blades slide. Machining tolerances are extremely tight.

In small turbo-blowers used for constant-speed diesel engines, such as diesel generators, the blade root can be a tight fit in its slot.

In large turbo-blowers with long rotor blades, particularly those used for propulsion engines where a wide range of operating speeds is met in service, the fit of the blade is such that it can move slightly. When blades are fitted loosely in this manner, a segmented binding wire is fitted through holes near the blade tips.

The blades are held in their slots by upsetting or raising a caulked edge on the rotor at the innermost part of the root slot; this edge prevents the blades sliding out. (Fig. 5.2)

■ **5.23   Why are turbine blade roots sometimes made to be a sliding fit in the rotor? What is the purpose of the binding wire?**

The sliding fit of the blade root in large turbo-blowers gives a large amount of damping to the blade to reduce the risk of blade vibration.

The binding wire is made in short lengths, each of which extends through four to six blades. All the blades are connected in this way. The binding wire passes through holes near the tips of the blades. It is not a tight fit in the holes. Each segment of binding wire is fastened by welding it to the first blade of the group in which it is fitted.

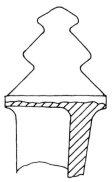

Fig. 5.2 Turbocharger turbine blade fastening, fir tree root type.

When the rotor is moving at high speed, the flexibility of the binding wire is such that the action of centrifugal force causes it to bear against the outside of the holes through which it passes. If the turbine rotates at a speed at which the blades would vibrate, the friction between the binding wire and the sides of the holes in the blade tips damps out the vibration and prevents ultimate failure.

Some turbo-blowers used in engines running at constant speed, such as diesel generators or alternators, do not have binding wires.

■ **5.24 What is the reason for using a fir-tree form of blade root for turbine blades?**

The fir-tree root form is superior to other blade fastenings because there is less stress concentration at the junction between the root and the blade. Because there is a more even distribution of stress at the blade root there is less chance of blade failure in this region.

■ **5.25 Describe how you would determine whether exhaust turbo-blower air filters were clean or dirty.**

Small engine filters of the renewable type must be visually examined. In larger blowers with removable filters, a vacuum gauge or U-tube gauge is fitted on the suction side of the blower, so that the air pressure in the inlet space between the filter and blower can be measured. As the air filter becomes dirty, its resistance to air flow increases and the pressure-drop across the filter is increased. This is indicated by a lowering of the air pressure downstream of the filter. A comparison of this pressure with the figures recorded at the engine test bed trial will indicate whether the filters are dirty. Sometimes specific pressures at which filters must be cleaned are given in the engine instruction book.

■ **5.26 What are the materials used for the filters on the air inlet side of an exhaust turbo-blower?**

The materials used for air filters on small engines may be special porous paper elements which are renewed when they become dirty. These elements are usually wound in a corrugated form, so that a large area of filter is obtained in a small space. This type is found on smaller engines, as used, for example, for emergency electrical generators. For larger engines air filters on the blowers have removable elements. The filtering medium within the element may be plastic fibre or a non-rusting metallic 'wool' material. Filters made with these materials can be cleaned when they become dirty.

■ **5.27 Describe the cleaning of turbo-blower air filters.**

Most engine builders supply a cleaning pan of the proper size to accommodate the air-filter element. The air filters are placed in the pan and soaked in alkaline solutions or special solvents. During the soaking period the dirt adhering to the filtering medium detaches itself and sinks to the bottom of the pan. After the filters are removed from the pan, they should be allowed to drain. Some filtering media may be back-washed with fresh water.

Note Care must be taken to use the cleaning material recommended by the engine builder; some solvents dissolve the plastic fibre material of which some filters are made. Care must be exercised in the use of many cleaning materials, to prevent skin infections and possible damage to eyes. This is particularly important when using compressed air for final removal of cleaning material or rinsing water.

■ **5.28 What effects will dirty air inlet filters in turbo-blowers have on the operation of the engine?**

Dirty air filters reduce the amount of air passing into the engine. As the quantity of air is reduced, so the exhaust gas temperature is increased. The degree to which this occurs depends on the restricting effect of the dirt on the passage of air through the filter.

5.29 Why are large amounts of lubricant supplied to turbo-blower bearings?

As the rotor revolves at high speeds large amounts of heat are generated in the bearings due to friction.

The friction may be caused by fluid friction as experienced in sleeve bearings or rolling friction as experienced in ball and roller bearings. The lubricant supplied serves both as a coolant and a lubricant. Generally larger amounts of lubricant are required in sleeve bearings.

■ **5.30 What indications will be given of a faulty lubricating-oil supply to the turbo-blower bearings?**

The lubricant supplied to turbo-charger bearings is supplied under pressure.

The pressure of the oil supply is monitored and any fall in pressure gives rise to an alarm condition and call out.

Turbo-chargers with sleeve bearings and external oil supply may also be fitted with thermometers on the oil-drain outlets from the bearings together with temperature sensors giving a bearing temperature reading at the control station. Oil flow indicators may also be fitted on the inlet oil line to the bearings.

In turbo-chargers fitted with ball and roller bearings and internal oil supply from reservoirs in the end covers, a sight glass is fitted in the reservoir to indicate fluid height. Visual examination and feel of the bearing casings or oil reservoirs also give an indication of satisfactory operation or possible trouble.

■ 5.31 What do you understand by the term 'two-stage turbo-charging'?

When air is compressed its temperature rises and the compression may approach adiabatic conditions. If air is compressed in two stages the air can be cooled between each stage of compression and the amount of work done in compressing the air is reduced. Compression then approaches isothermal conditions.

Two-stage turbo-chargers are generally built up from two separate matched standard size turbo-chargers. The rotor of each stage is on a common polar axis or centre line. The rotors are not mechanically connected and are free to move independently of each other. The exhaust gas stages are usually arranged back to back so that the gas passes direct from the outlet of the first stage into the inlet of the second stage. This reduces loss of heat from the gas in passing between each stage.

The second-stage turbine drives the low-pressure compressor. Air is cooled after passing from the low-pressure stage and then enters the high-pressure stage which is driven by the first-stage turbine. The air is again cooled after leaving the high-pressure stage.

Two-stage turbo-charging allows higher boost pressure ratios and imparts much higher overall efficiency to the turbo-charger unit. (Fig 5.3)

■ 5.32 Where are air coolers fitted on pressure-charged engines? What is their purpose? Name the cooling medium and explain why it is used.

Air coolers are fitted on turbo-charged engines to cool the air after it has been compressed in the turbo-blower. Cooling the air reduces both the temperature and the volume, and the mass per unit volume is therefore increased. Colder air also cools the internal parts of the cylinder more effectively during the scavenge period; a greater mass of air can therefore be present in the cylinder when compression commences.

Sea-water is the usual cooling medium. The temperature of sea-water is much lower than the temperature of the fresh water used for cooling the engine, and therefore has a better cooling effect.

■ 5.33 What causes fouling of the air-side of air coolers?

Fouling of the air passages in an air cooler forming part of a turbo-charging

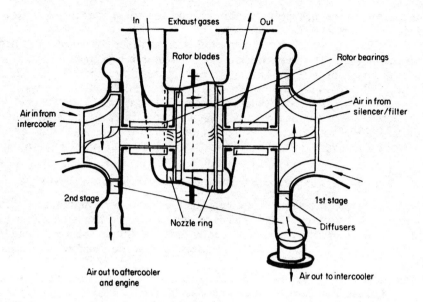

Fig. 5.3 Two-stage turbocharger. The lubrication system, labyrinth packings, cooling jackets and other details have been omitted for the sake of simplicity.

system is usually due to oil and oily-water films collecting on the sides of the tubes and tube fins. Lint and similar material adheres to these films of oil or emulsion. The presence of oil may be caused by faulty air filters, or by improperly placed filters which allow the air to pass by the side of the filter element. Sometimes the oil is drawn from the bearing at the blower end of the turbo-blower.

The presence of moisture is usually the result of high humidity, when the engine is operating in warm air temperatures in conjunction with low sea-water temperatures.

■ **5.34   What are the usual causes of oil being found in the air passages between the turbo-blower and the air cooler?**

If large amounts of oil are found in the air system, the usual cause is faulty oil seals at the blower bearing; another reason may be that the sealing air pipe to the compressor labyrinth packing has become choked or shut off, in which case oil or vapour will be drawn through the labyrinth packing.

The air breathers on the ends of the turbo-blower end covers must be kept clean. These air breathers connect through to the bearing oil drainage space and allow the air leak-off from the labyrinth packing to escape. The air breather can be used to check the air flow into the labyrinth packing, and the leak-off. If a strip of paper is held in front of the breather and is blown away, air is passing from the leak-off; if the paperstrip is drawn towards the breather it indicates that air is being drawn into the labyrinth packing owing to lack of sealing air.

Oil vapour will then be drawn into the blower and eventually pass and condense in the blower discharge lines.

The remedy is to clear the air supply to the blower labyrinth packing. If the air supply is normal, it is usually indicated by a small amount of oil vapour wafting from the air breather.

When turbo-blower casings are painted, care must be taken to prevent the air breather becoming sealed with paint.

■ **5.35 How will fouling of air coolers be shown on an engine during its operation?**

When air coolers become fouled, less heat will be transferred from the air to the cooling water. This is shown by changes in the air and cooling-water temperatures. Changes will also occur in the pressure drop of the air passing through the cooler. The amount of change will depend on the degree and nature of the fouling.

The symptoms of air-side fouling are as follows.

1 Decrease of air temperature difference across cooler.
2 Increase of air pressure drop across cooler.
3 Rising scavenge air temperature.
4 Rising exhaust temperature from all cylinders.
5 A smaller rise in cooling-water temperature across the cooler.

Fouling of the cooling-water side is shown by the following symptoms.

1 Rising scavenge temperature.
2 Reduction in the difference of the air temperature across the cooler.
3 Reduction in the temperature rise of the cooling water across the cooler if the fouling is general on all tubes.
4 Rising exhaust gas temperature from all cylinders.
5 Increase in the temperature rise of the cooling water if fouling or choking materially reduces the amount of water flow.

The temperature and pressure differences recorded should be compared with those obtained from the engine when on the test bed.

■ **5.36 What is the purpose of the drain cocks or valves on the air side of air coolers?**

The purpose of these cocks is to check the tightness of the air cooler against water leakage from the cooling side, and to drain off any condensation that may occur.

If a leak occurs it can be found by continuing the circulation of water when the engine is stopped, and leaving the drains open. If water is found to drain in any appreciable quantity, it indicates a water leak.

The cocks should be regularly used at sea and left open in port when the engine is shut down.

The presence of water may indicate that condensation is occurring in the cooler. This may happen when the atmosphere is humid, and the sea-water cools the air below the dew-point temperature. This condition is most likely to be met when the air is warm (warm air holds more water vapour than cold air) and the sea-water is cool. Condensation can be prevented by opening the cooling-water bypass on the cooler, or reducing the amount of water passing through the cooler.

Note Some engines are not fitted with a cooling-water bypass on the air coolers. If other heat exchangers are in a series circuit with the cooling water from the air coolers, care must be taken if the amount of water flow is restricted because it will also reduce the cooling-water flow to the other heat-exchangers.

■ **5.37   How are air coolers cleaned?**

The cooling-water spaces can be cleaned with the cooler in place. The connecting pipes to the cooler are removed, and the lower cooling-water branch is blanked off. The cooling-water space is then filled with the cleaning liquid. Sometimes a stand pipe is fitted on the upper branch to provide a pressure head on the cooler. The air spaces on smaller coolers can also be cleaned in place by fitting blanks, and filling the air space with cleaning fluid.

It is of the greatest importance to keep the air coolers clean, and engines are designed with a view to facilitating this job as far as possible.

Air coolers on large engines are designed so that the tube stack or nest can be quickly dismantled and lowered into a cleaning bath. The bath is made of glass-fibre reinforced plastic, which is light in weight and stores easily. After the stack is lowered into the tank, it is soaked in cleaning fluid to remove dirt. After the requisite soaking period the cooler should be rinsed clean. Care must be taken in the final examination to see that all the spaces between the tube fins are clear.

Note Engine builders usually give instructions for cleaning air coolers in the engine instruction book.

The cleaning materials or chemicals that can be used without causing damage are usually also listed. If no instruction is given, the cleaning material must be carefully selected, since some cleaning materials may actively corrode some of the metals used in coolers. Cleaning fluids having any content that may create flammable vapours must be used discreetly and any chance of vapour entering the engine at the first start must be obviated. This is done by carefully rinsing away all traces of the cleaning fluid.

Care must be exercised if the coolers are cleaned *in situ,* as the supports for the cooler may not be strong enough to support the cooler when it is full of cleaning fluid.

■ **5.38   After a period of service the gas inlet nozzles and blading of the turbine in exhaust turbo-blowers become dirty. What causes the dirt, and how would you know that the nozzles and blading were dirty?**

When combustion is clean, the deposits found within the nozzles and blading of the exhaust turbine are usually sodium compounds, in the form of sulphates, and vanadium compounds, which may be oxides. Some additives used in the cylinder lubricant may also be found in the ash.

If combustion is dirty, sooty deposits and carbonaceous materials may be found. These deposits may be due to impurities in the fuel, or to poor combustion. Sometimes poor combustion is caused by operation of the engine at low loads for lengthy periods.

When nozzle and rotor blades become dirty the engine symptoms are rising exhaust-gas temperature at the turbine inlet, accompanied by a fall in the speed (rev/min) of the turbine. The exhaust-gas temperature rises because more kinetic energy is given up by the gas before it enters the nozzle blades. This kinetic energy is converted into heat and consequently increases the temperature. Dirt in the nozzles and blades changes the velocity pattern of the gases passing through them; this prevents the turbine working efficiently, and the speed of rotation falls. When the turbine operates at a lower speed the air delivery is reduced and this in turn causes the cylinder exhaust temperature to rise.

■ 5.39   Describe how turbine rotor blading should be cleaned.

Turbine rotor blading can be cleaned by water-washing while the turbo-blower is in service, or the turbo-blower can be dismantled to clean all parts thoroughly. Water-washing of the blades while the turbo-blower is in service requires special apparatus and connections on the turbo-blower exhaust-gas inlet spaces. This apparatus consists of a probe which passes into the gas space. The probe is in effect a water sprayer which fits into the gas inlet space. The outer end of the probe is fitted with connections for compressed air and water, which are taken through flexible pipes from the supply points at the side of the blower. An automatic cock is fitted into the probe. When opened, it allows water and compressed air to flow to the sprayer. The compressed air atomizes the water as it is blown into the gas space. The usual cleaning time is 10 to 15 minutes, depending on the amount of dirt.

The ash found in the nozzle blading is usually water-soluble. The finely atomized water breaks down the ash formations, which then pass up the exhaust pipes with the exhaust gases to the atmosphere.

While the blades are being washed the turbine blower speed must be considerably reduced, and the water drain cocks from the exhaust spaces must be left open. After washing, the speed of the engine and the turbo-blower must be increased gradually, and the water drain cocks left open until all the water is cleared.

Some turbo-chargers are fitted with an arrangement to inject small particles of broken walnut shells into the exhaust-gas passages before the nozzle blading. The sharp edges of the broken shells have a good scouring action on the nozzles and blading without damaging the smooth surfaces required for the high-velocity gases to operate the turbine in an efficient manner, with only minimum losses from blade and nozzle friction.

When a turbo-blower is dismantled to clean the turbine rotor blades, the rotor is set up on a pair of wooden V-notched trestles that allow part of the rotor disc and blading to soak in a water bath. During the soaking period the rotor must be regularly and frequently turned, so that the deposits are completely removed. Incomplete removal may leave the rotor out of balance.

■ **5.40 What do you understand by the term 'surge', when applied to the centrifugal-type blower used for pressure-charging diesel engines?**

If the curve of the quantity of air discharged by a turbo-charger plotted against the pressure ratios is examined, it will be seen that a line is drawn from the intersection of the axes. From the right-hand side of this line curves are shown giving the quantity of air discharged against the pressure ratio. No points are plotted to the left of the line. The points plotted on the right-hand side are in the region of stable operation. Instability occurs if the turbo-charger operates on the left-hand side of the line. The line separating the two areas is referred to as the surge line. (Fig. 5.4)

The surge line passes through the points where the pressure ratio is very near a maximum value; as the amount of air discharged increases the pressure ratio is seen to fall away.

Under stable operating conditions, any change in the amount of air taken into the turbo-charger is accommodated by a change in pressure; one change is balanced by another.

When the turbo-charger operates under unstable conditions any reduction of air demand on the discharge side of the compressor causes the pressure to fall.

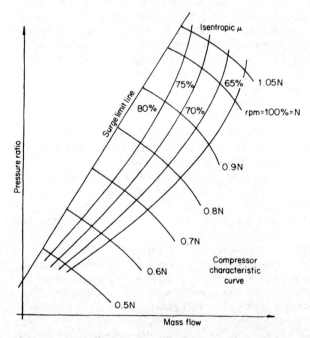

Fig. 5.4 Compressor performance curves showing the effect of speed changes on mass output and pressure. The surge line is also shown.

If this occurs the pressure falls rapidly and air flows back from the scavenge air trunk to the diffuser. The increase in pressure then causes an increase in the air flow. As the engine cannot use this air the pressure starts to fall again and the action is repeated. This condition will continue until the air demand is increased and the turbo-charger is allowed to operate under stable conditions again.

To prevent these conditions arising the scavenge trunk capacity should be large enough to reduce pressure fluctuations, and the capacity of the turbo-charger must be carefully matched to the engine. It will then operate in the stable range under normal conditions.

■ **5.41 What are the symptoms when a turbo-blower surges, and how would you rectify the condition? If you had a turbo-blower surge how would you prevent the trouble repeating itself?**

The symptoms of a surge condition are a repeated, irregular violent thud from the air intake to the blower, and rapid surges in the scavenge-air pressure. If you were standing by the air intake to the blower you would sense that the air was being taken into the blower in a series of 'gulps'. If surge occurs the engine speed must be reduced immediately; easing the scavenge-trunk relief valves will help to reduce the shocks from air-flow surges.

In an engine with a correctly matched blower, surge is caused by a combination of two factors. The first is dirty air-intake filters, which restrict the flow of air to the blower and the amount of air discharged. The second is the pressure pulsations, created by the opening and closing of the scavenge ports, and the irregular air flow from this cause.

Dirty air filters alter the discharge pressure ratio of the blower, and, in effect, change the blower characteristics. The pressure pulsations, when felt at the blower discharge, cause the blower to become unstable and consequently to surge.

On engines with two or more blowers discharging into a common casing, the air movement in the scavenge casing may bounce from one blower to the other, causing each blower to go into and out of surge conditions in an alarming manner.

To prevent the trouble recurring, the air filters must be immediately cleaned very thoroughly, so that the blower can work within the stable range.

■ **5.42 What are the main advantages of the impulse and the constant-pressure exhaust turbo-charging systems?**

The main advantage of the impulse system is that at low loads it is more efficient than the constant-pressure system. It does not require assistance from scavenge pumps, except at very low-power operation.

The advantage of the constant-pressure system is that the exhaust-pipe arrangement is more simple. At high engine loads, the constant-pressure system becomes more efficient than the impulse system. At low loads, scavenge pumps, under-piston pumps, or electrically driven blowers are necessary to supply the required air.

■ **5.43 What provision is made for running turbo-charged propulsion engines at very low power?**

When a propulsion engine is running at very low power, the air output from the turbo-blower is insufficient for the requirements of the engine. The air supply can be augmented by the following methods.

1   Some engines are fitted with an electrically driven blower that will give enough air for slow-speed operation, or for emergencies when the turbo-blower is out of service.

2   Other engines retain scavenge pumps, which are connected in series with the turbo-blower, and at slow speeds these scavenge pumps will supply sufficient air:

3   In crosshead, loop-scavenged two-stroke cycle engines the space between the underside of the piston and the diaphragm can be fitted with suction and delivery valves so that it acts as a reciprocating scavenge pump at low engine speeds. At higher engine speeds the pumping action reduces the mechanical efficiency of the engine, but it is possible to open a butterfly valve and, in effect, increase the volume of the clearance space. This reduces the pumping load, increases the mechanical efficiency of the engine and so lowers its fuel consumption.

■ **5.44 How is the efficiency of a turbo-charger obtained?**

The true efficiency of a turbo-charger is difficult to obtain. It is not easily adaptable to instrumentation, consisting as it does of a gas turbine and an air compressor combined into a single unit and driven by the power obtained from the exhaust gases of a diesel engine to which it is attached.

Turbo-charger manufacturers have their own test stands and testing facilities used both for research and testing for efficiency; due to commercial considerations regarding secrecy they do not usually publish details of their work.

A basic statement for efficiency is given by

Efficiency = (Energy in – Energy out)/Energy in

but this statement gives only the thermal efficiency; it will have to be multiplied by the mechanical efficiency to obtain the overall efficiency.

The energy input must cover both the kinetic energy and the heat energy contained in the exhaust gas. The energy utilized covers the difference between the energy input from the engine exhaust and the energy contained in the compressed air discharged from the compressor. There is no problem in dealing with the heat content in the exhaust gas and the compressed air because these values can be obtained from standard tables and charts, but a problem arises in deciding the datum point for the other forms of energy and whether stagnation enthalpy considerations should be brought in.

From this it is seen that an academic approach to finding the true efficiency of a turbo-charger by thermodynamic analysis raises many questions.

For simplification the efficiency of the gas turbine and the air compressor are considered separately.

The efficiency of the gas turbine is obtained from the values of the temperature and pressure of the exhaust gases entering and leaving the turbine.

These values can be set up on a heat-entropy chart for exhaust gases (also known as an enthalpy-entropy chart or a mollier chart). From this chart the adiabatic or isentropic heat drop is obtained; dividing this by the heat input gives the thermal efficiency of the gas turbine. The efficiency of the air compressor can be found in a similar manner by setting up the compressor pressure and temperature values on a heat-entropy chart for air.

The efficiency of the turbo-charger can then be found by taking the product of the turbine and compressor efficiencies, then multiplying the value found by the mechanical efficiency. This efficiency is not the true efficiency of the turbocharger but is accurate enough for comparative purposes.

An expression giving the isentropic efficiency of a turbocharger can be developed from basic thermodynamic statements as shown. The turbocharger efficiency is equal to the product of the turbine efficiency, the compressor efficiency, and the mechanical efficiency of the complete unit.

$$\frac{\text{Turbocharger}}{\text{efficiency}} = \frac{\text{Turbine}}{\text{efficiency}} \times \frac{\text{Compressor}}{\text{efficiency}} \times \frac{\text{Mechanical}}{\text{efficiency}}$$

$$\eta_{Tc} = \eta_{Tu} \times \eta_{Co} \times \eta_{Mech}$$

The turbine efficiency and the compressor efficiency are now treated separately.

Turbine efficiency η_{Tu}. The efficiency of the turbine may be obtained from a study of the heat given up by the exhaust gases on their passage through the turbine. The gases passing through the nozzles and blading are subjected to friction, this causes a temperature rise in the gases above the temperature rise expected if expansion had been isentropic (adiabatic). This is known as reheating and causes an increase in the entropy of the gases.

$$\eta_{Tu} = \frac{\text{mass flow} \times \text{sp. heat} \times \text{actual fall in temperature of gases}}{\text{mass flow} \times \text{sp. heat} \times \text{isentropic fall in temperature of gases}}$$

If the specific heat and mass flow are constant, they may be cancelled. Then:

$$\eta_{Tu} = \frac{\text{actual temperature drop in exhaust gases}}{\text{isentropic temperature drop in exhaust gases}}$$

$$= \frac{(T_1 - T_2)}{(T_1 - T_3)} \qquad \text{where:}$$

T_1 = temperature of gases entering turbine at pressure p_1
T_2 = actual temperature of gases leaving turbine at pressure p_2
T_3 = temperature of exhaust gases at pressure p_2 based on fall of temperature when expansion is isentropic.

The following equation may be used to find the value of T_3 :

$$\frac{T_1}{T_3} = \left(\frac{p_1}{p_2}\right)^{\frac{(\gamma_g-1)}{\gamma_g}} \quad \text{and } T_3 = \frac{T_1}{\left(\frac{p_1}{p_2}\right)^{\frac{(\gamma_g-1)}{\gamma_g}}} = T_1 \left(\frac{p_2}{p_1}\right)^{\frac{(\gamma_g-1)}{\gamma_g}}$$

γ_g is the ratio of the specific heats C_p/C_v for exhaust gas.

Compressor efficiency η_{Co} The efficiency of the compressor may be found in a similar manner, then:

$$\eta_{Co} = \frac{(T_6-T_4)}{(T_5-T_4)} \quad \text{where:}$$

T_4 = Temperature of air entering compressor at pressure p_4
T_5 = Actual temperature of air leaving compressor at pressure p_5
T_6 = Temperature of air at pressure p_5 based on isentropic compression

The following equation may be used to find the value of T_6:

$$\frac{T_6}{T_4} = \left(\frac{p_5}{p_4}\right)^{\frac{(\gamma_a-1)}{\gamma_a}} \quad \text{and } T_6 = \frac{T_4}{\left(\frac{p_5}{p_4}\right)^{\frac{(\gamma_a-1)}{\gamma_a}}} = T_4 \left(\frac{p_4}{p_5}\right)^{\frac{(\gamma_a-1)}{\gamma_a}}$$

γ_a is the ratio of the specific heats C_p/C_v for air. By substituting the expressions for the turbine and compressor efficiencies into the general equation given above an expression for the turbocharger efficiency may be obtained. Then:

$$\eta_{Tc} = \frac{(T_1-T_2)}{(T_1-T_3)} \frac{(T_6-T_4)}{(T_5-T_4)} \eta_{mech}$$

$$Tc = \frac{(T_1-T_2)}{(T_5-T_4)} \frac{(T_4)}{(T_1)} \frac{\left[\left(\frac{p_4}{p_5}\right)^{\frac{(\gamma_a-1)}{\gamma_a}} - 1\right]}{\left[1 - \left(\frac{p_2}{p_1}\right)^{\frac{(\gamma_a-1)}{\gamma_a}}\right]} \eta_{mech}$$

All pressures and temperatures must be absolute values. This expression may be further refined by substitution and simplified.

The efficiency of the turbo-charger can then be found by taking the product of the turbine and compressor efficiencies, then multiplying the value found by the mechanical efficiency. This efficiency is not the true efficiency of the turbo-charger but is accurate enough for comparative purposes. In studying technical literature covering efficiencies of various turbo-chargers, the method of finding the efficiency should be the same in each case otherwise a fair comparison will not be obtained.

Note The thermodynamic analysis of turbo-chargers is beyond the scope of this book. This matter is covered in books on advanced thermodynamics and specialist papers from technical institutions.

■ **5.44 How does the timing of the air-inlet and exhaust valves differ between naturally aspirated and pressure-charged four-stroke engines? What is the reason for the differences?**

In giving valve opening and closing times, it must be remembered that considerable differences occur in different makes of engine. These differences depend to some extent on the speed of the engine, and also on the degree of supercharging.

The following table gives the valve timings for four-stroke, medium-speed engines.

	Naturally aspirated	Pressure-charged
Inlet valve opens	up to 20° b.t.d.c.	up to 80° b.t.d.c.
Inlet valve closes	up to 25° a.b.d.c.	up to 45° a.b.d.c.
Open period	20° + 180° + 25° = 225°	80° + 180° + 45° = 305°
Exhaust valve opens	up to 45° b.b.d.c.	up to 50° b.b.d.c.
Exhaust valve closes	up to 20° a.t.d.c.	up to 60° a.t.d.c.
Open period	45° + 180° + 20° = 245°	50° + 180° + 60° = 290°

It will be noticed that the period for which each valve is open is much longer in the pressure-charged engine.

When examining the valve timing for each engine, it will be noticed that the inlet valve is opened before the exhaust valve is closed. The period when both valves are open is referred to as the overlap period. In the naturally aspirated engine, the overlap is (20° b.t.d.c. + 20° a.t.d.c.) = 40°; in the pressure-charged engine, it is (80° b.t.d.c. + 60° a.t.d.c.) = 140°.

The large overlap period in the pressure-charged engine allows the exhaust gases to be expelled from the cylinder, except for a small amount of exhaust gas that is mixed with the incoming air. During this period a large amount of air passes through the inlet valve, the cylinder, and the exhaust valve, cooling the parts in the process.

This cooling helps to keep the surface temperatures of these parts at a lower value, which in turn reduces thermal stresses.

After the exhaust valve closes, less heat is passed into the air that follows into the cylinder, and a greater mass of air is therefore present when compression begins.

The overlap period, when both the exhaust valve and air-inlet valve are open together, is sometimes referred to as the scavenge period in pressure-charged four-stroke engines.

■ **5.46   What pressure-charging system is used on modern slow-speed two-stroke uniflow-scavenged engines, fitted with exhaust valves? Give the crank angles at which the exhaust valves open and close, and the angles at which the scavenge ports open.**

All modern slow-speed, two-stroke cycle engines operate on the constant-pressure turbo-charging system. The number of turbo-chargers used depends on the size of the engine. Because the efficiency of a turbo-charger increases with an increase in its physical size and output, the number of turbo-chargers is kept to a minimum.

Scavenge ports open at approximately 35 degrees before b.d.c. and close 35 degrees after b.d.c.. The exhaust valve will open ahead of the scavenge ports to give a blow down period and close at some time to leave the correct amount of air in the cylinder for the combustion of fuel. The open period for the exhaust valve will be about 80 to 90 degrees of crank rotation and the valve will open about 45 degrees before b.d.c.. It should be noted that considerable variation may be found in valve timing figures for various engines. The figures used in any valve-timing checking work should be taken from the engine instruction book. These figures will take into account lag from hydraulic operated valves and valves with hydraulic lifters.

■ **5.47   What pressure-charging system is used on modern medium-speed, four-stroke V-engines? Give the timing angles for the air-inlet and exhaust valves.**

This type of engine is operated with the impulse pressure-charging system.

Air inlet valves open 50° b.t.d.c.
Air inlet valves close 45° a.b.d.c.
Exhaust valves open 40° b.b.d.c.
Exhaust valves close 60° a.t.d.c.

■ **5.48   What are the provisions for operating impulse and constant pressure exhaust turbo-charged engines with a disabled turbo-blower?**

Impulse pressure-charged engines can have the rotor locked, and the blower outlet blanked off with a blind flange, if the period of operation under emergency conditions is not long. The second blower can be left in operation, and the electric blower brought into use. In certain cases, reducing orifice plates must be fitted into the electric or turbo-blower, to balance the discharge pressures and prevent surge.

If the engine is to be operated under emergency conditions for a long period, the emergency exhaust pipes must be fitted, so that the exhaust gases bypass the turbine. This prevents damage to the labyrinth packings and the turbine rotor. It will still be necessary to blank the outlet from the blower, to prevent air backflow from the scavenge receiver.

When operating on one turbo-blower with the exhaust gases from the other bypassed, it is still necessary to use the electric blower as mentioned previously.

In constant-pressure turbo-blown systems, the gas inlet to the disabled turbine and the air discharge from the blower should be blanked off with blind flanges. If the engine is fitted with scavenge pumps it may not be possible to blank off the air outlet from the turbine, as this would shut off the supply air to the scavenge pumps. In this case the rotor must be locked, or the emergency air-inlet branches to the scavenge pumps opened. The engine will be operated on the remaining turbo-blower. The speed of the engine must be restricted so that the allowable turbine speed is not exceeded.

Generally, engines with two turbo-blowers can operate at 60% to 80% of full power when one turbo-blower is disabled.

When a blower is isolated from the exhaust by the fitting of blind flanges or emergency exhaust pipes, the lubricating oil and cooling-water services can be shut off.

■ **5.48 What has been the overall effect of increasing turbo-charger efficiency?**

The overall effect of increased turbo-charger efficiency has been and will continue to be shown as a decrease in fuel consumption.

This has fitted in well with the new generation of slow-speed, two-stroke cycle long-stroke engines. The increased turbo-charger efficiency has reduced the heat input requirement making it possible to open the exhaust valves later and get a greater amount of work from the fuel due to a greater expansion ratio of the combustion gases.

As about fifty per cent of the energy in the fuel passes through the turbine of the turbo-charger there is still plenty of heat available for exhaust gas boilers although in the future the heating surfaces of the boiler will have to be increased to accommodate the lower-temperature gases.

Note Schemes are available to utilize the energy in exhaust gases either to generate electrical power through an exhaust gas-driven turbo-alternator or generator, or to connect an exhaust gas turbine to the propeller shaft through a clutch and reduction gearing.

As a point of historic technical interest, similar schemes were used very effectively with reciprocating marine steam engines from the early 1900s until well into the post-war years, to utilize some of the waste heat in the exhaust steam before passing into the condenser.

One scheme connected an exhaust steam turbine on to a centre-screw shaft in twin reciprocating engine installations, making the ship a triple-screw ship. Another scheme connected the exhaust turbine to the screw shaft through a hydraulic clutch and double reduction gearing. A third scheme used the exhaust steam to drive a rotary compressor used to compress the steam between expansion stages. A fourth scheme used the exhaust steam to drive an electrical generator. The current produced was used in a set of resistance elements to reheat the steam between expansion stages.

■ **5.50   What precautions must be taken to minimize thermal stresses in modern, highly rated, turbo-charged engines?**

The modern, highly rated engine requires large amounts of cool air to pass through the cylinders during the overlap period of the inlet and exhaust valves, in four-stroke engines; and during the period that scavenging is taking place, in two-stroke engines. To maintain this air flow at a maximum it is necessary to keep a careful watch on the air suction pressure to the exhaust turbo-blower. Any reduction in this pressure from normal indicates that the suction filters need to be cleaned.

The turbine speed and exhaust temperatures must be carefully watched and any indication of fouled blading should be investigated and rectified.

The temperature of the air entering and leaving the cooler must be observed, and any decrease in the difference rectified by cleaning the coolers.

The grids fitted in the gas passages to the turbine may also require cleaning if combustion has been poor at any time.

Most engine instruction books give both normal values and the values which indicate that corrective action should be taken. Automatic valves fitted in scavenge receivers and under-piston scavenge pumps also require careful attention and regular cleaning, to prevent restriction of air flow.

■ **5.51   What attention do exhaust gas silencers require?**

The internal spaces and the baffle plates of exhaust gas silencers become fouled after a period of service. The dirt on the baffle plates increases the back pressure on the engine exhaust system, and must therefore be cleaned off at regular intervals so that the back pressure is kept at a minimum. The silencers are fitted with doors at the sides or bottom of the silencer outer casing to facilitate cleaning.

Accumulations of dirt are also dangerous. If the dirt ignites, burning carbon particles and sparks may be discharged with the exhaust gases from the funnel.

■ **5.52   Why are some turbo-charger manufactures dispensing with water jackets on the exhaust gas side of turbo-chargers.**

Water jackets hold the temperature of the castings of the turbine casings at reasonable values, but they have the following disadvantages also.

They cool the exhaust gases and reduce the turbo-charger efficiency.

They are more expensive to produce than non-jacketed castings due to the complications of coring, cost of cores, the possibility of defective casting due to core shift, and the costs of removal of core material when castings are cleaned.

Problems may arise when leakage occurs from the water jackets into the exhaust gas spaces.

These factors are prompting some manufacturers to consider dispensing with jackets; others are producing turbo-chargers with no water-jacketed spaces.

On the debit side for consideration, better materials will be required for casings without jackets in order to withstand higher temperatures without expansion; the casings will have to be very effectively heat insulated to prevent or reduce the risk of fires and accidents to engine-room personnel.

Note Small engine turbo-chargers normally did not have water-jacketed casings.

■ **5.52 State what can be done to continue operation with a turbo-charger having water leakage into the exhaust gas space.**

If the water jacket on a turbine casting leaks, the problem is usually due to corrosion on the cooling-water side of the jacket in the hotter regions of the water space. This is more likely to happen when sea-water is used as the cooling medium and proper attention has not been given to the corrosion protection anodes fitted in various locations in the cooling-water spaces.

When leakage occurs it has been the practice in the past to remove the plugs on the cover plates, or the cover plates covering the core holes in the cooling spaces. The aim is to set up a thermo-siphon action and cause air to flow through the cooling space to hold the casting at a safe operating temperature until a new casting is obtained and fitted.

In some engine rooms where ventilation around the gas side of the turbo-charger is inadequate, the cooling may have to be augmented with an air hose connected to an air supply, or by leading air through a temporary duct from a ventilator outlet.

This form of temporary operation has been carried out many times, but it should not be attempted without the cognizance of the Company's technical department and the turbo-charger's manufacturer.

The operation is more easily carried out where owners operate their engines at some value below the maximum continuous rating. (Fig. 5.5)

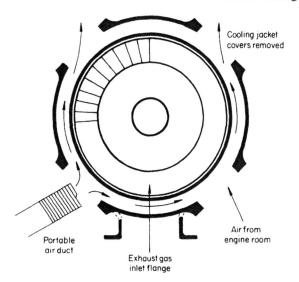

Fig. 5.5 Emergency air cooling arrangement for turbocharger turbine water jacket when leakage occurs across water/gas space boundary.

6

CONSTRUCTION MATERIALS, WELDING, MATERIALS TESTING

■ **6.1 What are the factors which influence the choice of materials used for the manufacture of the various parts of a diesel engine?**

The parts of a diesel engine play a different role, and are subjected to constant or varying stresses. Certain parts must withstand high temperatures and/or a corrosive environment, while other parts may be subjected to wear. The material chosen must be able to withstand the stresses to which it is subjected. If a part has to work at high temperatures or withstand corrosive attack, the material must not undergo physical or chemical changes under these conditions. If it has to resist wear, the surfaces must be sufficiently hard to give low wear rates, and it must be compatible to work in conjunction with the material of its mating part. Material and manufacturing costs are also factors to be considered when alternative materials are available.

■ **6.2 Name the various ferrous materials used in the construction of a diesel engine.**

The ferrous materials used are: cast-iron, alloy cast-irons, nodular cast-iron, steel of various grades (classified by their carbon content), and alloy steels.

■ **6.3 How are the various forms of cast-iron manufactured?**

Cast-iron is produced by reducing iron ore in a blast furnace. The resultant pig iron can then be further refined to reduce impurities, or remelted to make castings. In the form used for castings, cast-iron is essentially an alloy of iron, carbon, and silicon, in combination with smaller amounts of manganese, sulphur and phosphorus. If the carbon is in the form of flakes of graphite, in the presence of silicon, the grey irons are produced. When the silicon content is low the carbon will combine with the iron, if the metal is chilled, to produce white irons.

Alloy cast-irons are produced by alloying with other metals. Common combinations are chrome–nickel, nickel–molybdenum, and nickel–vanadium. Sometimes high-quality cast-irons are made in the electric furnace.

Nodular cast-iron is made by alloying iron with an alloy containing either magnesium and copper, or nickel and cerium, (a 'rare earth' metal). The alloy is added to the molten cast-iron prior to pouring. The added alloy causes the graphite to take up a spherical form, thus increasing the strength and ductility of the iron. Nodular irons also are modified by heat treatment. The processes for producing nodular iron, also known as spheroidal graphite iron, are protected by patents.

Meehanite is the name given to a group of cast-irons produced by inoculating the iron with calcium silicide in controlled processes; these processes are also covered by patents. Malleable iron castings are produced by heat-treating castings made from iron of a suitable composition.

■ **6.4   Name the steel-making processes and the types of steel produced.**

Open-hearth process. This process produces either acidic or basic open-hearth steel, depending on whether the furnace lining is acidic or basic.

Bessemer process. The Bessemer process uses a converter; air is blown through the molten charge which oxidizes the impurities and liberates heat. After the charge is blown, deoxidation and recarburization are carried out by the addition of alloys containing carbon, manganese, and silicon.

Electric furnaces are used to produce small amounts of high-quality steels or alloy steels.

Oxygen steel is produced by blowing oxygen on to the surface of the molten metal in a furnace similar to a Bessemer converter.

Killed steel is produced by deoxidizing molten steel, which is afterwards cast in special moulds to produce ingots. Pipes, blowholes, and slag inclusions are then reduced to a minimum. This produces an ingot that is less likely to develop faults later during the forging or rolling processes. The deoxidants used are ferro–manganese, ferro–silicon, ferro–titanium, and aluminium.

The type of steel used for important forgings for engine parts is usually killed open-hearth steel.

■ **6.5   How is steel affected by its carbon content and by impurities?**

The carbon content of a steel controls its internal structure. As the carbon content increases the steel becomes stronger but its ductility is decreased. Impurities also affect the internal structure and alter the strength and ductility of the steel.

■ **6.6   What do you understand by the term 'heat treatment'? Name the heat-treatment processes.**

Heat treatment is the name given to the process of heating metal above the critical temperature, at which internal changes occur, and then cooling the metal slowly or rapidly to give it more desirable physical and mechanical properties.

The heat-treatment processes are normalizing, annealing, hardening, and tempering.

■ **6.7   How is steel normalized and what is the purpose of normalizing?**

Steel is normalized by heating it slowly to just above the temperature at which changes in the internal structure take place; it is then soaked at this temperature. After the soaking period, the steel is cooled slowly in air. Air cooling is referred to as medium cooling.

Normalizing refines the grain structure of the steel.

■ **6.8   What does the term 'annealing' mean? How is steel annealed? What is the purpose of annealing?**

Annealing is similar to normalizing. Different rates of cooling are required for different materials. Steel is annealed by heating to the required temperature at which its internal structure changes, and then soaking at this temperature. After soaking, the steel is cooled slowly in a furnace. Furnace cooling is referred to as slow cooling. Annealing refines the grain structure, reduces hardness, and increases ductility.

■ **6.9   How is steel hardened? What effect does hardening have on the steel?**

Steel is hardened by heating to a specified temperature followed by rapid cooling. The cooling may be in air, or by quenching in oil or water: the method used depends on the type of steel and the properties required. A gain in hardness is accompanied by a loss of ductility. In the fully hardened state steel is brittle. Steels containing very small amounts of carbon cannot be hardened.

■ **6.10   How is steel tempered? What is the purpose? Give an example.**

Steel is tempered by heating hardened steel and then cooling it by air or quenching. The temperature to which it is heated and the quenching rate control the degree of hardness, strength, and ductility.

Steel is tempered to obtain special mechanical or physical properties. Important engine parts such as coupling bolts, bearing bolts, and parts subjected to alternating stress, are tempered to increase their strength and resistance to fatigue failure.

■ **6.11   What is surface hardening? Name the surface-hardening processes.**

Surface hardening is a process whereby the structure of the surface of a material is changed to give it better wearing properties. The process may increase the carbon content of the steel by carburization or, in the case of steels with higher carbon contents, by heat-treating the surface concerned.

The names of the various surface-hardening processes are as follows.

Carburizing processes   cyaniding – pack method
nitriding – gas method

Heat treatment processes   flame hardening
induction hardening

■ 6.12 Many diesel engine parts are subject to wear. Give examples of some parts that are surface hardened, and name the process used.

Part	Surface-hardening process
Medium-and high-speed engine crankshaft journals and pins	Nitriding
Special alloy cast-iron cylinder liners	Nitriding
Gear wheel teeth	Flame hardening
Gudgeon pins	Induction hardening
Cams	Flame hardening

(Different engine manufacturers may favour different processes for some of the parts given in the example.)

The method adopted must be suited to the steel used. Care must always be exercised during treatment, since if the treatment is not correct, exfoliation of the hardened surface may occur.

■ 6.13 What are austenitic steels and where are they used?

Austenitic steels are a group of alloy steels that exhibit a particular structural form. They have high resistance to the changes which occur in normal steels when subjected to high temperature. They are therefore used where strength is required at high temperatures. This type of steel may be used for exhaust valves, exhaust turbo-blower turbine nozzle blading, and turbine rotor blading, depending on the prevailing temperature and stress conditions that the part has to work under, whether corrosive conditions are present, and whether a high creep resistance is necessary.

■ 6.14 Where is Stellite used in diesel engines? What is it composed of and how is it applied?

Stellite is the trade name of an alloy consisting of cobalt, chrome, molybdenum, tungsten, and iron. Varying proportions of some or all the metals are used to produce alloys with various characteristics. These alloys resist loss of strength at high temperatures. Stellite is extremely hard and corrosion-resistant and is often used for metal-cutting tools. In diesel engines, Stellite is used to surface the valve faces and seats of exhaust poppet valves. It is applied by welding processes on to the valve face and seat, which must be preheated.

■ **6.15 How are the large, heavy, moving, steel parts of a diesel engine manufactured? Is hammering or pressing preferable for working large, heavy sections?**

The steel used for the heavy, moving parts of an engine is produced in the form of an ingot. In the cast state the steel ingot has a coarse crystalline structure and, being brittle, would be unsuitable for use.

The ingot is therefore hot-worked; this consists of heating the ingot in a gas furnace and working it under a hydraulic press, so that the part is gradually 'squeezed' to the shape and dimensions shown in the forging plan. The hot-working refines the structure of the steel so that the grain is less crystalline and has a finer and more uniform structure. The ductility of the steel is also greatly improved. The amount of hot-work is governed by the shape and sections of the forgings, but it must be above a minimum value.

Pressing is more suitable than hammering for large heavy parts, because the hydraulic press works the material right through to the centre of the ingot so that grain refinement occurs throughout the section. Hammering of large pieces disturbs or hot-works the material only on the surface. It does not produce a uniform structure right through the section of the forging.

After the parts are forged they must be annealed or normalized, and tempered.

■ **6.16 If steel ingots must be hot-worked to produce steel of the right quality for heavy parts, explain how it is possible to use steel castings for crank webs.**

When steel castings are used for crank webs, the moulds for the castings are made with large spaces that form bossings on the castings. These bossings form reservoirs of metal which are pulled into the main part of the casting as it cools. This reduces the internal stresses set up when the metal passes from the liquid into the solid state.

After the casting has cooled it is removed from the mould. The risers and discards are removed. The casting is grit blasted and examined for surface and internal defects. If it is found in order, the casting is heat-treated under carefully controlled conditions.

The heat treatment may consist of full annealing or normalizing, followed by tempering, depending on the size and proportions of the casting. The heat treatment refines the grain structure so that the crank web has the required strength and ductility. The test pieces are not removed from the casting until the heat treatment is carried out.

■ **6.17 Give the analysis of a cast-iron suitable for the manufacture of piston rings.**

Nickel-chromium cast-iron is often used for piston rings.

Analysis	Carbon	3.0%
	Silicon	1.5%
	Sulphur	0.1%

Phosphorus   0.3%
Manganese   1.0%
Nickel   1.0%
Chromium   0.4%
Balance iron to 100%

■ **6.18 Name the non-ferrous metals and their alloys that are used in engine construction (exclude bearing metals).**

The non-ferrous metals and their alloys are mainly copper, zinc, tin, aluminium, nickel, and chromium.

Copper is often used in pipe systems, while aluminium or aluminium alloys can be used for various castings.

Copper when alloyed with zinc and small amounts of tin, in various proportions, forms the range of alloys known as brass.

Copper alloyed with tin and small amounts of zinc forms the bronzes and gunmetals.

Aluminium bronze is an alloy with 80 to 90% copper and up to 10% aluminium, with small additions of other metals.

Phosphor-bronze is the name of a range of tin–copper alloys containing small amounts of phosphorus.

Manganese bronzes are produced by additions of up to 2% manganese to approximately 60/40 copper–zinc alloys.

■ **6.19 Name the metals used in the production of anti-friction bearing alloys.**

Anti-friction metals are also referred to as white-metal or Babbitt's metal. There are two classes of white-metal: one is tin-based, the other is lead-based. Tin-based white-metals contain tin, antimony, and various amounts of copper and cadmium. Lead-based white-metals are not normally used in marine diesel engine practice. Some tin-based white-metals do contain very small amounts of lead, either by addition or as an impurity.

■ **6.20 Name the metals used in the manufacture of thin-wall bearings.**

Copper, tin, zinc, lead, nickel, and manganese are used to produce alloys of varying proportions which are bonded on to a thin steel backing strip which forms the bearing shell. Selected metals are used to form the following alloys:

lead–bronze alloys
copper–lead alloys
tin–aluminium alloys
aluminium–bronze alloys

■ **6.21** Give the analysis of a cast-iron considered suitable for cylinder liners. State the mechanical properties of the cast-iron. What impurity must be kept to a minimum? How can the tensile strength be increased and the resistance to wear be improved?

Analysis   Carbon   3.0%
                 Manganese 1.0%
                 Silicon   0.7%
                 Phosphorus 0.4%
                 Sulphur   0.1%

The silicon content should be kept as low as possible: silicon promotes growth in cast-iron when subjected to continued heating. Cast-iron having the above analysis has a tensile strength of approximately $216\,MN/m^2$ (14 tons/in²).

The tensile strength of cast-iron is sometimes increased by the addition of mild steel scrap to the pig iron charge put in the furnace. Alternatively, small amounts of chrome and nickel may be used. The wear resistance can be improved by small additions of vanadium.

■ **6.22** Describe the behaviour of the constituent parts of bearing white-metals.

The tin content forms what is called the matrix. This matrix is sufficiently soft to accommodate the small changes in alignment between the journal and the bearing surfaces.

The antimony forms cubes or cuboids, which are very hard. These cuboids take the load from the journal or pin and transmit it to the supporting matrix. They also have a high resistance to wear. When the bearing is being cast and the white-metal is in the molten state, the antimony cubes tend to float and conglomerate, a process referred to as segregation.

The copper constituent prevents segregation. The copper has a high melting point and solidifies first, forming long needles which interlace in a criss-cross pattern through the liquid tin. The interlaced copper needles hold the antimony cubes in an evenly dispersed pattern through the tin matrix.

When the white-metal ingots are melted prior to casting, extreme care must be exercised to prevent overheating the metal. Care must also be exercised when bearings are centrifugally cast to prevent separation of the constituents.

Cadmium improves the toughness of the bearing metal and helps prevent fatigue.

■ **6.23** Give the analysis of white-metal suitable for diesel engine main and bottom-end bearings.

Analysis   Tin   85–88%
                 Antimony   8–10%
                 Copper   4–5%

Normally the white-metal used for diesel engine bearings is produced by metal refiners and sold under various brand names.

■ **6.24 What are the special requirements of the white-metal used in crosshead bearings?**

The loads placed on the crosshead bearings of modern engines are extremely heavy. The requirements of the white-metal depend on the design of the bearing. For example, very stiff crosshead assemblies with a thin white-metal layer require a different metal from a more flexible assembly with a thick layer. The requirements of each bearing must be formulated initially from previous experience, and then possibly modified following service experience to give the desired characteristics.

For modern, slow-speed, highly rated crosshead engines, a white-metal known to give reasonably satisfactory service is analysed as shown.

Analysis   Tin   88%
             Antimony   8%
             Copper   4%
             Cadmium trace

■ **6.25 What are plastics? Where is use made of plastics in diesel engines and ancillary equipment? Name the plastics used.**

Plastics, a product of the organic chemist, are made up of long chains of identical molecules, i.e., they are polymers. The constituent molecules are carbon and hydrogen compounds and may also contain a wide range of other elements. The raw materials used to produce plastics are obtained from coal or petroleum oils.

Plastics are not used in any of the major parts of diesel engines. Their use is limited to instruments and electrical fittings.

The materials mainly used are as follows.

Nylon. This material is used for bearings and small gearwheels in instruments. It is tough, has a low coefficient of friction, and can be machined from rod or plate sections. It is also made into fibres.

Polymethylmethacrylate (Perspex). This is a clear plastic used for instrument glasses, level gauge tubes (not boiler or steam service), and bearing-oil-well sight glasses.

Polytetrafluoroethylene (PTFE or Teflon). This material has a very low coefficient of friction and resists heat; it also has good chemical resistance in the presence of oil. It is used for gland packings or for coating conventional soft packings. Due to its low frictional resistance it may be used as a packing for shafts with high rubbing speeds. PTFE is supplied in moulded O-rings, or in sheet or shredded form. It gives good service in fuel pump and valve glands.

Phenolic resins. These resins are normally bonded with linen or other fibrous materials. They are used for bearings in pumps and on screwshafts. They swell in water, an action to be considered when calculating bearing clearances. One material which gives excellent service in water-lubricated bearings is Tufnol.

A wide range of other plastic materials are used to make paints and compositions, thermal insulation, sound absorbents, bonding cements, etc.

■ **6.26   What are elastomers, and where are they used?**

An elastomer is a rubber-like material. The term covers both natural products and synthetic materials which are elastic or resilient.

Elastomers are used to make O-rings, jointing material, and packings. For service at low temperatures, vulcanized or treated natural rubber can be used for joints, O-rings, and similar packings. Treated natural rubber will not withstand oil. Where there is contact with oil, the elastomer Neoprene is used. Neoprene is produced from acetylene, which is converted to chloroprene and then polymerized.

Nitrile rubber (a butadiene-acrylonitrile copolymer) is a synthetic material used for service at high temperatures. It is used for the sealing rings fitted between cylinder liners and jackets, and in the exhaust and scavenge port belt of a two-stroke engine cylinder liner.

■ **6.27   Name some of the materials known as technical or engineering ceramics.**

Some of the engineering or technical ceramic materials are listed below.

Aluminium oxide	Al_2O_3
Silicon carbide	SiC
Silicon dioxide	SiO_2
Silicon nitride	Si_3N_4
Zirconium dioxide	ZrO_2

Many of these materials are able to withstand very high temperatures before melting or change of state occurs. The list is by no means complete, but covers some of the more important materials likely to be met in diesel machinery practice.

The production of new ceramic materials and their use and application in the field of diesel engine technology must still be regarded as in its infancy, although progress in the last few years has been rapid, particularly in the automotive engine field.

■ **6.28   State where some of the ceramic materials listed in the previous question can be used.**

Aluminium oxide is used in the manufacture of electrical insulators, abrasive materials such as grinding wheels, grinding powders, pastes, and the like; it may also be used in the manufacture of refractory cement. In certain hydrated forms it can be used as a gas adsorbant. It is also used in abrasive blast cleaning processes (see Note). Silicon carbide is used in the manufacture of electronic circuit components such as light-emitting diodes and in the manufacture of abrasives. It is used in abrasive blast cleaning processes.

Silicon dioxide is used in glass manufacturing, abrasives, and foundry moulding work. Silicon nitride is sometimes used in the manufacture of gas turbine components operating at extremely high temperatures such as stator blading. It has a high resistance to thermal shock. It can sometimes be used for cutting tools as it is extremely hard.

Zirconium dioxide is also known as zirconia and is used in applications for the manufacture of special glasses, piezoelectric crystals and ceramic-metallic materials (cermets). Zirconium oxide can withstand very high temperatures. A considerable amount of research work has gone into trying to improve the efficiency of diesel engines by using it to coat the piston crown, cylinder head, valve heads and cylinder walls with zirconium oxide. The purpose is to increase the operating temperature range in the cylinder and raise the thermal efficiency of the engine. The oxide coating protects the metallic parts of the engine from the higher temperatures and reduces the amount of heat passing into the cooling water. The engine then operates on a higher range of temperature and so increases the thermal efficiency.

A secondary consideration is to increase the life of parts by protecting them from wear.

The coating of the parts is carried out by the plasma flame spraying process. In this process the ceramic is fed in powder form into a spraying torch, an arc is struck between a cooled central electrode and the nozzle, but is not transferred on to the work. The nozzle forms an orifice and constricts the arc, increases its temperature and causes ionization of the plasma gas. The plasma gas is usually argon. This is fed into the nozzle and emerges at an extremely high temperature and melts the powdered ceramic. The velocity of the gas carries the melted particles on to the work being coated.

Composites of different ceramic materials and ceramic materials and metals are made up to give the necessary properties of strength and make them isotropic, to increase resistance to mechanical and thermal shock loading, balance the thermal expansion between parts of different materials, improve corrosion resistance, and the like.

Note Extremely hard, sharp abrasives should not be used in abrasive blast cleaning processes on materials subjected to cyclic stress variations; the microscopic abrasions left on the surface will form nucleation sites for the commencement of fatigue failure.

■ **6.29 Name the processes by which two pieces of metal can be joined (excluding mechanical means such as bolting and riveting). Give an example of where the processes can be used.**

The processes by which two pieces of metal can be joined are soft soldering, brazing, welding, or with adhesive or bonding materials.

Soft soldering is mainly used for tinplate work or electrical connections.

Brazing is mainly used for non-ferrous pipe work such as joining slip-type pipe flanges to copper pipes, or steel flanges to steel pipes. The filler material can be a low melting-point brass or a silver solder; it is drawn into the joint by capillary action.

Welding processes vary considerably, from blacksmith's hammer welding and gas welding processes to the mechanized electric arc and resistance welding processes.

In diesel engine manufacture and repair work, the most common form of welding is the manual electric arc process. This process is used in the manufacture of fabricated bedplates, engine columns, steel pipework, and starting-air reservoirs or receivers.

Gas welding processes are used on small size pipework and thin sheet metal work.

When welding some metals by the electric arc method, it may be necessary to shroud, or envelop, the location of the arc with inert gases such as argon or helium.

Adhesive and bonding cements. Plastic adhesives and bonding cements are available, but their use is mainly limited to instruments, the fastening of nameplates, and the like. In some cases their use requires heat.

■ **6.30   Describe how two pieces of mild steel are joined by the manual electric arc welding process.**

After the two pieces of mild steel to be joined have been suitably prepared, they are brought together until they touch (or a gap may be left, depending on the thickness and length of the welded joint).

The electric current can be direct or alternating, with a voltage of 80 to 100 volts. The metal being joined is connected to one pole of the current supply, while a coated electrode or rod is connected to the other pole.

The electrode is used to make the arc, which is then kept as short as possible. The heat from the arc melts the metal being joined, the metal in the electrode, and the coating on the electrode, so that a small pool of molten metal forms. As the pool of metal is formed the electrode is moved along with a series of small sideways movements, extending the molten metal pool; the metal in the pool furthest from the arc cools and solidifies. The continued movement of the arc and the solidifying molten metal gradually create the joint. The electrode is held at approximately 30° from the vertical with the hand in advance of the electrode tip in the direction of movement along the joint.

The metal in the electrode is sometimes referred to as the filler. To create a good weld, the filler and the sides of the joint must melt and mix together, referred to as complete fusion. The coating must be allowed to float to the top of the pool before the metal solidifies; if the slag formed by the coating is trapped in the joint, a weakness will result which may cause the joint to fail in service. When joining thick metal sections two or more runs must be made to complete the joint.

When multiple-run welded joints are made, the first runs are sometimes cut out from the back and then rewelded.

■ **6.31   What is the purpose of the covering on the electrodes used in hand electric arc welding?**

The coating on an electrode serves several purposes. There are: the creation of gases to stabilize the arc, the formation of slags to protect the molten metal against oxidation, and the provision of fluxes.

■ **6.32 What type of welding equipment and tools are likely to be found on motor ships?**

Single-operator, transformer-type welding machines are used on ships with an alternating-current supply. They are connected to the high-voltage supply. The transformer can be used with a rectifier to supply direct current for welding. In older ships with direct-current supply a motor generator set can be used. Direct current is considered to be safer for use in ships, where welding may be done in hot, damp conditions. The tools used are a welding hood, a hand shield with coloured and transparent protecting glasses, insulated rod tongs, welding current supply and earthing cables, a slag hammer, wire brush, gloves, fireproof apron, and a welding bench.

The protective hood, hand shield, and gloves are to protect the operator from ultraviolet and infrared ray burns. The apron is to protect the clothing of the operator from being ignited by weld spatter. The hand shield protects the eyes of the operator, and enables him to see the welding arc and pool of metal. The slag hammer and wire brush are used to remove slag from the weld metal, and help prevent inclusions between the various runs.

■ **6.33 Name and briefly describe the types of joints made by electric arc welding processes.**

The joints made by electric welding may be of the following types.

Butt joints. In butt joints, the joint is formed by placing two pieces of metal with their edges side by side and then welding the edges together.

Lap joints. This type of joint is made by placing one piece of metal on the other so that they overlap. The edge of the upper piece is then welded to the surface of the lower piece on which it rests, forming a weld fillet. It is usual to weld lap joints with fillets on both sides.

■ **6.34 Which type of welded joint has the highest resistance to fatigue failure?**

Full-penetration, butt-welded joints give the best resistance to fatigue failure.

In fillet-welded lap joints, the forces acting on each section are not in line and a bending moment is set up in the joint. The edges of the fillets also act as stress raisers or notches.

In T and cruciform joints made with fillet welds, the lack of penetration between the fillets creates a stress raiser or notch.

T and cruciform joints subject to fluctuating stresses, such as in engine bedplates and A-frames, are made with full-penetration welds.

■ **6.35 Define weld penetration, weld reinforcement, throat, root, and toe.**

Weld penetration is the depth to which fusion occurs in the two pieces of metal being joined.

Weld reinforcement is the metal contained within the convex surface of a welded joint.

Throat is the term given to the thickness of the weld metal. In butt welds, it is equal to the thickness of the weld metal minus the reinforcements, or the thickness of the plate. In fillet welds, it is the diagonal distance between the corner of the joint and the surface of the weld, minus the reinforcement.

The *root* is the bottom of the welded joint.

The *toe* is the edge between the weld deposit and the parent metal in fillet joints.

■ **6.36 Name the positions in which welding operations can be performed. Which position is the easiest, and which the most difficult?**

The welding directions are downhand, vertical, and overhead. The downhand position is the easiest as gravity holds the pool of molten metal in place. The overhead position is most difficult as gravity tends to cause the pool of molten metal to drop out. It calls for considerable skill. Special electrodes are available for overhead welding. Vertical welding is usually done in an upwards direction, but special electrodes are available to make vertical welds with a downward direction.

■ **6.37 What is undercut, where does it occur, and what detrimental effects does it have?**

Undercut occurs at the toe in fillet welds and at the side of the weld reinforcement in full-penetration welds. It is most common on the upper toe in a fillet weld, and is caused by the parent metal melting at the fillet edge and running down into the fillet. Undercut, in effect, creates a notch, with a concentration of stress adjacent to the notch.

■ **6.38 How may a badly welded joint fail?**

Welded joints may fail in one of three ways.

1 Failure of the weld metal, due to internal stresses set up when the weld metal cools. The cause is usually excessive restraint.
2 Failure at side of weld metal, due to lack of fusion. Insufficient current, too rapid heat conduction, or too long an arc, are the usual causes.
3 Failure of base material in side of metal near joint, due to heat affecting the steel. Failure takes place at the boundary of the heat-affected zone.

■ **6.39 What do you understand by the term joint preparation (or edge preparation)?**

Joint or edge preparation is the name given to the profiles formed on the edges of the sections of butt joints. The profiles may be single or double Vs, or single or double Js. The included angle of the V should not be less than 60°, preferably

70°. The edges and profiles formed may be flame-cut or planed off in an edge-planing machine.

■ **6.40 Describe what occurs when a heated piece of metal is prevented from contracting during the period in which it cools. What is weld restraint?**

If a piece of heated metal is prevented from contracting when it cools, the metal will be increasingly strained as cooling proceeds. The total amount of strain will be equal to the difference between the length of the hot metal and the unrestrained cold length. From Hooke's Law we know that stress is proportional to strain; therefore as the restrained metal cools it will in effect be subjected to a tensile stress which increases as cooling proceeds. Similarly, if a welded joint is prevented from contracting when it cools, a tensile stress will be created within the weld and parent metal up to the anchor points forming the restraint.

Weld restraint sets up stresses within the welded joint and adjacent metal during cooling, which remain when cooling is completed. They are referred to as residual stresses. The residual stresses form a complex pattern, and increase as the number of runs required to make a joint increases. The stress in the direction of the joint is referred to as the longitudinal stress, and that across the joint as the transverse stress. Generally the number of runs to form a joint should be kept to a minimum by using the largest possible electrode; this reduces the complexity of the stresses in the joint.

The fact that restraint causes stresses is sometimes used to control distortion. Restraints are created by tack welds and clamps to hold the parts rigid in one direction. The locked-up stress created in one direction is used to control and cancel distortion in another direction.

■ **6.41 How is weld distortion prevented or controlled?**

The factors controlling distortion caused by welding are the initial design, the preparation of the joint, the correct root gap, and the sequence of welding the various parts forming the structure. Even when all the factors are correctly interrelated, some unavoidable distortion may occur.

Distortion is sometimes controlled by presetting the parts to be welded in what appears to be misalignment. As the parts cool after welding, the resulting shrinkage gradually pulls the misaligned parts into correct alignment. This minimizes the residual stress in the joint. Presetting, however, can be used only for simple assemblies.

Distortion can also be controlled by restraints as mentioned in Question 6.40.

■ **6.42 Why is a restriction put on the maximum carbon content of the steel castings and plates used in the construction of welded engine bedplates and other important engine structures?**

The reason for the restriction on the carbon content is that as the carbon content increases it becomes more difficult to make a sound and reliable weld, or in other words the weldability decreases. Generally the carbon content is not allowed to exceed 0.23% unless special welding techniques are used.

The thickness of the sections being welded also influences the weldability: thicker sections conduct the heat away faster, which has a quenching effect on the hot zone. Heavy sections are therefore preheated.

Note   The weld metal bears some resemblance to steel in the cast condition. The heat from the weld will also harden the parent metal adjacent to it; if this steel has a high carbon content, cracks are more liable to occur.

■ **6.43   Why are important welded engine structures, such as bed-plates and columns, heat-treated after all the welding is completed? Is the heat treatment carried out before or after machining?**

Welded engine structures such as bedplates through which the firing load is transmitted must be heat-treated. The heat treatment usually consists of heating the welded structure in a furnace; the heating is slow so that the welded structure is uniformly heated throughout all its parts. The structure is raised to a temperature of 580–650°C (1080–1200°F) and it is soaked at this temperature for a period dependent on the thickness of the thickest parts. After soaking it is slow-cooled in the furnace.

This type of heat treatment refines the grain structure of the metal in the welds and the adjacent metal, and also relieves the residual stresses within the welded structure. Heat treatment is carried out after welding is completed and prior to machining.

■ **6.44   What are the mechanical properties of a metal and how are they found?**

The mechanical properties of a metal are found by tests so that the behaviour of the metal under stress can be predicted. The mechanical properties which give these indications are: the tensile strength of the metal; the value of the yield stress; the elongation up to the yield point, and the total elongation up to failure; the contraction in the area of failure; the shear strength; and the hardness.

The mechanical properties of a metal are found by destructive testing of metal test pieces in testing machines, the most common being the tensile testing machine.

The mechanical properties are as follows:

Tensile strength
Yield strength
Elongation (%)
Reduction in area (%)
Hardness
Shear strength

Also included are Young's modulus, the modulus of rigidity, and Poisson's ratio, which may be obtained from the various tests.

Note Materials used in engine construction are tested to ensure that the material is of a quality known to give satisfactory results in service.

■ **6.45 How is the tensile strength of a metal determined? What other information is obtained when a tensile test is made?**

The tensile strength of a metal is found by breaking a specially shaped test piece under a tensile load. The test specimen is placed in a tensile testing machine which records the load applied and the extension, or strain, of the specimen under the load. The load may be applied through a system of levers or by a hydraulic pump.

The maximum load shown on the stress–strain diagram, divided by the original cross-sectional area of the test piece, gives the ultimate tensile strength of the material; this is usually referred to as the tensile strength. When the specimen is being tested, the yield point and the percentage elongation are also found.

In the design of many machinery parts the value of the yield point is of more importance than the tensile strength.

The ductility of the metal is characterized by the percentage elongation. The actual value of the percentage elongation figure for the various grades of steel used in important engine parts is carefully studied. If the figure is not up to the necessary requirements, the material is rejected.

■ **6.46 What is an impact test? What is its purpose?**

In an impact test a square section specimen is prepared and one side is notched with a slot cut to a particular form. A common form is a V-shaped notch with a radius at the bottom, the angle of the V being 45°, the bottom radius 0.25 mm, the depth 2 mm, and the specimen section square with 10 mm sides.

The specimen is subjected to a shock load in, for example, an Izod impact-testing machine or a Charpy pendulum machine. A weight on an arm swings through a certain distance and strikes a hammer blow on the notched test specimen, with a known amount of kinetic energy. The specimen is fractured, and the weight continues its movement until the swing is completed. The energy required to fracture the specimen is recorded on a dial.

Impact tests are made on test specimens to indicate whether their metallurgical properties are up to the required standard. The test is applied to test pieces taken from heat-treated castings, forgings, the weld metal taken from welded joints, and steel plates used for pressure vessels and important engine structures.

Note In the Charpy machine the test piece is supported at two points like a beam. The hammer blow is taken at the centre of the test piece, with the notch on the opposite side to the point of impact. In the Izod machine the test piece is held in a vice and projects upwards like a cantilever. The notch faces the hammer blow and is set at a fixed height above the jaws of the vice.

■ **6.47 How is the fatigue resistance of a material found?**

The fatigue resistance is found by subjecting a sample specimen of the material to periodic stress changes until fracture occurs. The number of cycles up to fracture and the stress pattern give an indication of the fatigue resistance of the material.

By conducting a series of tests with decreasing stress values, the number of cycles up to fracture increases. A point is eventually reached when the stress is insufficient to cause fracture irrespective of the number of cycles applied. This stress value is referred to as the fatigue limit.

The results of the various tests are plotted on a graph called a stress-number or S–N diagram. The S values denote the stress range and are plotted as ordinates; and N values denote the number of cycles, and are plotted as abscissae. The curve flattens as N increases. By plotting the values of log S and log N, the fatigue limit is shown as a well-defined turning point.

The results from fatigue tests require careful interpretation when deciding on the allowable design limits for a material.

A considerable number of machines are available for fatigue testing. One type uses magnetic means to create a fluctuating load cycle of tensile stress. Other machines use a hanging weight on the specimen, as in a cantilever; when the specimen is revolved, it is subjected to a stress cycle which alternates between numerically equal values of tensile stress and compressive stress.

■ **6.48 What is the purpose of creep tests and how are they carried out?**

Creep is a form of slip which occurs when metal is subjected to a tensile load at high temperature. Creep tests are used to find the safe working stresses for the materials working at high temperatures. Creep tests, to be reliable, must be conducted over long periods.

In creep-testing machines constant tensile load is put on the specimen by a system of weighted levers, while the specimen is kept at constant temperature. The strain is measured at set time intervals and a graph is plotted of strain against time. The test are repeated at the same temperature but with increasing loads, and further graphs are plotted. From these graphs, values are obtained for creep rates which are used to find the safe stresses. The results require careful interpretation.

Creep tests are mainly made on materials used for turbo-blower turbine blading.

■ **6.49 How is the surface hardness of a material measured? Name the machines used to test surface hardness.**

The surface hardness of a material is found by measuring the resistance of the surface to penetration by a very hard, specially shaped indenting tool, when it is pressed into the material by a known load.

The size of the indentation gives a measure of the resistance to penetration and therefore the surface hardness of the material.

The machines used to measure hardness are the Vickers Pyramid Hardness

Tester, the Brinell Hardness Testing Machine, and the Rockwell Hardness Testing Machine.

The hardness value is expressed as a number with a prefix indicating the machine used, such as BHN (Brinell Hardness Number) or DPH (Diamond Pyramid Hardness). The indentation made is very small, and in the Brinell and Vickers machines it is measured with a microscope. A formula is then used to calculate the hardness number from the load applied and the size of the impression. Rockwell machines show the hardness number on a graduated dial.

Tables are available which give approximate conversions from the hardness number on one machine to that of the others.

The Rockwell and Brinell machines use a hardened steel ball to make the indentation or impression. The Vickers machine uses a diamond shaped like a pyramid.

■ **6.50 Name the non-destructive tests used during the manufacture and service of diesel engine components and associated equipment. Give an example of where each test is used.**

Non-destructive testing methods include radiography, ultrasonic testing, magnetic particle testing, eddy-current testing and dye penetrant testing. These tests enable the engineer to decide whether a part is likely to be reliable in service.

Radiography is usually confined to the testing of welds in pressure vessels such as starting air reservoirs. Large reservoirs for high-pressure use may require complete examination of all welds, smaller reservoirs may require only spot tests to be taken. X-rays or gamma-rays are used to expose the emulsion on the radiographic film. Welding defects give a greater exposure of the film and show as darker areas. The use of radiographic equipment and interpretation of the negative requires considerable training and skill.

Ultrasonic testing is performed by equipment that transmits high-frequency vibrations through the material to be tested; the vibrations are reflected back from the opposite surface or from any discontinuity in the material. Ultrasonic methods can be used to measure the thickness of materials or to detect internal or surface defects in welds, castings, or forgings, either during manufacture or when in service. Defects are shown as extra pulses to the transmitted and reflected pulses, on a cathode ray oscilloscope.

Magnetic-particle testing methods can be used for detecting surface and near-surface defects in materials that can be magnetized. When a magnetic field is induced in the part to be tested, defects allow a flux leakage to occur. This causes the magnetic particles used in the test to congregate at the leakage, indicating the location of the defect.

Magnetic-particle testing is used mainly for checking the condition of engine parts and shafting, which are liable to fatigue failure. The use of this type of test equipment requires skill and experience.

Eddy-current testing methods are used mainly in production line work during the manufacture of small ferrous material parts for use in either large or small engines. A coil is used in the tester, any defects present causing a change in the impedance of the coil. The change of impedance is utilized in various ways to call attention to the presence of a defect.

Dye-penetrant tests are used to detect surface defects such a fatigue cracks in crankshafts and screwshafts. The part to be tested is throughly cleaned, and a dye penetrant is sprayed on the cleaned area. If any surface defect is present, capillary attraction draws the dye into the crack. The dye is cleaned off (but remains in the cracks), and absorbent material is spray-coated on the test area. The absorbent draws out the penetrant dye from the crack, showing a coloured line which indicates the presence and location of the defect.

■ **6.51   How can the actual working stresses on engine components be measured while the engine is operating? Name the type of instrument used and state the principle of its operation.**

The working stress on fixed or moving engine parts can be found by accurately measuring the strain and then calculating the stress by the application of Hooke's Law.

Stress/strain = a constant = Young's modulus E
Therefore
Stress = E strain

The instrument used to measure the strain on the engine component is an electrical-resistance strain gauge. The principle of its operation is associated with the phenomenon of the change in electrical resistance which occurs in an electrical conductor when it is deformed.

The electrical-resistance strain gauge, usually referred to as a strain gauge, consists of a grid of wire cemented to paper or thin card which acts as a dielectric. The card is fixed with adhesive cement to the engine component, so that the wire grid is shortened or lengthened by the same amount as the surface of the part to which it is attached. The linear strain on the engine component is thus exactly equal to the linear strain or deformation on the strain gauge.

By applying a voltage to the strain gauge the change in resistance that occurs may be related to the linear strain, from which the working stress can be calculated.

The circuits into which a strain gauge is wired vary according to whether the stresses are static or changing, and whether the effects of temperature must be compensated for. If transient stresses are present they can be shown on an oscilloscope or recorded using suitable instrumentation.

Note   An oscilloscope is a device which uses a cathode ray tube to produce in graphic form a representation of a variable electric quantity, usually with respect to time. A record can be made by photographing the oscilloscope display.

■ **6.52 Where are strain gauges used?**

Strain gauges of the electric-resistance type are used in research and development projects on prototype engines to find the actual working stresses on fixed or moving engine parts. They are also used by engineers to investigate operational difficulties or the causes of material failures. By wiring the strain gauges in various patterns, complex strains and stresses can be investigated. The gauges can be used in most places, provided that electrical connections can be made between the gauge and the instruments. For example, four strain gauges could be wired up in square formation and cemented on the surface of a shaft axis to provide a torque measurement. The electrical connections to the strain gauges for input and output would be made with slip-rings and silver–graphite brushes.

Strain gauges are usually incorporated into an impedance bridge, and operated with either dc or ac current. The latter offers some advantages, at the expense of some complexity in the circuitry.

In work involving moving parts where the changes in stress values must be studied, such as in a crankshaft or moving parts in similar locations, a battery is fastened to the moving part to supply current to the strain gauge circuits. The changes in current flow from the battery are monitored, amplified and changed into a radio signal which is transmitted from an aerial or antenna attached to the moving part. The signal is received by a stationary aerial and transformed into some form where it is usable for the analysis of the strains and the associated stresses to which the moving part is subjected.

The input signals are filtered and fed into a computer that will give plots of stress on a time base or the cyclic frequency and amount of the stress changes.

■ **6.53 When and how is polarized light used to determine the true stresses likely to be met in diesel engine components?**

Polarized light is used in photoelasticity experiments to analyse the action of complex stresses and find the true stress. The stresses may be parts of two- or three-dimensional systems. In photoelasticity experiments, use is made of a polariscope which produces polarized light to investigate thin transparent models or shaped forms of the part being tested. When the model is stressed and placed in the field of polarized light, patterns become visible; from these stress patterns the magnitude of the stresses can be determined.

Special plastics are available for three-dimensional studies. These plastics are held in a state of stress at a high temperature, and then allowed to cool slowly to normal temperature. At normal temperature the photoelastic properties are frozen in the plastic, allowing the model to be broken down to plane sections for testing.

Experiments using photoelastic methods are made during the design stage of prototype engines. They allow the engineer to determine if any stress concentrations are present. If any are shown, modifications are made to bring the ratio of maximum stress to nominal stress within acceptable limits.

The present knowledge available on fillet forms at bolt heads, surface changes made at the edge of shrink fits, transitional forms at changes in sections, etc. is derived from the results of photo-elasticity experiments.

Note The principles governing the photoelastic properties of transparent materials, and polarized light, can be found in any physics book dealing with optics under the subject headings of: double refraction, temporary double refraction (double refraction is sometimes termed birefringence), monochromatic light, the polariscope or polarimeter, polarization of light, circular and elliptical polarized light forms, and molecular optics.

7

BEDPLATES, FRAMES, GUIDES, SCAVENGE TRUNKS, CYLINDER JACKETS

■ **7.1** Describe briefly the construction of bedplates used in slow-speed propulsion diesel engines. Is there more than one type of bedplate, and are different materials used? What are the advantages of each?

There are two types of bedplate in general use in slow-speed engines, both of which are completely enclosed on the underside in order to form an oil sump, while the bearings for the crankshaft are supported by cross girders which form part of the bedplate. The trestle type sits upon two parallel stools or raised portions of the ship's structure located in a fore-and-aft direction. The other type of bedplate is the deep box pattern with a flat bottom which enables it to be bolted to the ship's flat tank top. Propulsion engine bedplates are usually fabricated from mild steel plates and steel castings, welded together; smaller engines may have cast-iron bedplates. The main advantage of the fabricated bedplate is lightness in weight and lower construction cost.

Bedplates made of cast-iron have a desirable quality of being able to absorb vibration better than fabricated bedplates, due to the good internal damping characteristics of cast-iron.

Generally large engine bedplates are of the deep box pattern with a flat bottom, those for smaller power engines being either of the flat-bottom type or of the type that sits on stools forming part of the ship's structure. (Fig. 7.1)

■ **7.2** Which parts of a modern, slow-speed propulsion engine bedplate are strength members? How are the strength members formed?

All parts of an engine bedplate contribute to its strength. The parts are made from mild steel plates and steel castings which are assembled and welded together so that the bedplate is strong longitudinally (fore-and-aft) and transversely (port and starboard) with good resistance to twisting along its length. The longitudinal strength is obtained by making each side of the bedplate in the form of a girder. These girders may be box formed with two flanges and two webs, or as a normal girder with one web and two flanges. The cross girders in

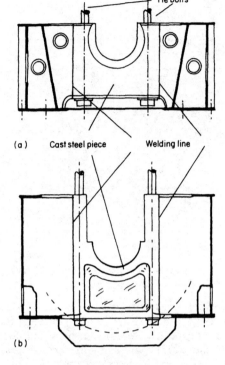

Fig. 7.1 Tranverse section through:
(a) older type slow-speed engine bedplate;
(b) modern slow-speed engine bedplate;

which the main bearings are housed give the bedplate its transverse strength. The cast steel cross girders are welded to the sides of the bedplate and the assembled unit has good resistance to twisting along its length.

The weight of the static engine parts is transferred to the bedplate through the A-frames, and it is therefore necessary to stiffen locally the sides of the bedplate in the way of the A-frame landings. Part of the stiffening comes from the cross girder. In bedplates with box girder type sides, diaphragms are fitted inside, between the webs and the flanges. In bedplates with a single plate web forming the side girders the stiffening is carried out by fitting brackets between the flanges in line with the junctions of the cross girders, additional tripping brackets being fitted to prevent localized buckling between the girder flanges and the web.

■ **7.3 Why are bedplate cross girders, cross members or transverse members now usually steel castings whereas previously they were usually fabrications?**

Fabricated cross girders have often shown failure of welding around bearing pockets and at the junction welds between the fabricated cross girder and the

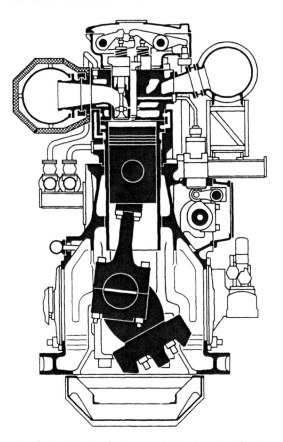

Fig. 7.1 (c) section through modern medium-speed engine showing the bedplate (MAN B&W)

side girders. This has caused costly engine repairs often involving removal of the crankshaft and consequent loss of earnings during repair time. Fabricated cross girders are difficult to weld due to difficult access, and problems sometimes arise in the assembly and junction welds. Generally cast steel cross girders have proved to be more reliable in service.

■ **7.4 How is the cylinder firing load transferred to the bedplate in single-acting propulsion engines?**

The firing load from the cylinder covers is transferred through the cover studs to the cylinder beams. The beam transfers the load through the tie bolt nuts and tie bolts to the bedplate cross girders.

The gas pressure on the piston acts downwards. The force created is transmitted through the piston, piston rod, connecting rod, crankpin, crankwebs and journals on to the lower halves of the main bearings supported in the bedplate cross girders. The upward acting force on the cylinder cover is transmitted through the cover studs into the cylinder beam. It is then transferred into the tie bolts and the bedplate cross girder where the downward acting force is balanced at the main bearings. The downward acting force is not identical in magnitude to the upward acting force, the small differences are shown as stresses in the engine structure. (Figs 7.2 (a) and 7.2 (b))

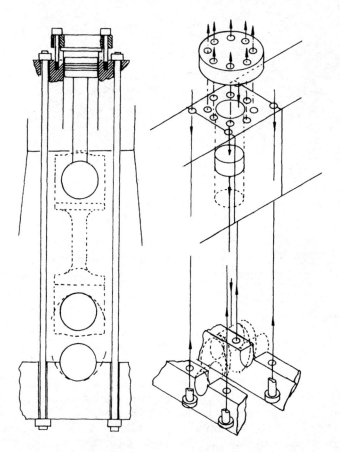

Fig. 7.2 Diagrams showing how the gas pressure acting upwards on the cylinder head is transferred through the cylinder head studs (tensile stress) and into the cylinder beam. The upward acting force on the cylinder beam is then transferred into the tie rods (tensile stress) and down to the transverse beam supporting the main bearings. The tensile stress in the tie rods tends to lift the transverse beam, but this is resisted by the downward-acting force from the gas pressure on the piston. This downward-acting force puts a compressive load on the piston rod; this is transferred through the crosshead bearings, down through the connecting-rod and into the bottom end bearing and crankpin. The downward-acting force on the crankpin is transferred down each crankweb and on to the main bearings; it is then balanced by the upward-acting force from the tie rods.
Note. The A-frames are in compression.

Some single-acting engines with cast-iron bedplates and A-frames are not fitted with tie bolts. In such cases the cylinder beam is held on to the A-frames by a transverse row of studs and nuts between the cylinders. Flanges on the base of the A-frame fit to corresponding flanges on the bedplate, and are bolted together. When firing occurs the firing loads are transferred to the bedplate through the studs holding the cylinder beam to the A-frame, through the A-frame and through the bolts holding the frames to the bedplate. At the time of firing all these parts are subjected to a tensile load. (Fig. 7.2)

7.5 Which parts of the structure of a medium-speed V-engine are strength members?

V-engines are made in two basic structural forms. One form has a conventional-type bedplate with the crankshaft main bearings housed in cross girders, and the cylinder beam is often pentagonal in section and houses the cylinder jackets, etc. The crankcase sides are integral with the cylinder beam and join on to the bedplate with flanged joints. In this form the strength members are the bed-plate sides, the main bearing cross girders and the cylinder beam.

The other structural form for the V-engine is fundamentally a pentagonal-shaped cylinder beam with heavy side plates forming the side of the crankcase, these heavy side plates being fastened to a heavy base-plate with areas cut out to allow crank rotation. The crankshaft main bearings are fastened into diaphragm plates forming part of the engine structure, and the crankshaft is under-slung in the housings and supported on the main bearing keeps. The heavy base-plate of the engine structure sits on fore and aft supports or stools built into the ship's structure. The lower side of the engine structure is made oil-tight by the oil pan which is light and does not make any strength contribution. The strength members of this structural form are the heavy inverted V plates forming the apex of the pentagon and the heavy base at the bottom which, in effect, form flanges of a girder, and the sides of the pentagon and the crankcase which form the webs. The diaphragms give rigidity to the sides forming the webs. Tie-bolts may be used in either of the structural forms. (Fig. 7.3)

7.6 Which other engine part, or parts of the hull structure, assist in supplementing the longitudinal and transverse strength of an engine bedplate?

The cylinder jackets are accurately machined on the ends and bolted together so that an assembly the length of the engine is formed. A considerable number of the bolts joining the jackets are of the fitted type so making the assembly rigid and preventing relative movement between jackets. This assembly of cylinder jackets is sometimes referred to as the cylinder beam.

The tie-bolts passing through the A-frames also pass through the sides of the cylinder beam at the corners of each jacket, and hold the jackets firmly on the A-frames.

The engine bedplate is supported by the foundation plate which forms part of the inner bottom plating or tank top of the hull double bottom. The hull structure under the engine is made very stiff, longitudinally by fitting additional intercostals on each side of the centre keel plate, and transversely by making all

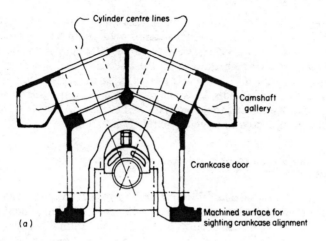

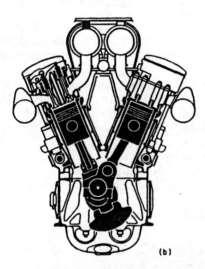

Fig. 7.3 Sections through V-engine frames and bedplates.
(a) Section through fabricated engine frame built from cast steel sections and steel plates joined by welding processes.
(b) Section through V-engine showing bedplate, frame, running gear and bottom of tie rods. (MAN B&W)

floors solid. The stiffness is also increased by making the depth of the double bottom under the engine to the maximum possible dimension. The localized stiffness under the engine is tapered off gradually towards the sides of the hull, and forward and aft, so that stress raisers are reduced to a minimum between the very rigid cellular structure under the main engine and other parts of the hull.

Note In ships with engines aft the localized stiffness under the engine is usually continued right through as far aft as possible.

The hull structure under the engine and the cylinder beam both add to the longitudinal strength of the bedplate. The transverse strength of the bedplate is increased by the hull structure and the A-frames. The increase in strength of the bedplate is considerably reduced if holding-down bolts, tie-bolts or A-frame bolts are allowed to become slack.

The stiffness of the hull is always greater than that of the engine structure and shafting. It follows, therefore, that the engine structure and the shafting will follow any changes in the hull which come about due to the vessel working in a seaway, or due to changing amounts and positions of cargo, bunkers or ballast.

■ **7.7 How are tie-bolts tightened? What control is exercised during tightening to ensure correct tension?**

Large tie-bolts are tightened with an adaptation of a hydraulic jack which loads the tie-bolt in tension. The tie-bolt nut is usually drilled to take a toggle bar or slotted to take a hook spanner, and when the correct pull is on the tie-bolt the tie-nut is pulled up hand-tight; the pressure in the jack is then released leaving the tie-bolt tight. The load placed on the tie-bolt by the hydraulic jack is controlled by the hand-pump pressure, which is indicated on the pump pressure gauge. The pump pressure to be applied is given in the engine instruction book and this pressure should not be exceeded.

■ **7.8 How would you find loose tie-bolts while the engine is in operation? How often would you make a search?**

If tie-bolts are not taking their proper load due to slackened nuts, the cylinder jackets adjacent to the slack tie-bolts can be seen lifting when the piston is nearing the end of compression, or on firing. If no movement is visible the thumb can be pressed on the cylinder jacket, with the thumb nail in contact with the tie-bolt nut. Small movements, too small to be visible, can be felt with the thumb nail. If washers are fitted between tie-bolt nuts and cylinder jackets, a washer may sometimes be twisted at the end of the firing strokes in each cylinder adjacent to the slack tie-bolt. Dial indicator gauges can also be used to detect relative movement between the tie-bolts and cylinder jackets.

In medium-speed engines the tie-bolts are sometimes hidden from view by valve rocker-arm covers and casings. Their locations should be found and tie-bolt nuts should be checked when changing exhaust or fuel valves.

Tie-bolts should be kept under observation at all times, particularly when running at reduced power in bad weather and following power increases during moderating weather.

Note When first making searches for slack tie-bolts, care must be taken to avoid mistaking the natural movement and 'spring' of the various parts during firing for slackness. Care is also necessary in the interpretation of dial-indicator gauge readings. Tie-bolts must be tightened carefully and close watch kept on crankshaft alignment or crankweb deflections (See Question 18.51).

■ **7.9   If fretting is occurring at landings of the cylinder beam on the A-frames and at top tie-bolt nut landings, what indications are usually shown?**

Fretting of a surface (particularly at the places mentioned in the previous Question) is often indicated by a rust-red powder being present at the outside of the faces that are fretting. If the location is dry and free of oil the red powder looks as if it has been dusted on the surfaces, being thickest at the joint. If the location is oily the powder mixes with oil and it looks as if a reddish coloured grease is exuding from the joint formed by the landing faces. The reddish, rust-coloured product from the fretting action indicates that the tie-bolt nuts have been slack for some time.

■ **7.10   What are the consequences of running an engine with slack tie-bolts?**

If an engine is run with slack tie-bolts the cylinder beam flexes and lifts at the location of the slack tie-bolt. In time the landing faces of the tie-bolt upper and lower nuts, and the landing faces of the cylinder beam on the A-frame, fret and the machined faces are eventually destroyed. The fitted bracing bolts between the cylinder jackets will also slacken and the fit of the bolts will be lost.

If fretting has occurred in an uneven pattern where the cylinder beam lands, and the tie-bolts are tightened, the alignment of the cylinders to the line of the piston stroke is destroyed. After fretting has occurred nut landing faces may be out of square, and if tie-bolts are tightened on faces which are out of square a bending moment will be induced in the tie-bolt. This, in turn, causes an uneven stress pattern in the tie-bolt, which could lead to early fatigue failure.

■ **7.11   How are A-frames or columns constructed?**

A-frames are fabricated from flat steel plates welded together with full-penetration welds in critical places. The plates are grit- or shot-blasted and given a coat of priming material before cutting. Various attachments are welded to the frames. These attachments form the mountings for the crosshead guides, main crankcase covers, piston cooling supply pipes, piston cooling return drains and the like. Flanges are mounted at the upper and lower ends of the A-frames. After fabrication and welding, the A-frames may be heat treated. Following heat treatment, machining operations are carried out on the upper and lower flanges and the guide-plate mounting face. The bolt holes for the upper and lower flange bolting are drilled undersize. When the A-frames are erected on the bedplate and brought to the correct alignment, the bolt holes in the flanges are reamed to the correct size for the fitted bolts.

■ **7.12   How is the bedplate fastened to the hull structure? What are chocks and holding-down bolts?**

The engine bedplate is supported on a series of chocks fitted around the underside of the periphery of the base of the bedplate. The chocks sit on the foundation plate which forms part of the inner bottom plating of the hull structure. The holding-down bolts pass through holes in the bedplate, chocks and foundation plate. When all the holding-down bolts are tightened the bedplate is held fast to the hull structure. The chocks are fitted in place after the engine has been aligned to the intermediate shafting. The spacing between the chocks is approximately 250 to 400 mm, and they are more closely spaced in the location of the cross girder so that good support is given to the area of the bedplate which is supporting the main bearings and the engine A-frames. This prevents any localized distortion or sag in the bedplate and so makes the support from the chocks almost continuous in effect.

The name 'holding-down bolts' is a carry-over from reciprocating steam-engine days; a more suitable name now would be stud bolts. The bolts used are between 25 and 75 mm diameter, they are threaded at each end, and the upper ends have two flats or a square machined on them to take a spanner. The plain part of the stud bolt is machined down on the centre shank to a diameter just slightly less than the bottom of the thread, thus providing a stretching length between the threaded sections and preventing a stress concentration at the bottom of the last thread. If a stretching length is not machined on the bolt it is more likely to fail in service.

The foundation plate is tapped to accommodate the thread on the lower end of the holding-down bolt, and the bolt is screwed into the tapped hole with a spanner applied to the square or two flats on the upper end of the bolt. Self-curing plastic sealing tapes are sometimes used on the lower threads, to prevent leakage of fluids from the tank top into the space below and also to protect the thread of the bolt and the foundation plate from corrosion. In other instances the locking nut fitted under the foundation plate is machined to take an O-ring or grommet, which then makes a seal.

In some cases long plates having a slightly tapered cross-section are tack and seal welded to the foundation plate supporting the engine. (Tack and seal welding prevents ingress of water and rust growth that may lead to later misalignment.) The long plates run the length of the engine. The engine is set up on chocks machined with the same angle of taper. The chocks and the plate then become in effect a set of folding wedges. This makes it easier to align the bedplate and saves a considerable amount of time for alignment and chocking. The hand fitting of chocks is obviated.

This form of chocking system will require careful examination at regular intervals to check for any possible rust growth between the foundation plate, the long plate, and the chocks, which could affect the engine alignment. (Fig. 7.4)

Another chocking system extensively used involves the use of synthetic resin materials cast with the engine *in situ* after alignment of the bedplate is completed (see Question 7.18 and Fig. 7.5).

■ **7.13 What are the consequences of running an engine when the holding-down bolts are slack?**

If an engine is operated with slack holding-down bolts it may quickly be apparent by a change in the vibration pattern and thwart-ship movement at the upper part of the engine. Sometimes, however, there may be no such indication.

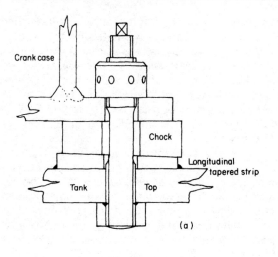

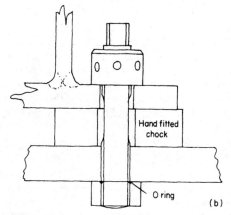

Fig. 7.4 (a) Section through a cast-iron or steel chock placed on tapered plates running the length of the engine. The taper is shown exaggerated for clarity.
(b) Section through hand fitted chock.

Continued operation with slack bolts allows fretting to occur on the mating surfaces of the bedplate, chock and foundation plate. If the fretting occurs in areas covering a number of adjacent chocks the crankshaft may be seriously damaged through misalignment.

■ **7.14 Many slow-speed engines have the thrust block housing incorporated with the engine bedplate. In such a case how is the propeller thrust transmitted to the hull?**

The propeller thrust is transmitted to the thrust-bearing casing and bedplate through the thrust pads or thrust-bearing ring. The thrust is then transmitted to the hull through the holding-down bolts and the foundation plate. As a safeguard against holding-down bolts slackening, which would allow the engine to slide backwards and forwards on the tank top with ahead and astern movements, the holding-down bolts at the aft end transverse of the bedplate and around the thrust block casing are made as fitted bolts. After the engine is chocked to the correct alignment the holes for the holding-down bolts are brought to the size of the fitted bolts with a reamer, and the fitted bolts are placed and tightened.

Normally, with all holding-down bolts tight the propeller thrust is taken up by the friction between the bedplate, chocks and foundation plate; if the holding-down bolts slacken, the fitted bolts around the thrust casing sustain a shear stress and the propeller thrust will be transmitted through the fitted bolts. The thrust from the propeller is then taken by the foundation plate which transmits it to the hull through the welded attachments of the foundation plate to the intercostals and floors.

■ **7.15   How often would you check holding-down bolts to ascertain that they are tight? What methods are used to check tightness?**

A requirement of classification societies (Lloyd's Register of Shipping, Norske Veritas, American Bureau of Shipping, B.V., etc.) is that holding-down bolts be checked by a surveyor within each survey cycle or four-year period. In practice this period of time is too great and the bolts should be checked at 6-monthly intervals unless there is a case history of the bolts going slack more frequently, in which case the period must be related to experience. In new vessels the bolts should be checked within one month of the commencement of the maiden voyage, or earlier if possible. The interval may then be gradually increased if all is found in order. After a vessel has been through bad weather the bolts should be checked as the weather moderates.

Holding-down bolts are hammer tested to ascertain if they are tight. The method is to hold the tip of the thumb on one side of the nut face and strike the nut on the opposite side. If the nut is slack, the nut and stud spring against the thumb and then retract. The movement can be felt against the thumb. If a holding-down bolt is of the fitted type this test is ineffective, and either a spanner or the hydraulic jack must be used. Due to bilge water and other slack water being on the tank top at various times, the holding-down bolt nuts may rust and seize on the studs; if this occurs care must be taken not to be misled by the seized condition of a nut when checking for tightness. The hammer testing method, however, stands up well in finding nuts which are not tight down on the landing face even when they are seized on a stud.

■ **7.16   What are the methods used to tighten main engine holding-down bolts? How are chocks fitted?**

Main engine holding-down bolts on modern motor ships are tightened with an adaptation of the hydraulic jack in a similar manner to the tightening of tie-bolts (see Question 7.7). In older ships they are tightened with hammer blows on to a heavy 'star' spanner supplied for this purpose. A hammer of 2 kg weight is normally used.

Chocks are fitted in place after the engine has been aligned to the intermediate shafting, and the bedplate has been brought to the same degree of flatness as it had during erection on the test bed. The engine at this stage is supported on a series of folding wedges around the periphery of the bedplate. The chocks are machined in the workshop to sizes taken from the ship (with some allowance for hand fitting). After machining, chocks are lightly driven into place so that 'hard' bearing marks are made on the machined surfaces of the chock. The chocks are then backed out and the hard spots are eased by draw filing. The process is repeated and the bearing surfaces on the chock faces increase. By the time the chock is about 75 mm from its final position there is a good bearing surface on the chock. At this stage it will be driven home with a 2 kg hammer, after which the holding-down bolt will be fitted.

■ **17.7** **If, soon after joining a motorship, you found a number of holding-down bolts slack, and fretting to have occurred in the areas of the slack bolts, describe how you would handle the situation.**

When chocks and their mating surfaces on the bedplate and tank top have fretted, the chocks cannot properly support the engine. If the holding-down bolts are tightened the crankshaft alignment may be seriously affected with lesser effects being felt on crosshead guide and cylinder alignment. The seriousness of the situation will be proportional to the amount of fretting that has occurred.

Before any tightening of the holding-down bolts is carried out, the alignment of the crankshaft should be checked by the crank deflection method with a dial gauge. If the crankshaft alignment is satisfactory the slack chocks can be removed and smoothed on the mating surfaces and then replaced. Slotted slip liners can then be made and fitted between the chock and bedplate. The bolts can then be tightened to harden the chock. After all the bolts and chocks have been tightened the crankshaft alignment must be rechecked. If the crankshaft alignment is satisfactory the engine can be operated.

If the crankshaft alignment is found to be unsatisfactory on the check made before tightening of bolts, it may be necessary to lift the bedplate in the region of the fretted chocks. This can be done by using slow-tapered folding wedges or box wedges, but not without difficulty. In order to get the necessary mechanical advantage it will be necessary to use three or more sets of wedges. When the crankshaft alignment is satisfactory the chocks can be treated as mentioned earlier and then fitted with slip liners or shims, after which the crankshaft alignment must again be checked.

After carrying out corrective work such as that described, the engine should be tested, preferably at a berth or by operating at low power, until it is established that no bearings are running hot.

Repairs of this nature could only be considered as temporary to get a vessel back to home port. Permanent repairs would then be necessary.

■ **7.18 Describe the work involved in making permanent repairs to main engines with badly fretted chock landings and chocks.**

In carrying out permanent repairs the aim is to bring the bedplate back to its original alignment to reduce stresses on welds, and then correct any crankshaft misalignment that may have developed. The method of repair is to discard the original chocks and replace them with a resin chocking system. The landing surfaces on cast-iron or steel chocks are approximately 30 to 36 mm wide over the length of the chock, which may be about 150 to 250 mm. The landing surfaces are also referred to as the fitting strips; there are two fitting strips on the upper face of the chock and two on the lower face.

When fretting occurs the surfaces of the underside of the bedplate in contact with the fitting strips become recessed, and in a similar manner the tank top becomes recessed and the chock becomes thinner. The sum effect of this is to allow the bedplate to sag in the areas where fretting has occurred and then the crankshaft becomes misaligned.

To effect repairs, the holding-down bolts are slackened off, in and on each side of, the damaged areas. The bedplate is lifted with a series of steel wedges, during which time the top surface of the bedplate (surface where A-frames land) is kept under observation with optical alignment equipment. The bedplate is lifted until the line of the bedplate is brought back to its original position. The damaged area of the bedplate and tank top is cleaned with chemical solvents down to bare metal. By using rubber or neoprene moulded strips a mould is formed between the wedges, the bedplate and the tank top, in a similar manner to the shuttering for a cement box. The holding-down bolts and the rubber strips are lightly greased so that the chocking resin does not adhere to them. The steel surfaces of the bedplate and tank top are sprayed with a very thin coat of chemical which prevents the resin adhering.

The volume of the mould is calculated so that the correct amount of pourable resin is mixed. The resin is supplied in two packs which when mixed begin to cure. As soon as the base and catalyst are mixed they are poured into the mould so that contact is made between the bedplate and the tank top; this is done by making the outside mould rubber strip higher than the thickness of the chocks. A slight head is then created so that the resin completely fills the chock cavity. The resin is then allowed to cure completely, taking about 48 hours when the temperature of the surroundings is 16°C (60°F) and proportionately longer if the temperature is lower.

After the resin has completely cured, the supporting steel wedges are removed and the holding-down bolts are tightened. The bedplate is then again checked. The crankshaft alignment is checked and any necessary correction is made by remetalling or machining the lower halves of the main bearings.

The area of the resin chocks supporting the engine must be greater than that of the cast-iron chocks. The area can be calculated from the weight of the engine and the initial stress in the holding-down bolts, and the load-bearing characteristics of the resin used. The resin used must be approved by the appropriate Classification Society. (Fig. 7.5)

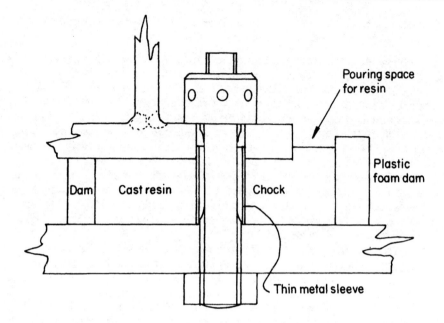

Fig. 7.5 Pourable continuous cast resin chocks.

■ **7.19 What is the purpose of resilient chocks?**

When an engine is operating, forces are set up in the moving parts and these forces are transferred to the engine bedplate. When the engine is on solid chocks movements from these forces are transmitted to the hull structure as a form of vibration. When resilient chocks are used the vibratory movements of the bedplate are damped out within the chocks and little or no vibration is transmitted to the hull. They are mainly used on medium- and high-speed engines where vibration may cause damage or be a source of annoyance in other parts of the ship.

■ **7.20 Some medium- and high-speed engines are set on resilient chocks. Describe how the holding-down bolts are tightened; what are the precautions that must be taken?**

A fairly common form of resilient chocking system is to use a bonded cork chock set between two steel plates. The cork is chemically impregnated to make it resistant to breakdown in the presence of oil. In some cases a binding strip of metal plate is fitted around the cork layer to help it resist compressive loads and prevent it 'barrelling'. In other instances an elastomer is used in place of cork.

With resilient chocks of this form the holding-down bolts are tightened with a torque spanner to some required load. The load placed on the bolt is fairly light. After tightening, the nuts must be carefully locked with lock nuts. Normally, once set, the holding-down bolts should not be tightened further.

When the nuts are initially hardened down the crankshaft alignment is kept under observation using the crank deflection method with the aid of a dial

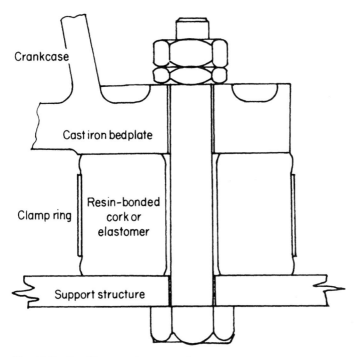

Fig. 7.6 Resilient engine mounting.

gauge. Should it be necessary to retighten holding-down bolt nuts, the specified torque must not be exceeded, and crankshaft alignment must be checked before and after tightening the nuts to see that it has not been adversely affected. (Fig. 7.6)

■ **7.21 How is the crankcase oil pan connected to the double-bottom drain tank?**

The lubricating oil in circulation drains through a pipe connection jointed on to the bottom of the oil pan or lower part of the bedplate and the tank top. Normally the pipe is of a large diameter and fitted with a normal flange at the upper end where it is jointed to the crankcase oil pan. At the lower end the flange is fitted within the pipe. The section of pipe and flanges then follows a Z form. Where the bottom of the crankcase is stiff one or two bellows convolutions may be fitted within the pipe length to give it the necessary flexibility for temperature changes and thermal movement.

The drain pipe is made after the engine is chocked: a pipe template is lifted and the pipe is made to match it. The flange distance then coincides with the distance between the oil pan and tank top joint faces. The drain pipe is jointed up and assembled to the tank top from within the crankcase. A perforated plate is fitted over the drain outlet to prevent rag and the like getting into the drain tank.

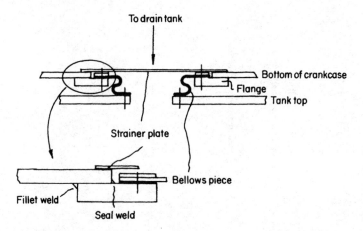

Fig. 7.7 Section through flexible crankcase drain. Drain connections of this form can be removed in order to renew the upper and lower joints or gaskets.

A flexible type of bellows piece is shown in Fig. 7.7. This type of bellows is fitted from inside the crankcase. It may be unbolted to renew leaking joints or gaskets.

The drain can be a source of lubricating oil loss and if the joint on the tank top leaks it may be a source of water contamination. In searching for points of oil loss and sources of water contamination, careful attention should be paid to the crankcase oil drain, including, if fitted, the bellows section.

■ **7.22   How is the oil for lubricating the cylinders of a crosshead-type engine prevented from draining into the crankcase?**

The oil draining down from a cylinder liner drains into the scavenge space around the bottom of the cylinder liner and is collected on the diaphragm. The diaphragm is sloped to one side of the engine so that the oil drains downwards, towards a scavenge space drain fitted on the lower side of the diaphragm, and the oil is discharged from the drain connection.

The piston rod metallic packing prevents the scavenge air blowing into the crankcase, and a set of scraper rings below the packing prevents oil from the crankcase being carried up into the scavenge space by the piston rod. The piston rod packing and scraper rings are fitted in a housing which is held in a boss, the upper surface of which is brought above the level of the diaphragm space, thus preventing any oil from the cylinder liner draining directly on to the piston rod.

■ **7.23   What material are guide plates made from, and how are the guides cooled?**

Guides plates, in some engines, are made of mild steel and form part of the structure of the A-frame. In other engines the guide plates are made of cast-iron and are bolted to the A-frame.

In engines where the guide plate is made of mild steel (forming part of the column) cooling is effected by the lubricant supply carrying heat away from the slipper and the guide face, and some of the heat generated is transmitted by conduction through the material of the A-frame.

Some cast-iron guide plates are cast with a hollow space in their back. When the guide is bolted on the A-frame a cavity is formed between the plate and the A-frame. This cavity is connected to the lubricating oil pressure lines and lubricating oil is circulated through it. The oil flow is controlled by a small needle valve fitted in the outlet line, and the flow is kept under observation by leading the outlet through a tundish with a hinged cover.

■ **7.24 How is the space around the scavenge ports in the cylinder liner formed and made airtight?**

The space around the scavenge ports is annular in uniflow-scavenged engines and approximately semicircular in cross- and loop-scavenged engines.

In uniflow-scavenged engines the space may be formed by continuing the sides of the cylinder jacket down below the cooling water space, or by steel plating fitted and welded to and around the tops of the A-frames. The bottom of this space is sealed off from the crankcase with a cover in which the piston-rod seals are mounted, and through which the piston rod passes. The cover is referred to as a diaphragm and the piston-rod seals as the diaphragm packing.

In cross- and loop-scavenged engines, the scavenge space about the scavenge ports forms part of the cylinder jacket casting. This section of the casting forms a passage through the cooling water space, or just below it.

■ **7.25 To what parts of a cylinder jacket would you give attention after removing an old cylinder liner prior to fitting a new one?**

The parts of the jacket coming into contact with the cooling water should be cleaned and examined for corrosion, and if any corrosion is present the corroded area should be cleaned out and filled with an epoxy filler. The inside of the jacket should be painted with one of the paints produced for this purpose, unless corrosion inhibitors used in the cooling water render painting unnecessary. In some engines the cooling water is given a high velocity at the upper parts of the cylinder liner. Where the high-velocity cooling water is in contact with the jacket, erosion sometimes occurs. Erosion is usually caused by air within the cooling water, and if erosion is present an examination should be made to locate the point of air ingress. The air bleed-off points at the upper parts of the cooling system should also be checked to see that they are not blocked.

The landing recess in the top of the jacket must also be cleaned and the joint face checked. Sometimes lapping plates are provided to lap the joint face in the recess to a smooth surface. The bell mouth entries for sealing rings often lose their taper due to sludge settling out and going hard. The tapers must be cleaned to their original shape.

If leakage tell-tale holes are drilled in the lower belt of the jacket in cross- and loop-scavenged engines, the hole must be cleaned out and blown through with a

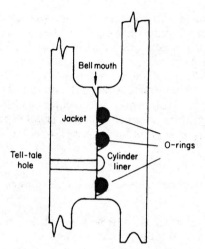

Fig. 7.8 Lower seals between the cylinder liner and water jacket showing the bell mouth entry for leading in the O-rings without causing damage. The bell mouth space usually fills with sediment from the cooling water treatment and must be carefully cleaned out before renewing sealing rings. The O-rings must be lubricated with a soap/water solution before fitting the liner in place. Many engines are not fitted with 'tell-tale' holes for indicating jacket water leakage through the upper sealing rings.

jet of compressed air. The surfaces in the lower part of the jacket which contact the copper and the rubber sealing rings must also be carefully cleaned and smoothed down with emery cloth. The surfaces in this area must on no account be chipped at with a chisel or chipping hammer as these tools may cause damage which could lead to water leakage across the rubber seal.

Prior to renewing or replacing a cylinder liner in its jacket, the bell mouth entries for sealing rings must be carefully cleaned to remove hard sediment build-up. If the sediment build-up is not cleaned and the bell mouth properly restored, the sealing rings will become damaged when the liner is lowered home. The sealing rings must also be lubricated with a soap-water solution prior to fitting a liner. If the rings are not lubricated they twist and become damaged when passing through the bell mouth. Oil should not be used as its presence in the cooling water may later lead to localized overheating and cause serious damage. (Fig. 7.8).

If the cylinder lubricating oil quills pass through the jacket, on the top face of the jacket will be found the engine builder's transverse centre-line markings which are put there for marking-out purposes and to line up the cylinder liner so that it comes into the correct position for fitting the oil quills. These marks should be found and chalked up or re-marked with a light centre-punch mark if necessary. The holes in the cylinder jacket through which the oil quills pass should have sediment removed and be throughly cleaned.

If the engine is of the cross- or loop-scavenge type the ports in the jacket should be cleaned to bare metal.

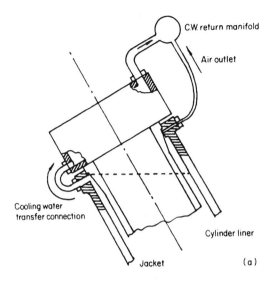

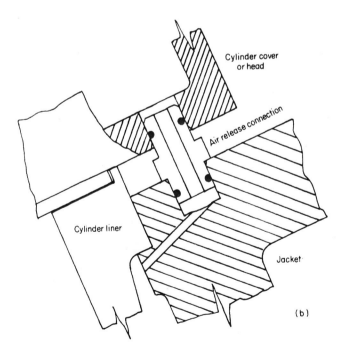

Fig. 7.9 (a) Sketch showing how upper part of V-engine cooling water space may become air locked.
(b) Sketch showing air release connection. The connection is held in place with two screws not shown.

■ **7.26   At what points on the jacket does the cooling water enter and leave?**

The cylinder cooling water enters at the lowest point on the jacket and leaves at the uppermost point. This arrangement prevents air locking or air pockets forming at the top of the cooling space.

V-engines require special air release holes in the uppermost point between the liner and the jacket to prevent air locks forming in any space above the main cooling water outlet from the cylinder. The air release arrangement may take various forms; it is essential that it be checked to ensure that it is clear when removing cylinder heads, or cylinder heads and liners. (Fig. 7.9)

■ **7.27   How can the cooling spaces within a cylinder jacket be examined without withdrawing the cylinder liner?**

Inspection openings are arranged round the bottom of a cylinder jacket, normally covered by blank flanges. To examine the cooling space when the engine is shut down, the cooling water supply and outlet valves on the jacket are closed and the jacket is drained through the drain connection. Drainage is facilitated if an air release connection has been fitted on the engine in which case it should be opened. When the cooling space has drained, the inspection covers can be removed to give access.

■ **7.28   Which parts of a cylinder jacket are machined to accommodate the cylinder liner?**

The uppermost surface of a cylinder jacket is first machined flat. A circular recess is then machined in this surface and the liner flange fits into this recess. In some modern engines a ring is fitted into this recess and the liner flange then fits on to the top of the ring.

In cross- and loop-scavenge engines a wide belt is cast towards the lower end of the jacket. This belt has passages passing through it which take the scavenge air and the exhaust gases. The belt is smooth-bored out to match the exhaust and scavenge belt on the outside of the cylinder liner. The upper end of the belt in the jacket is bored to a taper over a depth of 10 to 15 mm. This taper forms what is called a bell mouth. Its purpose is to give an easy entry for the sealing rings on the cylinder liner when the liner is fitted into the cylinder jacket.

The lower part of a cylinder jacket for a uniflow-scavenged engine is machined in a similar manner with a bell mouth for easy sealing ring entry, and in some engines the sealing ring is fitted into a stuffing box on the lower outside part of the jacket, a gland ring being used to tighten the sealing ring.

In those engines fitted with oil quills which pass through the cylinder cooling space, the bossings for the quills are bored and faced. Around the upper face of the jacket the holes are drilled and tapped for fitting the cylinder cover studs.

Modern design arranges for the cylinder head fastenings to be fitted away from the top face of the cylinder beam or jacket. This is done to reduce the stresses in the region of the stud holes which may become quite complex in nature in older-design engines, particularly if the engine has been improperly up-rated without sufficient attention being given to possible problems. In some cases failure to study this problem at the design stage has led to serious cracking of the cylinder beam and jacket area.

Some medium-speed engines have extended the cylinder stud or bolt fastening well down and away from the critical areas mentioned. (Fig. 7.10)

An arrangement of cylinder head fastenings is shown in Fig. 7.10(a). In this arrangement the downward acting force on the cylinder liner acting through 'B' is resisted by the upward acting force 'C' at the liner support in the jacket. As the two forces are not acting through a common line a bending moment is created. The cross-section at a–b has to resist a tensile load set up by the forces acting through 'A' and 'B'. Thermal stresses are also created in these parts when they come up to operating temperature and the location is said to be in a state of complex stress. If the radial clearance between the top of the liner and jacket is insufficient excessive tensile hoop stresses are produced in the top of the jacket.

Figure 7.10 (b) shows the way failures occurred in the cylinder jacket due to the older form of fastenings.

Figures 7.10 (c) and (d) show the designs adopted to prevent the failures shown in Fig. 7.10 (b). See also question 8.9. In some engines the cylinder jacket is made separate from the engine frame as shown in Fig. 7.10 (c).

■ **7.29 What periodic attention must be given to the scavenge air space and piston-rod packing at the bottom of a cylinder liner?**

The scavenge air space, although designed to be self-draining, does collect oily sediments on the diaphragm, which must be cleaned periodically. The interval between cleaning operations is a variable which will be dependent upon the amount of cylinder lubricant used, the effectiveness of the piston-rod packing and scraper rings, and the effectiveness of the air filters to the turbo-blower inlets. When the scavenge space is cleaned, the drain connection must be blown through with compressed air.

The attention given to the metallic packing consists of opening up and cleaning the packing. The clearance on the butts of the ring segments must be checked and corrected to take up the wear which occurs. As the rings wear the load on the garter spring around the packing is reduced and this sometimes calls for shortening of the spring length. The scraper rings are treated in a similar manner, and the width of the scraper surface will also require narrowing down if the design of the packing allows it. In other instances the scrapers must be renewed when the width of the scraping surface exceeds the figure given by the engine builder. The gallery round the scraper ring, and the oil drain holes, require careful cleaning. If drain pipes are fitted to the scraper packing drain space the pipes should be checked to see that they are clear.

The piston-rod packing does not normally require attention every time the scavenge space is cleaned. The frequency of overhauling packing is therefore much less than the frequency of cleaning the scavenge space. In some engines it is more convenient to attend to piston-rod packing at the time the pistons are removed for overhaul.

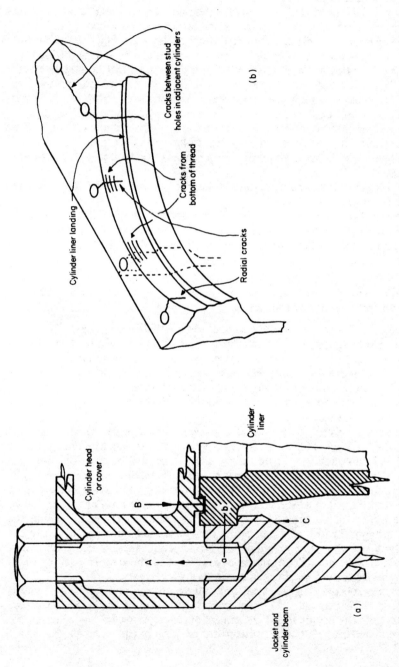

Fig. 7.10 (a) Sketch showing older arrangement of cylinder cover or cylinder head fastenings. The lines of action of loads B and C create a bending moment in the liner flange. The stresses arising out of this bending moment are additive to the thermal stresses in the flange. The stresses in the jacket are created by the upward force A in the jacket and the reaction in the threaded section of the hole. The location is in a state of complex stress due to the various forces A, B and C. (b) Sketch showing location of failures in older type cylinder cover fastenings.

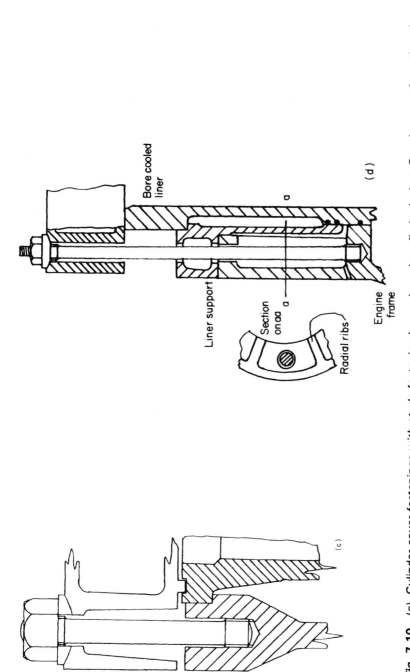

Fig. 7.10 (c) Cylinder cover fastenings with studs fastening lower down in cylinder jacket. Development from that shown in (a).

(d) Modern design of cover fastening with water jacket separate from engine frame and cylinder liner. The spherical washer at the cover nut prevents a bending moment coming on to the stud if the nut landing face is not square with the cylinder cover.

■ **7.30   How are the guides arranged on modern single-acting engines?**

On modern single-acting engines guides are fitted on the columns or A-frames to accommodate the slippers fitted at the forward and aft ends of the crosshead. Each A-frame, except those at the ends of the engine, has four guide plates, while those at the ends have two.

■ **7.31   What is the reason for arranging the guides of crosshead engines in the manner described in the answer to Question 7.30?**

The reason for arranging guides in this manner is in order to achieve a standardization that will enable engines to be arranged to run ahead in any direction without reversing the side of the engine to which the guides are fitted. Manufacture of right-handed and left-handed A-frames as would be required in twin-screw ships is then obviated.

Notes Controllable pitch propellers are usually made to turn left-handed (clockwise) when viewed from the after end of the ship, whereas solid cast propellers for use in single-screw ships are made to run right-handed when going ahead and left-handed to go astern.
   Controllable pitch propellers are unidirectional; when going astern with a left-handed controllable pitch propeller the stern of the ship will swing in the same direction as if fitted with a conventional, solid cast right-handed propeller.

■ **7.32   How are the guides arranged on opposed-piston engines fitted with crossheads?**

Each cylinder unit on an opposed-piston engine has three guides. The centre guide of the three accommodates the slipper thrust from the lower piston and lower piston or centre connecting-rod. The guides on each side of the centre accommodate the thrust from the upper piston side rod crossheads and connecting-rods. As the stroke of the upper piston is less than that of the lower piston the working length of the centre guide is noticeably longer than the side-rod guides. The guide plate of an opposed-piston engine is cast in one piece and bolted between the A-frame on each end of a cylinder unit. Each crosshead works within a slot which constrains any forward or aft movement of the crosshead. The crosshead is held in its slot by the guide bars which also take up slipper thrust.

■ **7.33   How are crosshead slippers and guides lubricated?**

In some engines the oil is fed to the guide by a line piped to the lubricating oil main. The line connects to a drilled hole (on the top of the guide plate) which is connected to a hole in the guide face. The oil flows on to the guide face. The slipper surface is white-metal lined, and upper and lower edges of the slipper are machined to a slight angle so that an 'oil wedge' is created between the slipper and guide when the engine is operating. In other engines the oil supply passages to the crosshead bearings are also connected through to the crosshead slippers.

The lubricating oil then flows from a hole in the slipper face. A lubricating oil groove is cut horizontally fore and aft from this hole. The edges of the groove are scraped off with a scraper and an 'oil wedge' is created above and below the groove according to the direction of the piston movement. The upper and lower edges of the slipper are also machined off to a slight angle to further lubricate the slipper.

■ **7.34 What is the amount of clearance between the guides and slippers on a modern single-acting crosshead engine?**

The maximum and minimum figure for the guide clearance is given in the instruction book supplied by the engine builder: the amount is very small when the engine is cold. When the jackets are warmed through by circulating warm water prior to starting the main engines, the guide clearances tend to increase, and are subsequently reduced again slightly when the engine is operating and the slipper comes up to its working temperature. On some modern engines no provision is made for adjusting the guide clearance. In engines where adjustments are possible it is good to have records of the change which occurs in the clearance under various conditions. In the absence of any ship records the guide clearance can be checked when the engine is cold, after preheating the jackets when preparing the engine for sea, and again at the end of a passage while the running heat is still in the engine. The three sets of figures will give a good indication for estimating minimum guide clearances.

■ **7.35 What is the purpose of the guides mentioned in earlier questions?**

When an engine is in operation and the connecting rod and piston rod are not in the same straight line (not on TDC or BDC) the angularity of the connecting rod to the line of the piston stroke causes a thrust to be set up. The thrust force created is transferred from the crosshead pins to the crosshead slippers, it is then accommodated by the guides.

The force acting on the guides may be quite large and causes or tends to cause the engine to rock or vibrate in a transverse direction.

■ **7.36 What is the direction of the guide thrust on the engine?**

The thrust from the guides acting on the engine frames is found from the triangle of forces drawn up from the force acting downwards from the piston rod along the line of the piston stroke and the reaction from the upper part of the connecting rod.

The force acting in the horizontal direction represents the guide and slipper load. This force completes the triangle and represents the magnitude of the equilibrant or resultant of the forces from the piston rod and the connecting rod reaction.

The free body or space diagram and the triangle of forces are shown in Fig. 7.11. In the triangle of forces a–b is the vector representing the amount and line of action of the force on the piston, a–c is the vector of the force in the connecting rod. b–c is the vector representing the value of the guide thrust. It should be noted that when the line of the connecting rod is identical with the line of stroke (when θ and ϕ are zero) the load on the guides is zero.

At these points in the piston stroke when the piston is on the top or the bottom dead centre the direction of the load on the side of the cylinder liner or the guides is reversed.

The thrust arising out of connecting rod angularity is exerted on the sides of the cylinder liner by the piston skirt in trunk piston engines.

When the piston is moving downwards under the action of the forces from the expansion of the combustion gases (the expansion or working stroke), the magnitude of the guide and slipper load will vary and act in one direction; when the piston is moving upwards during the compression stroke the guide load varies but acts in the opposite direction. This is illustrated in Fig. 7.11. Note that when the piston is on top and bottom dead centre θ and ϕ are zero and the slipper or piston side thrust is zero.

The magnitude of the guide load will be related to the gas pressure in the cylinder, the inertia of the moving parts and the angularity of the connecting rod to the line of the piston stroke. (Fig. 7.11)

■ **7.37 What parts take the place of the guides and crosshead slippers in trunk-piston engines?**

Whenever the centre line of the connecting rod is not coincident with the line of the piston stroke, a side thrust will be set up, irrespective of whether the engine is of the crosshead type or the trunk-piston type.

In trunk-piston engines the side thrust from the piston pin or gudgeon pin is transferred to the side of the piston skirt or trunk. The thrust from the side of the piston skirt is balanced by the reaction of the cylinder liner.

The piston skirt must be carefully designed and machined so that the effective area of the skirt coming into contact with the cylinder liner is adequate for the loads it has to meet. The design aim is such that when the piston comes up to its working temperature a circumferential arc of contact covering approximately 90° is made between each side of the piston skirt and the liner. The arc of 90° should extend over the length of the piston skirt.

■ **7.38 The bedplates, columns or A-frames and the cylinder beam on modern longstroke, slow-speed engines appear different from the older generations of engines. What are these differences and how did they come about?**

An examination of the details of the bedplates and A-frames on a modern engine show them to be generally lighter in construction than on older engines of the same power and speed. The cylinder beam may be made heavier in construction than on older engines.

These changes have come about due to a better understanding of the magnitude of the stresses placed on these parts and the stress variations that occur over a cycle of operations. This better understanding stems from finite element analysis studies made with the computer and the programs drawn up for these studies.

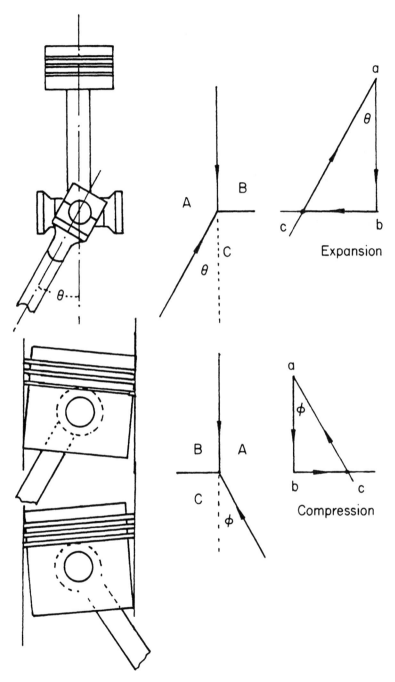

Fig. 7.11 Forces coming on to crosshead guides and the sides of a cylinder liner due to angularity of the connecting-rod.

With the use of a projected indicator diagram showing the cylinder pressures throughout the cycle, together with the masses of the moving parts to calculate the inertia forces, it is possible in a relatively short space of time to obtain the ranges of loads and their respective stresses. From the values found it is possible to make a more balanced design for these parts without the risk of fatigue failure. This in turn has made it possible to reduce scantlings and the weight of an engine.

It is usual to confirm the finite element studies by statically loading manufactured parts and finding the stresses in the critical areas with the aid of strain gauges. When manufacture of the prototype engine is completed, further studies will be carried out on an engine in service. Strain gauges will be fitted on the parts being studied and the testing will be carried out under actual service conditions.

A heavier and stiffer cylinder beam will spread the firing loads over a greater number of tie-bolts and allow them to be made smaller in diameter.

8

CYLINDER LINERS, CYLINDER HEADS, VALVES

■ **8.1 Why is it usual to cast the cylinder liners and cooling water jackets of large diesel engines separately?**

There are four main reasons for casting these large, heavy parts separately.

1 When the cylinder liner and jacket come up to their working temperatures the liner is at a much higher temperature than the jacket. The thermal expansion of the liner is therefore greater than the jacket. When the two parts are made separately, the increase in length and diameter of the liner is unrestricted. If the two parts were cast as one piece the stresses, set up due to restricting expansion of the liner, could cause fracture of the parts and consequent failure.

2 In normal service the cylinder liner wears and must be renewed, whereas the cooling water jacket normally lasts for the life of the engine. When the parts are made separately it is necessary only to renew the liner.

3 By casting the parts separately there is less risk of defects, and if there are any they are more easily found. The residual stresses left in a casting after solidification are much less when the liner and jacket are made separately.

4 The material required for the cylinder liner casting, due to the demands made on it, is more costly than that required for the cooling water jacket; by making the parts separately material costs are reduced.

Note The cylinders of small high-speed automotive-type engines are often cast in one piece with the cylinder block. In such engines the cylinder walls are comparatively thin and operate at a much lower temperature than thick marine engine liners; consequently the difference in expansion is small. Furthermore, the jacket walls of small engines are also thin and sufficiently resilient to take care of slight differences in expansion between cylinder wall and jacket.

In other cases it is desirable to fit renewable cylinder liners without increasing the costs by making watertight landings at the top of the liner and fitting elastomer sealing rings at the bottom.

In such cases, the casting of the cylinder barrel and jacket is in one piece. The cylinder barrel is machined out to suit the outside diameter of the renewable cylinder liner. The liner is then reduced in diameter by 'freezing' it before fitting it in the barrel. When the liner comes up to ambient temperature it becomes an interference fit or a shrink fit and has an acceptable heat transfer rate to the cooling water through the thickness of the cylinder liner and barrel.

This gives rise to the descriptive terms Integral Cast Cylinder; Dry Cylinder Liner (as just mentioned); and Wet Cylinder Liner, as is the normal practice in larger engines where the cooling water is in direct contact with the liner when circulating through the jacket.

■ **8.2   What forces are cylinder liners called upon to resist?**

The internal gas pressure within the cylinder liner causes a tensile stress, referred to as a hoop tensile stress. The cooling water pressure on the outside of the liner causes a hoop compressive stress which, however, is so small that it has no appreciable effect.

When the engine is in operation the mean temperature on the inside surface (gas side) of the liner is much higher than on the outside (cooling water side). This produces a temperature gradient across the section of the liner. As the temperature across the thickness of the liner is different, being hotter on the gas side and cooler on the water side, the hotter inside part of the liner tries to expand more than the outer part in contact with the cooling water. This temperature difference and expansion restriction causes a thermal stress. The thermal stress reduces the hoop stress (from gas pressure) on the inside of the liner (gas side) and increases it on the outside (cooling water side).

At the upper part of the cylinder liner in the region of the flange, bending moments are set up when the cylinder cover is hardened down.

As the gas pressure in the cylinder is changing throughout a cycle it follows that the hoop tensile stress is also changing, so the material of the cylinder liner is subjected to cyclic stress patterns. Care has therefore to be taken in the design of the liner to combat failure due to fatigue.

■ **8.3   Cylinder liners are generally thinner at the bottom than at the top. Why is this?**

The hoop stress on the liner is dependent on the gas pressure within the cylinder; as the piston progresses downwards the gas pressure decreases. Any section of liner has therefore to be strong enough to resist gas pressure above the top piston ring as the piston descends on the power stroke. The liner may therefore be made progressively thinner in sectional thickness to suit the lower gas pressures towards its lower part.

By making the liner progressively thinner, the rate of heat transfer from the thicker section is increased as the heat flows downwards to the thinner section which is cooler, both by reason of the gas temperature and because the thinner section transfers heat more easily to the cooling water.

■ 8.4 **Draw a simple sketch showing the lines of constant temperature through a cylinder liner from a four-stroke cycle engine.**

The lines of constant temperature are also referred to as isotherms.

The lines of highest temperature are experienced in the corner of the liner formed by its inside diameter and the landing of the cylinder head or cover. This line runs at an angle; as lower temperatures are found, the isotherms move outwards and become nearer to the vertical.

As heat flows from regions of higher temperatures to regions of lower temperatures it is seen that a quantity of heat flows downwards from the thicker section of the liner to the thinner sections lower down. (Fig. 8.1)

■ 8.5 **How is it possible for bending moments to be set up in the upper part of a cylinder?**

If a sectional view of the upper part of a cylinder liner, jacket and cover is examined it will be seen that the liner is supported by the lower face of the flange

L9OGB Cylinder liner isotherm

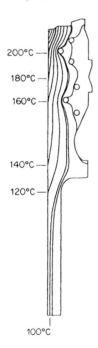

Computer plot based on measurements on
6L9OGB at MCR load

Fig. 8.1 Isotherms in a cylinder liner. (MAN B&W)

over a small width at the outermost radius. The joint face between the cylinder cover and the top surface of the liner is usually at or near the bore of the liner.

When the cylinder cover is hardened down a downward force is exerted on the joint face and an upward reaction is given on the outer landing face of the flange. As the force and reaction are not in the same line, a bending moment is created on the liner upper flange and the cylindrical part of the liner adjacent to the flange.

■ **8.6   What are the considerations that limit the diameter of diesel engine cylinders?**

The factors that govern the maximum possible diameter of a diesel engine cylinder are as follows.

1 The maximum working pressure within the cylinder.
2 The strength of the cylinder liner material. The maximum pressure and the strength of the material govern the thickness of the cylinder liner which must be sufficient to keep the hoop stress within acceptable limits.
3 The thickness of the cylinder liner governs the rate of heat flow from the hot side of the liner to the cooling water. The heat flow must be sufficient to keep the working side of the liner cool enough to prevent lubricant breakdown and hold the thermal stresses within safe limits.

■ **8.7   What factors limit the temperatures of the cylinder liner on the gas side?**

The working part of the liner, in which contact is made with the piston rings, must be kept at a temperature sufficiently low so that breakdown of the oil film does not occur. In the area of the liner above the piston rings the temperature must also be kept low enough to prevent high thermal stresses.

■ **8.8   How are cylinder liners held in place in four-stroke cycle engines, two-stroke cycle engines, opposed-piston engines?**

Four-stroke cycle engines. The upper end of the liner is formed in the shape of a flange which is cast integral with the liner. The underside of this flange rests on and is supported by the cylinder beam or jacket. The cylinder cover holds the liner in place (at its upper end) when the cover studs are hardened down.

At the lower end of the liner one section is made thicker to form a belt, this thicker section being located in the lower part of the cylinder beam or jacket which holds it in place. This is where the lower sealing rings are located.

Two-stroke cycle engines. The upper part of the cylinder liner is held in place in the same way as in a four-stroke cycle engine. The ported section of the two-stroke liner fits into the lower part of the cylinder beam which also forms the jacket in many two-stroke loop- and cross-scavenged engines.

In many uniflow-scavenged engines the thickened section of the liner is formed about halfway down its length and fits into the lower part of the cylinder beam which also forms the jacket. This thickened section houses the scavenge ports through which the scavenge air passes.

Opposed-piston engines. The combustion space is formed when the two pistons come to the inner dead centre. This section of the cylinder liner must therefore be stronger than the ends.

Reinforcing rings and a cast steel jacket are shrunk on to the centre part of the liner. The thickened centre part is machined with circumferential grooves and longitudinal slots to allow passage of the coolant from the lower to the upper cooling spaces. The lower section of the cast steel jacket has a flange which rests on the top of the scavenge trunk and is held in place with studs and nuts.

The upper section of the cylinder liner is machined circular in the way of the exhaust ports. A water cooled exhaust box fits over this section of the cylinder liner. The exhaust box is fitted with four corner lugs which are bolted to the top part of the engine framing.

■ **8.9 Why must cylinder covers be carefully hardened down? What is likely to occur if the cover is excessively hardened?**

Cylinder covers are held in place and hardened down by tightening the nuts on the cylinder cover studs. The nuts can be tightened by various means according to the size of the engine. They can be tightened by torque spanner, spanner and hammer, spanner and toggle bar of recommended length, or by hydraulic jacking.

The engine builder gives details of the allowable stresses which can be put on the cylinder cover studs, or specifies the forces which can be applied to the tools or hydraulic jacks. The instructions must be strictly adhered to when hardening down cylinder covers. If the cylinder cover studs are overstressed during tightening, excessive stresses will be put on the cylinder cover and dangerous stresses may be set up in the upper part of the cylinder liner. These high stresses, when combined with the thermal stresses, are a common cause of failure of cylinder liners, and may also lead to failure of the cylinder cover. Sometimes the deformation produced by the high stresses causes gas leakage from the cylinder cover and liner joint (see Question 7.28).

Modern engines of both slow- and medium-speed types use hydraulic tightening devices to harden the cylinder cover or cylinder head on to the gas-tight joint on the liner top. Errors in the pressure gauge used on the hydraulic pump may lead to over-tightening or under-tightening of the cover and the cover studs. It is therefore essential to regularly test the pressure gauge and protect it from mechanical damage when moving the pump about the engine room.

With older engines, torque wrenches may be used to harden cylinder cover studs; torque wrenches or controlled torque spanners should also be regularly tested to prevent possible engine damage when tightening a cylinder cover with a faulty wrench.

In one case where an investigation was being carried out by the author into problems with a highly rated engine, it was found that the torque wrenches gave excessively high torque values when being used and this led to some of the problems being experienced.

■ **8.10   What is a flame or fire ring?**

Flame or fire rings are fitted in the upper part of a cylinder liner which forms part of the combustion chamber space. The purpose of these rings is to protect the cylinder liner from the high temperatures which occur when injection and combustion of fuel take place. The upper part of the liner is recessed and the flame or fire ring fits loosely in the recess. The bottom of the recess is kept above the uppermost position of the top piston ring. Flame rings are made from alloy steel which will withstand the high temperatures without burning or scaling.

■ **8.11   How has it been made possible progressively to increase the diameter of diesel engine cylinders taking into account the factors mentioned in earlier questions?**

Research using modern instrumentation has made it possible to measure temperatures within the critical areas of the upper part of a single-acting engine cylinder liner.

The knowledge gained from this research has made design changes possible to increase both the cylinder diameter and the maximum firing pressure.

This has been done by increasing considerably the thickness of the upper part of the cylinder liner in order to decrease the hoop stresses due to internal pressure. The thickened section of the cylinder liner is drilled with a series of closely spaced holes around its circumference. The holes go from the bottom of the thickened section through to the top of this section and are drilled on a diagonal line a little away from the vertical. The holes pass relatively close to the working surface of the liner. It is then kept at a safe temperature to maintain effective lubrication of the upper working surface.

If the hot surface side of the liner is kept relatively cool, the thermal stresses in this critical area are reduced.

Due to the reduction in area for the cooling water flow through the drilled holes, the velocity of the water increases. The increase in water velocity gives much better rates of heat exchange between the surfaces being cooled and the cooling water.

This form of cooling has been given the name 'Bore Cooling' and is illustrated in Fig. 8.2. The outlet holes (not shown) usually come out of the side of the liner and then connect with the cooling bore in the lower part of the cylinder cover or cylinder head.

■ **8.12   Why are the cooling water holes in bore cooled cylinder liners drilled on a slight diagonal instead of the vertical?**

If the holes were drilled on the vertical there would be a considerable difference between the strength of the liner in way of the drilled holes and the undrilled

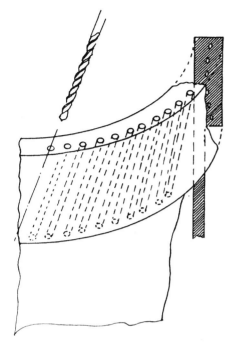

Fig. 8.2 Bore cooling holes in the top flange of a cylinder liner.

sections. The weakening effect would be considerable because of loss of strength in way of the drilled holes. The cross-section of the hole is a rectangle equal in area to the product of the height of the thickened section and the diameter of the hole.

When the holes are drilled off the vertical, the section through a drilled hole becomes an ellipse. The number of holes made in a section is dependent on the angle at which the holes are drilled. The loss of strength from material represented by three or four ellipses is less than that of the section through a vertical hole and the stress is more uniform around the circumference.

Figure 8.3 shows the geometrical development of angled and parallel bore-cooling holes. The horizontal projection of the angled bore-cooling holes shows the hole is tangential to the working surface of the cylinder liner and passes close to it. This is shown by the location of the elliptical holes on the right hand sketch.

■ **8.13 How is cooling water prevented from leaking from the cooling spaces around a cylinder liner?**

The flanged part of the top of a cylinder liner and the surface of the cylinder beam or jacket on which it sits is a metal-to-metal joint, and the mating surfaces are usually lapped with a tool or jig using grinding compound. In some designs varnishes or pastes are used on the metal surfaces. It is important that they be of

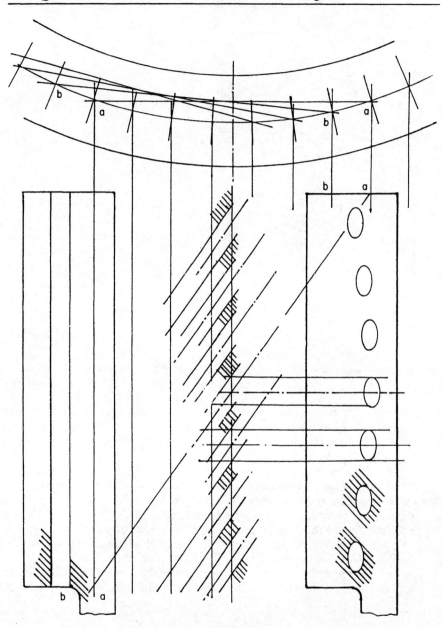

Fig. 8.3 Diagram showing how the bore holes pass very close to the cylinder liner inner surface. The loss of strength can be gauged by noting the section area loss from the vertically drilled holes when compared with the sectional area loss from the seven elliptical holes.

the slow-drying type, because if they harden before the liner is pulled home, severe stresses may be set up or leakage caused.

The lower part of a cylinder liner is thickened-up for some distance to form a belt, in which grooves are turned to accommodate nitrile rubber rings; the rings are compressed on their contact surfaces with the liner and the jacket so that a water-tight seal is formed.

■ **8.14** **What is the shape or section of the grooves used for fitting sealing rings at the lower end of a cylinder liner? What is the relationship of the cross-sectional area of the groove to the cross-sectional area of the ring?**

Different shapes of groove are used by different engine builders. The groove must, however, be such that the distance between the bottom of the groove and the jacket is less than the diameter of the nitrile rubber ring. When the liner is fitted in the jacket the rubber ring is then compressed and reduced in sectional diameter, so that a good seal results.

The cross-sectional area of the groove must be greater in amount than the cross-sectional area of the rubber ring. This is to allow distortion of the ring to take place and effect a good seal without the ring becoming volume-bound in the groove (see Fig. 8.4).

Note The bulk modulus of solid rubber is very high so it is virtually impossible to compress it and reduce its volume.

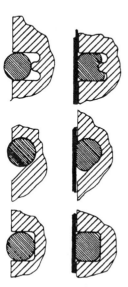

Fig. 8.4 Various sections through O-ring grooves. **Note.** In all cases the cross-sectional area of the groove is greater than the cross-sectional area of the O-ring to prevent the ring becoming volume bound.

■ **8.15   What is likely to occur if an oversize nitrile rubber ring is used when fitting a cylinder liner?**

It will be very difficult to pull the cylinder liner home due to the rubber completely filling the ring groove. If excess pull is placed on the liner drawing gear the load exerted on the rubber may fracture the jacket. In some instances when oversize rings have been fitted, fracture has occurred after the engine has been started and come up to working temperature.

■ **8.16   How are the jacket sealing rings protected from heat in way of the exhaust ports in cross- and loop-scavenged engine cylinder liners?**

Narrow grooves are machined circumferentially round the thickened section of the liner, one just above and one just below the exhaust ports. A copper ring is caulked into each groove and machined down so that it makes light contact with the jacket, or has a very small clearance between it and the jacket when cold. Exhaust gases are then prevented from making contact with the rubber sealing rings by the barrier created by the copper ring.

■ **8.17   If the copper sealing rings are found to have too large a clearance what can be done to rectify the fault?**

In minor cases the copper rings can be expanded with specially shaped caulking tools which deform the copper and swell it out, thus increasing the external diameter of the ring. If the clearance is too large to respond to caulking the copper ring must be removed and a new ring fitted. The old ring can be cut from its groove with a cross-cut chisel or machined out. A new copper ring is made by caulking a copper strip of the correct size into the groove. Sometimes the strip is machined to make a good surface.

If the belt in the jacket which fits against the new copper strip is corroded the liner can be put into the jacket and lowered down so that contact is made with the copper sealing ring, causing it to be marked. The marked areas are then hand dressed with a file until a good fit is obtained. The liner must be lowered in its correct position circumferentially so that the oil quills and ports are correctly located.

■ **8.18   What effects will leaking nitrile rubber seal rings have on engine operations?**

If leakage from rubber rings occurs into spaces external to the engine and the leakage is small it will have only a nuisance value from the mess it makes.

In some two-stroke engines if leakage occurs the water may be carried into the cylinder by the scavenge air and settle on the cylinder walls. The water has the effect of breaking down the lubricating oil film, and over-heated pistons and liners may result from the leakage.

In trunk-piston engines leakage from the rubber seal rings may go into the crankcase and cause deterioration of the lubricating oil. When the engine is stopped lubricant remains in bearing oil pockets and oil grooves. Water then separates from the retained lubricant and causes etching or pitting on the highly polished working surfaces of the pins and journals, and corrosion of the bearing metal leads to rapid bearing wear.

Cooling water leakage in trunk-piston engines may also be the start of degradation of lubricating oil through bacteria growth and build up (see Questions 3.15, 3.16, 3.17 and 3.18).

■ **8.19   How can leakage from cylinder liner rubber sealing rings be found?**

Often it can be found only by visual inspection when the engine is stopped. The cooling spaces should be kept under pressure during the examination. The design of the engine dictates the places that must be examined. The usual places are the crankcase and bottom of the cylinder liner in trunk-piston engines, and the scavenge spaces in two-stroke engines. In cross-scavenged engines the exhaust and scavenge ports can be visually inspected from the exhaust manifold or trunk. Some engines are fitted with leakage tell-tale holes so that leakage when it occurs in inaccessible places is indicated by leakage at the tell-tale holes.

■ **8.20   What precautions must be taken to prevent damage when lifting pistons?**

After removal of the cylinder cover and freeing the piston rod from the crosshead, the space above the top ring must be cleaned of all deposits to bare metal. This prevents the piston rings jamming the piston within the liner during the lift. The cylinder liner must also be fastened down with finger-plates fitted over the cylinder cover studs. The plates are held down with the cover nuts and pipes to make up the thickness of the cylinder cover.

In opposed-piston engines the centre part of the liner must be cleaned to bare metal before attempting to remove the lower piston. If the fuel valve tips or any of the cylinder valves extend into the combustion space these valves must also be removed.

■ **8.21   How is the cylinder oil fed into the cylinders of large diesel engines?**

The cylinder oil is forced in against the gas pressure in the cylinder by a small pump. The oil passes through a fitting which is called the oil quill or cylinder quill. The quill is usually screwed into a hole tapped into the cylinder liner. Some oil quills are fitted with a non-return or check valve within the quill, others have the non-return valve fitted external to the quill. (Fig. 8.5)

■ **8.22   What is the difference between wet and dry oil quills?**

Dry quills are fitted into cylinder liners external to the cooling water space, or they may pass through the cooling space but are jacketed so that cooling water

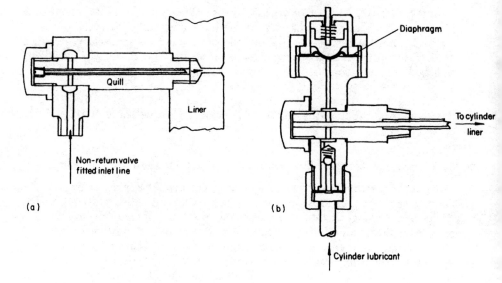

Fig. 8.5 (a) Cylinder lubricating oil quill.
(b) Cylinder lubricating oil quill with a diaphragm-type accumulator.

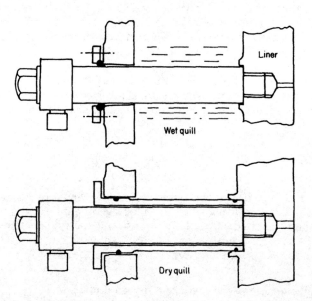

Fig. 8.6 Wet and dry oil quills used to carry cylinder lubricant on to the working surfaces of the cylinder liner.

does not touch the quill. Wet oil quills pass through the cooling water space and cooling water contacts the oil quill. It is not usual to find wet oil quills on modern engines due to the difficulties of making an oil and watertight joint at the junction between the quill and the liner. Non-return valves are fitted at the inlet to the quill. Many engines are, however, still operating with wet quills. (Fig. 8.6)

■ **8.23 What chain of events is likely to occur in engines fitted with wet quills if the joint face between the quill and cylinder liner becomes defective?**

If the joint face between a wet quill and the cylinder liner becomes defective, the results may be disastrous and the damage costly to rectify. When an engine is in operation the pressure exerted on the cylinder oil to cause it to enter the cylinder liner is greater than the cooling water pressure. If the joint face between the quill and the liner becomes defective cylinder oil leaks into the cooling water. The small oil globules which leak into the cooling water are carried around the cooling system and settle on or adhere to the various surfaces within the cooling spaces, coolers and piping.

When oil enters the cooling system some of the globules settle within cylinder covers and on cylinder liners. The minute oil film created within the cylinder cover and on the liner surfaces causes the heat flow from the gas space to be seriously affected so that the temperature of parts is greatly increased. These higher temperatures, which may be localized in way of the oil film, cause the cylinder cover and liner to distort so that difficulties are experienced in maintaining the cover joint gas-tight and the cylinder liner joint watertight.

If leakage persists the whole of the surfaces within the cover and liner is affected. This causes the cylinder liner to run at much higher temperatures on the gas side and difficulties are experienced with cylinder lubrication. In turn this causes extra heat to be passed to the pistons. The general effect is to cause deterioration of the nitrile rubber sealing rings within the piston and at the bottom of the cylinder liners.

In order to get the engine in proper working order again it is necessary to draw all liners and thoroughly clean them of the oil films and also to clean chemically the inside surfaces of cylinder covers. If overheating has caused warping so that difficulties are experienced in maintaining the cylinder covers gas-tight at the liner joint, the joint faces on the cylinder cover must be machined true again.

When water enters the quill during the time that the engine is stopped the water can affect the additives in the cylinder oil and cause them to solidify within the quill so that eventually blockage takes place in the quill. This in turn causes difficulties to be experienced with the pipe lines and indicator glasses associated with the cylinder lubricators.

The extra heat passing into the pistons usually causes failure of the sealing rings and consequent coolant leakage.

In loop- and cross-scavenged engines deterioration of the rubber rings at the lower end of the cylinder liner first occurs above the exhaust ports and causes water leakage at this point.

■ **8.24   Why do cylinder liners wear internally mostly at the combustion ends?**

The principal reason is that the combustion ends of liners are exposed to the high-temperature burning gases and lubrication of this part is not so efficient. Also it is at this part of the liner that the piston rings exert the greatest outward pressure. A further contributory factor is that the combustion ends of liners operate at a much higher temperature than the remainder of the liner and at the higher temperature cast-iron has less resistance to wear and the corrosive effects of very hot combustion products.

Note   The causes of cylinder liner wear can best be appreciated by imagining what takes place during the pressure period in the cylinder. Assume the piston is about to begin the compression stroke. The piston rings are then in the smallest diameter of the cylinder, which is well lubricated, and the radial pressure exerted against the liner is equal only to the spring in the rings. As the piston moves inward the cylinder pressure rises, and since the rings are resting on the bottom face of the grooves, gas will leak between the top of the piston ring and the upper face, and the pressure behind the rings increases and presses the rings against the liner with increasing force. The friction between the rings and the liner increases owing to this, and also owing to the fact that the lubricating oil film on the upper part of the cylinder, which has been exposed to high-temperature gas, has had its lubricating properties impaired.

■ **8.25   Name the tools required for measuring cylinder liner wear. State how they are used and the precautions taken to obtain accurate measurement.**

The tools required are a suitably sized internal micrometer and a micrometer locating gauge. The micrometer locating gauge is nothing more than a long flat bar covering the length of the liner. A series of holes is drilled in the bar. The holes are located at the points where the diameter of the liner is measured. As the maximum wear is usually found at the upper part of the liner, the holes are so arranged that the bore of the liner is first measured at the point where the lower edge of the top piston ring comes when the piston is on top centre. Two other measuring-point holes are located at approximately 50 to 75 mm and 100 to 150 mm from the first for large engines.

In smaller engines the distances between the upper gauging points will be less and are usually arranged so that the three upper measurements cover the distance between the positions of the top and bottom piston rings when the piston is on top dead centre.

There are two other holes in the micrometer locating gauge, one to locate the micrometer for gauging the liner at midlength and another to gauge the liner at the lower end of the piston travel.

Note   In opposed-piston engines the micrometer locating gauge is arranged to cover the travel of both the upper and lower pistons.

After the cylinder cover or upper piston has been removed and the liner has been cleaned, the locating gauge is placed on the port or starboard side of the liner. A series of horizontal chalk marks is made in the liner corresponding to the position of the micrometer locating holes. The locating gauge is then placed on the opposite side of the liner, the extension of the internal micrometer is placed in the top locating hole and the micrometer head is swung in a horizontal arc on the level of the chalk mark until the right 'feel' is obtained on the micrometer head. The micrometer is then read and the measurement recorded.

The process is repeated down the liner and the various diameters of the cylinder are obtained in the port and starboard or athwartships plane. Similar measurements are made in the fore and aft plane of the cylinder liner bore.

In taking measurements down to hundredths of a millimetre or thousandths of an inch a high degree of accuracy is required. In the first place this requires very delicate handling of the micrometer and adjustment of the head when it is being screwed out to measure the worn bore diameter. Other essential precautions are the setting up of the micrometer extension pieces and being aware of the relative temperatures of liner and micrometer.

The micrometer must be adjusted at the time the extensions are fitted to give a correct reading preferably in a standard cylinder bore gauge. If a gauge is not available a spare unused cylinder liner can be used, but it should be noted that some variation occurs in cylinder bore sizes of different liners due to the maker's allowable machining tolerances, and this may introduce discrepancies in measurements that are barely acceptable.

The temperature of the micrometer should be stabilized at engine-room temperature. For example, if the micrometer is normally stored in an air-conditioned office, it should not be brought from the office and immediately used in a hot engine room. Similarly during use it should not be placed in warm or hot places. During extended use the heat of the hand will also affect the accuracy of the measurements. Some micrometers have a plastic insulating tube which is slipped over the extension pieces for handling purposes. If this is not available the extension piece should be rag wrapped and taped where it is held.

Cylinder liners should always be at similar temperatures when measurements are taken, and this may be difficult to achieve due to the trading pattern of ships. As much as possible should be done to achieve this, such as circulating adjacent cylinders with cooling water at some temperature which can be easily obtained over widely differing ports or places.

If these precautions are conscientiously observed it will save the waste of time that comes about when a cylinder liner is measured and comparison with earlier measurements shows that it is apparently getting smaller in bore!

■ **8.26 After measurement of cylinder liners, how are the results put into a useful form?**

The actual bore diameters at the various locations are recorded on a form supplied by the Company's technical department or an engine builder. The differences in the latest measurements and those last recorded are noted in order to ascertain that maximum wear is still occurring in the same location. The difference in the latest maximum diameter and the original size of the liner is also noted.

These two figures give the maximum increase in diameter since the last recording and the total maximum increase in diameter since new.

In order to put these figures into useful form for comparison purposes, they are related to time, in the form of diameter increase per thousand running hours. This gives a time wear rate.

The following simple equation shows how the wear rate is obtained:

$$\text{wear rate since last recorded measurement} = \frac{\text{increase in diameter since last record}}{\text{running hours since last record}} \times 1000$$

$$\text{wear rate since new} = \frac{\text{total increase in diameter}}{\text{total running hours since new}} \times 1000$$

The units will be: wear mm/1000 hr or wear inches/1000 hr depending on the units of measurement used.

The values obtained are plotted as co-ordinates on a graph with the time in thousands of hours as the abscissa or base and the wear rate as the ordinate or vertical.

Note If measurements were taken during the early life of a cylinder liner it would be seen that the wear rates in the first few hours of use were relatively high. The wear rate would then fall away and remain fairly constant over the life of the liner until the wear amount became large. With large amounts of wear it becomes increasingly difficult to maintain correct lubrication and the wear rate increases.

The length of the bedding period will depend on the hardness of the liner material and the lubricant used during this time. It is followed by a uniform wear rate.

■ **8.27   Why is cylinder liner wear related to the maximum increase in diameter of a cylinder bore rather than to the amount actually worn off the surface of a cylinder liner?**

If the wear that takes place on a cylinder liner was always concentric to the centre of the cylinder bore, it would be an easy matter to divide the increase in the diameter by two. This would then give the amount worn off the cylinder liner surface. In actual practice the wear never takes place concentrically; further, the wear in the port and starboard plane is rarely the same as in the fore and aft plane.

These differences come about from various causes, such as heel and trim of the ship in service and the effective guide clearances. In trunk-piston engines the wear pattern of the cylinder liner can be completely different from that of a crosshead type engine.

It can easily be seen that for tankers and bulk carriers, where long ballast passages are made with the vessel trimmed by the stern, the maximum wear in a cylinder will be in the fore and aft plane and most of this wear will come on the after side of the cylinder liner. In any one engine, a study of liner wear is further complicated by the fact that no two cylinder liners have the same wear rates, even when all the factors influencing wear have been equalized as far as humanly possible. For these reasons cylinder liner wear is usually related to the increase in diameter.

■ **8.28** If a change is noted in the wear pattern or rate of the central and lower cylinder gauging sizes, what is it likely to indicate?

Any increase in the wear rate or pattern at the centre or lower end of a cylinder liner is usually indicative of the beginning of some piston alignment change, or the start of misalignment. It could also indicate slackness in the crosshead guides or the running gear of that cylinder unit.

■ **8.29** In service cylinder liners wear internally and the practice is to renew them periodically. What form does the wear take? What are the objections to reboring the liners when they become worn to the permissible limit and (a) bushing the liner to return it to its original internal diameter, or (b) boring to a larger size and fitting an oversize piston?

The maximum wear generally occurs at the upper end of cylinder liners, so that after a time the bore assumes a decidedly tapered form, the enlargement at the lower end being from one-tenth to one-fifth of the enlargement at the top. Bushing the liner is a costly operation and not altogether satisfactory from an operational point of view, because the bush must of necessity be relatively thin, and there is a tendency for it to distort due to the intense heat to which it is exposed in service. Merely to rebore the liner and fit an oversize piston is costly also, and even if the liner is strong enough to resist the internal gas pressure the results will not be satisfactory, because rebored liners wear at a greater rate than new liners owing to metal so far below the original surface of a casting being comparatively soft and therefore less wear-resistant.

■ **8.30** How is it possible to make a worn cylinder liner fit for further use?

Worn cylinder liners are sometimes brought back to original bore sizes by chrome plating processes. The worn liner is cleaned and examined for defects such as fractures which would render it unfit for further use. If the casting is sound, the internal bore of the liner is first plated with a 'sandwich' layer of a ferrous material that has good bonding properties with the parent material and with electro-plated chromium. Chromium plating follows, and is so controlled that more chromium is deposited on the more worn parts of the liner, thus converting the oval and tapered shape of a worn liner to the round and parallel shape of a new liner. Finally, the surface of the chromium is treated to give a matt finish that will retain an oil film on the working surface.

Note The economics of this operation would have to be carefully studied before proceeding with the process of chromium plating.

The cost will depend on the cost of the chromium plus the labour and overheads. This can then be compared with the cost of a new cylinder liner. Generally the number of running hours between overhauls with chromium-plated liners and plain cast-iron piston rings is greater than with a chromium-plated top piston ring together with plain cast-iron rings operating in a plain cast-iron liner. Provided that other problems do not arise necessitating removal of the pistons, the number of times pistons are lifted during the life of the ship must be brought into the economic study. Reducing the number of times pistons are lifted reduces the number of times an engine is put at risk following overhaul of cylinder units.

■ **8.31   What is the maximum amount of wear allowed on a cylinder liner? Is the liner renewed when this maximum wear is reached, or earlier?**

The maximum amount of wear normally allowed on a cylinder is one per cent of the diameter. For example the maximum amount of wear on a 600 mm bore liner would be 6 mm. Some of the ships' classification societies and engine builders recommend figures lower than this.

Cylinder liner renewal is normally a matter of Company policy based on the economics of ship operation. It is usually related to the expected life of the hull, cylinder liner life periods being made a whole factor of the expected hull life, or the period that a ship will be kept in service before sale.

For example, if the expected life of a ship is twenty years and the life of cylinder liners is eight years the liners might be renewed at eight years and sixteen years.

When the ship is subsequently disposed of at twenty years, the cylinder liners will have four more years of life remaining. Or, they could be renewed at just under seven years which gives the same number of liner renewals but better and less costly operation of the engine over the ship life.

■ **8.32   It is common practice to plot in a graphical form the wear of a cylinder liner against the number of hours it has been in operation. When this is done it is often noticed that some scatter exists between the plotted points after a few wear figures have been recorded. What is the reason for the scatter and how can the wear rate be shown in a more acceptable form? How would you forecast the length of life for a cylinder liner?**

When wear rate figures are plotted in graphical form it is common practice to record the wear value in millimetres as an ordinate and the running hours as an abscissa. In order to make the running hours a convenient figure to work with it is usual to make the unit of running time as one thousand hours. For deep-sea vessels this is the number of running hours spent at sea on a 'full away' to 'end of passage' basis divided by one thousand. For coastal vessels and ferries a different basis may be adopted to suit the operating conditions.

Steps to improve the accuracy of measurements are covered in the note at the end of Question 8.25, but it must be remembered that a high degree of sensitivity is required when taking cylinder liner measurements, and no two engineer officers have the same level of sensitivity in their fingers. Another cause of variation or scatter in the recorded results is due to the wide variations in the conditions under which the measurements are taken. This is most particularly so when working under conditions as far removed from metrological laboratory conditions as it is possible to get.

Although we may assume in prediction work that cylinder liners wear at a uniform rate, this is far from the actual truth as in any one engine the maximum wear in any two cylinders is never identical in amount or location.

Differences in fuel quality may also account for wide variations in the wear rate over any one period of time.

Some scatter must then be expected when recording the wear amount measurements against a time base in a graphical form. The method for finding the average value of the plotted points is referred to in mathematics as 'curve fitting'. This can be a quite complex process when trying to find the equation governing the values of experimental data plotted in graphical form. In order to simplify the finding of the equation governing the rate of cylinder liner wear, we can assume that the wear rate is uniform for all previously recorded measurements.

The plotted values of the wear against the corresponding running hours can then be assumed to form a straight line. From this we then know the equation or the graph will be of the form:

$$y = mx + c$$

or   wear $= m$.running hours$/1000 + c$

where m the coefficient of 'x' gives the rate of liner wear. As the initial high wear rate of liner during the bedding-in period occurs over a very short time-span the value of the constant 'c' can be made equal to zero. The plot of the values will then pass through the origin of the axes.

An example of the plotted values of wear against time is shown in Fig. 8.7; the method for obtaining the wear rate from the data given in the figure and in the table below is as follows.

Hours	4800	9500	14750	15900	19370
Wear (mm)	0.65	1.20	1.75	2.10	2.35

Let $x =$ total running hours$/1000$ and $y =$ wear measurements in mm. Record their values in a table as shown below with columns for the x values, the y values, the product of the x and y values and the value of x^2. Total each column.

x	y	x times y	x^2
4.80	0.65	3.12	23.040
9.50	1.20	11.40	90.250
14.75	1.75	25.8125	217.5625
15.90	2.10	33.390	252.810
19.37	2.15	41.6455	375.1969
64.32	8.05	119.2420	958.8594

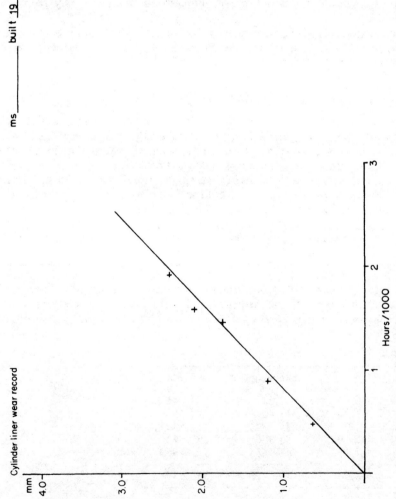

Fig. 8.7 Cylinder liner wear recorded in graphical form showing the values given in question 8.32. The straight line has been fitted to the data by linear regression, giving the most reliable means of forecasting liner life. The maximum allowable wear is sometimes plotted with a horizontal line at the maximum wear value.

$$m = \frac{\Sigma xy - \dfrac{\Sigma x \Sigma y}{n}}{\Sigma x^2 - \dfrac{(\Sigma x)^2}{n}}$$

where n is the number of pairs of liner gaugings plus 1. Since there are five pairs of readings, $n = 5 + 1 = 6$. To find m substitute the Σ values and the n value into the equation. This gives

$$m = \frac{119.2420 - \dfrac{64.32 \times 8.05}{6}}{958.8594 - \dfrac{64.32^2}{6}}$$

$$= \frac{119.2420 - 86.2960}{958.8594 - 689.5104} = 0.1223$$

For 25 000 running hours the total wear will be

$$25 \times 0.1223 = 3.06 \text{ mm}$$

The 'fair straight line' can be plotted by marking up an ordinate of 3.06 mm at 25 000 hours and drawing the line from this point through to the origin of the axes.

If the maximum wear is allowed to go up to 6 mm the life of the liner can be found by substituting the value of 6.0 in the equation

$$y = mx + c \qquad c = \text{zero}$$

giving

$$6 = m \times 0.1223$$

Therefore

$$m = 6.0/0.1223 = 49.0597$$

The expected life of the cylinder liner is nearly 50 000 hours. If the ship averages 6 000 running hours per year the expected life would extend over a period equal to 50/6.0 years which is approximately 8 years.

You would expect to renew the liners after about 8 years of operation on the basis of the figures recorded so far.

It is the practice of some companies to build up a reserve fund for cylinder liner renewal. This will depend on the policy of the company and the taxation laws of the country they are operating from.

The formula given to find the value of 'm' may form part of a programmed set of calculations found under the term 'Linear Regression' given with pocket calculators programmed for statistics. It is more convenient to use a calculator or a computer when many pairs of values must be used.

■ **8.33   When copper rings are fitted as cylinder cover joints or as joints for cylinder head valves, what attention do they require?**

Before new copper ring joints are fitted they should be annealed; similarly, if used joints are to be re-used they should be annealed immediately before they are fitted. The clearances between the edges of the ring joint are also important to ensure easy subsequent renewal.

For cylinder cover joint rings there should be adequate clearance on each edge of the joint ring. When the joint is hardened down after annealing it is squeezed, made thinner so that its outside diameter is increased and its inside diameter is reduced. Giving adequate clearance to each edge of the ring ensures that it never jams in its slot or fossette and is always easily removed.

When copper joint rings are used with cylinder head valves the clearance between the outside edge of the ring and the valve pocket is important. Giving good clearance to the outside diameter of the ring ensures that it never squeezes out to the point that it jams within the valve pocket.

If the clearance between the inside diameter of a joint ring and the spigot forming the end of the valve is kept small the joint ring squeezes on to the valve end and lifts out with the valve when it is removed. Exhaust valves must sometimes be re-hardened down after the engine is stopped.

Generally copper ring joint thickness should be kept minimal. The actual thickness will depend on the size of joint but it seldom exceeds one and a half millimetres.

■ **8.34   When air inlet valves, exhaust valves and air starting valves are being overhauled what particular care is required in the procedures to prevent difficulties in service?**

Air inlet valves

1   When a valve is dismantled careful examination must be made of all parts to note whether the visual appearance is normal. Anything abnormal should be followed up to ascertain the cause.
2   Careful attention *must* be given to the cleaning of the parts so that in the matter of cleanliness the condition is as new.
3   The lubrication points to valve spindles and valve-spindle guides must be thoroughly cleaned to allow unrestricted passage of lubricant.

Note   The lubrication channels to the valve spindle and guides may, in large valve bodies, be through a series of holes and an annular space, or in smaller valves a simple bell-mouth end at the top of the valve-spindle guide.

4   When grinding the valve and valve seat, any ridges, however small, formed during grinding must be removed either by machining, use of a small portable hand grinder, or hand tools. The tools used will depend on the size of the valve and the hardness of the valve materials.
5   Heavy parts must be handled carefully to avoid any risk of mechanical damage in the form of indents or burrs.

6 All locking devices must be made good and fast. The parts requiring lubrication must also be oiled with the correct grade of lubricant.

7 After overhaul, valves must be carefully stored to prevent damage from water or dirt. All open ports and gas passages must be plugged.

Exhaust valves. The precautions to be taken for exhaust valves are similar to those for air inlet valves. Additionally, special care is necessary to ensure that any gummy or hard deposits on the valve spindles are completely removed. The cooling spaces must be examined for cleanliness, water hardness deposits, and signs of erosion and/or corrosion.

If the spaces show water hardness deposits a check must be made on the cooling water condition, and the point of ingress of the contaminant must be discovered. This could be due to cooler leakage, poor evaporator and distiller performance, water condition in storage tanks, etc. The scale found must be removed by acid cleaning.

If corrosion and/or erosion is present the amount of additive in the cooling water must be checked; other points to check are de-aerating leak-offs in the upper sections of cooling-water lines, and ingress points if any parts of the cooling system are at subatmospheric pressure under operating conditions.

If erosion is present it is usually the result of some foreign object in the system causing cavitation and bubble impingement.

It may come from a badly cut pipe joint where the hole in the joint is too small, or if the hole is not correctly aligned in the flanges.

Air starting valves. The precautions for air starting valves are similar to those for air inlet valves. In addition, the bore of the air operating cylinder must be examined for ridges at the stroke ends. Any ridges found must be removed. If new piston rings are fitted the end clearance at the ring gap must be checked. Old piston rings must be radiused off on the edges of the working faces of the ring, and the ring compression checked if the valve has been leaking or become overheated. The compression of the closing spring must also be checked.

When air starting valves are fitted with wings on the valve end, the inside ends of the wings must be checked after valve grinding to ensure that the wings do not foul anywhere and prevent proper valve closure. After assembly the valve should be tested for tightness with compressed air.

Some air starting valves are fitted with bursting cartridges or discs which burst if a pressure build-up occurs in the starting air manifold. During overhaul of an air starting valve any safety devices fitted on the valve should be examined for corrosion which could lower the bursting pressure to that of the working air pressure in the system. This is important on a main engine: a burst cartridge or disc could make an engine inoperative at a critical period of manoeuvring or cause starting to be very sluggish with consequent early loss of starting air. The bursting disc or cartridge should also be examined to ensure that it is of the correct thickness and of a type that will burst at the correct pressure without damage to the starting air manifold and air lines.

■ **8.35** In one form of exhaust valve the burnt gases pass through the valve and leave by an opening external to the cylinder head, while in the other form the gases enter the valve and leave by a passage through the cylinder head. What are the advantages and disadvantages of each form?

In the form first described the necessary gas-tight joint between the valve and the cylinder head can be situated at or near the top of the cylinder head. This allows the part of the valve below the securing flange to hang, as it were, in the cylinder head; and, not being under stress due to excessive tightening of the securing nuts or expansion due to increase in temperature when the engine begins operating, the ground faces to the valve head and its seat are less likely to distort and do not require to be ground together so frequently. The disadvantage of this form of exhaust valve is that it is heavy for a valve which requires to be removed frequently for overhaul. The other form of valve is of smaller weight but has disadvantages due to the running hours between overhauls being somewhat less. The valve or cylinder head must consequently be removed for valve overhaul more frequently.

Note When the burnt gases enter the exhaust manifold direct, as in the first form of valve, there is no need for a large-diameter passage through the cylinder head, and the construction of this casting is simplified. It is also claimed that owing to the elimination of this passage, through which high-temperature gases pass while being circulated on the outside by relatively cold water, there is less likelihood of the cylinder head cracking owing to temperature stress. Foundry practice, however, has advanced so much during recent years that under normal operating conditions cylinder heads rarely, if ever, crack, irrespective of the form of exhaust valve used.

■ **8.36** Why is it usual to provide the form of exhaust valve which lands near the bottom of the pocket in the cylinder head with two exit ports, when there is always only one passage through the cylinder head?

The object is to preserve symmetry of the part of the casting exposed to the high-temperature gases leaving the cylinder, and avoid possible distortion of the seat, with consequent leakage between the valve seat and the head. If part of the valve between the landing and the securing studs was weaker than the remainder, distortion would result when the part becomes heated and expands.

■ **8.37** Since the exhaust temperature of a small oil engine is the same as that of a large engine, why should the exhaust valves of large engines only be water-cooled?

If the heads and seats of exhaust valves become overheated they soon begin to distort and burn. To prevent this, conduction of the heat away to the surrounding media is relied upon; consequently, with large valves there are greater thickness of metal, the rate of cooling by conduction is slower and the process must be accelerated.

Note The cooling of exhaust valve seats becomes very necessary in engines burning high-viscosity fuel. This is due to the combination of the sodium and vanadium content in the fuel. Under some circumstances at high temperatures the sodium and vanadium content in the exhaust gases is very corrosive. The cooling space for the seats is sometimes adjacent to the valve seat. (Fig. 8.8) Valve rotators also help in maintaining exhaust valves tight over longer periods of time.

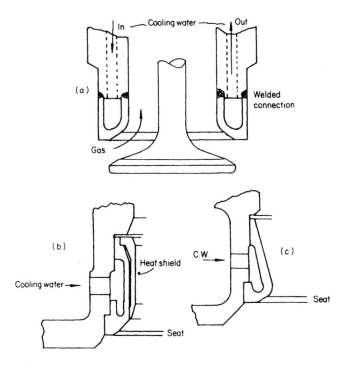

Fig. 8.8 (a) Slow-speed engine exhaust valve seat cooling.
(b) Highly rated trunk-piston engine exhaust valve seat with a heat shield.
(c) Highly rated trunk-piston engine exhaust valve seat without heat shield.

■ **8.38 What are the indications that an exhaust valve is leaking? How would you confirm that a valve was leaking?**

The first indication of exhaust valve leakage is a rise in the exhaust gas temperature measured at the exit from the exhaust valve. The temperature will be above that usually shown for that particular engine load. The leakage can be confirmed by taking an indicator card from the cylinder with the leaking exhaust valve. The card should be taken with the fuel on and off the unit. If the valve is leaking the compression pressure and maximum pressure will be low.

Air starvation due to very dirty scavenge ports will also give similar indications on indicator diagrams, but in the absence of any abnormal scavenge air or supercharge pressure increase, it will usually be found that the exhaust valve is at fault. In four-stroke naturally aspirated engines similar indications will be given on the indicator diagram if the air inlet silencers become dirty.

Indicator cards cannot normally be taken on medium- and high-speed engines. In these circumstances the maximum pressure indicator should be used and the maximum firing pressure and compression pressure checked and compared with normal values.

■ **8.39   If an exhaust valve burns on the ground and lapped sealing faces, what are the probable causes? Explain how the burning of the metal occurs.**

The most likely causes of exhaust valves burning are as follows.

1   Cylinder power in excess of design rating.
2   Poor combustion of fuel, which may be due to dirty fuel injection valves, incorrect injection pressure, incorrect fuel temperature, late timing, air starvation, badly tightened-down fuel valves, or impurities in fuel not brought out during fuel centrifuging or separation. Usually poor combustion leads to after-burning of the fuel.
3   Valve not closing completely, through incorrect tappet or roller clearance, or build-up of deposit on exhaust valve seat from fuel impurities or excess lubricating oil, or valve spindle becoming sluggish due to dirty oil and carbon build-up in valve spindle guide. The exhaust valve spring may then fail to close the valve sufficiently rapidly or the valve may remain slightly open.
4   In the case of water-cooled valves insufficient water flow or cooler malfunction may cause the valve to overheat, with consequent build-up from impurities in the fuel.
5   Incorrect hardening-down of exhaust valve in the cylinder head, which may lead to valve seat distortion.

Once a valve has started to leak, from any cause, the passage of hot gas over a small area of the seat causes distortion which then increases the amount of leakage. The temperature of the valve and seat in the area of the leak rises considerably, the strength of the metal is impaired, and its resistance to the erosive effect of the hot gases is seriously reduced. During the combustion period gases at high pressure in the cylinder expand across the location of the leakage at very high velocities. At this stage, the leakage across the valve seat (during the combustion period) increases very rapidly and large burned areas occur in the seat and valve.

■ **8.40   What do you understand by the term 'shrouding' in relation to cylinder head valves?**

Shrouding occurs if mitred valve seats are repeatedly ground and the consequent ridges are not removed. If the material of the valve is harder than the valve seat the shroud builds up in the seat. If the seat is harder than the valve, shrouding occurs in the valve. Taken to its extreme the shroud builds up in both valve and seat.

When a valve and seat are ground together with grinding or lapping paste, some material is removed and eventually the seating faces become lower than

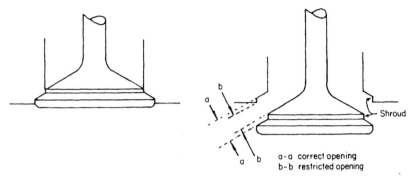

Fig. 8.9 Effect of excessive manual grinding leaving the exhaust valve seat and valve head shrouded. **Note** The effect of the shrouding is to reduce the effective area for gas flow even though the proper valve lift is maintained.

the surrounding material and a step is formed circumferentially. Although the actual distance that the valve opens remains constant the effective opening area across the valve and seat face is reduced by the amount of the shroud. Even small amounts of shrouding can seriously affect the air or gas flow across a valve, both by virtue of the reduction in effective opening area and the flow eddies created by the shroud causing a loss of streamlined flow. (Fig. 8.9)

■ **8.41** Where is shrouding most commonly found? How is shrouding removed?

Shrouding is often found in the smaller sizes of diesel engines where the air inlet and exhaust valves are non-removable and form part of the cylinder head. It may also be found on exhaust valve seat rings fitted in the cylinder head. Shrouding is removed from large valves in the manner described for removal of ridges in Question 8.3. In smaller valves a shroud formed on the seat body can be removed with a valve seat cutter that is larger than the valve. If special hard steel seat rings are fitted directly into the cylinder head the shroud must be carefully 'swept' away with a small hand-held portable grinder, or a new seat ring must be fitted.

When shrouds are formed in the head of a valve, it is usual practice to set up the valve in a lathe with the head on the back centre. The excess material forming the shroud can then be turned off, or if the valve is of hard material the shroud can be ground off with a grinding attachment on the tool rest.

Note When machining or grinding off the shroud on a valve it is better to use a fixed or solid back centre rather than one of the revolving type.

■ **8.42** State how the exhaust valves on modern slow-speed engines are actuated. How is the valve rotated and returned to its seat on closing?

The exhaust valves on these engines are opened by oil supplied from the camshaft lubrication system. The oil is supplied to a hydraulic pump operated

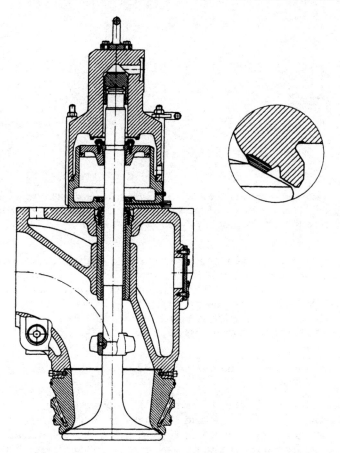

Fig. 8.10   Modern slow-speed two-stroke engine exhaust valve. The valve is opened by hydraulic pressure and closed by air pressure. (MAN B&W – MC exhaust valve)

by a cam fitted on the camshaft. The oil is delivered under pressure to a cylinder mounted on the top of the exhaust valve. Inside the cylinder is a piston which is actuated by the oil supplied under pressure from the exhaust valve hydraulic pump. The oil under pressure forces the piston down; the piston being in contact with the exhaust valve spindle forces it down and causes the exhaust valve to open.

Within the casing above the exhaust valve body is another cylinder arranged concentrically; this surrounds another piston fastened on the exhaust valve spindle. The space under the piston is supplied with compressed air at a pressure of the order of 6 bar (87 psi). The air under pressure causes the exhaust valve to close when the pressure in the hydraulic pump discharge drops on the downward or return stroke of the pump piston.

Within the outgoing exhaust gas space above the head of the valve a small fan is fastened to the valve spindle. The high-velocity exhaust gases passing through this space act on the fan; the fan then rotates the valve. (Figs 8.10 and 8.11)

■ 8.43   How are the seats and valves ground on modern slow-speed engines and what is the purpose?

Formerly valves were ground into their seats by putting grinding paste on to the seats, replacing the valve and then grinding both seat and valve together. The seat and the valve eventually had the same angle and a gas-tight surface was obtained.

Today the seats and valves are ground at slightly different angles by a high-speed grinding wheel adapted specifically for grinding the valve seat attached to the valve body and the seat landing face of the valve head.

The seat of the valve and the valve body can be considered as the curved surfaces of a frustum of a cone. The apex angle of the seat in the valve body is at a slightly greater angle than the seat on the valve head. When the valve is cold and in the closed position the outer diameter of the seating surfaces will be slightly open.

The combustion chamber side of the valve head at operating temperature is hotter than the opposite side. This causes the head to become slightly convex on

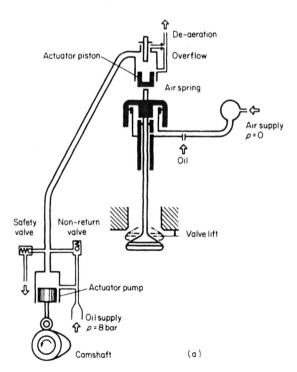

Fig. 8.11   (a) Hydraulically actuated exhaust valve system. Engine shutdown and air off system. The valve opens when air is shut off. (Sulzer)

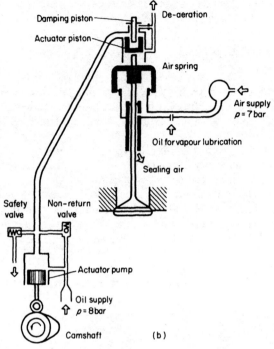

(b)

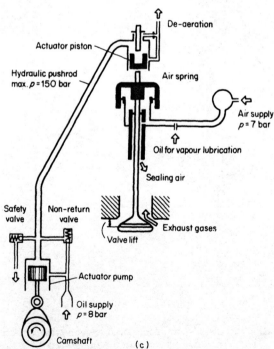

(c)

(b) Engine in operation, valve closed.
(c) Engine in operation, valve open.

the combustion chamber side and the open space between the seats closes; the effect of pressure in the combustion chamber space tends to open the space very slightly. The combined effect leaves a seat wide enough to give a good life period between valve overhauls. Valve rotation gives a more uniform temperature to the valve seats. The purpose of obtaining the combined effects of differing seat angles, valve rotation and cooling of seats, is to extend the period between valve overhauls. Up to as many as 10 000 hours of operation between overhauls has been obtained and longer times have been reported. The period between overhauls however, must be related to operating results and the related costs of valve overhaul. (Fig. 8.12)

■ **8.44 Older-type exhaust valve operation could be observed by visible moving parts. Exhaust valves operated by hydraulic and pneumatic pressure have no moving parts visible. How would you check the movement of the valve while the engine is in operation?**

Indicators fitted to the valve can be engaged during engine operation. These indicators when engaged contact the pneumatic piston used for closing the valve. The movement of the valve when opening and closing is shown by the identical upward and downward movement of the indicator which follows the movement of the pneumatic piston and hence the valve. The amount the

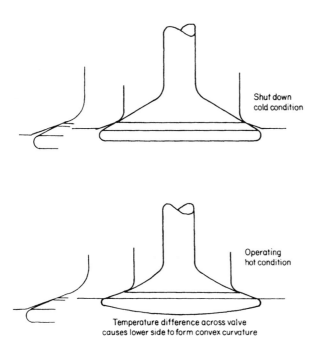

Shut down
cold condition

Operating
hot condition

Temperature difference across valve
causes lower side to form convex curvature

Fig. 8.12 Change in exhaust valve shape due to temperature variation across valve head. The effect has been exaggerated in the sketches.

valve is opening and closing can be measured. After use the indicators can be lifted off the piston and disengaged.

■ **8.45 Certain types of cylinder liner must be aligned circumferentially so that scavenge and exhaust ports are correctly positioned. How is this done? What checks and tests must be made at the time a new cylinder liner is fitted?**

Cylinder liners of loop- and cross-scavenged engines, and those fitted with quills that pass through the cylinder jacket casting, need careful alignment circumferentially to ensure that liner parts match up with jacket ports and to avoid difficulties in fitting oil quills. If the new liner is examined before fitting, a scribed line and centre-punch marks will be found on the side of the top flange. Sometimes lines are put in two places, diametrically opposite. Similarly, a scribed line and centre-punch marks will be found on the top of the jacket casting; this line is on the transverse centre line of the cylinder at the front of the engine. In some engines the transverse line will also be found on the top of the jacket on the back of the engine (see Note at end of answer).

The new liner, after preparation for fitting, is then lowered into the jacket until the bottom rubber sealing ring is just entering the bell mouth entry of the transfer port belt. At this point the liner is lifted slightly so that it can be revolved on the crane lifting hook.

A square is set up at the scribed mark on the jacket, and the mark on the flange of the liner is aligned with the square. The liner is then lowered again and pulled home with the drawing gear. While it is being pulled home frequent checks must be made on the marks to ensure that they match correctly with the square.

Prior to the fitting of a cylinder liner into a jacket, leakage tell-tale holes must be cleaned and proved clear with compressed air.

If any corrosion is found which could possibly allow leakage past the sealing rings, the corroded area must be completely cleaned out and the wasted space built up with a suitable plastic metal.

When the cooling spaces, transfer ports, rubber ring sealing surfaces, etc. are satisfactory the liner can be fitted.

After the liner is fitted, the edges of the cylinder liner ports must be checked with their corresponding transfer ports in the cylinder jacket. Any small steps should be carefully 'washed' away with a small portable grinder or rotary file.

The cooling water outlet from the jacket must be blanked off and the cooling space hydraulically tested. Care must be taken to see that the top of the cooling space is not air locked. If the cylinder has wet quills fitted they must also be carefully examined for tightness at the interface between liner and quill. Slight leakage is seen by a gradual build-up of moisture which forms a water globule at the cylinder oil outlet into the cylinder liner.

Note In earlier days, going back to steam reciprocating engines, the front of the engine was where the controls were located and the back of the engine was where the condenser was located. On marine diesel engines the back of the engine is usually accepted as the side on which the exhaust pipes are located and the front is naturally the opposite. In twin-screw installations the front sides of the engines are the inboard sides.

9

PISTONS, PISTON RODS, PISTON SKIRTS, PISTON RINGS

■ 9.1 Of what materials are diesel engine pistons made? Which material is most commonly used in marine practice? State the advantages and disadvantages of the various materials employed.

The materials used in the manufacture of pistons are pearlitic cast-iron, flake graphite cast-iron, spheroidal graphite cast-iron, and alloy cast-irons containing nickel and chromium. Forged or cast alloy steels, and aluminium alloys particularly aluminium-silicon alloys, are also used. The factors which determine the selection of a material are its strength at the maximum operating temperature, and its thermal conductivity. The cost of the material and its processing are also studied. None of the materials named is used universally throughout marine practice, but each material might be said to be commonly used with specific engine types and particular engine ratings.

The pistons in large, highly rated slow-speed engines are usually manufactured from steel forgings or castings of nickel-chrome steel or molybdenum steel. Pistons of this type require intensive and efficient cooling to enable the material to retain its strength and to hold the thermal stresses within satisfactory limits. As they are used in slow-speed engines the weight of the material is not a critical factor, but nevertheless it is considered.

Pistons used in highly rated, medium-speed and high-speed engines are of the trunk-piston type, where weight is an important factor, and a material commonly used is aluminium alloy, both by reason of weight and high thermal conductivity. In order to keep thermal stresses reasonable, and to reduce wear rate of piston ring grooves, cooling tubes are cast in the piston crowns and cast-iron or alloy-steel inserts are fitted in way of piston ring grooves. The coolant circulated through a piston of this type is lubricating oil.

Some medium-speed engines use composite-construction pistons having aluminium-alloy or cast-iron skirts with cast alloy steel piston crowns; in other types cast-iron may be used for the skirt material. Lightly rated medium- and slow-speed engines may use pistons made from flake graphite or spheroidal graphite cast-iron. If the thermal conductivity is insufficient for a particular purpose, an alloy of cast-iron containing nickel and chromium may be used, which has both better thermal conductivity and greater strength. The cast-irons have the advantage of low material and manufacturing cost; if an alloy cast-iron is used because the technical demands on the material are greater, this will only increase the cost of the material, processing costs remaining about the same.

Where weight is an overriding consideration because of engine speed, inertia loads, and balance, aluminium alloys must be used because of their light weight. As these alloys have excellent thermal conductivity the arrangements for cooling the piston crown can be simplified, and in smaller engines are often omitted.

In large, highly rated, slow-speed engines alloy-steel castings have advantages due to considerations of strength and cost. If forgings are used, the manufacturing costs increase considerably without a corresponding increase in technical advantage.

■ **9.2 During manufacture it is usual to anneal the forgings or casting from which pistons are made. Why is this done?**

Such castings and forgings generally have a combination of parts of varying thicknesses. When the hot forging is worked, or the castings cool in the mould, the thinner sections cool more quickly than the thicker parts and stresses are set up at the change of section. Annealing and similar heat-treatment processes reduce the internal stresses set up from uneven cooling between thick and thin sections. In castings, annealing also refines the grain structure in the material, as well as reducing the internal stresses (see Question 6.8).

■ **9.3 Why are the ends or crowns of pistons in contact with the burning gases made either concave or convex instead of flat?**

There are various reasons for shaping pistons in this way. Flat surfaces are not self-supporting if subjected to a pressure on one side, as occurs when an engine piston is loaded. An unsupported flat surface takes a concave form on the side which is loaded and as the load or gas pressure increases, the curvature of the concave surface increases, the amount of curvature being dependent on the pressure applied. Thus, if the piston top were flat, it would go through a pattern of varying concavity, increasing as the pressure in the cylinder increased, and decreasing as the pressure diminished. This regular and cyclic change of form which would occur in a flat piston crown would cause it to suffer from fatigue. Early failure would occur at the point where the greatest variations in stress occurred, no matter how well the other parts of the piston were designed.

Curved surfaces such as the curvature of convex or concave piston crowns are very nearly self-supporting when loaded on one side. As these curved surfaces are self-supporting they do not change their shape or move to any degree with the changes of pressure over a working cycle in the engine. Consequently the risk of fatigue failure is so reduced that it now rarely occurs.

The upper part or crown of a piston also forms the lower part of the combustion chamber. The concave or convex curvature must thus also be arranged in conjunction with the fuel-injector-nozzle spray pattern so that good combustion is encouraged within the form of combustion chamber produced.

Note The effectiveness of a curved or dished surface to be self-supporting is readily seen if the dished ends of a water-tube boiler drum are compared with the end plates of a Scotch marine boiler. In order to prevent the end plates of the Scotch boiler buckling when subjected to the pressure in the boiler, the ends are tied together by longitudinal and other stays, whereas the end plates or dished ends on a water-tube boiler drum have no such longitudinal stays.

Some large, slow-speed, highly rated two-stroke cycle engines have only a small amount of convex curvature on the outer periphery of the piston crown. This convex curvature is closed by a fairly extensive flat area supported by brackets which also assist in cooling and reducing the compressive load on the side walls of the piston. The brackets also support the bottom flange of the piston where it is mounted on the piston rod.

Some aluminium-alloy pistons have a flat crown on the upper surface but have a deep concave surface on the underside.

■ **9.4 Diesel engine pistons are subjected to various stresses. Name the nature of the stresses and describe how each arises when an engine is operating normally.**

The stresses to which a piston is subjected are compressive and tensile caused by bending action due to gas pressures, inertia effects, and thermal stresses. When the crown of a piston is subjected to gas pressures the top surface of the piston is under compressive loading and the lower surface is under tensile loading. The piston crown is then behaving somewhat like a uniformly loaded beam.

When the piston is moving upward towards the end of its stroke, retardation occurs and the inertia effect tends to cause the piston to bow upwards so that the top surface of the piston, together with the sides, is under tensile loading and the lower surface of the crown is under compressive loading. The pressure on the top of the piston nullifies the inertia effects when the piston approaches top centre position in the upward direction.

Note The inertia stress on a four-stroke cycle engine may not be nullified at the end of the exhaust stroke in the same manner as at the end of the compression stroke, due to the gas pressure at the end of the exhaust stroke being much lower.

When the piston is retarded on its approach downwards to bottom-centre, the piston crown tends to bow downwards, and its upper surface and the piston walls are in compression. The lower surface of the piston crown is then in tension. As the stresses from inertia effect are in the same direction as those caused by gas pressure on the piston crown the two stresses become additive; thus when the piston approaches bottom-dead-centre the inertia stresses increase the stresses caused by gas pressure.

The thermal stresses set up in a piston are caused by the difference in temperatures across a section. The free expansion of the hot side is resisted by the cold side which does not want to expand so much, as it is cooler. This section then sets up thermal stresses in the material of the piston, these stresses being greatest where the difference in temperature of the material across any section is greatest.

■ **9.5   The pistons of some engines are cooled internally by oil or water. Why is internal cooling necessary in some cases and not in others, in view of the fact that the combustion temperature is practically the same in all engines?**

All pistons, in effect, are cooled when heat flows from the combustion side of the piston crown to anything in contact with it which is at a lower temperature. In small engines the heat flows from the upper part of the piston crown to the lower side of the crown and the walls or sides of the piston. The heat is taken away partly through the cylinder liner by conduction, and partly into the air in the crankcase by radiation and conduction. Oil splashed off the bearings on to the underside of the piston removes a large amount of heat by conduction. The amount of heat removed by radiation is small. The heat passing into the piston crown is dissipated at a rate such that the temperature of the piston material does not rise to the point at which its strength is reduced below allowable limits, or cause breakdown in the lubrication of the piston and piston rings. It can be seen that a four-stroke engine piston will have to transmit only half the heat of a two-stroke engine piston having the same bore, stroke and speed.

As the size of engine is increased, the weight of the fuel-charge per working cycle increases; consequently the amount of heat liberated by the combustion of fuel increases although the maximum combustion temperature remains practically the same. When a greater amount of fuel is used during a working cycle, the heat liberated is greater and consequently in order to keep a piston at a safe working temperature, a larger amount of heat must be transmitted through it, and this heat transmission is governed by the difference in temperature across the section of the piston.

By the circulation of a coolant to contact the underside of the piston crown and its inside walls the temperature of the cooler side of the piston is reduced, thus increasing the heat transmission rate across the piston and preventing the hot side of the piston crown from rising to too high a temperature. Thus when the heat liberated by the fuel is so great that the temperature rise of the piston material would cause loss of strength and failure, the piston must be circulated with a coolant to keep its temperature within safe working limits. The strength of the piston material is thus retained and satisfactory lubrication of the piston and rings is maintained.

Note   Pistons that do not have an internal coolant are made with thicker crowns and side walls. As well as being stronger, the thicker material aids in increasing flow of heat away from the piston crown. Materials having good heat conductivity are favoured.

9.6   What are the media generally employed for cooling pistons? State the advantages and disadvantages of each.

The coolant used for removing and conveying the heat from a piston may be either fresh water, distilled water or lubricating oil. Water has the ability to remove more heat than lubricating oil. This can be seen from the fact that the specific heat of water (in SI units) is approximately 4 while the specific heat of lubricating oil is about 2. Further, the temperature range $(t_2 - t_1)$ of cooling water passing through a piston may be of the order of 14°C while for cooling oil it will be 10°C for a similarly rated engine. Let

Q = quantity heat removed in any given time

Then

Q = weight of coolant used in time $T \times (t_2 - t_1) \times$ specific heat

Assume the weight of water passing the piston in a time T is equal to unity. Then

$$Q_w = 1.0 \times 14 \times 4$$

Assume the weight of oil passing the piston in time T is W_o. Then

$$Q_o = W_o \times 10 \times 2$$

If the same amount of heat is removed from the oil-cooled piston as from the water-cooled piston in the equivalent time T then

$$Q_o = Q_w$$

and

$$W_o \times 10 \times 2 = 1.0 \times 14 \times 4$$

So

$$W_o = \frac{1.0 \times 14 \times 4}{10 \times 2} = \frac{56}{20} = 2.8$$

From this it can be seen that for the same cooling effect, the amount of cooling oil circulated will be nearly three times the amount of water. In actual design practice there are many other factors to be taken into account when designing and comparing the relative merits of water or oil for piston cooling systems.

Fresh water
Disadvantage
Contains hardness salts which could form scale on internal surface of pistons.
Advantage
Relatively easy to obtain, and does not require special reserve storage facility.

Distilled water
Advantage
Absence of scale-forming matter.
Disadvantages

1   Must be produced by evaporation of fresh or salt water and condensation of vapour produced.
2   Strict supervision of evaporation control necessary to keep the contaminants carried over to a low order – in parts per million.

Fresh- and distilled-water piston cooling systems
Advantage
The main advantage of cooling pistons by water is the ability of water to absorb large amounts of heat.
Disadvantages

1   The piston-cooling water conveyance pipes and attendant gear must be kept out of the crankcase as far as possible, because of the danger of

contamination of the crankcase lubricating oil by water leakage. Because of possible contamination of the jacket cooling water with oil, the jacket cooling-water system must be made separate from the piston cooling system. This necessitates duplication of cooling-water pumps, piping, motors, wiring, starters, coolers and control equipment.

2   When an engine has water-cooled pistons the piston cooling system should be drained of water after the engine is shut down for an extended period. A drain tank is necessary to hold the piston cooling water. This is often incorporated with a cascade type filter for separation of oil and scum from the piston cooling water.

3   Generally, water cooling of pistons makes for added complication, and a higher risk of contamination of the crankcase lubricating oil system.

Lubricating oil piston cooling systems
Advantages

1   The piston-cooling oil pump is combined with the lubricating oil pump and the piston-cooling oil cooler is combined with the lubricating oil cooler. This makes for overall simplicity in ancillary pumping and piping systems, and in the control equipment associated with these systems.

2   Internal stress within the material of the piston is generally less in oil-cooled pistons than in water-cooled pistons, but good design in a water-cooled piston can negate this advantage to some degree.

3   No risk of crankcase-system oil contamination, even when piston-cooling oil-conveyance piping is fitted in crankcase.

4   Simpler arrangements for cooling-oil conveyance piping with less risk of 'hammering' in piping and bubble impingement attack.

Disadvantages

1   Larger power requirements for pumping cooling oil.

2   Larger amounts of lubricating oil required in lubricating oil system, if oxidation is to be kept down.

3   Increased period of time to cool down after stopping main engine, if coking in piston is to be avoided.

Note   Each system coolant has its advocates; of the two largest main engine builders, one favours oil cooling and the other water cooling.

If simplicity is an objective, as it should be, then oil cooling of pistons is desirable.

■   **9.7   Why are some four-stroke cycle engine pistons fitted with extension pieces or trunks as they are called?**

The extension piece or piston trunk gives rise to the name trunk piston. The purpose of this extension piece or trunk in four-stroke cycle engines is to act in a similar manner to a crosshead. It takes the thrust caused by connecting-rod angularity and transmits it to the side of the cylinder liner, in the same way as the crosshead slipper transmits the thrust to the crosshead guide. With such engines, which are termed trunk-piston engines, the engine height is considerably reduced compared with that of a crosshead engine of similar power and speed. The engine-manufacturing costs are also reduced.

Note A common belief is that trunk pistons are used only when the speed of an engine is high, or when headroom is limited. Actually there is no reason why low-speed engines should not be fitted with trunk pistons provided that there is no objection to the disadvantages they possess. In engines with a large stroke/bore ratio, the reduction in height is not marked, since the connecting rod must be unusually long in order not to foul the piston trunk when the crank is passing the quarter position.

■ **9.8 Why are some two-stroke cycle engine pistons fitted with trunks or skirts?**

Two-stroke cycle trunk engine pistons have a trunk for the same reasons as do four-stroke cycle engines of the trunk-piston type.

In older loop- and cross-scavenged two-stroke cycle engines the piston trunk has a further purpose – to blank off scavenge and exhaust ports during the part of the cycle when the piston is high in the cylinder. Piston trunks or skirts are also fitted to some crosshead type two-stroke cycle engines for the same reason.

Note The terms trunk and skirt are synonymous, but the term trunk piston is usually used when referring to pistons of a specific type as used in an engine without a crosshead. The trunk or piston extension is usually called a skirt irrespective of its actual function.

■ **9.9 What may happen when wide load changes occur instantaneously or over a short time period in highly rated trunk-piston engines?**

The contact between the piston skirt and the cylinder liner where the piston side thrust is taken up can extend only over a limited part of the circumference of the cylinder liner. This is due to the fact that some clearance must be allowed.

If the piston skirt diameter were machined to a size equivalent to the diameter of the cylinder bore minus the required clearance, the contact between the piston skirt and the cylinder would only be a line contact. This would be insufficient to accommodate the side thrust and the piston skirt would overheat. Piston skirts, then, must be machined in a special manner so that an arc of approximately 90° extending over the depth of the skirt makes contact with the cylinder liner when at operating temperature.

When a large instantaneous load increase occurs the side thrust from the piston skirt on to the cylinder liner increases and the temperature of the piston skirt increases in way of the contact area (due to the increased side thrust load). The temperature of the skirt then ceases to be uniform around its circumference and the arc of contact changes. If the arc or arcs of contact are insufficient and leave the contact area between the piston skirt and cylinder liner inadequate to take the increased side thrust load, the result may lead to overheating of the piston skirt during the time the temperature of the piston skirt is stabilizing.

■ **9.10   Describe the changes designers and manufacturers have made to safely accommodate large instantaneous engine load changes in modern trunk-piston engines.**

The design changes made cater for increased combustion pressures, wear rate of upper piston ring grooves, ease of manufacture, and the selection of materials to ensure that materials having optimum characteristics are used in strategic locations.

This had led to making pistons in two-part form, sometimes called composite pistons. The piston crown is made of cast alloy steel to give the necessary strength to resist excessive distortion under load, good heat transfer characteristics, enabling the piston ring grooves to run at a satisfactory temperature, and absence of growth.

Careful attention is given to the location and design of the landing of the piston crown on to the piston skirt and the clearance between the outer part of the crown and the skirt to accommodate the thermal expansion of the ring belt.

The transfer of the gas pressure load on the piston crown through to the connecting-rod via the piston skirt and the piston or gudgeon pin is also carefully studied and often results in stepped gudgeon-pin bosses.

Piston skirts are now manufactured without internal brackets and supports. Their thickness tapers from a thicker section at the top to a lesser thickness at the bottom with a uniform thickness around any circumferential line. The material is usually an aluminium-silicon alloy. This has good strength characteristics, good heat transfer capability and is light in weight.

The shape of the skirt is made similar to a barrel having elliptical cross-sections. The minor axis is in the same plane as the gudgeon pin. This gives a good arc of contact between the skirt and the cylinder liner; the barrel shape accommodates the temperature differences between the upper and lower parts of the cylinder skirt, which gives a good contact line in the vertical direction. The combined effects of the shape give an adequate area of contact between the skirt and the liner when at operating temperature to take safely the side thrust load. The heat transfer capability of the aluminium-silicon alloy and the absence of internal stiffening maintains a more even temperature around the skirt, allows it to come up to a stabilized condition more rapidly, and takes care of transient conditions due to rapid changes in engine loads.

In the event of an aluminium-silicon piston skirt overheating the damage is usually only of a minor nature. In many cases it is not noticed until a piston is lifted for overhaul and renewal of piston rings. If a skirt of cast-iron overheats it can sometimes lead to engine seizure. Aluminium-silicon alloys do not scuff and score in the same manner as a cast-iron skirt in a cast-iron liner.

The fastenings between the piston crown and skirt are made with long studs. These are made as long as possible with a reduced diameter between their threaded ends to give a good resistance to fatigue failure. Belleville washers are sometimes used to maintain tightness of the fastenings over a wide temperature range. Locking plates are sometimes fitted to the studs to hold any broken part of a stud *in situ* if stud failure occurs; this prevents engine damage.

Today, many engine builders do not make pistons. They contract this work out to firms specializing in the design and manufacture of trunk engine pistons.

Some of these companies cover the full range of piston sizes from automobile engine pistons to the largest composite pistons as used in highly rated medium-speed propulsion engines burning heavy fuel.

Finite element analysis used in the studies of stresses, strains, and heat transfer, has made it possible to forecast the effect of design changes in the early stages of design and is used in the design of pistons. (Fig. 9.1)

Note The pistons in many small engines are still made in one piece. These are not designed in the same manner as larger pistons, but within certain ratios and proportions appropriate to their dimensions. In order to accommodate side

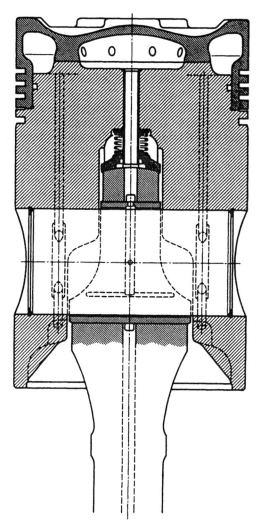

Fig. 9.1 Modern two-part piston for a four-stroke cycle trunk-piston engine of high output. (MAN B&W)

thrust the depth of the piston is made greater over an arc approaching 90° on both sides of the piston at a position 90° to the gudgeon pin. The parts of the piston over the other two arcs house the gudgeon pin and are of much lesser depth. The skirt on many of these pistons is split and is flexible to some extent, in order to accommodate the temperature changes due to changes in engine load.

Friction-reducing coatings may also be used on piston skirts of all sizes and different materials.

■ **9.11   Some trunk engines are built with pistons or piston skirts that revolve during engine operation. How is the revolving action obtained and what is its purpose?**

There are two different types of this form of piston. In one, the whole piston revolves, in the other only the skirt can revolve. The latter type has the skirt fastened to the piston crown in such a manner that it is free to float to any circumferential position it may wish to take. The hotter parts of the skirt increase in radius locally, the forces coming on to the skirt then tend to move the larger radius part around until a location with a lesser radius is taking the side thrust. The gudgeon pin is housed in an inner skirt which does not revolve.

The other type of revolving piston is made to revolve with a circular ratchet and pawl arrangement. The pawls, two in number, are spring loaded and fitted in the upper part of the connecting rod. The circular ratchet is fitted in the inside of the piston skirt.

When the connecting-rod moves through an arc due to the rotation of the crankpin, the pawls engage with the circular ratchet and cause the whole piston (crown and skirt) to rotate.

There is no gudgeon pin fitted in this type of piston. The upper part of the connecting-rod is made spherical in form. The spherical portion at the top of the connecting-rod is fitted into a spherical-shaped housing. This allows the piston to turn during engine operation.

The purpose of revolving pistons and revolving piston skirts is to give a uniform temperature around the circumference of the skirt and reduce the risk of overheating and possible damage.

■ **9.12   Some pistons have circumferential bands of a different metal fitted in them. What is the purpose of these bands? How are they held in place?**

The bands fitted round a piston, or a piston trunk, are referred to as wear rings or rubbing bands. They are usually made of a bronze alloy, but for engines operating on heavy fuel, steel or special cast-irons are sometimes used. Wear rings have a dual purpose in that they provide a rubbing surface with low frictional characteristics, and they prevent the hot upper side wall of the piston making contact with the working surface of the liner. This is achieved by increasing the radial clearance between the piston and liner and maintaining the working clearance in way of the wear ring. This gives a further advantage as the working clearance is on the wear ring which is always in a cooler section of the piston or skirt. As the wear ring is not subjected to heat in the same manner as the upper part of the piston it allows some reduction in allowable working clearance.

In trunk-piston engines the use of wear rings allows the detrimental effects of trunk distortion, caused by the interference fit of gudgeon pins in the piston trunk, to be minimized. They are fitted above and below the gudgeon pin location.

Wear rings are fitted into circumferential machined grooves. The edge of the groove is then lightly caulked or upset so that the ring is retained in the groove. They are usually fitted in two pieces with good clearance at the butts to allow for thermal expansion.

■ **9.13 How would you decide whether wear rings require renewal?**

When a wear ring is just about flush with its parent material in the piston it should be renewed. If the cylinder liner is also to be renewed in the not too distant future, no real damage should be caused to the old piston if wear-ring renewal is held over until the cylinder liner is renewed. If bad blow-past occurs and the cause is piston slackness in the bore of the cylinder, renewal of the wear rings must not be delayed.

Wear-ring renewal should always be reviewed when piston ring grooves are being reconditioned to bring them to original size.

■ **9.14 Why is it usual to taper that part of pistons above the rings, i.e. progressively reduce the diameter from the top piston-ring groove upwards?**

The portion above the piston rings is the hottest part of the piston and greater allowance must be made for expansion radially. It is at this point that the greatest mass of metal occurs, and as it is the least effectively cooled part of an internally cooled piston the expansion of this part is relatively great and due allowance for this must be made.

Note There is no material advantage in making that part of the piston above the piston rings a good fit in the cylinder. The body of the piston must be a reasonably good fit in the cylinder in order that the piston rings have a minimum of overhang. If the rings project too far out of the groove when all parts are in the working position, the gas pressure as well as the friction between the rings and liner will tend to tilt the rings, and such action causes excessive wear of the rings as well as of the piston grooves.

■ **9.15 State some values of the piston and cylinder clearance for crosshead-type engines.**

All crosshead-type engines have internally cooled pistons. The clearance allowed depends on the actual engine type and rating. A common figure for diametral clearance of piston and cylinder in way of piston ring location in two-stroke and four stroke cycle engines is as follows.

Two-stroke cycle engine
Piston clearance in cylinder bore = 0.2% of cylinder bore. For example, if
   bore diameter = 750 mm

$$\text{Clearance} \quad = 0.2 \times \frac{750}{100} = 0.2 \times 7.5 = 1.5 \text{ mm}$$

Four-stroke cycle engine
Piston clearance in cylinder bore = 0.1% of cylinder bore.

These figures may be increased when wear rings are fitted. The working
clearance for a new piston and new cylinder liner must be within the standards
set by the engine builder.

■ **9.16   State some values of the piston and cylinder clearance for trunk-
piston engines.**

The running clearances of trunk pistons are usually found by the engine builder
from previous experience and experimental work on a prototype engine, and
later confirmed by service experience. The actual diametral clearance then
found will be dependent on the operating temperature of the piston, which will
in turn depend on the way the piston is cooled. The piston and piston skirt
material and the amount it expands from cold relative to the cylinder liner
material will also have some influence on the clearance. For trunk-piston
engines the diametral clearance of the piston at the upper part in way of the top
piston ring will be approximately 0.4% to 0.5% of the bore for cast-iron
pistons, while for aluminium-alloy pistons a clearance of double this amount
may be necessary.

   At the top of the piston skirt the clearance will be approximately one quarter
of that already given; thus it will be 0.1% to 0.125% of the bore diameter for
cast-iron piston skirts and up to three times this figure for aluminium-alloy
skirts. At the bottom of the skirt the clearance will be less again. For cast-iron
piston skirts this will be approximately 0.075% of the bore diameter and double
this figure for aluminium alloys.

Note   The engine builder's recommendations in this respect should never be
departed from without a full assessment of all pertinent and relevant factors.

   The figures given in the answer are representative of those found in most
engines, but wide departures from these figures are not uncommon.

■ **9.17   What determines the distance from the top of the piston to the
uppermost piston-ring groove?**

Piston rings give the best results when their working temperature is the lowest
practicable. It is necessary therefore that the uppermost piston ring, which has
the severest duty to perform, should be well clear of the hottest part of the
piston or the large mass of metal at the junction of the piston crown and cylindrical wall. This then imposes a limit to the minimum distance from the top of
the piston to the uppermost groove and consequently all other piston grooves.

On the other hand, the further the uppermost groove is from the top of the piston the greater must be the overall length of the piston in order to accommodate the requisite number of piston rings. The objections to increasing the length of pistons are that the length of cylinders and overall height of the engine are correspondingly increased and the reciprocating mass is increased.

There is also an objection to fitting the top piston ring low down. This is because the space formed between the side of the piston crown and the cylinder liner above the top piston ring is an area where carbon and ash from lubricating oil may encrust and build up.

If the carbon and ash deposit flakes away it can cause rapid abrasive wear between the upper piston ring landings in the piston ring grooves and on the piston ring landings.

From this it can be seen that the selection of a suitable cylinder lubricant, the cooling of the piston, and the location of the piston ring grooves are very important factors. (Fig. 9.2)

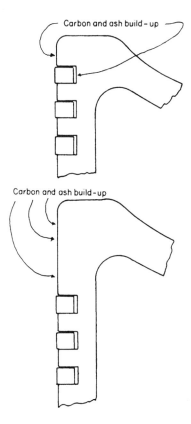

Fig. 9.2 Effect of upper piston ring position on carbon and ash build-up.

■ **9.18   Some pistons have circular ribs cast on the inner side of the crown. What is the object of these ribs, which are not attached to any other part of the piston and cannot, therefore, give any support to the crown?**

The object of such ribs, which extend from the crown to the interior of the piston, is to facilitate the transfer of heat from the crown to the cooling medium. The heat flows into the ribs which, being of thin section, quickly give it up to the cooling water or oil.

■ **9.19   What is the flow pattern of the coolant through an internally cooled piston? Why is the flow direction arranged in this way?**

The flow pattern is generally such that the coolant enters at the lowest part of the cooling space and leaves from the uppermost part. It should move in such a manner that the upward movement of the coolant is uniform on opposite sides of the piston to give even cooling without causing distortion due to unequal expansion. The flow direction is arranged in this manner so that the piston is always full of coolant and the underside of the piston crown is always in contact with it. This is particularly important in slow-speed propulsion engines, because when the engine is running at dead slow speed the coolant in the piston is not 'shaken up' as when running at full speed. If the coolant flow took place in the opposite direction, it would be possible at very slow speed for the coolant to drain from the piston, and lose contact with the piston crown; the piston could then become overheated even though the engine was operating at low power.

Some water-cooled pistons have the outlet for the water at approximately half the cooling-space height. When running slow, the piston is half full of water and piston movement agitates the water in the piston so that it is splashed on the underside of the piston crown and up the inside wall of the piston; heat is transferred to the water droplets. The 'cocktail shaker' action may be further increased by a ported inverted cone baffle fitted within the cooling space. The baffle acts in a similar manner to the air vessel on a reciprocating pump.

When the engine is stopped a jet action from the piston cooling pipe nozzle directs cooling water on to the piston crown, thus removing residual heat and catering for an emergency stop at full speed. The splash method of cooling is sometimes referred to as 'cocktail shaker cooling'.

■ **9.20   What attention do the internal parts of an internally cooled piston require?**

The internal surface of the piston, which transmits heat to the coolant, must be kept clean and free of scale formations and oil films in water-cooled pistons. Oil-cooled pistons must have their internal surfaces kept free of carbon build-up. The internal fittings within the cooling space, and their fastenings, must be carefully examined when exposed after removing a piston from the piston rod. The points to look for are corrosion and erosion of the coolant directors, funnels and similar parts used to direct the coolant flow through the piston. The internal joints and fastenings must also be carefully checked, because a bad joint or slack or unlocked fastening could allow the cooling media to be by-passed and lead to an overheated piston.

In water-cooled pistons solid material similar in appearance to very fine gravel or coarse sand may be found in the lower part of the cooling water space. It will usually be more in evidence in fresh-water-cooled pistons than in those cooled by distilled water, particularly if sludges formed by chemical additives to the cooling water are not regularly cleaned out.

■ **9.21 What types of piston ring are generally used on diesel engines?**

1 Rings for sealing the gases above the piston and preventing gas leakage, called compression rings or pressure rings, those most commonly used being of the Ramsbottom type. This is a ring made of cast-iron with a uniform section, either square or rectangular, the inside of the ring being 'hammered'.
2 Rings for controlling the amount of lubricating oil passing up or down the cylinder wall, or spreading the oil evenly around the cylinder, called oil-control rings or scraper rings.
3 Rings used for spreading oil evenly around the circumference of a cylinder, called oil-spreader rings.

Some medium- and high-speed engines use piston rings having a different cross-section from the rectangular section of the Ramsbottom ring. (Fig. 9.3)

■ **9.22 Sometimes piston rings are made eccentric, i.e., the radial thickness is least at the gap and greatest at a point diametrically opposite. What are the advantages of this form of ring?**

Concentric rings exert the greatest pressure at each side of the gap and at the point diametrically opposite, and the effect of thinning the ends or making a ring eccentric is to reduce the outward pressure at these points and thus produce a more uniform pressure all round.

Note This can be proved as a mathematical exercise. See any textbook on strength of materials dealing with curved uniformly loaded beams. Eccentric form piston rings are not normally used today.

■ **9.23 How is the pressure or load between the rubbing surface of a piston ring and a cylinder liner made uniform around the circumference of the ring?**

Piston rings are 'hammered' to give them the required outward spring so that they exert a pressure on the cylinder wall when placed in the cylinder. By varying the amount of hammering in any particular location the amount of outward spring will also be varied, producing a piston ring of uniform section having similar characteristics to an eccentric ring. If the amount of hammering is gradually increased from small amounts on each side of the piston ring gap to a maximum amount opposite the gap, the ring when sprung into the cylinder will exert an even pressure on the cylinder walls.

Some piston rings are machined in what is known as the 'cam turning' process. In this process the ring is machined so that it has varying radii at different angular positions around its circumference relative to the location

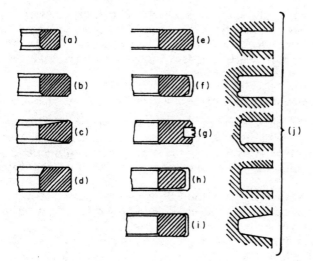

Fig. 9.3 Various types of piston ring.
(a) Square section iron piston ring.
(b) Rectangular section piston ring.
(c) Wedge-shaped piston ring (sometimes referred to as a Keystone section ring).
(d) Rectangular section piston ring with inside bevel.
(e) Rectangular section piston ring with elliptical or barrel-shaped working face.
(f) Chromium-plated barrel-shaped piston ring with plating on working face.
(g) Piston ring with copper or bronze insert to give rapid bedding-in.
(h) Piston ring with chromium plating on working face and lands.
(i) Rectangular section piston ring with chromium-plated working face.
(j) Piston-ring grooves showing various methods of leaving a fillet at the bottom of the ring groove. The bottom sketch shows a groove for a wedge-shaped piston ring.
Note. Chromium-plated pistons rings are fitted only in the top piston-ring groove.

where it will be split. The outside circumference of the ring then has a shape somewhat similar to a hammered ring. When it is in place in a cylinder it gives a uniform pressure against the cylinder wall around its circumference.

Note If a new piston ring is examined around its inside, the 'hammering' will be seen as a series of marks similar in appearance to those that would be made by a blunt cold chisel. These marks do not go all the way across a section.

The hammering action on the ring material sets up a localized internal compressive stress adjacent to the hammer mark inside the circumferential surface of the piston ring. This internal stress is greatest opposite the ring gap and least adjacent to it. The initial internal compressive stress caused by hammering on the inside of a piston ring aids in preventing permanent distortion and breakage when the ring is sprung open to pass it over a piston and into the piston ring groove.

Cam-turned piston rings can be identified by the absence of hammer marks (Question 9.38).

■ **9.24 How much force does a piston ring exert on the cylinder liner wall? Is this force variable and how is it measured?**

The force exerted by a piston ring on the cylinder is measured as a specific pressure in kN/m^2 (or lbf/in^2 or kgf/cm^2). The value for this specific pressure when the ring is inserted in the cylinder will vary over a wide range, being high for small-bore engines and low for large-bore engines. This pressure is referred to as wall pressure.

Wall pressure – compression rings
For 150 mm (6 in approx.) bore cylinders it will vary between 98 and 147 kN/m^2 (1 to 1.5 kgf/cm^2 or 14 to 21 lbf/in^2).
   For 900 mm (35 in approx.) bore cylinders it will vary between 35 and 46 kN/m^2 (0.36 to 0.47 kgf/cm^2 or 5 to 7 lbf/in^2).

Wall pressure – oil-control rings
The wall pressure for oil-control rings will vary according to the type of scraper ring used, and whether it is used in a trunk-piston engine or crosshead engine. In trunk-piston engines, scraper ring wall pressure may be 70% more than that for compression rings, while for large crosshead engines it may be the same as for the compression rings. The specific wall pressure exerted by the ring when it is sprung into the cylinder is fixed during manufacture by the material of the ring, the proportions of the ring section, and the degree of hammering. The actual wall pressure can be measured only in special test equipment used by piston-ring manufacturers. The wall pressure is, however, related to the forces which would have to be exerted on a piston ring at diametrically opposite points (90° round the circumference of the ring from the gap) to close the gap. This force is referred to as the diametrical closing load or gap-closing force, and is more easily measured than the wall pressure. For 150 mm bore engines with rectangular-section rings the gap-closing force will vary between 62 and 89 N (14 and 20 lbf) while in a large-bore slow-speed engine the gap closing force could reach 785 N (176 lbf).

Note It is virtually impossible to find the actual wall pressure exerted by a piston ring during its operation in an engine, either by test apparatus or calculation. In any calculation the range of the variables would be such that the degree of accuracy in the final result would be in doubt.

■ **9.25 Should a piston-ring material be harder or softer than the cylinder material in which it works?**

Generally piston-ring materials are made considerably harder than the material of the liner in which they work. This has the advantage of giving the piston ring a long working life by reducing the radial wear rate consequent upon reduction of the amounts of wear debris, which acts as an abrasive. An exception to this is the case of hardened or chromium-plated cylinder liners.

■ **9.26** Give details of the checks or controls made when fitting new piston rings during a cylinder unit overhaul. Why are these checks necessary and where is failure likely to occur if the unit overhaul and checks are not properly carried out?

The piston, piston-ring grooves, piston rod, cylinder liner, cylinder-liner ports and combustion-chamber belt must be brought to a 'clean-as-new' condition and all loose dirt must be removed from the engine and working areas. Wear ridges formed in the cylinder liner, particularly at the top of the stroke, must be removed by carefully grinding with a portable grinder. In a similar manner wear ridges on piston skirts and piston rods from scraper-ring action must be carefully filed off. These points must be checked before piston rings are fitted.

Any new piston rings to be fitted should first be rolled round in the groove where they will be fitted. This gives a check that the radial width of the groove is deeper than the radial width of the piston ring (see Figs. 9.4(a) and (b)), and an indication that the clearance between the upper surface of the ring and groove is adequate. Any jamming found in the 'roll-round' test must be investigated and the cause removed or the ring changed.

If the working face of any new piston ring is found to be proud of the side of the piston as shown in Fig. 9.4(c) the radial width of the piston ring should be checked for correctness or the cause of the ring being proud must be found and removed.

The piston ring is then entered into the unworn portion of the cylinder liner and the gap checked. After checking the gap and making any necessary rectification the piston ring will be ready for fitting on to the piston, using a piston-ring expander. If an expander is not available, narrow width tin plate or brass strips can be used to pass the lower piston rings over the upper piston-ring grooves.

When all the rings are fitted each ring should be pushed sideways to the inner side of its groove and rolled around the groove. The working face of each ring should be below the side of the piston or flush as shown in Figs. 9.4(a) and (b) for its full circumference. If any used ring is found to have its working face proud of the side of the piston as shown in Fig. 9.4(c) the groove should be re-examined and the cause removed. The cause is often found to be hard carbon left behind during cleaning. When each ring is pushed over in its groove the vertical clearance must be checked with a feeler gauge at the location shown in Fig. 9.4(d). The check is repeated in four or more positions for small-bore engines and six or more for large-bore engines. The feeler gauge should be set at the requisite clearances and used as a 'go/no-go' gauge. For record purposes, the actual clearance should be measured at two places diametrically opposite. The process is then repeated for each ring irrespective of whether the ring is new or being re-used.

Failure is likely to occur in various ways if a fault is missed in checking. If the ring gap is inadequate the ring may break due to a restriction on its free expansion when coming up to working temperature. Whether a ring breaks or not, it may scuff the liner due to the very heavy wall pressures causing the oil film to break down, or at least causing a large increase in liner wear. Scuffing may be such that the liner must be condemned.

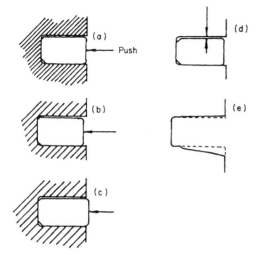

Fig. 9.4 Testing piston rings in their grooves.
(a) and (b) show grooves with the correct amount of depth for the radial width of the piston ring.
(c) shows a piston ring with excessive radial width causing the working face to be proud of the side of the piston. Do not use ring.
(d) Checking top clearance of piston ring with feelers.
(e) Shows a common pattern of piston-ring groove wear.

Piston rings fitted with insufficient vertical clearance will stick when they come up to working temperature. In turn they may break, jam in the groove on one side, and eventually allow blow-past which could lead to piston seizure on trunk-piston engines and two-stroke cycle crosshead engines with large piston skirts. It can also lead to scavenge fires in two stroke-engines of any type. The common pattern of piston ring groove wear is shown in Fig. 9.4(e).

■ **9.27** What factors would you consider in deciding whether to open up a cylinder unit for overhaul? After opening the unit and preparing for assembly, how would you decide whether to renew or re-use piston rings?

If the engine is operated under a planned maintenance programme, cylinder units should not be opened up until the required number of operating hours has been achieved. If, after taking indicator cards, or checking cylinder pressures with a maximum pressure indicator, it is established that the compression pressure is low due to leaking piston rings, overhaul of the unit should be arranged at the earliest opportunity. Failure to overhaul may lead to scavenge fires, smoking exhaust, and increased cylinder-liner wear rate.

In deciding whether to re-use piston rings removed from a piston, the wear rate of the piston ring must be known. It is then compared to the maximum wear allowed in the piston ring to establish the expected life remaining. If this is found to be less than the programmed hours between overhaul of the cylinder units, the ring must be scrapped.

Scrapping is also necessary in the following instances.
1   If a piston ring is found stuck in a ring groove and, as is likely, it is worn more on one side than another.
2   If the axial height of the ring is reduced so that a large clearance is found.
3   If preformed piston rings having pressure variations around their circumference are fitted in any groove and micrometer measurements show variations in the radial width of the ring.
4   If the chrome layer on chromium-plated piston rings is worn through or worn very thin.

These remarks show that a considerable number of man-hours could be spent in deciding whether to re-use a piston ring or to scrap it. As very few man-hours can be spent before the cost of a set of piston rings is covered, it is generally better to renew piston rings when a piston is lifted by ships' staff outside of a programmed overhaul period. If the programme time for unit overhaul has been reached the rings must be renewed, since if the correct period between overhauls has been established they will be almost worn out.

Some people generally consider that only the top two rings should be renewed. If it is found in practice that this is all that is necessary, it could well be that the period established for the running hours between overhaul is insufficient and further savings could be gained by extending the period.

■ **9.28   The bottoms of piston-ring grooves are generally rounded at the corners. State the reason for this, and mention how the fitting of the piston rings is thereby affected.**

If the corners of the grooves are not rounded there is greater possibility of the metal between the grooves breaking off.

The material between the piston-ring grooves forms a cantilever and is subjected to cyclic stresses. If the bottom of the groove has a square corner it acts as a stress raiser and may lead to eventual failure of the material by fracture. The radius at the bottom of the groove acts in the same way as a fillet. The shape of the fillet can take various forms.

When fitting piston rings the inside edges of the rings must be rounded off to coincide with the shape of the bottom of the groove. Neglect to do this may cause the rings to jam in the piston see Fig 9.3j).

■ **9.29   Why is an adequate radius necessary on the upper and lower corners of the outer circumference of a compression ring?**

In order better to lubricate the piston ring and the cylinder liner the radius is necessary to give an oil lead-in. The oil is then forced between the piston ring and the cylinder by the build up of an oil wedge and provides better lubrication for both the piston ring and the cylinder liner. If the corners were sharp, without a radiused profile, lubrication would be impaired because the ring would act somewhat like a scraper ring.

Note It is essential for this radius to be restored to a partially worn piston ring if it is to be replaced in a piston for further use.

■ **9.30 Give some values for the vertical or axial clearance, and gap openings of diesel engine piston rings.**

The vertical clearance of a piston ring in its groove must always be sufficient to allow the piston ring to remain free in the groove from the time an engine is started cold through to working on overload conditions. The upper piston rings work at a higher temperature than those in lower positions. Consequently in large slow-speed engines the vertical clearance of the upper two piston rings is made larger than the remaining lower compression and scraper rings. For large, slow-speed two-stroke engines the original vertical clearance of the upper two compression rings or pressure rings will be up to 0.2 mm (0.008 in) and for the lower compression rings up to 0.125 mm (0.005 in). For smaller engines the original vertical clearance of all compression rings will be of the order of 0.065 mm (0.0025 in approx.).

The vertical clearance of oil-control rings will vary according to their location relative to the top of the piston, the minimum clearance usually not being less than 0.0625 mm (0.0025 in) in any type of engine.

A common formula for the closed gap of a new piston ring when fitted in an unworn portion of the cylinder liner is

gap (closed) = 0.4% cylinder bore

Example

If cylinder bore = 600 mm
then

$$\text{closed gap} = 0.4 \times \frac{600}{100} = 0.4 \times 6 = 2.4 \text{ mm}$$

The value of 0.4% may sometimes be increased, perhaps to 0.5%, in some engines where hardened or chromium-plated piston rings are used, or whether the top piston ring is near the top of the piston.

However, the clearances and gaps recommended by the engine builder or specialist piston-ring manufacturer should be adhered to.

The free gap will vary according to ring material, the proportions of the ring section, and the amount of hammering received by the ring. It is usually about three to four times the radial width of the piston ring.

■ **9.31 How can the effective life of piston-ring grooves be extended?**

In large, slow-speed crosshead engines the piston-ring groove landing face is subjected to heavy wear. In order to increase the piston-ring groove life 'false' landing rings are omitted and the piston-ring groove is machined in the solid material of the piston wall. A layer of chromium is then electro-plated on to the lower landing face of the piston-ring groove or, preferably, both the upper and lower faces of the groove are plated. The chrome layer has excellent resistance both to the corrosive action of combustion products and to abrasive wear.

Aluminium-alloy pistons do not have good properties to resist abrasive wear in the piston-ring landings, so steel or cast-iron inserts are placed in the piston mould and the piston is cast round the insert. Sometimes an insert is used for the top ring only.

■ **9.32   It is noted in engines burning heavy fuel that the top of the piston crown may waste and gradually burn away. What is done to rectify this condition?**

When a piston is removed from an engine during the overhaul of a cylinder unit the shape of the piston crown is examined and compared with a gauge showing its proper shape and its height above some datum.

If the shape and height are not within the recommended limits the piston is taken out of service and sent to one of the specialist firms dealing with piston repair.

The firm will gauge the piston, examine it for fractures or cracks, examine the piston-ring grooves and note the general condition of the piston. A decision based on estimated repair costs will be made as to whether it warrants repair.

If it is decided to repair the piston, the crown will be welded up with a deposit of weld metal to bring it to its correct shape and height above datum. The piston-ring grooves will also be welded up. After the welding is completed it will be examined again and if in order it will be heat treated, fully machined and the piston-ring grooves will be chromium plated again.

A repaired piston is similar in appearance to a new piston and has the same expected life. The specialist repair work usually costs considerably less than a new piston. Engine builders and their accredited repairers usually hold stocks of repaired pistons for exchange with pistons that require repair. This shortens the delivery or lead time for the repair of a piston.

■ **9.33   Where are oil-control rings or scraper rings fitted? What is the sectional form of a scraper or oil-control ring?**

Oil-control or scraper rings are fitted in pistons, piston skirts, around piston rods and at the bottom of cylinder liners in some types of engines. The pressure on the scraping surface is usually obtained from the resilience of the material and hammering, although sometimes the load on the scraping surface is obtained from springs behind the rings. In four-stroke cycle and two-stroke cycle crosshead engines it is usual to fit at least one scraper ring in the bottom of the piston, arranged so that any oil coming down the liner is carried up again by the ring, the ring being said to scrape upwards. To achieve this the scraping edge of the ring is fitted uppermost and the bevel downwards, so that the upper portion of the liner is better supplied with lubricant.

Some modern crosshead engines are not fitted with a scraper ring in the bottom piston-ring groove.

The packing around a piston rod in two-stroke cycle crosshead engines has to perform a dual function; preventing oil from the scavenge air space under the piston draining into the crankcase, and oil from the crankcase coming into the scavenge space. An intermediate set of packing, seals the scavenge air in the scavenge space and prevents air leakage into the crankcase.

The top set of scraper rings in the piston-rod packing box is arranged with the sharp edge uppermost and the bevel towards the bottom. The scrapers then scrape oil up the piston rod as it descends into the crankcase, and they are fitted with drain holes which either drain the scraped oil into the scavenge space or into a gallery where it is drained away through a drain pipe to some point outside the engine.

The lower scraper rings act in the opposite manner and are fitted in the reverse direction to the upper scraper rings. They are also fitted with drain holes which often allow the oil to drain back to the crankcase, or in other cases it may be drained away through a drain pipe to some point outside the engine.

Note   Leak-off oil from the upper scraper rings of piston rod packing boxes should not be mixed with the leak-off oil from the lower scraper rings. The latter should be stored and saved. If analysis shows it to be satisfactory the oil may be returned to the crankcase lubricating oil system.

Scraper rings have many variations on three basic forms, which are the bevel type, the hook type, which might also be called a self-sharpening variation of the bevel type, and the twin or double-edge type.

Bevel-type scraper rings start with a basic square, or rectangular section, in which the radial width is greater than the vertical or axial height. Outward-springing rings have a bevel machined circumferentially on an outside corner of the ring section and inward-springing rings have the bevel machined on an inside corner. The angle of the bevel from the cylinder wall is between 70° and 75°. The section of a hooked scraper ring is, as its name implies, hook shaped. It is similar in shape to a bevel-edge ring but has an additional hook-shaped relief on the lower flat-face surface of the ring opposite to the bevel. Double-faced rings are rectangular in section with a peripheral U-section slot in the outer circumferential face of the ring. Two scraping lands formed by the slot make contact with the cylinder liner surface, and drain holes or vents are made through the ring across the thin section at the bottom of the U-shaped slot.

In trunk-piston type engines with splash-lubricated pistons and cylinders, there is always some difficulty in controlling the amount of lubricant passing up the cylinder to lubricate the piston. Some of this lubricant is splashed from the crankcase, but a large amount may come from the gudgeon pin by leakage through the fastening of the pin in the piston skirt. In such engines the scraper rings are usually fitted in the piston skirt and scrape the lubricant down the liner. The ring is then fitted sharp-edge downwards and bevel uppermost, therefore scraping downwards. Drain holes are usually drilled through the skirt so that excess lubricant passes through the drain holes back into the crankcase.

The various shaped sections of oil control rings are shown in Figs 9.5(a) through (e). When lubricating oil is scraped from the surface of the cylinder liner, drainage slots in the control ring and drainage holes in the piston skirt allow the oil to return back to the crankcase. In control ring (b) the drainage

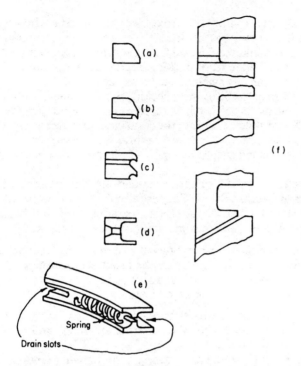

Fig. 9.5 Various sections of oil control rings or scaper rings.
(a) Normal older type of oil control ring.
(b) Hook-shaped oil control ring with radial drainage slots in bottom land of ring.
(c) Double hook oil control ring with drainage holes and drainage slots.
(d) Oil control ring with double scraping edges. Drainage hole is between scraper edges.
(e) Spring-loaded oil control ring with drainage slots between scraper edges.
(f) Various forms of oil control ring drainage holes.

slots are situated on its base. In the control rings (c), (d), and (e) the slots are located in the central section of the ring. Various forms of drainage holes and control ring grooves fitted in a piston skirt are shown in Fig. 9.5(f).

■ **9.34 How can the life of a piston ring and cylinder liner be increased?**

Piston ring life can be increased, with usually an increase in cylinder liner life, by fitting the top piston-ring groove with a piston ring which has been chromium plated on its outer circumference. This increases the hardness of the surface that makes contact with the working surface of the cylinder, thus improving anti-friction conditions and consequently increasing piston-ring and liner life.

■ **9.35 How much increase of piston-ring vertical clearance can be allowed before reconditioning the piston-ring grooves?**

In large slow-speed engines the greatest amount of wear takes place on the lower side of the upper piston-ring groove. When the original piston-ring clearance is slightly more than doubled it is common practice to machine the top and bottom faces of the groove so that it is 0.5 mm larger than the original size. An oversize piston ring having an increase of 0.5 mm in the axial height over the original height is then fitted and this restores the original vertical clearance.

As the major part of the cost of reconditioning piston-ring grooves by machining is in piston removal, and setting up the piston in the lathe or vertical boring machine, overall savings can be effected by machining all grooves, even though the increase of clearance in lower grooves may be considerably less than in the top groove. This has the further advantage that all piston rings on any piston are the same size enabling the number of different-sized rings carried in the ship to be reduced. Oil-control ring grooves must also be reviewed at this time, to decide whether machining is necessary.

In trunk-piston engines a figure of double the original vertical clearance in the groove may not be feasible due to the slack rings pumping lubricating oil and so increasing lubricating oil consumption. When this occurs it is sometimes possible to extend groove life by changing from single-edged bevelled scraper rings to 'fiercer' scraper rings having a 'hook' section, or to rings with double edges and narrow landing widths. Double-edged scraper rings can be used only if drainage holes are provided through the piston skirt.

In trunk engines with splash-lubricated pistons and cylinders, care must be exercised in changing scraper rings to another type, as the pistons may then become insufficiently lubricated. When scraper rings are changed in this way, the lubricating oil consumption should not be below the engine builder's recommendation. Also, all the pistons must be fitted with 'fierce' scraper rings. If only one or two pistons were fitted with 'fierce' scrapers these cylinders could be insufficiently lubricated to the point where seizure could occur without warning, since no indication would be given of a low lubricating-oil consumption; because of excessive amounts being used in other cylinders where the scraper rings had not been changed.

■ **9.36 List the causes of piston-ring failure which may result in gas leakage or ring breakage.**

1 Insufficient piston-ring and groove clearance. This may cause the ring to jam in the groove when the engine comes up to working temperature and allow blow-past.

2 Insufficient lubrication. Although there may be sufficient cylinder lubrication to prevent ring scuffing and an excessive rate of ring and liner wear, a borderline may be reached where there is insufficient lubricant available to keep the piston rings free in their grooves. A small excess amount of lubricant over that necessary to lubricate the liner is required so that additives in the lubricant can clean sludges, varnish and gum formations out of piston-ring grooves. The rings then work effectively. This is

particularly important in new engines or when a new cylinder liner and piston are fitted in an old engine.

3 Large amounts of wear in cylinder liners.

4 Excessive diametral clearance between the piston and cylinder.

5 Excessive wear on piston-ring landing face in the piston-ring groove, particularly if associated with (3) and/or (4). This is often the cause of piston-ring troubles in cross- and loop-scavenged engines, the gas pressure built up behind the piston ring causing the ring to bow outwards slightly as it rides beyond the top edge of the port. It is then supported only by the port bars, and the pressure behind the ring causes this bowing action.

6 Ring gap too small. This usually leads to ring breakage, but could ultimately lead to disastrous or very serious consequences.

7 Incorrect preparation of ends of piston ring adjacent to gap, particularly in loop- and cross-scavenged engines with relatively wide ports. The usual preparation in this case is to give the ends of the piston ring a near semi-circular profile which is washed into the corner radius of the ring, on the face making contact with the cylinder. For example, the radius on a piston ring of 12 mm height would start at 5 mm. This 5 mm radius would be reduced moving around the circumference away from the gap so that at 25 mm away the radius is reduced to 2.5 mm, and at 50 mm away it would be reduced to the point that it merged with the radius at the corners of the ring section.

8 Radius at top and bottom of exhaust and scavenge ports in cylinder liner inadequate, the ring then receiving a shock when sliding past the port edges. As the design width of the port increases relative to the cylinder circumference, the port edge radius becomes increasingly important. Insufficient radius leads to ring breakage.

9 Wear on the port bars relative to cylinder-liner working surface such that surface of bar is below the surface of the liner. This condition is checked with a straight edge held vertically on the port bar, the hollow surface being checked out with a feeler gauge. This is important after the first few thousand hours of liner life. The marks and slight surface roughness from machining have now been worn away, and as the port bar is worn smooth its ability to hold lubricant on its surface is impaired and the wear on the port bar increases. To rectify the condition the surface of the port bar should be scratched with the corner tip of a small square file so that a trellis pattern of scratches is put on the surface of the bar. The unscratched square areas of port surface between the trellis pattern should have sides 3 to 6 mm depending on the size of the engine. This problem is found to be more prevalent on the exhaust port bars. If this pattern of wear develops on an old liner which has not given trouble previously, it may be indicative of the cooling space in the exhaust port bar becoming dirty or scaled.

■ **9.37 What do you understand by the term 'scuffing' as related to pistons, piston rings, piston skirts and cylinder liners? What are the causes of scuffing?**

Scuffing is a form of damage occurring between two sliding surfaces, when there is a breakdown of the lubricating film separating the surfaces. It may vary from something relatively minor in nature, appearing as a slight roughening of the surfaces, to a heavy abrasion rendering the damaged parts unfit for further use.

When scuffing occurs the breakdown of lubrication in way of the high points or surface asperities is caused by very high localized pressure on the high points. The heat generated by the friction between the high points on the cylinder liner and the piston ring causes the high points to weld together. The movement of the piston and piston ring causes the weld to break away as soon as it is created; a roughened surface then results.

Piston rings that have been subjected to scuffing action can be identified by subjecting the rubbing surface to a hardness test. If scuffing has occurred the ring surface shows a large increase in hardness value.

In appearance it shows as a slight roughening, brownish to very dark brownish-grey in colour, with light score marks emanating from the scuffed area.

Scuffing may occur at any time, but is more likely when cylinder liners, piston rings, etc. are new, such as when a new cylinder liner is fitted or when a new engine is undergoing trials on an engine works' test bed. Any factor which causes or allows the lubricating oil film between the working surface to become disrupted will be a likely cause of scuffing. Some of the more common factors are as follows.

1 A surface finish, in a new cylinder liner, which is too smooth, and thus has inadequate oil-retention properties. The surface of the liner must be smooth, but the honing marks at 45° to its axis should be deep enough to retain some lubricant and allow it to spread.

2 Sections of piston ring may have a high localized wall pressure on the liner during the initial period of bedding, leading to oil-film breakdown in the region of the highly loaded piston-ring part and possible subsequent scuffing.

3 Bad selection of cylinder lubricant, i.e. viscosity too high or too low. Using cylinder lubricant with too high viscosity and running engine with cooling-water temperature too low. Or, using a lubricant with too low a viscosity and holding cooling-water temperature too high. Incorrect proportions of additives in lubricant.

4 Absence of an upward-scraping scraper ring in some crosshead engine pistons, or wrongly fitted scraper ring.

5 Insufficient number of lubricating oil entry points in cylinder liner.

6 No oil grooves in cylinder liner.

7 Defective cylinder lubricator, or defective lubricating oil pipe to cylinder from lubricator.

After the piston rings have become bedded into the cylinder liner and lubrication is kept at the correct level, scuffing rarely occurs. In two-stroke cycle crosshead engines fitted with piston skirts scuffing sometimes occurs on the cylinder liner and piston skirt surfaces after the cylinders are run-in. This is due to broken or defective rings allowing blow-past which then overheats the skirt on one side, causing the skirt to distort and rub on the cylinder liner.

■ **9.38 What are preformed piston rings?**

Preformed piston rings are made to suit the difficult operating conditions found in some types of highly rated engines. For example, in certain types of cross- and loop-scavenged two-stroke cycle engines it was found that the temperature differences across the radial width of the top piston ring set up thermal stresses which distorted the ring, causing the ends of the ring to warp radially outwards; the ends then break off if they contact the upper or lower sides of the scavenge or exhaust ports.

The radial outward pressure pattern of the ends of the piston ring can be modified by giving them a residual stress when they are in the cold condition. The outward-acting thermal stress causing the ends to warp outwards at operating temperature is then nullified by the residual stress. The ring is then able to work without contacting the ports and end breakage is prevented.

Preformed piston rings are those rings which have the wall-pressure pattern of the ring modified to suit particular conditions found in particular types of highly rated engines.

Note The edges of the scavenge and exhaust ports in cross- and loop-scavenged engines must have the radius at the edges of the ports rounded off because the liner wears in way of the ports.

The radius of the scavenge ports in uniflow-scavenged engines is not so critical because the ports extend for the full circumference of the cylinder liner. The ports are much smaller in size and the material between the ports does not wear at the same rate as the relatively narrow port bars in loop- and cross-scavenged engines.

10

CRANKSHAFTS, CAMSHAFTS, CONNECTING-RODS, CROSSHEADS, SLIPPERS

■ **10.1 Name the various manufacturing methods of producing crankshafts for the various kinds of diesel engine used in marine practice.**

Small, medium, and high-speed engines, such as those used for lifeboats, emergency start-up air compressors, emergency lighting sets and the like, may have steel crankshafts produced by drop-forging methods for engines having up to two cylinders, or some times four cylinders. Engines of three cylinders or more may have the crankshaft cast with special cast-irons, or it may be made by normal forging methods.

In larger medium- and high-speed engines of the type used for electrical power generation, and up to the sizes used for ship propulsion, the crankshaft is normally forged in one piece from a steel billet.

Large, slow-speed engines have crankshafts produced from steel castings or forgings. The various parts making up the crankshaft are machined and assembled by shrink-fitting processes or welding processes and after assembly the crankshaft is finish-machined.

■ **10.2 Explain, with reference to crankshafts, the meanings of the terms 'fully built' and 'semi-built'?**

The parts forming a crankshaft are the journals, crank webs (commonly called webs) and the crankpins.

In fully built crankshafts the journals and crankpins are made from forgings, the crankshaft webs may be forged or cast, and each part is made as a separate piece. The holes in a pair of crank webs are accurately machined to a suitable size for shrink-fitting on to a crankpin. After two webs have been shrunk on to a pin, a web assembly is formed, and after the requisite number of web assemblies are made the remaining holes in the crank webs that take the journals are sometimes line-bored so that they are in correct alignment with one another. In other cases a pair of webs is bored together so that the hole centres are coincident. The various web assemblies are then shrunk on to the journals and the crankshaft is built up.

In semi-built crankshafts the web assembly is one piece, made from a steel casting or a forging. It is machined and the crankshaft is put together by shrink-fitting the webs on to the journals.

■   10.3   How are solid-forged crankshafts made?

Solid-forged, crankshafts are made from a suitably sized steel billet. The heated billet is held in a porter bar and worked down under a steam hammer or hydraulic forging press. Small crankshafts will be worked under a hammer, while larger crankshafts will be worked under a press, which for larger work gives superiority in quality, cost and output. The end of the heated billet is first worked to form an end coupling flange and journal. The next section worked will be that part forming two crank webs and a crankpin, which is in the form of a block. The third section to be worked will be the second journal, followed by the second block forged at the correct angle to the first block, the angle between each block being equivalent to the crank angles of the finished crankshaft. The forging is then continued through the crankshaft until it is completed. The colour of the material is carefully watched during forging; the heating of the billet is done slowly and is carefully controlled so that the steel is worked at the correct temperature. If the heating is done too quickly it would be possible to have the surface of the forging up to welding temperature and the inside of the material too cool, thus leading to faulty forgings due to insufficient working of the interior material of the forging.

The forging is heat-treated and after cooling is set up on the marking table for marking out. Generally the journals are roughed out, then each block forming two webs and the crankpin is machined so that the profile of the outside of the crank webs is produced. The crankshaft at this stage is rough machined all over except for the inside surfaces of the webs and the crankpin which still remains in the solid block. The journals and the sides of the webs adjacent to the journal are finish-machined. Material is then cut away from the block (not flame cut) to form the shape of the crankpin and a pair of webs. A pair of webs and a crankpin is produced in the rough stage.

The crankpin and the inside of the webs are then machined. The crankpin is usually machined by setting up in a special crankpin machining tool, in which the crankshaft is held stationary and the cutting head holding the cutting bits revolves around the crankpin. The oil passages and holes are also drilled. After final finish-machining, the crankpins and journals are often hardened (which is necessary if thin-wall bearings are used) and then ground to the finish sizes within close tolerances. If the crankshaft is for a highly rated engine in the medium- or high-speed range the inside surface of the oil holes in the crankshaft may be ground to a very smooth finish to remove surface blemishes which could act as stress raisers, the fillet radius at the ends of pins and journals being sometimes ground for the same reasons.

■ 10.4 What do you understand by the term 'grain flow' when related to crankshafts?

For many purposes steels can normally be regarded as homogeneous, but in actual fact the crystalline structure and impurities in steel cause it to have grain, resembling the grain in wood. This is shown when sulphur prints are taken from split ingots and forgings, the grain following the direction of the polar axis of the ingot. A knowledge of this grain structure is sometimes used in the forging of solid forged steel crankshafts. The forging is made so that the grain structure follows a path like the flow of water round the bends in a stream. This grain path would be along and parallel to the axis of the first journal and then bend into a curve and follow along the line of the crank web, round into and along the axis of the crankpin, down the next web and into the next journal again, and so on throughout the length of the crankshaft.

To achieve this grain flow, the forging will be carried out in a similar manner to that described in the previous question except for the crank webs and crankpins, which are forged with formers so that the material forming the webs is turned at 90° to the axis of the journal (and ingot), to cause the grain to flow from journals through webs to crankpins. Machining is done in the same way as described in the previous question.

The various stages of forging required to produce a forged crankshaft having the grain flow pass axially along the crankjournals and crankpins and then round the crankwebs are shown in Figs. 10.1 (a) through to (e).

After the ingot is forged down to its proper diameter sections of the forging are heated as shown in (a). Pressure is then applied axially along the shaft to swell up one or two sections which later form the crank webs. The 'jumped-up' section or sections are shown in (b) or (b-1) together with the grain direction along the forging. After the third heat pressure is applied with dies as shown in (c), the finalized grain flow is shown in (d) and the final forging is shown in (e).

Note Die-forged or drop-forged crankshafts usually have grain-flow characteristics.

■ 10.5 What advantages do crankshafts forged to give grain flow have over those where the crank webs and pin are forged as a solid block?

When the crank webs and pin are forged as a solid block the grain of the steel runs along the journal and *across* the crank web. In grain-flow forging the grain runs *along* the throw of the web and this gives the crankshaft better fatigue resistance.

■ 10.6 How is the shrink-fitting process carried out for assembling the parts making up a large engine crankshaft?

The holes in the crank webs are bored very slightly smaller than the crankpins and journals which they have to accommodate. The pins and journals are left at workshop temperature, while the web bore in which a pin or journal is to be

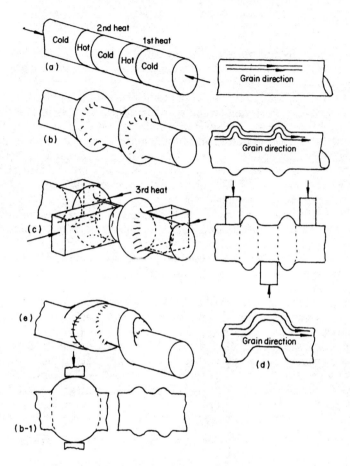

Fig. 10.1 Various stages in the forging of grain flow forged crankshafts.
(a) The first stage may be carried out with one or two heats as shown in sketch
(b) (two heads) or sketch (b-1) (one heat).
(b) and (b-1) Pressure on the ends of the section causes it to make either of the shapes shown.
(c) Producing the cranked section with dies and a hydraulic press.
(d) and (e) Direction of grain flow and finished forging.

shrunk is heated slowly with gas burners both inside the hole and round the web, every care being taken with the heating to avoid distortion. When the temperature of the web reaches approximately 377°C the hole will, by thermal expansion, be large enough for the pin or journal to be entered into it, and this is done fairly quickly; usually the web is in a horizontal position and the cold pin or journal is lowered into it.

Crankshafts may be assembled in either the vertical or horizontal position. Vertical assembly is usually done over a pit; as the parts are built up the crankshaft is lowered into the pit so that work goes on always at floor level. When crankshafts are built up horizontally they are often built up from assemblies that have been built in the vertical position.

After the crankshaft parts are shrunk together they are set up in a large lathe and the journals are checked for throw. Throw errors are machined out so that the polar axis of each journal is in the same straight line.

■ **10.7   How are crankshafts manufactured by welding processes? What are the advantages of welded crankshafts?**

Development in the manufacture of large crankshafts by welding together forged or cast components is relatively new. Approval by the ship classification societies for this form of construction is now being given.

The form of the components being joined together by a welding process will be dictated by the final cost for the construction of the crankshaft. The form of a component for a large crankshaft could then consist of a web and one half of a crankpin and one half of a journal. For smaller crankshafts the component could consist of a pair of webs joined by the crankpin with one half of a journal being formed on the outer surface of each web.

Narrow gap welding could be used to join the parts together. This is not a welding process but adaptations of existing welding processes such as MIG (metal inert gas arc welding process) or SAW (submerged arc welding process).

Narrow gap welding is ideally suited for crankshaft construction because the amount of weld metal deposited is minimal when compared with the normal 'V' and 'J' preparation used when welding thick material. Costs are lowered because of reduced welding time; the heat input is also lowered, which has a beneficial effect in the heat affected zone; the diameter of any crankshaft likely to be made comes within the limits of the weld thickness the system can handle; the materials likely to be used in crankshaft manufacture from normal carbon steels to higher carbon steels through to alloy steels can also be handled.

The weld metal can be put down with a multipass weld, either from two wire electrodes in the first-mentioned system with an inert gas shroud protecting the weld, or by a single wire electrode protected by an inert gas shroud or by submerging the arc in a suitable flux provided that its weight allows it to float easily on the pool of molten metal and its coefficient of expansion allows it to break away cleanly and quickly without any fear of slag inclusions in the weld. With the submerged arc system, alloying elements can be incorporated in the flux when welding alloy steels.

With single wire electrodes the weld can be made with a pass as wide as the groove, or the groove can be made up in two passes to give better side penetration.

One of the requirements of narrow gap welding is for an accurate set up so that the width of the gap is the same over the depth of the weld. The sections of shafts being welded are set up with the polar axis of each section concentric on a horizontal plane, the welding being carried out in the downhand position. The shaft rotates to give a suitable relative travel speed for the welding head. The weld pass is then continuous and in the form of a series of spirals.

The welding is automatically controlled and quality control is made by observation of the bead.

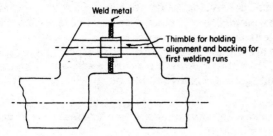

Fig. 10.2   Narrow gap-welded crankshaft fabricated with cast steel sections.

Another requirement of a welded crankshaft is for the journals and pins to be hollow. Backing rings are then used to obtain alignment and for the start of the weld. When welding is finished the ring and the initial passes are machined out to remove any doubtful material from the inner part of the weld.

The usual heat treatment and testing required for high quality welding are carried out prior to finish-machining.

The main advantages of welded crankshafts are savings in weight when compared with crankshafts constructed in the normal manner by shrink-fit assembly, lower cost arising out of the reduced size of parts being forged, and ease of forging the pieces to obtain good grain-flow characteristics in the finished components.

The sectional profile of a welded crankshaft can be made similarly to that of a 'cast-iron' crankshaft. That is, the main masses of metal are located in those regions where the highest stresses would normally occur. (Fig. 10.2)

■ **10.8   What state of stress exists in a web and crankpin or journal of a built-up crankshaft?**

The state of the stresses can be visualized if we consider the sizes of these parts before they are heated and shrunk together and then relate these sizes to those found when shrinking has taken place and the parts have cooled to normal temperature.

After the shrink fit of the pin or journal is made in a crank web the diameter of the pin or journal will be reduced in the way of the shrink; similarly the diameter of the hole in the web will be increased.

During the cooling process the heated web will decrease in size and start compressing the surface of the journal; this will continue until a balance point is reached, when the force per unit area between the journal surface and the surface of the hole in the web will be the same and will be acting radially. Thus there will be a radial compressive stress acting on the mating surfaces of the pin and the hole.

If we were to take a cast-iron washer and drive a steel taper pin in the hole, the washer would eventually split along a line radial in direction and then open up, failure of the washer being due to tensile loading. In driving the taper pin into the washer we are simulating the conditions existing in the shrink fit, so in a similar manner we create a hoop tensile stress in the web after it has cooled down. As the radial thickness of the web around the hole is large, the radial compressive stress and the hoop tensile stress are more concentrated at the surface of the hole and they both diminish gradually in an outward direction.

Note For a mathematical treatise on the stresses in crank webs due to shrink fits, see any textbook on the strength of materials which deals with the theorem of Lamé, and pressures in thick cylinders and compound tubes.

■ **10.9 What would be the result if the shrinkage allowance of crankshaft crankpins and journals were (a) too great, and (b) too small?**

If the shrinkage allowance is too great there is a danger of the crankpins and journals becoming slack in the webs, while if the allowance is too small the material of the webs around the holes may be overstressed and the elastic limit of the material reached. Steel in such a condition will quickly fracture.

Note The elastic limit of steel or similar materials is reached when the relationship between stress and strain ceases to follow Hooke's Law. Beyond the elastic limit steel begins to yield under the action of the load (see also Questions 6.44, 6.45 and 10.8).

■ **10.10 What is the usual shrinkage allowance for the shrink fits of built-up crankshafts?**

The shrinkage allowance for both the pins and journals of built-up crankshafts is the same and is between 1/570 and 1/660 of the diameter of the pin or journal.

Example. Find the bore diameter to which a web must be machined to accommodate a crank pin or journal of 400 mm diameter.
The size of the hole (lower limit) is

$$400-400/570 = 400 - 0.70 = 399.30 \text{ mm}$$

and the size of the hole (upper limit) is

$$400-400/660 = 400 - 0.61 = 399.39 \text{ mm}$$

The mean bore size is therefore

$$\frac{399.39 + 399.30}{2} = 399.345 \text{ mm}$$

and the bore-size hole in the web is

$$399.35(\pm 0.02) \text{ mm}$$

■ **10.11 What are the proportions of crank webs relative to the pin or journal diameter in built crankshafts?**

The proportions of crank webs must be such that the resultant shrink fit has enough grip without overstressing the material in the web.

Generally the thickness of the web axially is made greater than 0.45 of the diameter of the pin or journal; common values now are between 0.5 and 0.575 of the diameter. Formerly this proportion was made greater. The radial thickness of the material of the web forming the eye is normally calculated from various design formulae which take into account the axial thickness of the web and the minimum yield stress of the material. In some cases the axial thickness of the web may be fixed first and then the radial thickness calculated later.

In semi-built shafts there is a restriction on the smallest distance between the side of the hole in the web and the surface of the crank pin; this distance is usually kept in excess of 0.10 of the journal diameter.

Note The bore sizes in the webs for shrink fits are calculated and take into account the yield stress of the crankshaft material, the thickness of the web in the axial and radial directions, together with other factors. Further details on this matter are covered in the rule books of the classification societies.

■ **10.12 Is it necessary to fit dowel pins in way of the pins, journals and webs of built-up crankshafts? State reasons.**

When reciprocating steam engine crankshafts were first built up by shrink-fitting the parts together, the craftsmen of those days did not have the advantage of using accurate measuring tools such as internal and external micrometers. It was then necessary to use made-up pin gauges, inside and outside calipers, and feelers, and the accuracy of measurements depended almost entirely on the 'feel' of the craftsman. In consequence dowel pins were fitted to make up for any deficiences which might arise from a 'light' shrink.

As engine builders changed over from building steam reciprocating engines to diesel engines the practice of fitting dowels was continued.

When some crankshafts fractured in service, investigation into the failures showed that the dowel pin and hole (half in the web and half in the pin or journal) acted as a high stress raiser which initiated a fracture. This, together with more accurate methods of measurement brought about by the common use of the micrometer, produced more reliable shrink fits and caused the practice of fitting dowels to be discontinued. Their use is now generally prohibited by the engine construction rules of the various ship classification societies.

■ **10.13 If dowel pins or keys are not fitted in the parts of a built-up crankshaft how is it possible to drive the propeller without the various parts slipping?**

When the parts of a built-up crankshaft are assembled with shrink fits, the mating surfaces of the pins, journals and webs are subject to a very high radial compressive stress. This radial stress sets up an enormous frictional resistance which prevents the webs slipping around the journals and on the crankpins when the engine is operating. The radial compressive stress on the contact surfaces of the shrink fit is approximately 77 MN/m^2 (5 tons/in^2).

■ **10.14   Do the parts of a built-up crankshaft ever slip relative to one another?**

With normal service usage the various parts of a built-up crankshaft do not slip, but many instances of shrink-fit slip are known. The most common cause comes about in attempting to start an engine when water or fuel leakage has occurred and partially filled the space in the cylinder above the piston. When the engine is first started the piston comes up and the water or oil being incompressible causes the engine to stop; the momentum of the moving parts is then almost immediately absorbed as the shrink-fitted parts slip. The consequences of this are usually disastrous and the extensive damage is enormously costly to repair. In other instances (outside the control of the ship's engineer officers) damage of this nature has come about when the propeller has struck some fixed or heavy object.

Note   Before putting starting air on to an engine after a shutdown or long standby, the engine must be given a full turn with the turning gear, while the open indicator cocks are observed to ascertain that no water or oil is discharged while the engine is being turned. This may present difficulties, when an engine is kept on standby while the ship is at anchor. With good communication and understanding between the Chief Engineer Officer and the Master, arrangements can be made for the engine to be checked using the turning gear before 'swinging' the engine on starting air. In the case of diesel engines driving auxiliaries such as generators, the engine must be 'barred' round a full turn before starting, even if a solid-forged crankshaft is fitted in the engine concerned.

■ **10.15   How would you establish whether slip has occurred between any parts of a built-up crankshaft?**

After a crankshaft has been built the protrusion of the journal beyond the flat surface on the inside of the web is locally flushed with a grindstone so that a small rectangular flat surface is created half in the web and half in the journal. A light chisel-cut in a radial line is then made with a very sharp flat chisel so that half of the cut is in the end of the journal and the other half in the web. If slip occurs between the web and the journal the line cut by the chisel will separate. Each end of the crankpin and the outside of the crankwebs are treated in a similar manner on fully built crankshafts.

These reference marks, as they are called, should be sighted each time a crankcase internal examination is made and their correctness should be recorded with any other remarks made in the log-book entry covering the crankcase examination.

■ **10.16   Assume that a large, slow-speed main engine crankshaft journal has slipped in its web. Would it be possible to effect a permanent repair without removing the crankshaft from the engine? If such a repair is possible, how would it be carried out?**

Once the cause of the slipped journal has been established and careful examination shows that no other parts of the engine are seriously damaged, the savings that can be made in both time and money by making a corrective repair to the crankshaft without taking it out of the engine are obviously very attractive. For this reason repairs of this nature have been carried out on many occasions, and with satisfactory end results.

To make such a repair, the various parts of the engine such as the main bearings, connecting rods, etc. adjacent to the damage are removed from the crankcase so that good access is obtained to the repair site.

A hole is then drilled into the end of the journal that has slipped, far enough to meet with the oil hole which supplies oil from the main bearing to the crankpin. In some engines there will already be a central oil hole, in which case only the end plug will need removing.

The hole is drilled in the centre of the journal and follows its polar axis, its size being calculated so that it can be made as large as possible without seriously impairing the strength of the journal, or reducing the radial compressive stress on the shrink surfaces.

Sections of the oil hole are blanked off so that there is a clear passage only from the newly drilled hole to the oil hole in the journal. Thermocouples or resistance elements are cemented on to the ends of the journal adjacent to the web and on the curved surface of the journal near the web. The ends of the journal, the web and the circular part of the journal are then well covered with fireproof blankets or similar non-flammable heat-insulating material. A pipe is led from the oil hole in the journal to some point outside the engine room. Liquid nitrogen is then piped to the newly drilled hole in the end of the journal and bled through a throttling or reducing valve into the space behind the hole. The liquid nitrogen then flashes to gas in the journal and exhausts itself away to atmosphere. In flashing to gas, the latent heat required for making the change of state is taken from the journal and the journal is cooled down. The thermocouples or resistance elements connected to instruments give a temperature read-out of the parts being cooled.

The crankshaft must be supported at suitable places, and hydraulic jacks are placed in such positions that a torque can be applied to the journal from adjacent webs, the direction that the torque is applied being such that it will move the web in the opposite direction from the slip. Two jacks are normally used, one to supply the correcting torque and the other as a form of control. Calculations and graphs are made to cover the temperature to which the parts being cooled will be reduced. When the temperature is sufficiently reduced the hydraulic jacks are pumped up with separate pumps so that each jack can be individually controlled.

The insulation covering the web to be heated is removed and propane or similar burners are then used to heat the web over as large a circumference as possible so that localized overheating does not occur. The witness or reference mark on the journal and web is kept under observation together with the pressure gauges showing the hydraulic jack pressure. When the pressure on the jack supplying the corrective torque drops it indicates that movement between the journal and web has commenced. With the witness mark under observation further torque is carefully applied until the witness mark is in line again. If the mark is passed, the control jack must be activated to reverse the movement. When the witness mark is correctly re-aligned, the nitrogen supply is shut off, the heating is stopped and the heated parts are allowed to cool down again while the journal warms up; the shrink then restores itself.

Prior to fitting the parts back in the engine, the journal in way of the repair is checked for any errors which, if large in extent, must be corrected.

After a repair of this nature a trial of the main engine must be carried out, and if the engine operates satisfactorily the ship would be allowed to return to service. The crankshaft and bearings would then be examined at frequent time intervals which would be gradually extended. After a period of satisfactory operation the repair would be considered permanent.

Note A repair of this nature must be undertaken with great care. In a book of this scope it is impossible to go fully into all the details that must be planned to cover the exigencies that might arise.

The safety aspects must also be carefully studied as it can be seen that it would be a simple matter to fill up the lower part of the engine room with nitrogen. Liquid nitrogen at normal temperature is under high pressure so the method of piping the liquified gas to the parts to be cooled must be carefully studied. The method of blanking the holes in the journal and the outlet pipe from the journal is also most important.

The heating of the web must also be carefully controlled or distortion will occur.

An inert and non-toxic gas must always be used. Nitrogen is probably the best gas for the purpose as it has a very low boiling point ($-195°C$) at atmospheric pressure, with a latent heat of vaporization of 200.1 kJ/kg.

Other materials have been used to cool the journal; these include circulation of methyl alcohol cooled by 'dry ice' (solidified CO_2) and liquid CO_2. When any forms of alcohol are brought into the engine room great care must be taken against possible formation of explosive vapours.

■ **10.17 What loads and stresses is a crankshaft subjected to while in operation? State any causes from inside or outside the engine that could increase these stresses.**

The loads to which a crankshaft is subjected come initially from the compression of air and combustion of fuel, which cause rotation of the crankshaft. In order to gain some idea of the loads coming on to a crankshaft it simplifies matters if we consider only that part of the crankshaft associated with one engine cylinder, and consider it as static and subjected only to a constant load at this point. In this way we can more easily analyse what is actually happening.

If we consider the crankshaft on top-dead-centre we can see it is in effect a beam supported on the main bearings. The load it is subjected to will be the product of the piston area and the gas pressure in the cylinder at this time. This load will act downwards, and as the crankshaft will be supported by the bearings, the bearing reaction will be upwards. From this we can see that the load on the crankpin will give it a curvature downwards which will put the top part of the pin in compression and the lower part in tension. The two journals will receive a similar curvature and they will then be supported only on the inside edge (adjacent to the crank) of each main bearing.

The webs of the crank will be deflected similarly, and the inside surfaces of the webs will be in tension and the outside surfaces in compression. The whole of the cross-section of the web will also be subjected to half of the total load acting on the piston, which will be compressive.

If the crank is put at 90° from the top-dead-centre position the gases will have expanded and the load on the piston will be between 11% and 14% of the load when it was at top-dead-centre. The crank mechanism will still be acting as a beam, supported on the edges of the bearing, but the crankpin, webs and journal will have been turned through 90°, and the parts of the pin and journal which were subjected to the maximum tensile and compressive stress when the piston was at top-dead-centre will now be lying in the plane of the neutral axis. The action of the piston load will be downwards and so also will the curvature of the pin and journal, which will consequently be subjected to tensile and compressive stresses due to the bending moments set up by the piston load and bearing reaction. The stresses on the crank web will be greatly changed, since the load on the piston and the journal reaction from the twisting moment will cause the web to act as a beam and the bearing reaction will cause it to be subjected to twisting at the same time. The angularity of the connecting rod will also induce a tensile load on the crank webs.

When the crank is turned to bottom-dead-centre the gas load on the piston will be a mere fraction of that at top-dead-centre position, the stresses on the pins, journals and webs will be reversed, and consequently stresses that were tensile in the top centre position become compressive in the bottom centre position, and vice versa. The magnitude of the stresses is, however, greatly reduced due to the reduction in piston load.

This analysis serves to show that the stresses on a crankshaft are changing both in magnitude and direction as the engine rotates and that the torsional stresses brought about by the shaft reaction or the load being driven are additive to those mentioned. It can also be seen that the action of the loads on adjacent cranks induces further stresses into the crankshaft. Stress patterns in crankshafts are extremely complex.

When bearings wear we can see that if the wear-down between adjacent bearings is unequal the alignment of the crankshaft will be altered, this alteration further increasing the stresses to which a crankshaft is normally subjected.

Causes outside the engine may similarly produce misalignment, due to different loading patterns in the cargo spaces causing changes in the hull deflection. This in turn causes changes in the crankshaft alignment, since the engine bedplate and crankshaft, being less rigid than the hull, follow the hull movement.

In heavy weather as waves pass along a ship the increased buoyancy from the wave crest causes hull movement which again alters the alignment of the crankshaft and increases the stresses on it. When causes outside the engine produce crankshaft misalignment, this is the result of changes in the positions of the bearings relative to some norm. The loads on the main bearings can then be greatly increased, and this is one of the causes of the main bearings wearing down at different rates.

■ **10.18 What are the objections to drilling oil-ways in crankshafts? How are the effects minimized?**

It has been seen (Question 10.17) that the parts of a crankshaft are subjected to stresses which are constantly changing both in magnitude and direction throughout an engine cycle. When oil holes are drilled and channels are cut in crankshafts they act as stress-raisers which reduce the ability of the parts concerned to resist fatigue. The objections to drilled oil-ways are minimized by situating the oil holes in the regions of the crankshaft where they will be subjected to the least stress. An example of bad practice is seen in some older engines where the oil supplied to the crankpin bearing entered the bearing from four exit points in the crankpin obtained by drilling two holes, across the crankpin at 90° to each other, one of which came through the crankpin surface where it was subjected to the maximum tensile loading.

Nowadays the holes in the crankpin may be positioned horizontally across the diameter when the crank is at top dead centre. The oil hole then goes across the plane of the neutral axis of the crankpin when it is subjected to maximum bending. In other cases it is located on the top of the crankpin when at top centre, and goes radially downwards into the pin for half the diameter, where it joins a hole following the polar axis of the crankpin. When the hole is in this location the material of the pin is subjected to compressive loading when the pin is at top centre, to a low tensile load when it is at bottom centre, and at approximately half-piston stroke the hole is along the neutral axis of the crankpin.

The holes must be left smooth inside their bore; they must be made with sharp drills so that drill swarf does not score the surface of the holes, particularly near the working surfaces of pins and journals.

Where a hole and the circular surface of a pin meet, the corner must be carefully radiused with a clean smooth surface. Oil holes in journals are also given a radius in a similar manner.

In some cases no oil holes are drilled in the crankshaft, which is left solid. Lubrication of the crankpin bearing is obtained from the crosshead which is supplied with lubricant via a set of telescopic pipes. The lubricant then passes through a hole bored through the connecting rod.

■ **10.19 Describe some of the difficulties which arise in crankshaft design when the bore/stroke ratio approaches unity or when bore/stroke is over-square. How does this affect the engine operator?**

In engines with ratios of bore to stroke around unity the diameters of the crankpin and journal become relatively large in relation to the web dimensions. The crankpin-surface circle then overlaps the journal-surface circle when they are viewed end on. If the elevation of a drawing of a crankshaft of this type is studied it can be seen that the overlap of the crankpin and journal is such that the thinnest section of the web is across a diagonal running from the lower pin surface of the crankpin across the web and upwards to the top surface of the journal. The rigidity of the crank web across this diagonal is much less than in the other parts of the web, due to the influence of the lower portion of the crankpin and the upper portion of the journal. If the bore/stroke ratio is over-square the effect is more pronounced. The less-rigid section of the crank web makes for problems in design to minimize its effect as a stress-raiser or notch.

In turn, this affects the ship's Engineer Officer, who must watch more closely the alignment of this type of crankshaft. It is obvious that the amount of misalignment allowed will be much less than in a medium-speed engine with a lower bore/stroke ratio, as in such engines the notch effect across the webs is much less pronounced.

■ **10.20   What examinations would you carry out on the main bearings and crankshaft before, during, and after removal of the bearings from the engine?**

The bearing nuts, oil pipes, locking devices and split pins must be examined to see that nothing is abnormal prior to dismantling. If it is possible to measure the bearing clearance with a feeler gauge before dismantling the upper half of the bearing, the clearance should be measured and recorded. After removal of the upper half of the main bearing, the wear-down of the lower half should be taken with a bridge gauge and a feeler gauge and recorded.

The lower half of the main bearing is then removed and brought out of the crankcase. The white-metal of the bearing should be examined for cracks, which may be developing in the wearing surface of the bearing, and if present the white-metal around the cracks should be lightly tapped with the ball pein of a small hammer or with a large copper coin to check that the white-metal is still bonding with the shell of the bearing. The remaining parts of the bearing should also be checked to establish that the bond remains solid.

Attention must be given to any oil grooves which may be cut in the main bearings; if the sides of the groove are sharp they must be lightly scraped with a scraper to establish an oil lead-in. If the bearing has slightly wiped at any time (and not been previously noticed) any debris should be chipped out of the oil grooves and the sides of the groove scraped. Generally the lower half of the bearing requires a much more careful examination than the upper half. The outer surface of the bearing shells should also be examined for any fretting which may have taken place if the bearing was left or worked slack.

On the crankshaft the working surfaces of the journal are examined for signs of corrosion or pitting caused by water or acid contamination of the lubricating oil. The area around oil holes is a critical part and must also be carefully examined for cracks, which usually emanate at about 45° to the axis of the crankshaft, and in some cases may be up to four or more in number, forming radial lines from the oil hole. The surfaces of the crank webs are also examined for any signs of cracks, particularly in the higher-stressed regions midway between pin and journal. With built-up crankshafts the shrink fit reference marks should also be sighted.

For medium-speed engines it is most likely that the main bearings will be of the thin-walled type and the crankshaft solid-forged. In such engines the treatment and examinations are similar. The examination of bearing shells for cracking should be carried out in a similar manner as for white-metal bearings. The oil lead-in pockets or chambers at the butts of the bearing should have their edges examined and if necessary they should be eased lightly on the edge. Medium-speed solid-forged crankshafts must also be carefully examined around oil holes for any signs of cracking. The fillets between the journal and web area require careful examination, particularly so in the arc of the fillet between the position corresponding to 10 o'clock and 2 o'clock when the piston is on top-dead-centre.

The location where fatigue failure may commence is shown in Figs. 10.3 (a) or (b). Any crack which may be present will be more difficult to find in crankshafts having re-entrant or negative fillets. This type of fillet is more common today; it is often necessary to use a mirror and a flashlight to view the surface of the fillet. Cracks usually start at the bottom of re-entrant fillets. With normal fillets any crack will usually start at the edge of a fillet run-out. These locations are shown with arrows in the figures. Cracks may also start at the fillets of oil holes, this form of crack usually runs at 45 degrees to the shaft axis. Once initiated, any fatigue crack will grow until complete fracture occurs. See question 6.50 dealing with non-destructive testing.

Before replacing the lower half of the bearing it is useful to take a bridge gauge reading to establish the natural sag of the crankshaft with the bearing

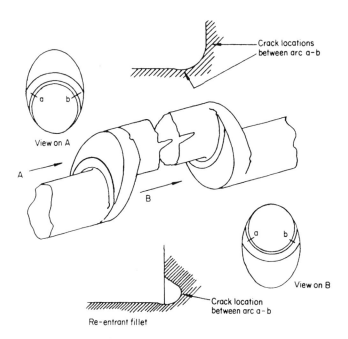

Fig. 10.3 Usual locations of crankshaft failures. Note the two forms of fillet shown in the sketches. The upper fillet form is a normal form of fillet. The lower form of re-entrant or negative fillet is often used today as it is possible to obtain a shorter length to the crankshaft without a reduction in bearing surface length. Another location of possible failure is around oil holes with cracks running at 45 degrees to the axis of the crankshaft.

removed, if the figure is not known or recorded. This is a useful figure to have when taking clock gauge readings of crankshaft deflections (see questions dealing with this matter in Chapter 14). (Fig. 10.3)

■ **10.21 Describe the examination carried out before dismantling main bearings. How are main bearings removed from the bearing pocket?**

Before dismantling main bearings an examination must be made to check that the marks locating the position of the keep and bearing shell are correct. The marks, usually cut in with letter and figure type, signify the bearing number and the way it is located. It may be that the spare main bearing has been fitted in an engine, in which case the direction in which the bearing is located may not be indicated, and some location identifying mark should then be made with a letter or figure stamp.

Various means are adopted to facilitate removal of the lower half of the bearing, a common method being to fit a special clamp on the crank web adjacent to the bearing which has to be removed. The engine is then turned slowly with the turning gear and the bearing is pushed round out of its pocket by a projection on the clamp. When the bearing has been turned for a half revolution it will be sitting on the top of the journal free of the pocket. It is then lifted off the journal and swung clear of the crank webs with a set of chain blocks rigged above the bearing and another set rigged near the crankcase door.

If the bearing bore has been machined slightly eccentric to the outer shell of the bearing, the engine must be turned in one direction to remove the bearing, and in the opposite direction to replace it.

With some very large engines the support for the bearing in the pocket of the bedplate is made with a small hollow area so that a pocket is formed. This pocket is then filled with oil and the oil is put under pressure with a hand pressure pump. The oil under pressure lifts the bearing in the pocket and reduces the turning effort required to turn the bearing out of the pocket.

In other engines it is necessary to place hydraulic jacks under the webs on each side of the bearing to be removed. Wood blocks are placed below the jacks to distribute the load evenly over the bottom of the crank pit, and the jacks are pumped up so that the weight of the crankshaft is transferred from the bearing to the jacks. The bearing is then free in the pocket. An eye bolt is fitted into the side of the bearing and the bearing is pulled round in its pocket with a chain block. When the bearing is half out of its pocket the eye bolt is transferred to another position so that the bearing can be completely turned out of its pocket.

■ **10.22 What examinations must be carried out when a crankpin bearing of a large, slow-speed engine is opened up for survey?**

When the bearing has been exposed the white-metal is examined for cracks, signs of wiping, and condition of oil grooves which should be dealt with as mentioned in Question 10.20. The foot of the connecting rod must be checked for cracks at the sides of the bolt holes and the underside of the foot must also be examined for cracks running across under the line of the run-out of the fillet, particularly so if a compression plate is fitted between the connecting rod and the bearing. The contact surface between the connecting rod and the top flat surface of the crankpin bearing is also checked for fretting.

The threads of the bolts must be examined for deformation and stretching; the bolts should also be hung up and sounded with a light hammer for detecting cracks. The fillets in the reduced sections of the bolt must be examined for the start of cracks which could lead to eventual bolt fracture. If any bolts are suspect, crack-detection tests should be applied and if any cracks are confirmed new bolts must be fitted. The crankpin should be examined for cracks, particularly in the region of the oil hole in the crankpin. The condition of the working surface of the pin and the reference marks between crankpin and crank web are also important points to note.

It is good practice to place a straightedge on the top of the crankpin when on top centre, and at 90°, 180° and 270° from this position. The profile of the wearing surface of the pin can then be checked and some indication will be given if high spots or areas are being left at the centre of the crankpin in way of the circumferential oil groove (if fitted) of the bearing and at the sides of the crankpin. Although no action may be taken to remove the high areas on the unworn parts of the crankpin, it is as well to be aware of their presence in case of difficulties arising in the alignment of the running gear or operation of the running-gear bearings.

■ 10.23   How does failure in a crankshaft usually occur?

In Question 10.17 we attempted to analyse the stresses in a crankshaft and found that they were highly complex. It was also noted that crankpins and journals were subjected to bending action, which results in push-pull forces and in consequence makes a crankshaft prone to fatigue failure. When a crankshaft is designed for a prototype engine, a careful study is made of the stress-concentration factors at oil holes, the ends of fillets, and across crank webs. By altering the geometry of the crankshaft the stress concentration factor can be improved, but stress concentrations will always remain. If an engine crankshaft should fail, it usually does so due to fatigue at some point where concentration of stress has been high. These points are oil holes in crankpins, oil holes in journals to a lesser extent, and at the run-out of fillets on the underside of crankpins in solid-forged crankshafts and semi-built-up crankshafts.

Although the design of a crankshaft may be good, misalignment of the crankshaft journal bearings will increase the stress range during operation and may cause fatigue failure at one of the points where a stress concentration occurs. Hardened crankshafts may also fail due to overheating of bearings from lack of lubricating oil or bad fitting. The overheated area may cause surface cracks to appear in the hardened layer, which then progress inwards until failure occurs. The crack, once initiated, acts as a stress raiser.

Failure of crankshafts has also occurred in the flange connecting the engine to the intermediate shaft or electrical rotor. This is likely to occur if the radius of the fillet between shaft and flange is large, making it necessary to make deep facings into the fillet for the coupling-bolt heads or nut landings.

Failure may also occur due to excessively large stresses caused by torsional vibration. This type of failure is dealt with in Chapter 17.

■ **10.24 How can the fatigue resistance of a crankshaft material be increased?**

The locations of the highest stress ranges are the locations where fatigue failure may occur in a crankshaft. These locations are in the regions of the fillets at the junction of the crankpins and webs, and the journals and webs. (Fig. 10.3)

Normally fatigue failure occurs at some point where local damage or a flaw becomes the nucleation site of the failure. When the material in this region is subjected to a cyclic stress that is tensile during some part of the cycle, a crack may start to grow from this point and will proceed until fracture occurs or the crack is found during a survey.

If the material in the region of the highest fluctuating stresses is given a residual compressive stress by cold working the surface, the possibility of fatigue failure is reduced.

If the surface is locally hardened to give a higher tensile stress a similar result can be expected.

Apart from normal hardening processes applied to the surfaces of crankshaft pins, journals and webs, as mentioned in Question 6.11, there are another two methods for surface hardening but these are not suitable for working surfaces.

The fillets on crankpins and journals may be work hardened by rolling the fillets with heavily loaded hardened steel rollers or by shot peening with hard shot. A residual compressive stress in the fillet surface is then obtained.

The depth of the compressed layer is of great importance. In the case of rolling treatments the depth cannot be easily measured. Procedures are established to determine the required amount of rolling by measuring the depth of the effect in test samples after their destruction.

In the case of shot peening, test strips can be subjected to the same treatment as the fillets, the degree of curvature of the test strip giving an indication of the depth of the induced stress.

Valve springs are often shot peened to increase their resistance to fatigue failure, and shot peening treatment is sometimes applied to the root fillets in gear teeth.

One of the problems related to these treatments is fading of the surface effect and the greater problems that may arise if this loss occurs.

It is pertinent to remark that some builders of automotive engines use shot peening treatment for crankshafts in engines that operate mainly under part load condition and are only subjected to maximum load at infrequent intervals.

Generally rolling treatments are applicable to large and small crankshafts whereas shot peening can only be considered suitable for the smallest crankshafts.

Before any of these treatments are used on crankshafts for marine engines the treatment must be approved by the classification society concerned.

Note   This subject can be further studied in books dealing with the strength of engineering materials and their fracture, and books on the subject of fracture mechanics.

■ **10.25 How would you deal with a suspected crack in a main engine crankshaft of (a) a slow-speed, direct-driven single-screw installation or geared, medium-speed engine installation with one engine; (b) a twin-**

engine, medium-speed installation geared to a single screw; (c) a twin-screw installation?

Before any real course of action can be taken it is absolutely essential that the suspicion is fully confirmed and a crack is found to exist. This can be done only by careful examination and use of dye penetrant tests or magnetic particle tests (see Question 6.50). Once the suspicion of a crack is confirmed the situation for the Chief Engineer Officer and the Master changes because the ship cannot function normally and may possibly be classified as a casualty. The shipowner's technical and operating departments must be informed immediately so that the technical department can attend to repair arrangements, and the operating department attend to the commercial and legal liabilities of the owner.

(a)   In single-screw engine installations the ship should not proceed until the following matters have been attended to. First, the depth of the crack must be ascertained. If it is shallow, it may be ground out so that its effect as a stress-raiser is minimized. The alignment of the bearing journals must then be checked in case misalignment is the cause of the failure, and other likely causes of failure should also be checked. Any likely causes found must be remedied.

In conjunction with the classification society's surveyor (Lloyd's Register, A.B.S., etc.) a certificate can be obtained which would permit the ship to proceed at reduced power. In the case of deep cracks some strengthening will have to be arranged before the ship can be allowed to proceed to a repair port.

(b)   In the case of twin-engined installations the conditions generally will be as for (a) except that it is an easier matter to get a classification certificate to allow the ship to proceed on one engine.

(c)   The conditions are similar to (b).

Note   In the case of such a serious matter as a defective crankshaft, once the crack has been absolutely confirmed it is imperative that the Master and Owners be notified. No repair work other than preparatory removal of parts should be attempted until the confirmation for other work is received from the Owner's technical department, who will be in a much better position to call for expert advice and assistance, deal with classification and insurance matters, and arrange for temporary repairs if necessary and then the permanent repair.

The nature of any strengthening or reinforcing work to effect a temporary repair would depend on where the crack was found; for example, a crack in a web would require a totally different type of treatment from a crack in a crank-pin or journal. Cracks at the after end of a crankshaft are generally more critical than at the forward end. The amount of available space in a crankcase may be a critical factor: any reinforcing or strengthening must be within the limits of crankcase space to allow free rotation.

■ **10.26** What special importance is placed on the side-rod crankpin and crosshead bearing bolts of opposed piston engines? Why are they considered in a different manner from the crosshead and crankpin bolts of normal crosshead engines?

The side-rod crankpin and crosshead bearing bolts are subjected to tensile loads from the gas pressure under the upper piston. The tensile load on the bolts from the gas pressure is further increased by the tightening action of the bolt nuts holding the bearings and running gear together. Together with loads from inertia effects, the range of tensile stress on side-rod bolts is such that they are much more liable to fatigue failure than crankpin and crosshead bearing bolts on normal slow-speed engines.

In view of side-rod bolts being prone to fatigue failure they are given a 'life' of a certain number of running hours. Special importance is put on this life period and side-rod bolts must be renewed when they have operated for the number of running hours making up the life period. Side-rod crosshead and crankpin bolts on opposed piston engines are given this recommended running-hour life-span due to the much greater possibility of fatigue failure, whereas normal slow-speed engine crosshead and crankpin bolts are not normally given a running-hour life.

■ **10.27   Why is it particularly important that the total thickness of the shims or liners for crosshead and crankpin bearing adjustment be kept the same on both sides of the bearing? What happens if the thickness varies on each side of side-rod crosshead and crankpin bearings on opposed piston engines?**

When bearings and bearing-keeps are machined for bearing bolts, bolt-head and nut-landing faces, they are machined with equal shims and the bolt-head and nut-landing faces should be absolutely parallel to one another and at 90° to the bolt hole. If at some later time during bearing adjustment the bearing is assembled with unequal thicknesses of shims or liners on opposite sides of the bearing, the landing faces for the nuts and bolt heads will be slightly out of parallel when the bolts are hardened up. This will cause the bolt to bend slightly in the hardened-up or tight position. In effect this causes a variation in the tensile stress across the bolt cross-section and seriously reduces the bolt's resistance to fatigue failure. For this reason it is imperative that the thickness of shims or liners be kept absolutely the same on opposite sides of the bearing.

In opposed piston engine side-rod crosshead and crankpin bearings, if the shim or liner thickness is not kept equal on opposite sides of the bearing the bolt life can be greatly reduced, so that failure occurs before the bolt becomes due for renewal.

While bearing-bolt breakage can have serious consequences in any engine, side-rod bolt failure in opposed piston engines can be, and usually is, particularly disastrous.

■ **10.28   If the crosshead and crankpin bolts are insufficiently tightened what effect does this have on the bolt?**

Insufficient tightening can have serious consequences even if the bearing continues to operate without becoming slack. When a bearing bolt is tightened it is subjected to tensile stress.

During operation in normal engines, changing inertia loads on the bearing keeps causes a variation in the bolt loading which is additive to the loading from tightening the bolt nuts. When a bolt is insufficiently tightened the stress range in the bolt is greater in service even though the maximum stress may be less. As fatigue is a function of stress range rather than maximum stress, an insufficiently tightened bolt is more liable to fatigue failure than one correctly tightened.

In opposed piston engines the correct degree of tightening of the side-rod top and bottom ends is of paramount importance if premature bolt failure is to be avoided.

■ **10.29 What particular attention is required for opposed piston engine side-rod crosshead and crankpin bearing bolts during the survey of the bearings?**

The side-rod bearing bolts must be examined for the start of fatigue cracks at the ends of the fillet run-out on to the smaller diameter of the bolts and at the fillets adjacent to the bolt heads. The bottom of the threads must also be carefully examined, particularly towards the end of the thread nearest to the bolt head and at the reduced section of the bolt at the thread run-out. The bolts should be hung up and checked for soundness by their ring. Dye-penetrant tests can also be used to make doubly sure.

If there has been any water contamination of the lubricating oil, corrosion pits often start at the critical areas described above. If the pitting is deep, renewal of bolts must be considered because pit holes often become the stress raisers that initiate fatigue failure.

The landing face of the heads of bolts must also be examined for burrs and minor damage which could prevent the head of the bolt landing properly; any such defects must be cleaned off before the bolts are refitted. In a similar manner feathers or dowels fitted in the bolt head should be examined and any surface defects on the feather, which could cause the feather to bind in its slot, and slot defects, must be dressed and cleaned off, thus enabling the bolt head to be pulled home square on its landing face.

When the bolts are refitted during assembly of side-rod bearings, the nuts should be pulled up lightly with a skeleton spanner. After the bolts are hand-tightened the landing face of the bolt head should be checked with the thinnest feeler-gauge blade to confirm that the head is fitting correctly on the landing face of the bearing. When new bolts are being fitted at the time of bolt changes, the diameter of the bolt head should be checked against the diameter of the head on the old bolts so that there is no fear of the head of the new bolt being too large for the spot facing on which it lands.

Note . The checks mentioned in the answer may appear troublesome, but they are necessary to prevent premature failure of side-rod bolts in service and the disastrous consequences of failure.

■ **10.30 What particular attention is required for top- and bottom-end bolts of medium-speed engines and bottom-end bolts of high-speed engines?**

The examinations carried out on top-end and bottom-end bolts for the engine types named in the question should be as set out in Question 10.29. In certain engine types the length of the bolt is measured when new, and the tightening stress is controlled by measuring the bolt during the tightening process. When the required amount of strain is measured, the bolt has been brought to the correct tension by the tightening of the nut. With materials used in certain bolts some permanent strain occurs in the bolts during their working life. When a specified amount of permanent strain is reached the bolt is considered to have reached the end of its useful life and is scrapped.

When bolts are given a life amounting to a certain number of running hours it is essential that accurate records of the running hours of individual bolts be kept so that they can then be renewed at the correct time. When bearing bolts that have a specified life are replaced with new bolts, the old bolts must be defaced or cut up so that they are unusable; this prevents inadvertent re-use of an old bolt.

■ **10.31   What is the purpose of cams and where are they fitted?**

Cams are used on diesel engines to operate the mechanisms connected with opening and closing exhaust valves and air inlet valves, driving fuel pumps, air starting valve pilot valves, etc. They are fitted to (or are part of) camshafts which are driven by the engine crankshaft.

■ **10.32   How are camshafts driven?**

Camshafts are driven by power transmitted from the engine crankshaft through roller chains or gear trains.

■ **10.33   What is the camshaft speed relative to the engine speed?**

In four-stroke cycle engines the operational cycle is completed in four piston strokes or two engine revolutions, so the camshaft must turn one revolution in this time. It, therefore, runs at half the engine crankshaft speed. In two-stroke cycle engines where the operational cycle is completed in one revolution of the crankshaft, the camshaft operates at the same speed as the engine crankshaft.

■ **10.34   Is the torque on a camshaft constant or varying?**

The torque on a camshaft varies considerably through an engine cycle due to the action of the cams on the valves and fuel-pump operating mechanisms. When a cam is opening an air inlet valve or an exhaust valve, the action of the cam in opening the valve compresses the valve springs. When it is time for the valves to close, the reaction of the compressed spring closes the valve. During the period that the valve is closing the action of the spring forces the roller on to the cam and some of the work done in compressing the spring is returned to the engine through the roller, cam, camshaft and camshaft drive. The return stroke of the fuel pump has a similar effect.

The same action occurs with exhaust valves opened by the pressure from a hydraulic pump and closed by the action of a pneumatically operated piston.

If the torque transmitted by the camshaft to a cam is plotted on a polar twisting-moment diagram, it will be seen that during the period a valve is opening the torque will be positive and when the valve is closing it will be negative. When the summation of the torque requirements for each cam is plotted it will be seen that the torque requirement for driving the camshaft varies considerably throughout the cycle. As the number of cylinders increase the variation in camshaft torque decreases.

Note Torsional vibration in the crankshaft can affect the camshaft. This matter is covered in Chapter 17.

■ **10.35 What effect does camshaft torque variation have on the camshaft drive?**

As the torque requirement to drive the camshaft varies, the load on the camshaft drive also varies. This load variation may cause problems in the driving gears which may chatter within the limits of the gear backlash, or cause the roller chains to vibrate and swing. Gear-wheel chatter and roller-chain vibration or transverse swing increase the load on the camshaft drive and may cause difficulties in service.

■ **10.36 The roller chains used to drive engine camshafts eventually increase in length while in service. How is this increase in length accommodated and how is increased transverse vibration catered for when the chain stretches?**

At regular intervals the stretch or length increase of roller chains is taken up by adjustment of the chain tensioning device. This usually consists of an idler sprocket wheel mounted at one end of a lever. At the other end of the lever is a nut fitted in a yoke. A screw passes through the nut. Turning the screw causes the lever to move about a fulcrum so that the idler sprocket wheel moves towards or away from the chain – increasing or reducing the tension.

In some engines correct chain adjustment is controlled by measuring the transverse displacement of the chain at its mid-position between two designated sprockets.

In other engines a helical spring loading device is fitted on the lever, nut and screw. As the chain stretches, the amount of compression on the spring is reduced, and restoring the spring compression to its proper figure gives the chain its correct tension.

The tensioning sprocket is fitted to the slack side of the chain. As the chain stretches and is retensioned, the camshaft sprocket is gradually retarded. This can affect valve timing and fuel-injection timing. The camshaft must then be re-positioned relative to the engine crankshaft so that the retarded timing is corrected. In some engines the camshaft couplings cannot be moved angularly and relatively to the camshaft sprocket wheel, and in such cases the individual cams must be re-timed.

Transverse vibration of the roller chain is caused by torque variations in the camshaft and torsional vibratory movement of the crankshaft. The amplitude of chain transverse vibration is controlled by guides and dampers which may be elastically or spring mounted. The fastenings of these chain dampers and guides must be carefully checked during crankcase examinations to ensure that they do not slacken in service.

■   **10.37   How are cams mounted on a camshaft?**

There are various methods of mounting cams on camshafts, but the most common method is to make the bore of the cam of such size that it is an interference fit on the camshaft, and is pressed on to it. The sections of the camshaft on which the cams actually fit are made progressively larger in diameter so that the cam fitted first has the largest bore diameter. This facilitates the fitting of the various cams, which easily slide over the smaller section diameters until they come up to the spot where they are to be fitted.

In some engines the bore of the hole in the cam is ground out to a very fine taper, the camshaft is ground to the same taper, and within the bore of the cam an annulus is ground. The cam is then slid along the camshaft up to the location at which it is to be fitted and the annular space in the bore of the cam is pressurised with oil from a high-pressure hydraulic pump. The oil under pressure expands the cam and allows it to be correctly positioned, both angularly and laterally, on the camshaft. When the oil pressure in the annulus is released the cam contracts and firmly grips the camshaft. Any later angular adjustment for timing purposes can be made by simply putting the annular space under oil pressure and turning the cam to its new position. When pressure is released the cam again grips the camshaft.

Camshaft couplings can be constructed in a similar manner for adjusting the position of all cams when the driving chain stretches and its tension is readjusted. An advantage of this method of fitting cams and shaft couplings is that the friction set up between the shaft and the cam or coupling is sufficient to drive the cam without fear of slipping. This obviates the necessity for keys and keyways and the stress raisers they produce.

■   **10.38   Why is the clearance between the cam and the roller important on air inlet and exhaust valve cams? What precautions must be taken when setting roller clearances on engines fitted with hydraulic loading devices on the push rods?**

The roller clearance setting is important because, if the clearance is insufficient, exhaust valves and air inlet valves will not seat properly and will quickly burn on the valve seating faces. The roller clearances are set with feeler gauges when the engine is cold. In some engines the clearance is measured between the rocker arm and valve spindle, and in others between the end of the push rod and rocker arm. Adjustment is made with tappet screws or other similar arrangements.

In some engines hydraulic loading devices are fitted to the bottoms of push rods. The purpose is to take up the clearances mentioned previously but at the same time to allow the expansion, due to heating of the various parts, to occur freely and without holding valves off seats. The design of these loading devices is such that the required clearances are arranged within the hydraulic loader and when lubricating oil is put on the engine the piston in the loader is activated and all the mechanical clearances are taken up. This takes away the shock loading in the valve gear and reduces wear and tear, and results in a quieter-running engine. When setting tappet or roller clearances on engines fitted with hydraulic loading devices on the push rods, the lubricating oil supply to the engine must be shut off and the loading devices must be properly drained before adjustments are made to the clearances.

■ **10.39 In some engines the camshaft driving gearwheel or chain sprocket wheel is made in two pieces with the joint on the diameter, and then bolted on to the crankshaft or crankshaft flange. What very particular examination does this type of gearwheel require when carrying out a crankcase examination?**

When this type of gearwheel or sprocket wheel is used, slackening of the wheel fastenings causes the pitch of the teeth to change at the mating surfaces of the two halves of the gearwheel or sprocket wheel. This in turn causes high loadings to occur on these teeth and leads to their early failure.

When carrying out a crankcase examination the wear pattern of the gear or sprocket teeth on each side of the joint should be compared with the wear pattern of teeth away from the joint. Any differences noted must be investigated and the causes (usually slackening of the fastenings) rectified. If differences in the wear pattern are noted there is every reason to suspect that fatigue cracks may have commenced in the teeth nearest to the joint, and they should be carefully examined and the roots of the teeth checked for cracking with dye penetrants. If cracks are found in gearwheel teeth, the teeth of all the other gearwheels in the train should be checked, particularly the teeth in idler gears if their diameter is small in relation to the driving and driven gears in two-stroke cycle engines.

■ **10.40 Name the types of bearing used for camshafts. What attention do these bearings require?**

Slow-speed engine camshaft bearings are usually of the journal type, operating in white-metal lined or bronze bearing shells or bushes, and lubricated by oil supplied from and returning to the engine lubricating system. In cases where fuel leakage from fuel pumps could contaminate the engine lubricating oil, the lubrication for the camshaft bearings is sometimes arranged as a separate system. Apart from attention to the lubrication system for camshaft bearings it is also important that the bearing keep fastenings and locking devices be regularly inspected. This is particularly important when the camshaft is underslung and the cam loads are placed on the bearing keeps. If a bearing keep slackens it can alter the engine timing and valve-opening periods, and ultimately lead to camshaft failure. The fastenings holding the bearing housings and fuel-pump foundations also require regular daily inspection, and should these fastenings slacken and damage the locating dowels, the housing will need careful re-alignment. In medium-speed engines the camshaft bearings may be similar to those found in slow-speed engines, or they may be roller or needle-roller bearings. The attention they require is similar to that for slow-speed engines. In high-speed engines the bearings are usually needle-roller type and require similar attention.

■ **10.41 Connecting-rods on slow-speed engines are round in cross-section while on medium- and high-speed engines the cross-section is found to be I or rectangular-shaped. Why is this so?**

The cross-section adopted for connecting-rods follows from study of the loads on connecting-rods and the cost of manufacture. Round-section connecting-rods are naturally cheaper to produce, but in medium- and high-speed engines the loads on the connecting-rods are such that the higher cost of production of I-and rectangular-section connecting-rods must be accepted due to technical necessity.

The basis of connecting-rod design is to consider the connecting-rod as a strut, pin-jointed at each end, and subjected to lateral loading from inertia (in V-engines additional lateral loading comes from slave rods) combined with the thrust from end loading through the pins. The loads coming on to a connecting-rod are the result of gas loads on the piston and inertia loads from the piston, piston-rod, crosshead or gudgeon and piston skirt. The inertia loads from the reciprocating parts are applied as a correction (depending on their action) to gas loads. The thrust- or end-loading on the connecting-rod may be 'push-pull' in single acting four-stroke cycle engines, the rod being subjected to high compressive loads in the 'push' stage and low tensile loads in the 'pull' stage. In a single-acting two-stroke cycle engine the load from end thrust is usually purely compressive.

In service the connecting-rod is swinging about the crosshead bearing or gudgeon pin bearing, the swinging movement being constrained by the bearing on the crankpin. This action sets up inertia loads on the connecting-rod in a transverse direction due to the mass of the connecting-rod and its swinging or whip movement. In slow-speed engines, round-section connecting-rods of normal length are strong enough to sustain the corrected gas loading; and the inertia whip loading is not of sufficient consequence to cause departure from the lower manufacturing cost of circular cross-section connecting-rods. With medium- and high-speed engines the transverse inertia loading is of such magnitude that the weight of the connecting-rod must be considered together with its strength to resist these loads. Rectangular- and I-section connecting-rods fulfil this function in the best manner, and this is the reason for their use, in spite of higher manufacturing costs.

■ **10.42 Name the causes of failure in connecting-rods.**

Connecting-rods may fail from fracture or cracking in vulnerable areas, or they may buckle. Failure of connecting-rods in service is extremely rare in slow-speed engines, but cases of slight buckling have been found in some instances when an attempt has been made to start an engine after water or oil has leaked into the cylinder space. In medium- and high-speed engines fatigue cracks have been found in locations subjected to wide stress ranges. With thin-walled steel-shell bearings seizure more easily occurs than with white-metal lined bearings, and crankpin bearing seizure usually results in the connecting-rod buckling in the transverse direction.

■ **10.43 How are the forces on the piston transmitted to the crankshaft in V-engines?**

In V-engines the gas load on the piston is transmitted through to the connecting-rod as in any other reciprocating engine. There are, however, different configurations in which the load or force is transmitted from the connecting-rods to the crankshaft. If the engine designer wishes to keep the centre-line of the cylinders in the same transverse plane, which keeps engine length to a minimum, articulated connecting-rods may be used. These have a boss on one side just above the bottom end. One connecting-rod, referred to as the master rod, is connected to the crankshaft through the bottom-end bearing in the normal manner. The other connecting-rod, referred to as the slave rod, is connected through a pin at its lower end to the boss just above the bottom-end bearing as mentioned above.

In other V-engines one connecting-rod with a forked end grips a bottom-end bearing bush which revolves around the crankpin in the normal manner. The other connecting-rod is built with a thin normal-shaped bottom end which fits between the fork end of the other connecting-rod and swivels on the outside of the bearing bush gripped in the fork.

In some V-engines the bottom-end bearings follow normal practice, but fit on to the crankpin adjacently. This arrangement, however, increases the crankpin length and the span between the main bearings. The cylinder centres must also be staggered longitudinally resulting in some small increase in engine length.

■ **10.44 How has the conventional bottom end of connecting-rods been modified in highly rated medium-speed engines? Why has this modification been made?**

Failure of connecting-rods at the lower end, housing the bottom-end bearing shells, led to studies being made using transparent models in photo-elastic testing experiments. As a result of these experiments, departures were made from conventional marine-type connecting-rods with flat lower palms, and those having a semi-circular end and bearing keep joining the rod on a face at 90° to it. The bottom-end bearing housings and rods are now made in three pieces, so that the lower end of the connecting-rod has a wide-angled wedge form; the bearing keep is made in two pieces, one portion nearly semi-circular and the other in the form of a sector to complete the circular eye for the bearing bushes. The mating surfaces of the three parts are serrated with V-shaped slots, the tops and bottoms of the slots being well radiused. The three parts are held together by two hydraulically tightened bolts set horizontally in the transverse direction.

Other designs have been developed in which the bottom-end bolts are set angularly to one another, and in another form the split between the connecting-rod and the bearing keep is at 45° to the connecting-rod centre-line; this design also has serrated mating surfaces. These designs give a much stiffer bearing housing and reduce the effects of stress raisers normally found at the foot of conventional marine and automotive-type bottom-ends and thus reduce the chance of fatigue cracks developing. The stiffer housing is much better suited to thin-wall bearings, and it reduces fretting between the bearing-shell and its housing and prevents bearing-shell distortion under load. The serrated mating surfaces of the keep parts and the connecting-rod take up shear loads and prevent transference of these loads to the keep bolts. They also reduce fretting of the mating surfaces.

In some older trunk-piston engines, pistons had to be removed by lowering them through the liner and out through the crankcase door because the bottom-end was too large to pass up through the liner. When connecting-rods have the bottom bearing keep angled at 45° to the centre-line, removal of the piston and connecting-rod from the engine is made easier because the connecting-rod will pass up through the cylinder liner.

■ **10.45   How is the clearance of slow-speed top-end and bottom-end bearings measured?**

These clearances can be measured with long, narrow-width feeler gauge blades of sufficient length to go across the width of the bearing. The blade is inserted under the top of the keep and then moved round towards the sides of the bearing. Slow-speed bottom-end bearing clearances can be measured using telescope feeler gauges applied between the side of the lower part of the bearing and the crank web. The blade of the feeler gauge is then inserted between the keep and the crankpin. When adjusting clearances on top-end bearings the feeler gauge method of measuring the clearances gives foolproof results, but during bottom-end bearing adjustment, use of thin lead wire gives better and more positive indications of clearance than telescope feelers.

■ **10.46   Why are the actual clearances of opposed piston engine side-rod crosshead bearings extremely important?**

It is imperative that the crosshead bearing clearances of the two bearings on either side of any one crosshead be kept equal. Wear and loading on a side-rod crosshead bearing takes place on the bearing keep, and if the bearing clearances become unequal the upward loading comes first on to one bearing and then the crosshead assembly distorts until the load is taken up by both bearings. This effect puts a bending moment on the lower end of the tie-rod so that the stress pattern over the cross-section of rod is unequal. A higher tensile load is then set up on one side of the rod, which tends to initiate fatigue cracks. It also causes unequal wear on the side edges of the crosshead slipper. It is also necessary to keep the bearings on both the forward and after sets of side-rod crosshead bearings the same, to prevent a similar action taking place on the portions of the tie-rods just below the upper piston yoke. It also prevents excess piston and cylinder-liner wear in the fore and aft direction.

. In some medium-speed opposed piston engines the crosshead bearing design has reverted to a form used in stream reciprocating engines and commonly known as the 'single top end'. With this type of bearing the end of the connecting-rod is forged with a fork, the sides of which are bored, and a hardened steel pin extending from one side of the fork to the other is shrunk in the bored holes. The bearing shells are housed in a slotted space in the crosshead block. With this type of bearing the problem of fracture of the lower end of the side-rod is minimized. It is, however, still necessary to keep the clearance of the forward and aft side-rod crosshead bearings similar to prevent troubles developing at the top of the side-rods and in the upper-piston york piece.

■ **10.47 How are the clearances of the bottom-end bearings, gudgeon-pin bearings, top-end bearings and slave-rod pins measured in medium-speed engines?**

The most convenient time to check the clearances on these bearings is when the cylinder covers are removed for piston removal and piston ring renewal. A hand-operated chain block or lift is fitted on the hook of the crane or main lifting gear. The piston lifting bar or eyebolts are fitted to the piston and rigged to the hand-operated chain block before the bottom-end bolts are slackened. A dial gauge with a magnetic base is placed on the connecting-rod with the dial spindle touching the top of the crank web. When the piston is given a lift with the chain block the keep of the bottom end lifts and contacts the underside of the crankpin so that the clearance is taken up. The amount of lift is registered on the dial gauge and the reading given is the bearing clearance.

By reversing the dial gauge so that the needle spindle touches the bottom of the piston skirt the clearance of the gudgeon-pin bearing can be measured using similar methods. Centre and side-rod crosshead bearings in medium-speed opposed piston engines can also have clearances checked in this manner.

Note The chain blocks or lifts must be of sufficient lifting capacity, and the lifting operation should be carried out by hand pulling. Excess force should not be applied nor should air or electric hoists be used in case of overloading the lifting gear or causing other damage.

If there is any chance of piston or other parts slewing, two dial gauges must be used and the mean of the two readings gives the bearing clearance.

■ **10.48 What effect does the gas load on the piston have on the crosshead pins? How is this effect catered for in the crosshead bearings?**

In large slow-speed engines the gas pressure acting on the piston and through the piston rod puts a load of approximately 600 tonnes on the crosshead pins when the pressure in the cylinder is at the maximum value. Loads of this value cause some deflection in the crosshead pins of large engines, even though the crosshead block is designed to give the pins maximum rigidity. The crosshead pin deflection is allowed for in most modern engines by providing a small amount of flexibility in the base of the lower half of the crosshead bearing. This allows the bearing to follow the shape of the pin under the high load, thereby reducing bearing operating troubles.

The latest generation of slow-speed crosshead engines now have a very stiff crosshead pin. This has been brought about by increasing the diameter of the crosshead pin to approximately the same size as the cylinder bore. Crosshead bearings in these engines have thin-wall bearing shells with either a thin layer of tin-based bearing alloy or a thin layer of one of the harder bearing alloys.

In some cases the upper end of the connecting-rod has been made heavier and more rigid. The axial length of the crosshead bearing is then made the same as the upper part of the connecting-rod so that the crosshead pin is supported over its entire length by the bearing shell and the upper part of the crosshead.

The piston rod is fastened to a relatively narrow landing on the crosshead pin. The whole assembly is very rigid when compared with the designs used in earlier engines. (Fig. 10.4)

In older engines the lower half of the bearing has a semi-elliptical area scraped hollow (the minor axis of the semi-ellipse is adjacent to the inside edge of the bearing against the crosshead block holding the piston-rod fastening). The depth of this hollow area is a maximum at the centre and is approximately 0.05 mm. The hollow area is scraped after the bearing has been machined and bedded in the normal manner. When the engine is in operation the crosshead pins deflect downwards and the whole of the bearing surface supports the pin. As the gas pressure falls away the pins straighten, and the part of the bearing surrounding the hollow area is sufficient to take the pin load. During this period of lesser loading when the pin is straight, lubricating oil flows into the hollow space so that an adequate amount of lubricant is available during the period of extreme loading.

With some other types of older engines the bearings are bored in the normal manner, after which the bolts holding the bearing to the boring-machine table are slackened off and a thin liner is placed under one edge of the bearing shell. The bearing is hardened down again on the table and the thin liner cants the bearing, and the semi-elliptical hollow area is machined in. In other cases the bearings are set up on the boring table with the thin liner under the side of the bearing; they are then machined with the bearing base canted.

The bearings are set up on the connecting-rod and bedded to the crosshead by scraping, a semi-elliptical hollow area at the inside edge of the bearings adjacent to the crosshead block being left as the bedding proceeds.

■ **10.49   When white-metal-lined crosshead bearings fail in service what are the usual types of failure?**

Various types of failure occur in crosshead bearing linings. Some, if found early, can be rectified so that the bearing can satisfactorily continue in service, in other forms of failure the lining must be condemned and the bearing remetalled. Failure of the white-metal in some less severe forms usually progresses so that eventually the bearing must be remetalled.

The forms of failure are as follows.

1   Cracking of white-metal.
2   Fatigue failure of white-metal.
3   Squeezing of white-metal so that oil grooves are partially blocked or obliterated; oil holes may be partially blocked, or wholly blocked in extreme cases.

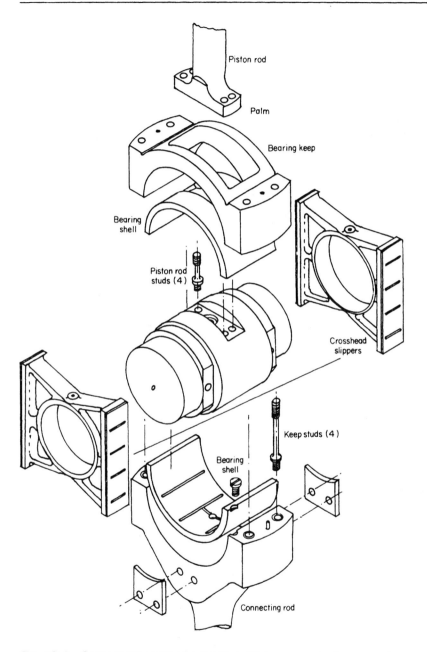

Fig. 10.4 Component parts of a modern, high-powered, slow-speed, two-stroke cycle engine crosshead bearings and slippers.

4   Failure by overheating, when the bearing surface of the white-metal becomes plastic or even melts.
5   Corrosion.

Failure of the white-metal by cracking or fatigue does not usually manifest itself under normal operation. If it is present it is usually found at the time the lower half of the bearing is exposed for examination. Failure by squeezing or overheating may become apparent during engine operation; it is often noticed when checking bearing clearances or during crankcase examination. Failure by corrosion is usually found when examination is carried out on the bearings, corrosion of the white-metal usually being accompanied by corrosion of the crosshead pins. The form of the corrosion will depend on the nature of the contamination of the lubricating oil.

■   **10.50   What are the usual causes of crosshead bearing failures?**

As there are many and widely varying causes of failure, each case must be considered individually and the cause ascertained from the findings of careful examination. Bearings have been known to fail on new engines when the engine is being worked up to full power for the shop trials, or after many years of satisfactory service.

Bearing failures may result from any one or combination of the following causes.

1   Misalignment of engine running gear.
2   Deterioration of surface finish of crosshead pins.
3   Quality of white-metal.
4   Insufficient supply of lubricant.
5   Impure lubricant, or water contamination.
6   Excessive firing pressure.

In opposed piston engines fitted with spherical crankpin bearings, locking or binding of the spherical portion of the bearing in the connecting-rod may lead to overheating of the bearing or squeezing of the white-metal. In some four-stroke cycle engines pounding of the bearing and consequent hammering out of the white-metal may occur if there is too much clearance in the bearing.

■   **10.51   How can damage occur to the crosshead bearings in slow-speed propulsion engines when the engine is operating under manoeuvring conditions?**

Many ships that operate on medium- and high-viscosity fuel at sea use diesel fuel to manoeuvre the main engines when proceeding in coastal waters, rivers and in and out of port. Medium- and high-viscosity fuels require much earlier injection than diesel fuel, to obtain good combustion. When an engine which has been timed for the cheaper grade slower-burning fuel uses diesel oil which is faster burning, the rate of pressure rise increases and the maximum combustion pressure may well exceed that for which the engine was designed. Normally the lower operating power of the engine caters for this pressure rise and keeps the maximum pressure within safe bounds. It is, however, good practice to take the maximum cylinder pressures whenever a new supply of diesel oil fuel is being used for the first time under manoeuvring conditions.

When ships with bridge control of the main engines are being controlled from the navigating bridge, the Master must be made aware of the maximum speed at which the engine can be safely operated on diesel fuel so that the engine speed is kept within a safe figure. Excessive combustion pressures will be shown by cylinder relief valves lifting, but operating at just below the relief valve lift pressure over extended periods can easily cause damage to bearings and possibly to other engine parts. If there were any malfunction of the relief valves, the high maximum combustion pressures could go unnoticed until bearing damage was found.

■ 10.52 If, on opening up a crosshead bearing it was found that the white-metal was cracked, how would you judge whether the bearing was fit for further service?

First, the white-metal in the cracked area must be examined to ascertain if any metal has lost its bond to the bearing shell or become detached from surrounding metal. Loss of bond can be ascertained by tapping the surface of the white-metal with the edge of a large copper coin. There is a distinct change of note between solid-bonded metal and detached metal. If no loose metal is found it is generally safe to continue using the bearing provided that the cracked areas are localized and do not cover too large an area. Loose metal may extend from a few small fragments or chips at the boundary of a crack to quite large detached pieces, of thickness up to the full thickness of the white metal lining, or lesser thicknesses where the crack has extended below the surface of the metal but not down to the bearing shell. When loose metal is present it must be examined to note how oil flow to the bearing will be affected if the detached metal should move.

In engines with non-drilled crankshafts, crankpin bearing lubrication is effected by oil flow which passes through the crosshead bearing. If loose white-metal should move in such engines, it may block off oil flow to the bottom-end bearing, resulting in damage to the bearing and most likely to the crankpin too.

Generally, if detached metal is adjacent to oil holes, and movement might block oil flow, the loose metal should be removed. If the amount of white-metal removed is not too large the bearing can continue to be used, until such time as it can be changed. While it continues in service the clearance of the bearing must be carefully checked and kept under observation. Any large clearance change must be investigated, particularly if a change occurs between the clearances in the forward and aft bearings of a crosshead.

If any doubt exists as to the ability of the bearing to sustain its load the remaining area of the bearing should be measured. The timing of the fuel pump can be slightly retarded to lower the maximum firing pressure, or the load on the bearing reduced by reducing the power on the engine unit. The power developed can be related to the remaining effective area. This should only be done when it is essential to keep a ship in service; normally it is better to change doubtful bearings.

With some engines in service it has been found that crosshead bearings have a 'life' of a certain number of running hours, and after this life is exceeded some breakdown in the crosshead bearing white-metal can be expected.

■   **10.53   What kind of surface finish is required for crosshead pins?**

After crosshead pins are turned to a smooth finish they are ground and may then be 'super-finished', an operation consisting of setting up the ground cross-head block and pins in a lathe and polishing the working surface of the cross-head pin with hones. The hones are held in a tool holder which is pneumatically or hydraulically loaded on to the surface being finished. The force holding the hones on to the surface remains constant if the hydraulic or pneumatic pressure remains constant. The crosshead pin is revolved and the hones are fed slowly along the length of the pin while being oscillated in the direction of the feed. Paraffin oil or cutting lubricant is used to prevent scores and wash away debris from the hone. This honing action produces an extremely smooth accurate finish on the working surface of the crosshead pins.

In some cases, with the grinding machines and special grit grinding wheels now available, the finish produced is sufficiently smooth to render super-finishing unnecessary.

There is a distinct difference in the 'lay' of the surface finish or texture between ground pins and super-finished pins. In ground pins the lay is seen as a series of circumferential marks from the grinding wheel, while in super-finished pins it is seen as a series of wavy lines running circumferentially round the pin. The wavy line (in appearance similar to a sine wave) comes from the combined revolving of the pin and the oscillating movement of the hone.

■   **10.54   How is the surface finish of crosshead pins measured?**

Special instruments are available which measure the microscopically small grooves or valleys and peaks of the finished surface. This part of the measure-ment shows in effect the roughness of the surface. Some instruments also meas-ure how the median point of these microscopic peaks and grooves undulates; this measurement then shows the 'waviness' errors in the surface. 'Waviness' errors are often seen to be repeated in cycles, and this cyclic error is referred to as 'form error'.

Due to the size of a crosshead and the fact that some of the surface-finish measuring instruments are not portable, a plastic strip may be employed which is pressed on to the crosshead pin surface to be measured. The strip reproduces the valleys as peaks and the peaks as valleys, and after it has been allowed to set it is lifted from the pin, and may then be taken to the laboratory for checking the surface.

Small surface profile measuring instruments are available but their use is mainly confined to production engineering.

Generally for crosshead pins, surface 'roughness' (as mentioned above), is the criterion by which the working surface is judged. The average depth of the peaks and grooves is referred to as the centre-line average or CLA. In the field and in many repair establishments special instruments are not available, and in such cases a comparator is used consisting of a set of prepared surfaces finished to various CLA values. The index-finger nail or thumb-nail is drawn across the surface of the crosshead pin and the feel of the roughness is compared with that of the surfaces of the comparator; the comparator surface that is closest to the crosshead pin surface gives the CLA value.

The index-finger nail or thumb-nail is very sensitive and there is no difficulty in making tactile comparisons. Some people, however, prefer to make the comparison with the edge of an unmilled copper coin.

Note 'Surface finish' and 'surface texture' are synonymous terms. The matter of surface finish becomes increasingly important in the operation of heavily loaded bearings similar to the crosshead bearings in a diesel engine where, in spite of the difficulties, maintaining an adequate oil film is an absolute necessity.

The load-carrying capacity of a super-finished bearing surface is approximately twice that of a very finely ground bearing surface.

Terminology, standards and the methods of measurement used in this subject are covered in the various standards published by such authorities as the British Standards Institution, the American National Standards Institute and the International Standards Organisation.

It is pertinent to note that the terms 'AA' (arithmetical average), 'Cla' or 'CLA' (centre-line average), and 'Ra' are identical.

■ **10.55 Some ships have, in their engine spares inventory, spare crosshead blocks. What attention must be given to this important spare part?**

The surface finish of crosshead pins requires the highest degree of smoothness and if this smoothness is lost the crosshead becomes useless for future use until it is re-finished on its working surfaces.

It is therefore necessary to store this item in a safe place where it cannot be damaged mechanically and where it will not be subjected to water drips from leaks or build-up of moisture from atmospheric condensation. The finished working surfaces of the pins must be adequately protected against rusting and also protected with suitable wrappings before the part is brought on board. If the wrapping is not one which hermetically seals these finely finished surfaces, it must be removed periodically to examine the pins for deterioration, and then carefully replaced.

If deterioration is present its degree must be noted and reported. Usually rust spots show as excrescences and they should be cleaned off carefully to bare metal, taking care not to damage surrounding good areas. After this cleaning a shallow hollow area will be left. The boundaries of this area can be smoothed off by rubbing down with the finest abrasive such as 'Crocus' paper. If the rough boundary edge is smoothed off, the crosshead should be good for further use provided that only a small amount of the total surface area of the pins is affected. Sometimes crosshead pins are chromium-plated on the working surfaces to protect them during storage.

■ **10.56 Name some of the improvements that have been made in the design of the shafting in modern, two-stroke cycle slow-speed engines.**

In many modern engines the camshaft drive is taken from a gearwheel or a chain sprocket mounted on the thrust collar. This has shortened the engine length. When the torsional vibration characteristics of the shafting system are studied attention must be given to the position or the location of any antinodes. When the drive for the camshaft is located at or near an antinode its amplitude should be studied in case of problems arising from vibration which may lead to drive-chain 'snatch' and heavy loads on sprocket teeth in chain drives, or heavy teeth loads on both sides of the teeth in gear drives.

The other design advance in crankshafts is the elimination of bored or drilled lubricating oil holes. This in turn has eliminated the stress raisers which come about when holes are bored in journals and crankpins. The elimination of oil holes when hollow crankshafts are used becomes a necessity if the fitting of oil piping is to be avoided.

Crankshaft manufacturing costs are lowered when oil holes are eliminated.

11

STARTING AND REVERSING

■ 11.1 **How are the various types of marine diesel engine started?**

Small engines such as those used for lifeboats, emergency fire pumps, and for driving small 'start-up' air compressors are usually started manually. To facilitate starting, many of these small engines have a decompression device which is operated when the engine is being swung, and which allows a faster swing to be made by reducing the work done during compression. When a good speed of rotation is reached using the starting handle, the decompression device is released, and the compression of the engine returns to normal and should be sufficient to ignite the injected fuel.

Small- to medium-size engines may be started by hydraulic starters. These consist of three essential parts. The first is a hand-operated hydraulic pump, similar to a test pump or the pump used for tensioning tie-rods, bottom-end bolts and the like. The pump discharges into the second part which is a compressed air-loaded or inert-gas-loaded hydraulic accumulator. This accumulator is nothing more than a vertical cylinder, closed at each end and fitted with an internal piston. The hydraulic fluid from the hand pump is discharged into the cylinder space below the piston, which is forced up, compressing the air above it in the cylinder. The third part of the hydraulic starter is the mechanism attached to the engine. The engine is started by this mechanism, which consists of a pair of hydraulic cylinders connected by piping to the lower power part of the hydraulic accumulator. Within each cylinder is a ram-type piston which is connected to a toothed rack with teeth cut on the skew. The racks drive a small pinion which couples through dogs to the engine crankshaft. The skew on the teeth of the racks causes the pinion to slide up to the crankshaft to engage the dogs and compress a spring. When the engine starts the spring pushes the pinion clear.

To start the engine, hydraulic fluid is pumped into the accumulator through connecting pipes. The hydraulic fluid under pressure in the accumulator is then allowed to flow into the hydraulic starter by opening a valve in the connecting pipework. The fluid drives the rams, and the pinion driven by the racks engages the engine crankshaft and turns it at sufficient speed to cause ignition of the injected fuel. The three parts of a hydraulic starter are connected together by pipework to a hydraulic fluid reservoir.

Emergency electrical generators driven by diesel engines are usually started by electrical starters, powered by a starter battery of the pattern used in automotive practice. The starter is actuated by a voltage drop in the ship's d.c. supply, or in ships with a.c. electrical power by a reduction in the frequency.

High-speed diesel generators with engines of the automotive type may have similar starting arrangements to emergency diesel-generating sets. In some cases an air turbine may be used instead of the electrical starter motor. Medium-speed diesel generator sets are usually started with direct compressed air supply through a starting air system. The air flows into the cylinders during the firing or expansion stroke in four-stroke cycle engines, or during the expansion portion of the stroke in two-stroke cycle engines. The air flowing into the cylinders sets the engine in motion and brings its speed up so that it is sufficient to promote combustion when the engine is put on to fuel.

Medium- and slow-speed propulsion engines are invariably started with compressed air through the starting air system.

■ **11.2   What are common causes of difficulty in starting small manually started diesel engines?**

The most common causes of difficulty are loss of compression, and fuel loss from the engine fuel system.

Loss of compression in four-stroke cycle engines is usually caused by leakage through the cylinder-head valves, or sticking of piston rings. In some engines with a variable compression ratio, the small double-seated valve used to connect or disconnect the small ante-chamber to the cylinder combustion space can be a source of air leakage causing compression loss. Loss of compression from binding or stuck piston rings is often caused by rain, sea-water or condensation draining back down the exhaust pipe and into the engine cylinder. For this reason, if the engine exhaust pipe outlet is in a position exposed to rain and sea-spray, care must be taken to secure the outlet properly against the ingress of water when the engine is stopped and shut down. Any drain cocks at the lower end of the exhaust pipe must also be put in the open position when the engine is stopped.

Loss of fuel from the engine fuel system can be rectified by priming the fuel system up to the fuel-injection valve prior to starting the engine. The location of the leakage or drain-back point in the system should be found and corrected.

Some other common causes of starting difficulties may be associated with maladjustment of decompression devices fitted to the cylinder-head exhaust valve or valves. Dirty fuel filters, water in fuel, and blocked gauzes in the fuel tank air pipes which prevent ingress of air into the fuel tank, are also causes of starting or running difficulties. If the exhaust outlet has been tightly secured and has not been opened up prior to starting, this will also cause difficulty.

■ 11.3 Name the main parts of a compressed air starting system for main propulsion engines. What is the function of the various parts?

Air-storage reservoir or starting air tanks. There are usually two such tanks on ships fitted with reversible engines and fixed pitch propellers. These tanks hold in reserve the compressed air supplied from the starting air compressors.

Starting air lines to automatic valve. These lines are fitted between the air storage reservoir and the engine automatic valve.

Automatic valve. This valve automatically opens when the engine controls are put into the starting position and compressed air can then pass from the air storage reservoir to the starting air valves fitted in the engine cylinder covers. The automatic valve is opened by pneumatic means with air passed from the pilot valve.

Automatic valve controller or pilot valve. This valve is controlled by the starting lever either directly or indirectly. Operation of the pilot valve supplies compressed air to the actuating piston in the automatic valve causing the automatic valve to open. Action of the automatic valve controller or pilot valve also controls the supply of compressed air to the starting air distributor.

Starting air distributor. This consists of a series of valves (one valve for each cylinder starting valve) mounted adjacent to the camshaft. The action of the pilot valve sets the air distributor valves on to the starting cam or cams on the camshaft. When these valves are opened in turn they supply air to the pistons within the starting air valves mounted on the engine cylinders, thus opening the starting air valves, in turn, and allowing compressed air to flow into the cylinders and rotate the engine.

Note   In some engine manuals the distributor valve may be referred to as the pilot valve.

Starting air valves. These valves, fitted in the engine cylinder covers, supply starting air to each cylinder. The timing for the opening and closing of starting air valves is controlled by the cam or cams (on the engine camshaft) which actuate the distributor valves. The supply of starting air from the automatic valve is led through a common starting air rail fitted along the length of the engine with separate branch connections to each cylinder starting air valve. The design and operation details of the various parts of the starting air system differ according to the engine manufacturer, but their function within the starting system is essentially the same in all makes of engine.

■ 11.4 How are engine-starting air valves operated? What attention do they require?

The body of a starting air valve is a casting, the upper part of which is considerably larger than the lower part which fits into the engine cover. This larger part forms an air cylinder, and the piston working in this cylinder is fitted to the starting air valve spindle. The cover of the air cylinder on the starting valve is connected through small-bore piping to the distributor valves. When a distributor valve opens, compressed air flows through the piping to the upper side of the piston in the starting air valve, forcing the piston downwards and opening the starting air valve, which is connected to the starting air rail. A spring fitted within the valve under the piston closes the valve when the distributor valve closes.

The underside of the actuating piston and cylinder in the starting air valve is open to the atmosphere so that any air leakage from the actuating piston leaks off to atmosphere; air pressure cannot build up under the piston and cause sluggish action of the valve. The upper part of the valve spindle is sealed with small rings similar to piston rings and any air leakage from this part also goes direct to atmosphere.

It is good practice to lubricate the actuating piston in a starting air valve prior to manoeuvring the engine and at the time of engine shut-down, but care must be exercised to avoid over-lubrication. The lubricants used may be either grease or lubricating oil, as specified by the engine builder. After certain periods of service starting air valves are changed and overhauled. Care must be taken to ensure that piston rings are free in their grooves. Should it be necessary to fit new rings, the end clearances of the rings must be carefully checked because they are usually made of brass which has a larger coefficient of expansion than the other parts of the valve. The valve and valve seat are ground with grinding paste and finished to a fine surface with lapping paste. It is essential to ensure that all parts of the valve are scrupulously clean before reassembly.

After overhaul the tightness of the seat must be tested. In valves that are removable from the cylinder cover it is usual to connect the valve to an air supply and immerse the end of the valve in paraffin oil or clean water; valve leakage is then shown by a flow of air bubbles through the liquid. When the starting air valve parts are integral with the engine cylinder heads and the heads are off the engine, the valve seat tightness (when under air pressure) can be checked by the application of a solution of soapy water.

A modern type of starting air valve is shown in Fig. 11.1. This valve is opened by air supplied from a pilot valve. Instead of having a single piston as in the valves described above, the valve shown has three pistons. The air from the pilot valve passes down through the axial hole in the valve spindle. The air then passes through the radial holes shown and creates a force which acts on the three pistons and opens the valve. The spring situated round the lower part of the spindle causes the valve to close when the automatic valve shuts off air supply to the pilot valves. The engine starting air after leaving the automatic valve passes through the ports and across the valve seat into the cylinder when the valve opens. The starting air entering the engine cylinders exerts a force on the pistons and causes the pistons to move and set the engine in motion.

■ 11.5 What are the indications of leaking starting air valves?

When an engine is in operation leakage of starting air valves is shown by overheating of the branch pipe connecting the starting air valve to the starting air rail. The heating occurs due to the leakage of hot gases from the engine cylinder into the starting air line connected to the starting air rail. During periods of

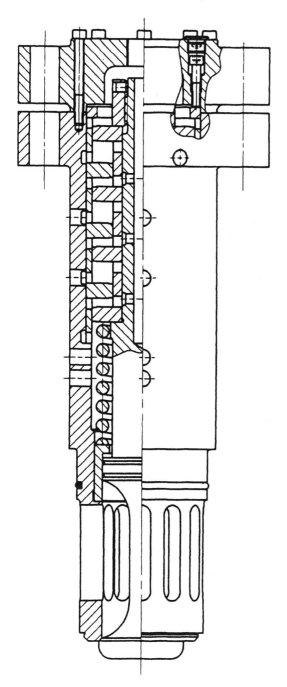

Fig. 11.1 Cylinder starting air valve with a multiple number of pistons to reduce its size. (MAN B&W)

manoeuvring the temperature of each supply pipe from the air rail to the starting air valve should be checked by feeling the pipe – as close to the valve as possible.

■ **11.6   What are the usual causes of starting air valve leakage?**

Leakage of a starting air valve is usually caused by sluggish valve action preventing fast closure of the valve, or by dirt or foreign particles from the starting air supply lodging on the valve seat and so preventing the valve from closing fully. Sluggish valve action can be caused by dirty pistons or valve spindle guides and the like. In newly overhauled valves sluggish valve action may be caused by parts fitted with inadequate clearances.

■ **11.7   What are the opening periods for starting air valves? How is the opening period controlled and what is overlap?**

Starting air valves are open only during the expansion stroke of the engine. In two-stroke engines, opening will occur some time after the engine piston has passed top-dead-centre, and closing should be fully completed before the exhaust ports, exhaust valves or scavenge ports open. In four-stroke engines the starting valves open after the engine piston has passed top-dead-centre and close before the exhaust valves open.

In order to conserve starting air, starting valves are designed to close as early as possible consistent with good starting, and some expansion of the starting air then takes place. The opening and closing of starting air valves is controlled by the cam (operating within the starting air distributor) actuating the distributor valve.

To enable a propulsion engine to be started from any crank position, overlap is necessary in the timing of the starting valves on an engine. Overlap occurs during the period that any two valves are open, it being understood that one valve will be opening whilst the other valve is closing. One valve will then always open when air is put on the engine to start some manoeuvre. If there were no overlap in the valves it would be possible for the engine to stop in some position where all the valves remained closed when air was put on the engine.

The opening of starting air valves usually occurs around 5° after top-dead-centre, with open periods sometimes up to as much as 130° of crank rotation, but usually less depending upon the make and type of engine and number of cylinders.

■ **11.8   What is the usual engine-starting air pressure in marine practice? There are advantages and disadvantages in using higher pressures; what are they?**

Starting air pressures vary with the various engine builders and engine types and lie between 25 and 42 bars (350 and 600 lbf/in²). They are within the same range for both main propulsion and auxiliary engines. The pressure of the starting air must be sufficient to impart enough speed to the engine pistons to compress the combustion air during the compression stroke sufficiently quickly for it to reach a temperature at which combustion of the injected fuel is initiated. If the starting air pressure is increased, the physical size of the parts in the system can be made smaller, but it will not necessarily follow that there will be any large savings in weight of the parts, as they must be made heavier to withstand the higher pressures.

■ **11.9 Is there a definite minimum starting air pressure at which any marine diesel engine will start?**

No. The minimum pressure at which a marine diesel engine will start varies from engine to engine. The pressure requirements will also vary according to whether the engine is at working temperature, or is comparatively cold as when first starting. An engine at working temperature will start with a lower starting air pressure than a cold engine. The temperature of the lubricating oil supplied to the engine bearings will also have an influence on the minimum required starting air pressure. When the lubricating oil is at working temperature the engine will swing more easily than when the lubricating oil is cold; hence a lower air pressure will start an engine with lubricating oil at working temperatures.

Badly leaking piston rings may also be the cause of an engine failing to start. Leakage of air from the piston rings lowers the compression temperature; it is then insufficient to ignite the fuel and start the engine. Preheating the cooling water to a temperature higher than normal may be all that is required to get the engine to start, particularly in very cold climates where the engine room temperature has been allowed to fall. After the engine has started and warmed up, the cooling water temperature is brought back to its normal value.

Note During trials of newly built ships it is a part of the main engine manoeuvring tests to put one starting air reservoir on to the starting system and continue manoeuvres ahead and astern until the pressure falls to the point that the engine cannot start. This pressure is recorded with the ship's trial data.

■ **11.10 If the automatic valve should fail during manoeuvring of the engine, what provisions are made so that manoeuvring of the engine can be continued?**

The arrangements vary according to the designs produced by the engine builder. In some designs the automatic valve (under a failed condition) can be operated by an emergency hand control consisting of a special lever made to engage with a collar on that part of the automatic valve spindle which protrudes through the valve casing cover. In other designs the automatic valve spindle (operated by hand-wheel control at the manoeuvring platform) can be put into one position for automatic control, or screwed into another position which positively opens the valve and allows it to be closed when screwed back. In other cases the positive hand-control wheel may be fitted locally on the automatic valve.

Note When newly joining a ship, the method of emergency control for the automatic valve should be studied. Portable equipment, if required to operate the valve, should be checked to see that it is in its correct location and in such a condition that it is ready for immediate use.

■ **11.11   Briefly describe the most common method of reversing the direction of rotation of four-stroke cycle engines**

In four-stroke cycle engines (which today are mostly of the medium-speed type) reversing is carried out by the use of duplicate cams for the air inlet valves, exhaust valves, and fuel pumps.

To engage the correct cams for ahead or astern operation with the cam rollers, the camshaft slides axially in its bearings. This axial movement is controlled by the camshaft reversing gear, which is usually a piston operating in a cylinder: motion of the piston is transmitted directly to the camshaft. In some engines the motion of the piston is transmitted to the camshaft via levers. In V-engines a single piston is made to move the camshaft on each cylinder bank by the use of a pair of levers, each operating its own camshaft.

The camshaft reversing gear consisting of the pistons and levers is sometimes referred to as the camshaft servomotor. It may be actuated by hydraulic or pneumatic means. In some cases the hydraulic system is fed from the main engine lubrication system, the pressure being sufficient to operate the servomotor piston. Pneumatic servomotors are usually supplied with air from the starting air system.

A reversing gear of the type used in four stroke cycle medium speed V-engines is shown in Fig. 11.2. The servomotor is pneumatically or hydraulically

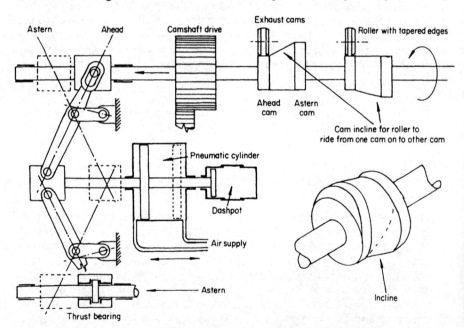

Fig. 11.2   Reversing servomotor gear on a four-stroke cycle medium-speed V-engine. **Note**. It is common to fit cylinder starting air valves on only one cylinder bank.

operated, an oil dashpot smooths out the motion of the parts and prevents shock.

The incline between the ahead and the astern cams allows the cam rollers to transfer from one cam to the other when the adjacent cam is in an open position. Some engine builders put a corresponding chamfer or a good radius on the cam roller edges.

■ **11.12   How is it possible for the camshaft to slide axially (in reversible engines) when the rocking lever roller on one cam happens to be adjacent to the peak on its duplicate cam?**

In older slow-speed engines the shaft carrying the rocker arms had eccentrics on it, the eccentrics being the fulcrums for the rocker arms. Rotation of the rocker shaft lifted the rocker arms and rollers clear of the cam peaks so that axial movement of the camshaft could be completed. When the camshaft was in its new position, continuous rotation of the rocker shaft lowered the rockers and rollers on to the duplicate cams.

In modern, medium-speed four-stroke cycle reversible propulsion engines the axial distance of each pair of cams is made sufficiently large so that the side of the peak of one cam inclines off slowly to the idle part of its adjacent duplicate cam. These inclined planes between pairs of cams enable the rollers to slide from one cam to another, even in the most unfavourable position of the cams.

■ **11.13   How are two-stroke cycle propulsion engines reversed?**

Some cross- and loop-scavenged diesel engines have a fuel-pump camshaft with duplicate cams for each fuel pump and starting air distributor, and reversing is carried out by moving the camshaft axially as in four-stroke cycle engines.

In other cross- and loop-scavenged engines reversing is carried out by circumferential translation of the fuel-pump camshaft relative to the engine camshaft. This circumferential movement is accomplished by the use of a servomotor in one of the gear wheels forming the fuel-pump camshaft drive. The servomotor consists of a pair of vanes, fitted on the fuel-pump camshaft, which move between another pair of vanes fitted within the gearwheel rim. By putting lubricating oil under pressure between opposite pairs of vanes the fuel-pump camshaft is moved relative to the gearwheel and the engine crankshaft. This relative movement changes the fuel-pump timing for ahead and astern operation, and in so doing changes the firing order of the cylinders. A similar arrangement is fitted in the exhaust-valve and fuel-pump camshafts of some uniflow-scavenged two-stroke cycle reversible engines.

A reversing servomotor of the type described is shown in Fig. 11.3. The valve controlling the oil flow to the servomotor is actuated through a pneumatically controlled valve forming part of the engine control system; it is shown in two positions (a) and (b) for ahead and astern operation of the engine. The oil passages through the camshaft are also shown in line form together with arrows showing the direction of oil flow during the reversal operation.

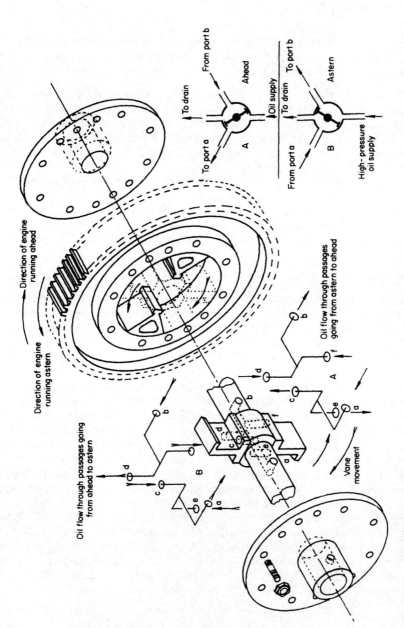

Fig. 11.3 Reversing servomotor fitted in the camshaft drive gear on some slow-speed two-stroke cycle engines. The changeover valve controlling the flow of lubricating oil to the servomotor is pneumatically actuated by the reversing gear. The pneumatic actuator is not shown.

Direction of engine running ahead

Direction of engine running astern

Oil flow through passages going from ahead to astern

Oil flow through passages going from astern to ahead

Vane movement

To drain   From port b

Ahead

To port a   Oil supply   A

From port a   To port b

Astern

To drain   B

High-pressure oil supply

■ 11.14 What do you understand by the terms 'interlocks' and 'blocking devices' in starting and reversing mechanisms?

Interlocks and blocking devices are provided so that the engine can be started or reversed only after certain conditions have been fulfilled. The starting interlocks prevent the engine being put on to fuel before all the sequences of the starting system have been completed. During the reversing of engines the interlocks ensure that the reversing mechanism and gear have completed all operations before air can be put on to the starting valves, thus preventing the engine from starting with the wrong direction of rotation.

The construction of the interlocks in the starting and reversing gear varies considerably between different designs of engines. With systems controlled by the operation of hand levers, the interlocks may be cams or pins which lock and prevent hand-lever movement. In engines started by hand-wheel controls, the interlocks are often slotted discs (fitted on the wheel shafts) and small levers which engage or clear the slots in the discs.

Blocking devices are mechanical, pneumatic, electrical or hydraulic devices used to make for safer operation of the engine. Some engines have a blocking device connected with the ship's engine room telegraph which prevents the engine being put astern when an ahead order is given, and vice versa. In other engines, blocking devices are fitted to the engine turning gear so that the engine cannot be inadvertently started with the turning gear in.

■ 11.15 Why is it important to know how the interlocks and blocking devices operate on an engine over which you have control?

When fuel-valve or fuel-pump timing is being checked the interlock devices must be cleared so that the fuel lever can be put into the position at which the timing is checked. Similarly the fuel lever must be put into various fuel-setting positions when the cut-off points on certain types of fuel pumps are being checked.

In the event of engine-room telegraph failure, any interlocking and blocking devices operated from the telegraph would prevent the engine being manoeuvred during this time of emergency. It is therefore important to know how the interlock and blocking devices may be overridden so that the engine can be manoeuvred under emergency conditions with orders via the bridge to engine-room telephone.

Note The instruction books from the engine builder give details of the interlocks on engine starting and reversing systems; sometimes the engine telegraph interlock details are covered in the engine instruction book and sometimes in the engine-room telegraph instruction book.

■ 11.16 Why are brakes sometimes fitted on the intermediate shafts of some diesel-engined vessels?

Some types of propulsion diesel engines will continue to turn after the fuel is cut off to stop the engine. If the engine is reversed and an attempt is made to start the engine it can happen that the engine operates in the reverse direction to the control setting. Operation of the engine on fuel under these conditions usually causes the cylinder relief valves to lift.

In order to make for precise and quick manoeuvring of the engine, a brake is fitted to the intermediate shaft, so that the engine can be brought to a standstill with the brake before an attempt is made to reverse and operate the engine in the other direction. Braking systems are sometimes fitted in geared propulsion systems where friction clutches are used. Their function then is to prevent high loading and possible burning of clutch friction surfaces when going from ahead to astern and vice versa. In some cases the brake is brought into use automatically through the engine's manoeuvring controls; in other cases it is applied as an emergency operation. Usually this form of brake is actuated by pneumatic loading. While braking systems may be fitted in any ship where their use is justified, they are more commonly found in high-powered ferries operating in confined waters. In other cases they have been fitted in passenger ships with diesel engine propulsion (see also Question 12.7).

■ **11.17   If it became necessary to operate an engine with a piston or part of the running gear for a cylinder unit removed, how would you deal with starting and reversing of the engine?**

If a piston has been removed from a propulsion engine it is obvious that the starting air valve (on the cylinder where the piston is removed) must not be allowed to fuction.

In some cases of breakdown a cylinder cover may not have been replaced before the engine is run, in which case the starting valve would be disconnected. To enable the engine to be manoeuvred the starting air branches on the air rail would have to be blanked off.

When manoeuvring an engine with one or two starting air valves removed, it sometimes happens that the engine comes to a stop in a position where it cannot be restarted. If this occurs the reversing control can be operated and the engine given a small amount of starting air to get it to another position, after which the engine reversing gear is again used to give the rotational direction required. The engine can then usually be started in the correct direction of operation.

If two starting air valves have been removed, the chances of the engine failing to start are naturally increased. If the engine still fails to start, the starting air must be shut off the engine and the engine-turning gear applied. The engine can be turned into some position where a starting air valve will be effective, by putting one of the pistons a few degrees past top-dead-centre in the direction of the required rotation.

Note   When circumstances make it necessary to cut out the use of any starting air valves, every effort must be made to make the machinery safe for manoeuvring. At all times the Master must be notified of the situation so that he can take the necessary precautions for the vessel's safety.

■ **11.18   In certain cases of main-engine breakdown it is sometimes necessary to remove a connecting-rod and hang the piston so that the** engine can be operated to bring the ship into port. How would you deal with the starting arrangements for that cylinder?

It is absolutely vital either to properly blank off the air supply to the starting air valve on the unit where the piston has been hung-up, or to remove the starting air supply branch pipe to the valve and blank off the branch piece on the starting air rail.

It can be seen that if starting air should get into a cylinder where a piston is hung-up, the load on the piston would most likely be sufficient to carry away the piston hanging bar and its fastenings.

Note In a 740 mm-bore engine with starting air pressure at 30 bar, the load on the piston in tonnes would be

$$\text{Load on piston} = \text{piston area} \times \text{air pressure}$$
$$\text{Load (tonnes)} = \frac{0.7854 \times 74.0^2 \times 30}{1000}$$
$$= 129 \text{ tonnes}$$
$$(\text{assuming } 1 \text{ bar} = 1 \text{ kgf } 1 \text{ cm}^2 \text{ approx})$$

It can then be seen that a hanging bar and its fastenings designed to support a load of 3 to 4 tonnes would be totally inadequate to sustain a load of 129 tonnes.

■ **11.19** Show in diagrammatic form the various parts of a starting air system for a slow-speed main engine. Show the piping connecting the various parts.

An air starting system supplying compressed air for starting main and auxiliary engines is shown in Fig. 11.4.

Compressed air is supplied into the system from either of the two main air compressors used when the engine is being manoeuvred; if an auxiliary compressor is fitted it will be used at sea under 'full-away' conditions. For starting up a 'dead' engine room plant the start up compressor will be used.

Air discharged from the compressors is passed through separators to remove moisture and oil, the air then passes into the compressed air receivers.

The main engine starting air lines, the diesel generator starting air lines and the auxiliary air lines are arranged as separate lines. The air to the main engine starting air lines passes through the automatic valve, a non return valve and then on to the starting air manifold fitted on the engine. The manifold has a connection to each starting air valve fitted in the cylinder cover. A starting air valve is shown in the diagram. A separate air line passes from the automatic valve to the pilot air distributor. A stop valve is often fitted in this line for use when testing starting air valves for possible leakage.

The starting air system for the auxiliary engines is similar to that for the main engines but the air lines are arranged with separate branches to each generator set.

Compressed air lines are shown branched to the ships whistle or syren, instrument air, control system air, exhaust valve air, workshop and tool air, lifeboat hoist air and deck air. In some ships the instrument air and control system air is separated from the systems as shown, the system is then supplied with compressed air from oil-free compressors.

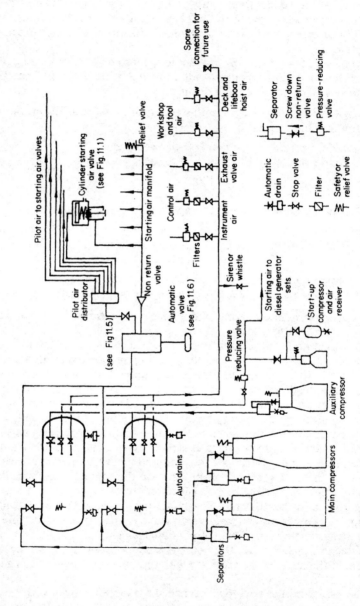

Fig. 11.4 Diagrammatic arrangement of a main and auxiliary engine starting air system.

■ 11.20 Show by sketches the parts of an air distributor or the pilot valves used for controlling the operation of air starting valves.

A common form of a pilot air valve is shown in Fig. 11.5, three valves are shown, each being in a different operating position.

When an engine is in operation, stopped, or shut down the pilot valve takes up the position shown in sketch (a). In this position the spring in the pilot valve lifts the roller off the cam and the valve remains inoperative.

When an engine is started the automatic valve supplies compressed air to the pilot valve, this acts on the spring loaded piston and forces the roller on to the cam. Air cannot be passed through the valve to operate the engine air starting valve as the roller has engaged on the idle sector of the cam as shown (b).

If the roller engages with the 'negative peak' of the cam as shown in (c) compressed air passes through to the engine air starting valve piston and causes the valve to open. The engine is then turned with the compressed air supplied from the starting air manifold. As the engine turns the roller comes on to the idle sector of the cam and the pilot valve reverts to the position shown in (b). Any compressed air trapped above the piston in the engine starting valve is quickly released to the atmosphere. The air passes through the holes drilled in the central piston of the pilot valve and through a silencer or muffler before

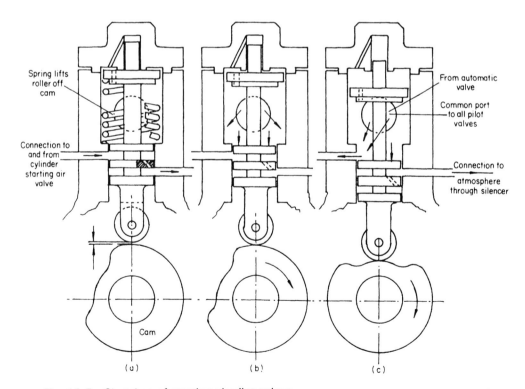

Fig. 11.5 Sketches of starting air pilot valves.

escaping. As the air pressure above the piston (in the engine starting valve) drops the spring rapidly closes the valve.

■ **11.21   Show by sketches the parts of an automatic valve.**

An automatic starting valve is shown in Fig. 11.6. The purpose of this valve is to supply air quickly to the engine starting system when the engine is started and to shut off the air supply automatically after the engine is working on fuel. This makes for safer engine operation and conserves starting air thereby saving energy and 'wear and tear' on the air compressors.

When an engine is brought into operation after a shut down the hand wheel on the automatic valve is moved to the position shown on the valve indicator as 'automatic'. In this position the piston, valve and the valve spindle assembly can move up or down freely and without hindrance from the handwheel and screw. When an air reservoir is opened compressed air passes through to the central part of the automatic valve which is initially held shut by a spring (not shown). The piston is greater in area than the valve below it. Compressed air acting on the top of the piston is obtained from a small pneumatic controlled valve (shown on the right hand side of the automatic valve). The downward acting force being greater holds the valve in the closed position.

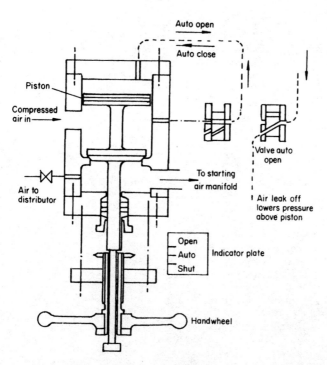

Fig. 11.6 Diagrammatic sketch of arrangement of automatic starting air valve.

When an engine is started the small pneumatic controlled valve moves upwards to the position shown in the extreme right-hand sketch; in this position the air above the piston quickly leaks off to the atmosphere. This unbalances the forces acting on the piston and valve, the upward acting force is now greater and causes the valve to open quickly against the load from the spring.

Compressed air is then allowed to flow from the open air reservoir through to the starting air manifold on the engine.

After the engine has commenced to operate on fuel the small pneumatically controlled valve is caused to revert to its original position and the automatic valve closes.

In an emergency the automatic valve can be manually operated by the handwheel.

In some engines the automatic valve is manually operated by a portable lever during an emergency. The lever is supplied by the engine builder.

The emergency operating lever is usually located in a position adjacent to the valve. Prior to a 'stand by' for manoeuvering an engine, it is good practice to check that the manual control lever is in its correct location and readily available for immediate use.

12

REDUCTION GEARING, CLUTCHES, COUPLINGS

■ **12.1 What are the advantages and disadvantages of geared diesel propulsion machinery installations?**

Geared diesel propulsion systems have many advantages and some disadvantages, and what might appear as an advantage for some specialized type of ship could be a real disadvantage for a different type.

The main advantages of geared propulsion are a saving of total machinery weight compared with direct propulsion, and a reduction of the headroom requirement, for lifting gear to remove cylinder heads and pistons for overhaul.

The main disadvantage is the number of cylinder units which have to be kept in good condition, though this is offset to some extent if it is possible (within the service commitments of the ship) to shut down one engine at sea. Overhaul of cylinder units can then be carried out when the weather is suitable for engine shutdown.

The fuel and lubricating oil requirements for geared diesel machinery installations are generally higher than for slow-speed, direct-coupled machinery of equivalent power. The fuel consumption is higher because the overall mechanical efficiency is less than for equivalent direct-coupled installations. However, this can to some extent be offset by designing for a lower propeller speed – which gives higher propeller efficiency. Operational experience with medium-speed geared installations shows that the total lubricating oil consumption is usually considerably higher than with conventional diesel machinery.

Total fuel and lubricating oil consumption figures for the whole machinery installation, with slow-speed direct-coupled installations and conventional diesel generating sets, still remains lower than for geared installations, even where electrical power is generated with an auxiliary drive from the reduction gearing.

In ships with machinery aft and a two-engine installation the extreme gear casing width can cause problems, which may be overcome by moving the machinery forward or increasing the fullness of the aft body of the hull or a combination of both. This can itself create a problem of having part of the machinery space difficult to utilize, and hull trim problems when the ship is fully loaded. These problems become less difficult to deal with when three engines are grouped round the reduction gearing.

In specialized types of ships such as fast ferries, car ferries, 'Ro-Ro' ships and the like, the low headroom requirements of geared medium-speed diesel engines makes this form of machinery most attractive, to the point that it is almost a prerequisite.

No overwhelming case can be made for geared diesel propulsion machinery except in the type of situations mentioned above.

Generally the total operating costs of geared engines and their auxiliaries is higher than that for conventional slow-speed direct-coupled machinery installations, which can be offset to some extent by capital cost reductions or the possibility of increased earnings. Naturally, increased earnings can come about only if cargo is available to fill the extra deadweight and hold cubic capacity obtained from the use of geared diesel installations.

■ **12.2 How are the engines in a geared diesel plant grouped around the reduction gearing?**

With single engines the engine is fitted forward of the reduction gearing. The engine shaft is connected to the reduction gear pinion which is at the top of the gearwheel. All the parts, consisting of the engine, gear pinion, gearwheel and intermediate shaft, lie on a common vertical plane which is on the centre-line of the ship.

In two-engine installations, the engines are most commonly fitted forward of the reduction gearing. The pinions, which are on the same centre-lines as the engine crankshafts, are placed diametrically opposite on the main gearwheel, and the axes of the crankshafts, pinions and gearwheels are on the same horizontal plane.

When three engines make up the installation, two engines are fitted forward of the reduction gearing in a similar manner to that described in the two-engine case. The third engine is fitted aft of the reduction gearing but coupled into the gearing in a similar way to the single engine. This brings the aft engine above the intermediate shaft, which extends aft from the main gearwheel, to the propeller or screw shaft. The aft engine foundation is arranged so that it allows the intermediate shaft to pass through it.

With four-engine installations the grouping of two of the forward engines is as in the two-engine case, and the other two engines are fitted aft of the reduction gearing. The pinion for the engines on the starboard side is common to both starboard engines and the arrangement on the port side is similar.

■ **12.3 Name the various types of gears and state which types may be found in marine diesel engine practice, either on engines or within propulsion reduction gearing.**

There are many types of gears; broadly they may be classified as gears used when the axes of the shafts are parallel, and those used when the axes of the shafts are at an angle or when their lines of projection intersect.

Gear types used when shafts are parallel

1   Spur gears with external teeth, or mating pairs of spur gears where the teeth on one gear are external, and on the other internal, within a hollow rim. The line of the gear teeth is axial.

2   Helical gears; in this type of gear the tooth line follows a helix; helical gears may be of the single or double type. Single helical gears set up a thrust in the shafts when transmitting power, due to the helix angle, but in double helical gears the thrust created by the right-handed helix is balanced by the thrust from the left-handed helix.

In a pair of gears meshing with each other (sometimes referred to as a matched pair) the smaller gearwheel is called the pinion and the larger one is referred to as the wheel.

Gear types used when the shafts are not parallel

3   Bevel gears.
4   Spiral bevel gears.
5   Worm gears.
6   Spiral gears. Spiral gears are similar to helical gears but used only when the shafts are not parallel.

Another type of gear is the rack and pinion; in this form of gear a rotary motion can be converted to a sliding motion or a sliding motion can be used to create a rotary motion.

In marine diesel engines spur gears are used for camshaft and fuel pump camshaft drives, cylinder lubricator drives and the like.

Bevel and spiral bevel gears are sometimes used for governor drives and indicator shaft drives.

Worm gearing is commonly used for engine-turning gear drives.

Speed-reduction gearing for propulsion drives from medium-speed engines is generally of the single helical type, though double helical gears are sometimes used.

For engines of very low power, spur gearing can also be used for speed reduction between the engine and propeller shafts.

Rack and pinion gear is used to connect and transmit the power from an electric motor in the automatic (Gyro Pilot) steering control to the control on the steering gear.

■ **12.4   What do you understand by the terms 'couplings' and 'clutches'; are the terms synonymous?**

We often speak of couplings and clutches in a loose sense as being the same thing, so they may appear to be synonymous; in fact there are essential differences.

The term 'clutch' is commonly applied to equipment which allows an operator *quickly* to connect or disconnect an engine into or out of the power train while the engine is working. The term 'coupling' is used for items of equipment which connect an engine to a power train but do not allow it to be connected or disconnected easily and quickly (*see also* Question 12.6, which deals with hydraulic couplings).

■ **12.5 How would you define the terms 'primary element' and 'secondary element' of a coupling or clutch?**

The primary element of a coupling or clutch is that part which is connected to the prime mover, i.e. the diesel engine in the case of geared diesel propulsion machinery. The secondary element is that part which takes the power transmitted from the primary element. The secondary element is indirectly connected to the propeller through the gearing and shafting.

Note There may be a considerable difference in the form that the primary and secondary elements of a coupling or clutch take. The form will depend on the actual type and operation of the coupling or clutch.

■ **12.6 Name and briefly describe the various types of couplings or clutches used with geared diesel propulsion machinery?**

Solid flange coupling. This is the simplest form of coupling and consists of two flanges, one being the primary element which is fastened to the engine crankshaft and the other the secondary element which is fastened to the pinion in the gearing train. The two flanges are bolted together with coupling bolts in a normal manner.

This type of coupling is used only in very low-power installations. Its use usually requires a flywheel heavier than normal so that the cyclic variation in the output torque from the engine is reduced. The latitude of allowable misalignment between the engine and gearing with this type of coupling is very small. The allowable misalignment can be made easier to handle by connecting the engine to the gearing pinion through a quill shaft (see Question 12.8).

Friction clutch. This form of clutch consists of both the primary and secondary sections connected in the normal manner to the driver and driven sides of the power train. The elements making up the internal parts of the clutch consist of a series of primary plates with splines which engage with and are driven by the primary element, and another series of plates, with splines on the opposite edge, which engage with the secondary element of the clutch. Between the primary and secondary clutch plates friction plates are fitted. When the plates are unloaded the primary section of the coupling can revolve but no torque is transmitted; but if the plates are forced together friction causes the primary plates to drive the friction plates, which in turn drive the secondary plate, and torque is transmitted through the clutch. The plates are usually forced against one another by hydraulic pistons fed from the lubrication system.

Another form of friction clutch has a toroidal elastomer ring having an elliptical cross-section. The ring is fastened inside a rim which is attached concentrically to the primary shaft. A series of friction pads are fitted on the outside surface of the ring within the inner circumference. The pads are connected with links to the rim holding the elastomer ring.

When the hollow elastomer ring is inflated with compressed air the friction pads move radially inwards and engage on to a drum forming the secondary part of the clutch. The torque from the primary shaft is then transmitted through the friction pads and into the secondary shaft.

The clutch is disengaged by releasing the air pressure on the elastomer ring, when the centrifugal force acting on the friction pads causes them to move radially outwards and deflate the ring. A clearance is then created between the friction pads in the primary section and the drum of the secondary section.

Slip coupling. This may be of the hydraulic or electro-magnetic type.

Hydraulic coupling. The main components of a hydraulic coupling are a pair of similar discs. Each disc has a semi-elliptical annulus, with radial vanes in the annulus, on its 'working' face. One disc is the primary element, attached to the engine crankshaft; the other is the secondary element, connected to the gear pinion. On the outer circumference of the secondary element a dome-shaped cover plate is bolted, and this plate envelops the primary element. The primary shaft passes through a centrally located hole in this cover plate. The clearance space between the hole and the primary shaft is sealed with bronze labyrinth packing. The edges of the labyrinth seals are knife edged. When the coupling is filled with oil, centrifugal force acting on the oil causes it to circulate in the pocket formed by the radial vanes and the semi-elliptical annulus in each of the primary and secondary portions of the coupling. The circulation of the oil causes torque to be transmitted across the faces of each section of the coupling.

When there is no oil in the coupling the primary section can revolve freely in the space formed by the secondary element of the coupling and the domed cover plate. When oil is pumped in the drive is gradually taken up.

If a dumping valve is fitted on the secondary element of the coupling so that oil can be discharged quickly from the coupling (due to centrifugal action), the engine can be clutched into and out of the power train. A hydraulic coupling with this refinement becomes a hydraulic clutch.

Electro-magnetic coupling or clutch. In construction this type of coupling or clutch is similar to a salient pole alternator, or an induction motor. One coupling element consists of a number of poles which are magnetized by current flowing through coils wound round the pole cores. The coils are connected with a direct current supply through slip rings.

The other element consists of a series of windings which may be of the cage form permanently short circuited, or similar to the windings on an alternator.

If the magnetic poles are not energised the primary section of the coupling can revolve freely, but when the magnets are energised there is relative movement between the magnetic field and the cage windings. This relative movement between the two causes current to flow in the cage windings which in turn creates another magnetic field. The interaction between the two magnetic fields links the primary and secondary sections of the coupling so that torque is transmitted magnetically across the gap between them.

Different manufacturers arrange the primary and secondary elements differently. Some arrange the magnetic poles and slip rings on the primary element, while others arrange them on the secondary.

If the slip coupling can also be used as an alternator the coupling becomes a salient pole machine and the direct current field poles are always arranged as the primary element and attached to the engine.

Note In hydraulic and electro-magnetic couplings some slip must always occur between the primary and secondary elements.

In the hydraulic coupling the slip is necessary to create circulatory conditions for the oil to transmit torque. In the electro-magnetic coupling the slip gives a relative velocity between the magnetic field of the poles and the cage windings. If this slip did not occur no current would flow in the cage winding, there would be no magnetic effect, and the coupling could not transmit torque.

By varying the amount of oil contained in a hydraulic coupling or by varying the excitation current in an electro-magnetic coupling it is possible to increase the slip between the primary and secondary elements of the coupling so that the propeller can be run at a speed lower than its normal speed relative to the engine speed.

Elastic coupling. In this form of coupling an elastomer is fitted between the primary and secondary sections of the coupling. Some damping of the cyclic variation in the engine torque takes place so the input to the gearing is smoother. No slip takes place between the primary and secondary elements.

Note Hydraulic couplings and electro-magnetic couplings also have the capability of damping out torque variations from the engine.

■ **12.7 Why and where are couplings and clutches fitted in geared diesel propulsion systems?**

Clutches are fitted in multi-engined geared diesel propulsion installations so that any engine can be connected with or taken off the power train easily and quickly.

In some multi-engined installations it is the practice when manoeuvring to have one or more engines operating ahead and one or more engines operating astern. Ahead direction of propeller operation can be obtained by engaging the clutches of the ahead running engines and leaving the astern running engines free, and vice versa. A brake in sometimes fitted on the intermediate shaft when the clutch is of the friction type. The brake then stops the propeller, shafting and gearing so that its inertia is destroyed and clutch plate wear is reduced.

Clutches are also fitted in gearboxes with reverse gearing. The clutches are used to engage the ahead or astern sections of the gearing train.

Clutches and couplings are almost always fitted on the high-speed side of a gearing train, for control of individual engines. The physical size of a clutch or coupling on the higher-speed side, and therefore the lower-torque side, is much smaller and lighter than, say, a coupling or clutch transmitting the same power on the lower-speed side.

A very important attribute of hydraulic, electro-magnetic, and elastic couplings is their ability to dampen torque variations from the engine under normal operating conditions, and shock loads coming back from the propeller as is experienced in heavy weather. It is for these last two reasons that these types of couplings are so often employed in geared diesel plants.

Another reason for fitting hydraulic couplings arises from the relatively large mechanical clearances between their primary and secondary elements: this large clearance can accommodate minor misalignment between engine crankshaft and gearbox.

■ **12.8   What is a quill drive? What is its advantages and where is it used?**

A quill drive is made up of two parts, a pinion and its adjacent journals made from a single forging bored right through its polar axis, and a shaft small enough to pass through the hole with a substantial clearance. One end of this shaft has a flange coupling on it and the shaft is long enough to pass right through the hole leaving the flange clear of the pinion journal. The end of the shaft opposite to the flange is fastened by a key to the bore in the journal.

It can be seen that if torque is applied to the coupling it will be transmitted along the shaft which in turn will transmit it to the pinion through the key.

It can also be understood that, due to the length of shaft passing through the

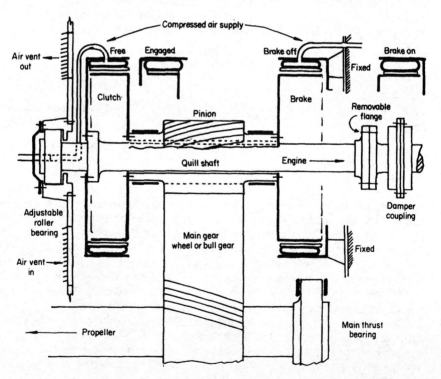

Fig. 12.1 Diagrammatic sketch of pneumatic clutch, pneumatic brake, and quill shaft of a geared diesel propulsion system.

pinion and the natural elasticity of the shaft, the arrangement will be very flexible; and this flexibility is obtained in approximately the same length as a normal solid arrangement of pinion journals and flange.

If the end of the shaft which is keyed inside the pinion journal has the key fastening replaced by a claw-type flexible coupling, the arrangement will not only have elasticity but will also be able to sustain a relatively large amount of misalignment between the flange at the input end and the connected prime mover. It must be understood, however, that as misalignment increases so also does the rate of wear on the teeth in the claw coupling.

Quill drives are often used to connect the prime mover through to the reduction gearing pinions. They may also be used to connect the engine to the primary sides of friction clutches when they are fitted on the aft sides of the pinions.

In reduction gearing installations which have a set of step-up ratio gears for driving an alternator or generator, it is common practice to use a quill drive between the pinion and the alternator or generator.

A main propulsion reduction gearing system consisting of a quill shaft, pneumatic clutch and pneumatic brake is shown in Fig. 12.1. The damper coupling is connected through to the coupling on the engine crankshaft. The air vents (located in the casing supporting the adjustable roller bearing) are placed vertically one above the other; cool air may then enter the lower vent, the air after cooling the clutch sets up a convection current and leaves through the upper vent. It is important to keep the vents clear of obstructions at all times.

The illustration shows the arrangement for one engine, for a twin engined propulsion system the arrangement will be duplicated.

■ **12.9 What is the effect of changes in rate of pressure rise on the pneumatic clutch shown in Fig. 12.1?**

The summation of the mass of the gearing, shafting, propeller, and the mass of entrained water within the bounds of the propeller together with the frictional effect, amounts to a considerable sum. When the clutch is loaded by supplying air under pressure the large total mass has to be accelerated while bringing it up to the desired propeller speed.

If the rate of pressure rise is too fast the effect is the same as putting a brake on the engine and the engine will possibly stall.

If the rate of pressure rise is too slow the heat generated by friction between the pads and the clutch ring will cause the pads to be badly overheated and the clutch ring to be damaged.

The rate of pressure rise in the clutch is adjusted to suit the power output of the engine and the mass of the parts that have to be accelerated while the clutch is being loaded.

If the engine is low in power and the mass of the parts being accelerated is high, the rate of pressure rise must be low to prevent stalling.

If the power of the engine is high and the mass of the parts being accelerated is low, the rate of pressure rise may be high to prevent overheating.

■ **12.10   If the ring of a pneumatic clutch were found to be cracked, how would you deal with the problem?**

There will always be some heat generated during the loading period in pneumatic clutches. The heat generated on the surface causes a crazed pattern of cracks in the hard surface of the clutch ring. If the cracks are only on the surface of the ring no action need be taken. If there is reason to think the cracks have some depth, the depth must be ascertained.

The depth of the cracks can be found by ultrasonic testing. If the cracks indicate the possibility of parts of the clutch ring becoming detached or the ring fracturing, the clutch ring must be renewed.

Spare clutch rings are usually carried in the ship or held in some location where they are readily available for transport to the ship.

■ **12.11   What do you understand by the terms 'single- ' and 'double-reduction' gearing? What is the purpose of each form of gearing and where may these forms of gearing be found in motor ships?**

In single-reduction gearing a set of gears consists of one or more pinions driving a gearwheel, the ratio being fixed by the number of teeth in the pinion relative to the number of teeth in the wheel. The gear ratio or reduction ratio is designed to suit the relation between the speed of the prime mover and the speed of the output shaft. If the speed reduction required is not too great, single-reduction gearing will be satisfactory.

If the speed reduction is large, double-reduction gearing must be used in order to keep the physical dimensions of the gearing within bounds. A set of double-reduction gearing consists of a pinion at the input end of the gear train, referred to as the first-reduction pinion, with its teeth meshing with the first-reduction wheel. The first-reduction wheel is mounted on a common shaft with the second-reduction pinion, so the rotational speed of the first-reduction wheel and second-reduction pinion is the same and is fixed by the speed of the input shaft and the gear ratio between the first-reduction pinion and wheel. The second-reduction pinion drives the second-reduction wheel, which is mounted on the output shaft.

Single-reduction gearing is used in geared diesel propulsion systems to utilize efficiently the available speed in medium-speed diesel engines with the more efficient propellers, which turn at much lower speeds.

If the ship has a steam turbo alternator coupled in with an exhaust gas boiler the steam turbine will drive the alternator through a set of single-reduction gearing.

The most likely place to find double-reduction gearing is in the engine room crane or hoist, on steam and electric winches, in lathe gearboxes and the like.

■ **12.12   Give some values of reduction gear ratios for propulsion reduction gearing driven by medium-speed engines.**

Medium-speed diesel engines for marine propulsion purposes are designed to operate at some speed within the range of 300 rev/min to 850 rev/min. Propeller speeds may be between 85 and 150 rev/min depending on the size and loaded draught of the vessel, or because of limitations imposed on the machinery weight.

The largest ratios will be for the higher-speed engines geared to the lower-speed propellers. The speed of an engine might for example be 850 rev/min geared to a propeller having a speed of 85 rev/min. Speed of driver divided by speed of driven equals ten. The gear ratio would therefore be 10 to 1.

Taking the other extreme we could have an engine speed of 300 rev/min driving, through gearing, a propeller at 85 rev/min. In this case the driver speed is 300 and the driven speed 85 so the ratio is 300 to 85, or 3.529 to 1. These figures are at the extremes of the range.

In actual practice high-powered, medium-speed engines operate at speeds considerably lower than 850 rev/min. Fairly common figures are in the range 250 to 600 rev/min, and for propellers 90 to 110 rev/min, this providing the best compromise between propeller size, cost, and efficiency. This gives a middle value of 425 rev/min for engine and 100 rev/min for propeller. These values give a gear ratio requirement of 4.25 to 1.

Note Design of geared propulsion machinery has taken a distinct trend towards using increased engine speeds and lower propeller speeds, where the draught conditions will allow a larger-diameter propeller.

Higher-speed engines have been available for some time, and now that their reliability has been well proven they tend to be selected more often. When lower-speed propellers are used the increased propeller efficiency offsets some of the other losses in geared propulsion machinery.

■ **12.13** When examining gear ratios or the number of teeth on matching pinions and wheels in reduction gearboxes, it is always noted that the number of teeth on the pinion is never an arithmetical factor of the number of teeth in the wheel. Why is this, and what is a hunting tooth or teeth?

Gear ratios for propulsion gearing always have a number which includes a decimal fraction, i.e. the number of teeth in the pinion is never an arithmetical factor of the number of teeth in the gear. By designing the numbers of teeth in this manner a particular tooth in a pinion does not repeat its contact with a particular tooth on the gear until many revolutions later. This equalises the wear between the teeth in the gear train. This can be demonstrated as follows.

An arithmetical factor (a whole number) would be obtained if there were for example 50 teeth on the pinion and 150 teeth in the gear. The factor would be 3 and the gear ratio 3 to 1. When the gears are rotating it will be shown that for 3 revolutions of the pinion a particular tooth on the pinion will make contact again with a particular tooth on the wheel. First find the LCM of the numbers of teeth.

$$50 = 2 \times 5^2$$
$$150 = 2 \times 3 \times 5^2$$
$$\text{LCM} = 2 \times 3 \times 5^2 = 150$$

The number of pinion revolutions for re-establishment of contact between a particular tooth on the pinion and a particular tooth on the gear is

$$\frac{150}{50} = 3$$

Let the teeth on the gear be increased to 152. The gear ratio is 152:50 which is 3.04:1. The ratio of the gearing has hardly been changed. We can now examine how many revolutions the pinion must make between re-establishment of contact between pinion and gear teeth. As before,

$$50 = 2 \times 5^2$$
$$152 = 2^3 \times 19$$
$$\text{LCM} = 2^3 \times 5^2 \times 19 = 3800$$

Revolutions of pinion

$$\frac{3800}{50} = 76$$

Before a particular tooth on the pinion repeats its contact with some particular tooth on the gear, the pinion makes 76 revolutions.

If the gear was given 151 teeth the ratio would be 151:50 or 3.02:1.

$$50 = 2 \times 5^2$$
$$151 \text{ is a prime number}$$
$$\text{LCM} = 2 \times 5^2 \times 151 = 7550$$

Revolutions of pinion is now

$$\frac{7550}{50} = 151$$

In the examples, the odd tooth or extra two teeth put on the gearwheel increase the number of revolutions before the pinion and gear repeat some particular position. They are called hunting teeth, and their purpose is to equalize tooth wear and prevent problems which might arise if increased wear occurred in localized areas. Hunting teeth make for longer gear life.

■ **12.14 What is an involute, a cycloid, an epicycloid and a hypocycloid?**

If a plank is placed on the curved portion of a cylinder lying on its side, and rocked so that it is in effect rolling round the cylinder without sliding, a mark on the edge of the plank would trace out an involute.

Involute. In geometrical terms, an involute is the locus of a point on a straight line which is rolling without sliding round the circumference of a circle.

Cycloid. Similarly a cycloid is a curve generated by a point on the circumference of a circle when it rolls, without slipping, on a straight line.

Epicycloid. An epicycloid is a curve traced by a point on the circumference of a circle rolling on the exterior of another circle.

Hypocycloid. If it rolls on the inside of the other circle the curve traced is a hypocycloid.

The construction or development of each of these curves is shown in Figs. 12.2(a) through (d).

If during the construction of a hypocycloid the diameter of the rolling circle is made half the diameter of the base circle the hypocycloid becomes a chord of the base circle equal the diameter of the base circle.

■ **12.15 The terms 'involute' and 'cycloid' have some relation with gear tooth profiles. What do you understand about this relationship and why is one form of gear tooth preferred to another?**

Cycloidal tooth forms are based on the geometry of the cycloid; involute tooth forms on the geometry of the involute. The involute tooth form is favoured for use in reduction gearing as used for ship propulsion purposes in connection with diesel engines. For two gears to operate satisfactorily it is essential that the tooth form be such that the velocity ratio is always uniform. In practice it is extremely difficult to maintain the theoretical centre distances between matched gears; with involute tooth forms relatively wide deviations from the theoretical centre distance can arise without affecting the velocity ratio, whereas variation in cycloidal tooth form centre distance immediately affects the uniformity of the velocity ratio.

With involute teeth, within limits, variation in centre distance increases the backlash but has little effect on the velocity ratio. It is on this ground, together with some other advantages related to manufacture, that involute tooth forms are favoured.

■ **12.16 Define the following terms when related to spur gears: pitch circle, addendum, dedendum, tooth face, tooth flank, backlash, clearance.**

Pitch circles. The pitch circles of a matched gear and pinion are those circles which if in rolling (without slipping) contact with one another give the same motion as the actual gear and pinion. The ratio of the pitch circle diameters is the same as the ratio of the gearing.

Addendum is that part of the tooth above the pitch circle measured radially from the pitch circle; the *dedendum* is that part of the tooth below the pitch circle measured radially from the pitch circle to the bottom of the tooth profile.

Tooth face is the working or contact surface of the tooth above the pitch circle.

Tooth flank is the working or contact surface below the pitch circle.

Backlash is the minimum open distance or clearance measured between the back faces of adjacent teeth on a pinion and gear.

Clearance is the gap measured radially between the tip of one tooth and the bottom space between two teeth opposite and on each side of it. The names of the various parts making up gear teeth are shown in Fig. 12.3.

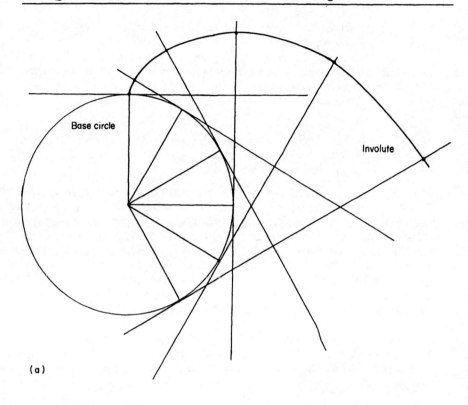

(a)

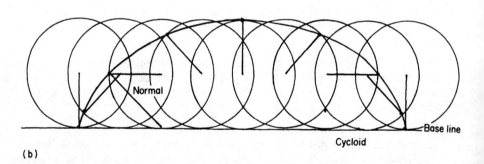

(b)

Fig. 12.2 (a) Development of an involute.
(b) Development of a cycloid.
(c) Development of an epicycloid.
(d) Development of a hypocycloid.
Note. If the rolling circle is half the diameter of the base circle the hypocycloid becomes a straight line equal in length to the diameter of the base circle.

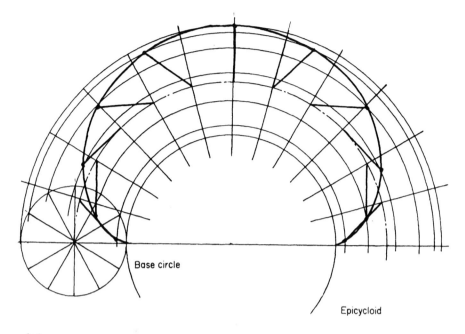

Base circle

Epicycloid

(c)

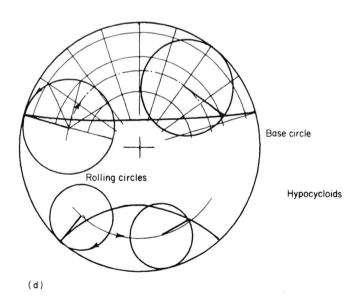

Base circle

Rolling circles

Hypocycloids

(d)

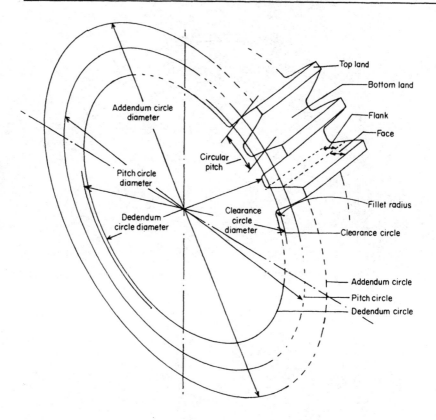

Fig. 12.3 Sketch showing the names of the various parts forming a gear tooth.

■ **12.17 Why is helical gearing favoured for propulsion reduction gearing; what are the advantages of double helical gearing over single helical gearing?**

Helical gearing is favoured because it is quieter and smoother in operation than normal spur gearing. This quieter and smoother action comes about from the fact that the left- and right-hand helices of a matched pinion and wheel give a gradual engagement of teeth, which in turn gives an increase in the allowable load which can be put on the teeth; with larger helix angles two or more teeth on the pinion engage with two or more on the wheel at the same time.

When helical gearing teeth are operating they set up axial thrusts which are related to the helix angle. The axial thrust from the main gear can be accommodated within the main thrust bearing, but the thrust from the pinion must be balanced by additional thrust bearings or some other means.

Double helical gearing, being cut with opposite handed helices, does not have this problem of thrust as the thrust from the right-handed helix is balanced by the thrust from the left-handed helix. It has a disadvantage in that the main gearwheel, being constrained axially by the main thrust bearing, must engage with a pinion which is free to float axially, so that the load transmitted by the pinion is balanced between the forward and aft sections of pinion teeth. This entails the use of claw couplings or similar devices to provide the floating action required.

In reduction gears for diesel-engined ships single helical gearing is most favoured. The axial thrust is kept to a minimum by using small helix angles, no problem arises from the axial thrust in the main gear wheel and no serious problem arises in balancing the thrust from the pinion. Single helical gearing is less costly to produce.

However, for high powers associated with low propeller speeds, where the width of the gears becomes large, double helical gears will usually be used.

■ 12.18 When two V-engines are used in conjunction with a two-input single-output gearbox, how are centres of the input shafts kept far enough apart to give adequate working space between the engines?

It would be possible to design gearing which consisted of two input pinions and one main gearwheel but it would be unnecessarily large and heavy. To keep weight and physical dimensions within economic bounds intermediate wheels are fitted between each engine input pinion and the main gearwheel on the output shaft. This increases the distance between the centres of the input shafts and gives the working space required.

In some twin-screw ships the working space between the two engines is obtained by an asymmetric arrangement wherein the gearing (moving outboard from the inboard input shaft) consists of an input pinion meshing directly with the main gearwheel. Between the outboard input pinion and the main gearwheel an intermediate wheel is interposed. This provides the necessary space between the engines, and forms the asymmetric arrangement.

■ 12.19 What is an epicyclic gear and where is it sometimes used?

The parts forming this reduction gear consist of the input shaft on which the input pinion, referred to as the sun pinion, is mounted. The sun pinion drives a planet wheel which is mounted on an axle attached to the planet wheel cage carrier. The carrier is attached to the output shaft and revolves with it. Surrounding the planet wheel is a gear rim with teeth cut on its inner circumference, and these teeth mesh with the teeth on the planet gear. The gear rim is fixed in the gear casing and does not move relative to the other parts. This gear is referred to as the annulus wheel.

When the input shaft and pinion rotate through a displacement of one tooth the planet wheel is displaced a similar angular distance around its axle, which is part of the planet wheel carrier. When the planet wheel is displaced angularly one tooth by the input pinion, the tooth on the planet wheel meshing with the annulus is displaced an angular distance of one tooth round the annulus wheel. As the annulus wheel is fixed the displacement of the planet wheel tooth causes the planet wheel axle to be displaced a similar angular amount. This in turn displaces the planet wheel axle carrier which, being part of the output shaft, causes the output shaft to revolve.

In practice, to reduce tooth loading, two, three, or four planet wheels may be used depending on the reduction ratio required, the space available and the power to be transmitted.

A considerable torque is placed on the annulus wheel so the casing is of necessity robust enough to withstand this torque together with the load transmitted from the thrust bearing. The gear casing fastenings to the foundation are also robust enough to transmit the propeller thrust into the hull for propelling the ship.

The annulus wheel is flexibly connected to the casing by means of a tooth-type flexible coupling which is sufficient to absorb any inaccuracies in the machining of the gears.

This type of gear is used in high-powered, single-screw motor ships driven by a single V-engine. The arrangement is very suitable for a fine aft body as is necessary for high powers and speeds. The power-to-weight ratio for this arrangement makes it very attractive in specialised ships where a high cubic capacity and low machinery weight is required.

In order to keep total weights down it is usual to utilize small-diameter propellers having rotational speeds between 125 and 150 rev/min. With such propeller speeds the reduction gear ratios are of the order of 2.75:1 to 3.25:1.

Double helical type gears are used for the sun pinion, planet, and annulus wheels.

Note   The definitions given for epicyclic gear arrangements vary and it could be argued that under some definitions the gear described is not of the epicyclic type. The reduction gear described has, however, been used in many diesel propulsion installations. It is often called a sun and planet gear, or planetary reduction gearing.

■ **12.20   What is the form of friction when involute profile teeth come into and out of contact?**

When the contact between a pinion tooth and a wheel tooth is at the pitch circle line, the friction is due to pure rolling action, because the rolling speeds of the wheel and pinion teeth are equal. At other points of tooth contact the rolling speed of the pinion teeth and wheel teeth will not be equal so the friction will be due to sliding action, and relative to the rolling speeds of each element. As sliding between tooth surfaces will give higher amounts of wear than pure rolling, the gear designer aims to keep the sliding speeds to a minimum. This is done by altering the proportions of the addenda in the pinion and wheel. Generally the sliding action is a maximum at the time the flank of a pinion tooth makes contact with the tip edge on the face of a wheel tooth.

Note   The action of gear teeth is covered in most books dealing with Theory of Machines.

In practice the major gear manufacturers producing the high-duty gears used for marine reduction gearing have their own designs of tooth profiles which for marine propulsion gearing are based on involute forms. But spur gear teeth on more lightly loaded gearing will be cut with standardized gear cutters which may follow B.S.S. designs or the designs of the specialist gear cutter manufacturers.

■ **12.21 What types of bearing are used to support the rotating elements of propulsion reduction gearing?**

Some reduction gearing installations use white-metal-lined bearings throughout the gearbox.

In other cases white-metal-lined bearings may be used for the main gearwheel bearings, with a combination of roller bearings or ball bearings for the pinion shafts. If ball bearings are used, arrangements must be made to allow for axial expansion of the pinion and shaft.

For reduction gearing used in lower-power propulsion installations ball or roller bearings may be used throughout.

Epicyclic gearing bearings for the gearbox described in Question 12.19 are white-metal-lined throughout.

White-metal-lined pinion journal bearings sometimes have the bearing shells split on an angle from the horizontal. In such cases the split is arranged at 90° to the line of action of the bearing load or at some angle near to this.

■ **12.22 How are reduction gearing teeth and bearings lubricated?**

Reduction gear teeth are lubricated by directing jets or sprays of oil on to the teeth of the pinion and wheel at the point where they are beginning to make contact when running in the ahead direction. Sometimes the oil jet is directed on to the opposite side also, for better cooling of the teeth. The lubricating oil sprayers are spaced approximately 100 mm apart.

The sprayers may be formed from hollow pockets cast in the gear case and drilled on the inside. Oil is fed into the pocket from the outside and issues from holes in the pocket drilled at the correct angle to direct the jets on to the desired areas.

In some instances, with cast gear casings, and with most fabricated gear casings, the sprayer consists of a straight pipe with holes drilled along its length to form the jets. In other instances nozzles which produce a fish-tail oil spray are fitted into the feed pipe.

Bearings are lubricated by forced lubrication, an annulus being machined round the bearing pocket which is connected to the oil supply. Oil is fed into the bearing through two holes on opposite sides of the bearing shells which are in way of the joint between the upper and lower halves. The holes feed the oil into the side pockets machined in the sides of the bearing at the shell joint.

The lubricating oil drains into pockets at each end of the bearing and drains away through external or internal drain pipes.

■ **12.23 State some values for the overall efficiency of a set of reduction gearing as would be used in geared diesel propulsion machinery (a) with slip couplings, (b) without slip couplings. Why is it important that the efficiency curve remain constant over a wide range of powers?**

The efficiency referred to here is the mechanical efficiency which is equal to the output power divided by the input power. The quotient when multiplied by 100 gives the mechanical efficiency expressed as a percentage.

Gearbox efficiencies at full load are of the order of 97% to 98%, which may decrease to 96% to 97% at half load.

Slip coupling losses, either hydraulic or electro-magnetic, are of the order of 3% to 4% of the load transmitted at full load. These losses expressed as a percentage efficiency will be equal to 100–3 or 4, or approximately 97%.

The efficiency of a set of reduction gearing combined with slip couplings will then be the product of the coupling and gearing efficiencies.

Coupling efficiency × gearing efficiency = 97% × 98% ≈ 95%

If reversing of the propeller direction is carried out with reversing gear incorporating reverse gears and clutches in the gearbox, the gearing efficiency will be reduced by about 1% so gearing efficiency is then of the order of 96% at full power.

Note   Electro-magnetic slip couplings require an electrical power supply for energizing the magnet poles and this power input must be included in overall efficiencies.

It is important that the efficiency curve plotted as the relationship between load transmitted and efficiency be as flat as possible, so that at low power output from the engine the power loss will be minimal. The curves are plotted following tests which may be carried out at various speeds and tooth loading related to the propeller law, or at constant speed with varying tooth load.

In ships with a controllable pitch propeller the efficiency curve will be plotted with the relationship of the tooth load and speed to suit the changes made in the propeller pitch for various rotational speeds.

■   **12.24   How do the power losses in reduction gearing manifest themselves and how are they dealt with? What causes these losses?**

The losses of power are mostly due to friction in bearings and from gear tooth contact friction, with a very small part due to windage losses.

These power losses manifest themselves in the form of heat, a very small part of which is lost by radiation and convection from the gear case. For practical purposes in calculating the quantity of oil to be circulated, it is assumed that all the heat generated is removed by the lubricating oil used for the bearings and gear oil sprayers. The heat taken away by the lubricating oil is in turn transferred to the water circulating through the gearing oil coolers.

Hydraulic coupling losses manifest themselves as heat energy. The fluid used in the coupling is taken from the gearing oil system and part of the oil used is continuously lost from the coupling through small holes drilled in the coupling casing expressly for this purpose. This oil finds its way back to the system by mixing with the oil from the sprayers and bearings and heat is eventually removed in the gearing oil coolers.

Electro-magnetic coupling losses also manifest themselves as heat which is removed by circulating air through, and around the coupling, in a similar manner to the cooling for large electric motors of the open or drip-proof type.

Friction couplings have virtually no power loss except when running idle or being engaged or disengaged. Wet friction couplings have oil passing through them which removes any heat generated under these conditions. Heat generated by dry friction couplings during engagement and disengagement is dissipated by conduction through adjacent parts and air cooling.

■ **12.25** Certain forces are set up in hydraulic couplings when they are transmitting torque. How do these forces come about and how are they balanced?

A common arrangement for the supply of lubrication oil (which acts as the hydraulic fluid) is to supply the oil through an axial drilling, which passes through the pinion and journals. The secondary element of the coupling is attached to the pinion shaft and the oil passes through its centre directly into the space separating the two elements.

Air is expelled through the small radial holes in the outer circumference of the coupling and a small amount of oil circulates through the holes for cooling purposes when transmitting torque. Oil completely fills the internal parts of the coupling between the primary and secondary elements and the dome-shaped casing.

In the static condition there is an out-of-balance force which tends to separate the coupling elements, this force being equivalent to the product of the oil pressure in the coupling oil supply and the cross-sectional area of the primary shaft to which the primary element is connected.

When the engine is started, oil contained in the primary element annulus is subjected to centrifugal force which acts radially outward, causing a circulation to be set up which is radially outward in the primary element and radially inward in the secondary element. This circulatory action causes vortices to be created which in effect transmit torque. As the secondary element gathers speed the slip reduces to approximately 3% and the speed of circulation is reduced. All the forces acting on the oil in the coupling cause further axial force which again tends to separate the couplings and is an increase on the forces mentioned previously.

These axial forces are used in the balance of single helical gear axial thrust on the pinion side of the coupling, and small out-of-balance forces are balanced by tilting pad thrust bearings or thrust collars and locating rings built into the bearings or casings. Axial thrust on the primary element shaft is usually balanced by a tilting pad thrust bearing fitted to the primary shaft between the engine and gearbox.

In other cases the thrust on the primary section is taken up by a thrust collar and bearing fitted within the aftermost section of the engine crankcase, thus keeping overall machinery length to a minimum.

■ **12.26** How are tests conducted to measure the efficiency of reduction gearing?

It is common practice to carry out tests on reduction gearing by connecting two sets of reduction gearing back to back. The two output shafts are connected together through a short section of shafting.

The input shafts are connected by a flexible shaft. Load can then be put on the surfaces of the teeth of each set of gearing by putting a twist in the flexible shaft connecting the pinions while the wheels are held together by the other shaft. The tooth loading is directly related to the angle of twist put in the torsionally flexible shaft, and by varying and measuring the angle of twist, the tooth loading can be controlled and known.

By coupling one of the pinion journals to an electric motor and running the gearing up to speed the power required to drive both sets of gearing can be measured. The power measured is the actual power loss for the two sets of gearing. The power losses at various speeds and tooth loading can be plotted on a graph in a manner coinciding with the power requirements of the propeller.

As the power losses in gearing are very low it would be extremely difficult to measure them with any accuracy using normal power testing methods with a prime mover and dynamometer.

■ **12.27   What is a hob? Where and how is it used?**

A hob is a cutting tool used in a hobbing machine for the manufacture and production of gearing. The hob itself is a specialized form of cutting tool which in appearance looks like a worm with sections cut out axially so that cutting edges are formed. The material behind the cutting edge is backed off as is normal with metal cutting tools. Hobs are made with two or three separate helices. Single-helix hobs are used for finishing cuts where a high degree of accuracy is required.

The hobbing machine consists of a horizontal circular table on which the gear blank is set up. The circular table is made so that it can be rotated. Gearing connects the table with the hob spindle, and the ratio of this gearing is controlled to suit the number of teeth being cut on the gear blank and the number of helices on the hob. The angle of the hob spindle, which can be swivelled, is adjusted so that the tangent of the helix on the cutting side of the hob is at the angle corresponding to the helix angle on the gear being cut. The swivelling hob spindle head is mounted on a vertical arm or stanchion, and guides on the arm allow the hob spindle head to be raised or lowered.

The gear blank and the horizontal table are rotated. The hob rotates in synchronism with the table (through the gearing) and is brought down on to the top edge of the blank so that cutting starts. The cutting of the gear then follows on as a continuous process, with the hob and table continuously rotating.

The cross-section of the cutting face of the hob is similar in profile to a rack and is normally straight-sided. The tooth profile on the gear being cut is therefore generated. Large hobbing machines have two cutting heads set at 180° to each other.

Pinions are cut in a similar-type machine, but smaller in size, and the pinion blanks are set up horizontally as in a lathe.

This method of cutting gears does not require a dividing head, index plates and the change wheels necessary if gears are cut in a milling machine with standardized gear cutters.

■ 12.28 What do you understand by the following terms when related to high-precision gearing as would be used for propulsion reduction gearing: 'shaving', 'lapping', 'grinding'?

Shaving is a process carried out on accurately hobbed gears to improve the accuracy and the surface finish of the profiles of the gearing teeth. Rotary shaving cutters are used in special machines in which both the gear and shaving cutter rotate. The shaving cutter is fed in at an angle which causes the cutter to slide sideways and produce a shaving action which is not unlike the action of a hand scraper. Shaving gives the tooth profiles a mirror finish.

Lapping of gears is carried out in a similar manner to a normal lapping action which can be likened to the action of lapping the high-pressure jointing surfaces of fuel valve parts on a lapping plate. When gears are lapped the gear is set up in bearings and meshed with an accurately matching cast-iron dummy. The dummy is rotated and a lubricant containing the lapping abrasive is fed between the teeth of the dummy and the gear. The keeps on the gear journal bearings are spring loaded so that some resistance is set up to increase the cutting rate of the lapping media.

The pinion is then meshed with the wheel and they are finally lapped with each other. In production of standard sets of gearing the pinion may also be lapped with a special lapping dummy made for this purpose.

Grinding of the profiles of gear teeth is carried out in special gear grinding machines. Surface- and through-hardened gearing is usually ground after hardening so that any discrepancies occurring from growth or warping during the heat treatment process are removed.

Grinding the profile of helical gears and pinions is carried out with a grinding wheel thin enough to go between the gear teeth. The motion of the side of the grinding wheel relative to the motion of the tooth profile is such that the profile form is generated. Obviously it is not generated in the same way as a hobbing machine, but follows the action of a gear rack planer. Grinding gives the gears a mirror finish.

The capacity of gear grinding machines is limited when compared to main gear wheel hobbing machines, but it can cover most of the reduction gearing requirements for geared diesel propulsion installations.

The terms shaving, lapping and grinding cover the methods used for finishing gears. When these processes are used to improved the surface of the profile of the teeth, the allowable surface loading on the profile can be considerably increased provided the stresses in the root of the tooth induced by bending are not outside allowable limits.

■ 12.29 In what direction do the engines run in a geared twin-engine propulsion installation?

If the propeller is of the solid type and right-handed, and the reduction gearing consists of a main gearwheel meshing with pinions on opposite sides (each engine connected to its own pinion), it can be seen that the main gearwheel must turn in the same direction as the propeller, which will be clockwise for right-handed propellers when looking from the aft end of the engine room towards the gearing. The pinions must therefore both rotate in the opposite direction which will be anti-clockwise, and each of the engines will turn in the same. direction as its pinion.

If intermediate wheels are fitted between the pinions and the main gearwheel the engines will run in the same direction as the main gearwheel and the propeller.

If an intermediate gear is fitted between the outboard pinion (as in some twin-screw ships) and the main gearwheel, the engines run in opposite directions.

■ **12.30   How is the direction of the load on pinion and mainwheel journal bearings found?**

The direction of the load on pinion and wheel journal bearings in helical gears will be the resultant of the direction and magnitude of the tooth load in the transverse direction and the weight of the parts that the bearing supports.

Referring to the first case in Question 12.29, i.e. two pinions and one main-wheel, it was seen that both engines rotated anti-clockwise. Consider first the starboard engine. With the gearwheel on the left-hand side of the pinion, the action of the tooth loading is to lift the starboard pinion journals upward so that the load is on the upper part of the bearing (i.e. tending to lift the bearing keep and putting the bearing keep studs in tension).

The port engine also rotates anti-clockwise, but now the gearwheel is on the right-hand side of the pinion. The action of the tooth loading on the port engine pinion journals is such that the journals are forced down into the lower part of the bearing.

The line of action of the force or the pressure line between two involute gear teeth is shown in Fig. 12.4(a). The method of finding the amount and line of action of the forces acting on reduction gear bearings is shown in Fig. 12.4(b)

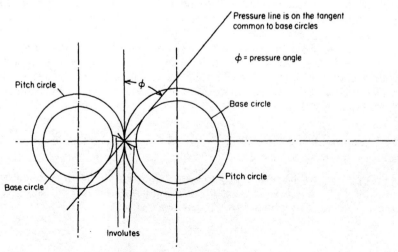

(a)

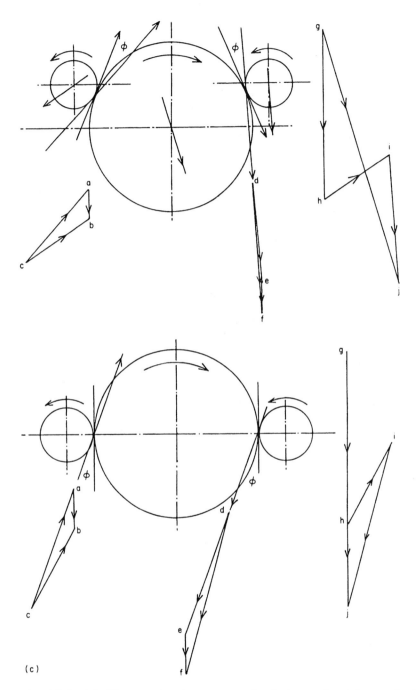

(c)

Fig. 12.4 (a) Diagram showing pressure line between involute gear teeth.
(b) Diagram showing the line of action of forces on bearings.
(Centre line of main gear shaft below that of pinion shafts.)
(c) Diagram showing the line of action of forces on bearings.
(Centre line of main gear shaft on same level as that of pinion shafts.)
Note. In many cases the weight of the pinions is neglected when finding the line
of action of the forces and loads on pinion and main gearwheel bearings.

and (c). Vector diagrams based on the gear tooth loads required to obtain the designed shaft torques are constructed as shown in the diagrams. If the weight of the pinion is small relative to the weight of the main gear wheel, the weight of the pinion is often neglected when constructing the diagrams.

Some gearing manufacturers arrange for the line forming the junction between the upper and lower halves of pinion bearings to be arranged at or near 90 degrees to the line of action of the force acting on the bearing.

■ **12.31   What effect does a relatively heavy coupling of the hydraulic or electro-magnetic type have on the pinion journal and bearing loads?**

Generally the weight of a pinion is a lot less than the weight of the coupling. In such cases the pinion shaft will be canted due to the weight of the coupling. The bearing nearest to the coupling becomes a fulcrum, and the weight of the coupling tips the pinion shaft about the edge of this bearing. The bearing furthest from the coupling restrains the tipping action, and the end of the journal furthest from the coupling contacts the end of this bearing. It can therefore be seen that in the static condition a heavy coupling can lift the end of the pinion shaft so that weight comes on to the upper half of the bearing furthest from the coupling.

Note   The tipping action will be dependent on the tipping moment caused by the weight of the coupling and the righting moment from the weight of the pinion. The moments will be taken about the nearest edge of the nearest bearing to the coupling. If the tipping moment is greater the pinion shaft will cant and the end furthest from the coupling will lift.

■ **12.32   How is bearing wear-down measured? Why must bearing wear-down figures be carefully examined?**

Some reduction gear bearings are provided with bridge gauges, and bearing wear-down can be measured by removing the bearing keep, placing the bridge gauge in place and taking feeler gauge readings between the bridge gauge and the journal.

In other cases a hole is drilled through the bearing keep and bearing shell. The top of the bearing keep is machined flat. A depth micrometer gauge is then used to measure the distance between the flat surface and the journal. Normally the hole is blanked to retain oil, and a screwed plug is removed to take the measurements.

By comparing the measurements taken with the previous and original measurements, the wear since the last record, and the total wear, can be found.

Bearing wear-down readings, when taken as mentioned previously, must be analysed with care bearing in mind that they will give no indication of developing trouble if it is the upper half of a bearing that is wearing, as mentioned in Questions 12.30 and 12.31.

The original bridge gauge readings are used in making the initial tooth alignment after bearings are remetalled.

It can be seen that after a set of gearing has been in service, changes in bearing clearance are more useful as a record of wear than bridge gauge readings.

■ 12.33 What effect does bearing wear have on the tooth loading of gears?

When gearing is new it is carefully aligned so that the whole length and working parts of the tooth profiles contacting each other have their surface loading evenly distributed.

If the two journal bearings of a pinion wear an equal amount in the same direction no detrimental effect is likely to be felt within normal limits of wear, but if one pinion bearing wears more than another the alignment of the gear teeth will be affected. The pinion teeth will bear more heavily on the wheel teeth at the end where the bearing wear is greater. The wear on the teeth will then increase at the end where heavier loading takes place so that eventually the tooth profiles are impaired. If the bearing wear increases over a short time and goes unnoticed breakdown of the tooth surface could take place.

■ 12.34 How can the clearance of reduction gearing bearings be found?

Some gear cases are arranged with all the bearings outside the casing so the bearings are easily accessible. In other arrangements it is sometimes necessary to remove part of the casing to get access to the bearings. In nearly all gear casings the mainwheel bearings can be removed without disturbing the casing.

Sometimes the access space is sufficient to measure the bearing clearance with a feeler gauge but when feelers are used care must be taken to see that the weight of overhanging parts does not cause false readings to be obtained (Question 12.31).

In other cases it will be necessary to remove the bearing keep and bearing shell so that the clearance can be measured with lead wire or plastic gauging thread.

The soft lead wire or plastic thread is placed on the journal and the bearing is reassembled and tightened down so that the wire is compressed and squeezed out to the same thickness as the bearing clearance. The bearing keep and shell are dismantled, the wire or thread is removed and the clearance measured either by using a micrometer or comparing the width of the squeezed plastic with a comparator from which the clearance is directly given.

■ 12.35 Some gears have what is called 'tip relief'; what is this and what is its purpose? What is 'end relief'?

Tip relief is carried out to prevent damage to the contact surfaces of the teeth at their extremities (face tip and flank root) where the sliding action is greatest (see Question 12.20).

Tip relief is a modification of the involute form at the tooth tip. The actual work involves some removal of material on the face of the tooth at the tip, and it is often done in a gear grinding machine. The small amount of material removed relieves or lowers the contact pressure at the tooth tip, which, if not relieved, would become damaged early in service due to the combination of pressure and sliding.

End relief is carried out by removing a small portion of the ends of the tooth profile extending over the depth of the tooth. It is seen as a chamfer or smoothing away of the profile at the ends of the teeth for a distance of 7 to 10 mm. End relief also relieves pressure at the tooth ends.

Note Teeth with coarse pitch are more prone to troubles if tip relief is neglected. Sometimes relief is made on the flanks of the teeth instead of the tips. When the flank is relieved it may be carried out as a grinding or shaving operation.

■ **12.36 How would you ascertain whether reduction gearing tooth alignment is satisfactory or departing from some acceptable norm?**

In order to find out if the tooth alignment is acceptable or otherwise it is necessary to examine the contact surfaces of the face and the flank on the tooth profiles. If the width of the contact surface is the same along the length (axially) of the teeth and the contact width is adequate and extends nearly across the depth of the tooth (radially) covering a major part of the addendum and dedendum, the alignment can be considered satisfactory. If the pattern of wear is wedge-shaped, being wider on one end of the tooth than on the other, it indicates misalignment.

If the change in width is only small it should not be necessary to take immediate action but a record of the contact surface should be made, and the gearing kept under observation.

If the contact surface does not extend to the full length of the tooth and diminishes to a point, action should be taken to correct alignment. The allowable amounts of misalignment vary with the gear material and original heat treatment. The amount that can be allowed is very small.

Generally, gearing noise levels give some indication of changes taking place in alignment, but in motor ships this cannot be relied on because engine noises often drown the noises coming from the gearing.

Note For engineers who have sailed with highly loaded steam turbine gearing it must be pointed out that the proportions of propulsion gearing in diesel installations are such that it is not necessary to make any compensation for twisting or bending of the pinions.

■ **12.37 If, in checking tooth contact patterns, it is found that, due to the finish of the teeth, the contact surface is difficult to follow, what can be done to give the pattern better definition?**

An area covering a span of four to six teeth is cleaned of all oil and staining. The teeth are then coated with a thin layer of spirit lacquer or blue tool-room marking dye, which must be oil-resistant. The engine is then operated at low power so that some load is put on the teeth, after which the engine is shut down and the lubricating oil shut off the gearing. An inspection of the coated teeth is made. The contact area will now be bright and bounded by the blue dye. Generally it is better to carry out this test with both engines operating at full power, but this is seldom practicable unless the teeth are marked in one port and inspected at the next.

In some cases pinion teeth are coated with a blue dye and mainwheel teeth with a red dye. Teeth which have been lapped have a matt surface left from the lapping that delineates the contact surface boundaries very clearly.

In new ships it is common practice to copper-plate some of the teeth and make frequent regular observations of the tooth contact. Each time an observation is made the results are recorded for future reference.

Some shipping companies maintain tooth-contact-surface records for diesel propulsion reduction gearing similar to the records kept for steam-turbine reduction gearing.

■ **12.38 What are the Hertz equations? Where are they used and what do they show us?**

The Hertz equations are a mathematical method for obtaining the surface compressive stresses and surface strains which occur when spherical or cylindrical surfaces of elastic materials are under load.

It can be seen that if one steel ball supports another so that they are in contact the area of contact will be infinitesimally small. If one of the steel balls is loaded with some weight the unit stress on the contact surfaces will be exceedingly high from the fact that load divided by area is equal to unit stress. If the area of the contact surfaces is infinitesimally small, the unit compressive stress at the surface becomes infinitely large. As the localized compressive stress is large it follows that the localized strain will also be large.

The Hertz equations are used by designers to calculate the high localized compressive stress and strain which occurs in elastic materials when two curved surfaces are under load and in contact with each other. The equations are the base from which were derived some of the equations used in the design of gearing teeth. The equations are also used in the design and evaluation of material for fuel-pump cams and rollers.

Note The Hertz equations were the work of H. Hertz and go back to the years 1891 and 1895 when they were published in Leipzig. They have since been shown to be correct by photoelastic methods (see Question 6.53). For interested students the mathematical methods devised by Hertz are covered in text books on the Theory of Elasticity or the Strength of Materials and dealt with under 'Pressures of Spherical and Cylindrical Bodies in Contact'. There are many equations derived from the Hertz equations which are of specialized form for design work. While it is not necessary for the student to be familiar with the equations or their derived forms, it is desirable to have knowledge of the problems that brought them about as such knowledge provides a better general understanding.

■ **12.39 List the troubles which sometimes occur with reduction gearing teeth.**

Gear manufacture has become so good that propulsion reduction gearing runs with remarkably little trouble. When the occasional trouble does arise investigation has often shown that it was caused by misalignment due to working of the ship in bad weather or due to some condition of cargo loading which distorts the hull in the region of the reduction gearing so that the gearing alignment is disturbed.

Troubles which may beset gearing teeth vary from minor pitting of the teeth found during the time the gearing is being run-in, to such faults as fracture of teeth with parts detaching themselves from the parent material.

Pitting is a form of fatigue failure which occurs more commonly with new gearing during the running-in period. When the gearing teeth are newly in service the very minute imperfections in the surfaces of the teeth profile sustain very high loads so the surface breaks down under fatigue failure and small pit holes develop. As the teeth bed themselves with one another correction naturally takes place and the pitting ceases.

If misalignment occurs, the surface loads in localized areas may rise so that pitting starts again and may go on to the point where tooth profile is destroyed if the misalignment is not corrected. In appearance pitting is seen as a series of irregular-sized holes usually near the pitch line.

Flaking may occur on tooth surfaces; some experts consider this to be another form of surface fatigue failure while others consider it to be due to the action of rolling and sliding. This form of failure is not necessarily serious. Evidence of it is found in magnetic filters and it often stops when the teeth are fully bedded-in. It is more common with non-hardened or non-surface-treated steels.

Other tooth surface failure can occur with surface-hardened teeth, in which the hardened surface breaks away from the softer sub-surface material. This is usually a metallurgical failure caused by internal stresses in the material, which are set up during the surface hardening process. This type of failure shows itself in the form of holes or craters where quite large pieces may detach themselves from the parent material. This kind of surface failure is sometimes referred to as sub-surface fatigue or spalling.

Scuffing is another form of surface defect which occurs when lubrication breaks down and small areas of pinion and gear teeth weld themselves together (under loading) and then break away. In appearance the tooth profiles are seen to have score marks running in a radial direction (see Question 9.37).

Tooth fractures may come about from a variety of causes such as: a row of pitting holes creating a stress raiser; damaged or bruised teeth causing high bending loads; incorrect design in the form of insufficient root fillet radius; minute cracks formed during quenching when teeth are being hardened; overloading when working on one engine (provided that engine characteristics allow it). In most cases these causes decrease the fatigue resistance of the material and ultimate failure is due to fatigue.

Wear debris and other small foreign particles which will accumulate in the lubricating oil if it is not properly treated, will cause surface scratch marks. If the parts embed themselves in the tooth profile quite a lot of surface damage may occur together with accelerated wear.

■ **12.40 If you were instructed to carry out an examination of a set of gearing, how would you go about it?**

Prior to making an examination of a set of propulsion reduction gearing, the inspection hole covers and nuts must be cleaned off so that dirt, paint chips, and foreign matter will not fall into the gearing when the covers are opened up. It is good practice to clean up the area around the gear casings and remove any loose material lying around.

The inspection cover nuts are then removed from the covers leaving each cover closed and held in place with two nuts; the nuts removed must be stored in a safe place away from the inspection openings.

The turning gear is put in, the usual precautions having been taken at the engine control station and prior to turning the propeller with the turning gear.

Inspection covers should only be removed one at a time and then replaced and fastened down with the two nuts left on the cover.

The first or last part of the examination will be to check gear oil sprayers and oil flow from bearing ends; if they are not fitted with drain pockets, either before shutting the oil pump down or by running the pump at the end of the examination.

If a grid or perforated plate is fitted in the run-down connection between the gear casing and the drain tank, it should be inspected for any debris, white-metal flakes and the like.

The profiles of the pinion teeth should be examined, noting particularly the wear pattern markings and the contact surface; both ahead and astern sides should be examined.

If the contact surfaces are normal the alignment will be in order (see Question 12.36). Mainwheel gear teeth are examined in the same way.

If the pinion teeth are hardened, defects in alignment will most likely show up in the gearwheel teeth first, especially if the gearwheel teeth have a softer surface than the pinion teeth.

The root fillets in *all* teeth must be examined for the start of any fatigue cracks, even though they are more usual in pinion teeth.

If any bearings are fitted within the gear casings their fastenings and locking devices should be checked, together with any wiring connected to temperature sensors.

The fastenings, clips and connections on lubricating oil pipes to bearings and oil sprayers must also be checked out. Prior to replacing covers the gear teeth, where cleaned during the examination, should be coated with oil.

Notes should be made of the findings so that they can be written up in the log-book or work reports, without omissions or inaccuracies.

At the end of the examination when the inspection covers are being replaced all nuts removed, spanners, wrenches, inspection hand-lamps and material used should be accounted for.

Note Any tools such as feelers, calipers, pencils, note-books, etc. should be taken out of the top pockets of boiler suits or overalls as they can be relied upon to fall out of the pocket when leaning over looking through inspection holes.

■ **12.41 How often should gear casing inspection covers be opened up to examine the gearing and internal parts of the gear case?**

Once the gearing teeth are properly run-in after the initial operating period and if there is nothing abnormal to cause concern, three- to six-monthly inspections should be adequate depending on the service conditions. But if there is any unusual noise or an increase in debris in the magnetic filters, the cause must be promptly investigated. Similarly, overheated bearings, which might have wiped and could lead to misalignment in tooth contact, must be investigated.

■ **12.42   How would a hot bearing be found in a set of reduction gearing?**

In unmanned and part-time unmanned engine rooms, reduction gearing bearings are connected through to the machinery alarm system. If any bearing develops a fault which causes a temperature rise to occur it is called out on the alarm system and visual indicators locate the bearing. In other cases reduction gearing bearings may be fitted with a thermocouple-type sensor in the bearing keep or housing. This is connected to an electrical meter which gives a temperature read-out. A selector switch is used to find the running temperature of each bearing. It is not usual for this type of instrument to be coupled into an alarm system. In older reduction gearing part of the oil circulating in each bearing is passed into an individual lubricating oil return line. Thermometers fitted into thermometer pockets in the return line show the temperature of the oil leaving each bearing. Portable cover plates on the line may be opened to check amounts of oil flow. During normal watch keeping, bearings external to the gear casing may be felt with the hand at regular intervals to check temperatures.

■ **12.43   How are gear casings fastened down? What attention do the fastenings require?**

Gear casings are fitted with a heavy flange which runs around the casing bottom periphery. When the gear casing has been lined up to the propeller shafting, chocks are fitted between the flange and the tank top. The chocks support the weight of the gear casing and the internal parts. Bolts are fitted through the flange, chock and tank top. The tank top is usually tapped with a thread so that the holding-down bolts can be screwed into the tank top. A nut and a sealing grommet is usually fitted to the bolt on the underside of the tank top plating. A large number of the bolts will be of the fitted type to ensure that the gear case is always correctly located.

The thrust block is usually incorporated in the gear casing; when this is so, a high percentage of fitted bolts may be used to hold the gear case down and for the transmission of the propeller thrust into the hull. A chock is fitted at the forward side of the base flange to further assist in the transmission of thrust and also to prevent movement of the gear casing from thrust action should holding-down bolts slacken. In some cases the thrust chocks are fitted with a wedging arrangement which is adjustable.

In other gear casing arrangements the space under the thrust bearing is left as an open hollow space accessible through lightening holes. The chocks fitted to the tank top for holding the gear case against thrust are then welded direct to the tank top and make hard contact with the lightening hole edges in the casing base plate. With this arrangement it is usual to make all the holding-down bolts at the aft end of the casing of the fitted type.

Some gear casings are fastened down on to poured resin chocks (see Question

7.18). The chock is usually cast to support the whole of the mounting flange of the gear case.

The holding-down bolts for the gear case require periodic examination to see that they are maintained in a tight condition and do not allow movement, which causes chock fretting.

Apart from periodic examination, holding-down bolts should be examined after a ship has had a heavy weather passage or during a lengthy period of heavy weather.

■ **12.44   If something is dropped on an open set of gearing so that the teeth became bruised, burred or locally deformed, what would you do to correct the situation? What will be the likely result if repair is not carried out correctly?**

While every care must be taken to protect open gearing, the occasional accident can happen. With non-hardened gearing the damaged teeth must be filed up with a smooth file so that their profile follows the original shape as closely as possible. When this stage is reached the teeth on the mating gear or pinion are thinly coated with marking blue and turned into and out of mesh with the damaged teeth. Any high spots shown up by the blue marking in the damaged location are then removed with a very fine tooth file or an oil stone. This is continued until good bedding is obtained, with all the high spots removed. Care must be taken to prevent filings from falling into other gear tooth spaces or into the casing. When the gearing is in operation at slow speed a check should be made to ensure that there is no 'tick-tick', which would indicate high spots remaining on the damaged teeth. If there is any unusual sound its source must be located and the cause rectified.

If teeth are bruised or their profile is upset, indication of this is given by unusual noise. If the gearing is kept in service without correcting the damage the high bending loads set up on the tooth root will eventually lead to fatigue cracking and fracture of the tooth in way of the higher loaded locality.

Hardened and surface-hardened teeth if damaged should be treated in a similar manner. In this case it may be necessary to use carborundum hand hones or very small, portable, lightweight grinders if the material is too hard for filing. After the tooth profile is corrected, the surfaces of the teeth in the damaged area must be checked for surface cracks with a dye-penetrant test.

After making a voyage, or in a lesser period if necessary, the damaged teeth must be examined again and if any hard bright high spots remain they must be removed with an oil stone.

■ **12.45   What are magnetic filters and where are they used?**

A magnetic filter consists of a casing which houses a cartridge in which two, or up to a large number of permanent magnets are fitted. Lubricating oil is passed through the filter casing and ferrous particles contained in the lubricating oil are attracted towards the magnets which hold the particles. The casing has a cover which when removed allows the magnetic cartridge to be withdrawn from the casing for cleaning. The magnets used in a magnetic filter have a high flux density, and they are fitted in such a manner that the north pole of one magnet is opposite the south pole of an adjacent magnet.

Magnetic filters used in a lubricating oil system are sized to take the full flow capacity of the lubricating oil pump, with which they are connected in series. It is normal to fit a magnetic filter in the lubricating oil system of propulsion reduction gears so that the lubricant passes through the magnetic filter before it passes into the oil sprayers and bearings. Sometimes small magnetic filters are fitted in the lubricating system of exhaust gas turbo-blowers when the rotor is held in ball- or roller-type bearings; in other cases a magnetic plug may be fitted in the oil sump. Some motor ships have turbo alternators (operated by waste heat boilers), and it is common practice to fit a magnetic filter in the lubricating oil system of such steam turbines and their reduction gearing.

In some Companies it is the practice to fit a magnetic filter on the main engine lubricating oil system during the early life of the engine. After the engine has been operating for some period the filter is removed from the system and stored for use in the next new ship. Many engine builders use magnetic filters in the lubricating system to which new engines are connected when on the test bed.

■ **12.46   Are magnetic filters fitted as duplex or simplex filter units?**

It is not usual to fit magnetic filters in duplicate or as duplex units, so that one can be in use while the other is shut down for cleaning, or as a stand-by unit. It is usual to fit a by-pass in conjunction with a single magnetic filter unit so that it can be shut down for draining and cleaning.

When a magnetic filter is fitted only temporarily for use during the early part of an engine's life no by-pass is fitted.

■ **12.47   How often should magnetic filters be cleaned? What attention must by given during the cleaning?**

No set time period of frequency can be given for the cleaning of magnetic filters; the period will be dictated by the build-up of ferrous wear debris seen on the magnets.

When a new ship with a geared diesel propulsion installation is brought into service, it is common to open up the magnetic filters and clean them every watch until the working faces of the reduction gears have stabilized. During this early operational period, which may last over some months, the amount of ferrous debris attached to the filter magnets should be carefully observed and commented on in the log-book. After the working faces of the teeth in the gears have stabilized, it is common practice to clean magnetic filters weekly. The cleaning should be carried out under the direction of a senior engineer who should carefully observe if any changes have occurred in the amount of wear debris attached to the magnets. If any increase is noted it must be investigated. Magnetic filters, when fitted temporarily for the early life of an engine, are opened up for examination and cleaning when the engine is shut down in port.

13

LINE SHAFTING, SCREWSHAFTS, PROPELLERS, THRUST BEARINGS

■ **13.1   How is the torque transmitted from the prime mover through to the propeller?**

A heavy flange is forged on the last section of the engine crankshaft, which may also be the thrust shaft. If the propulsion installation consists of geared diesel engines, the heavy flange will be on the main gearwheel shaft and be an integral part of it. The main gearwheel shaft often incorporates the thrust collar for the thrust bearing. In some cases the thrust collar may be forged on the main gearwheel shaft; in other cases it is made separately and fitted on to a cone machined on the forward end of the shaft and held in place with a large nut (see Question 13.38).

The intermediate shafting is slightly smaller in diameter (but see Note later) than the crankshaft (direct coupled engines) and the screwshaft. Each section of the intermediate shafting has a flange forged at each end. The screwshaft also has a flange forged on its inboard end.

The flange on the output shaft from the engine or gear case is coupled to the intermediate shaft with coupling bolts. Each section of the intermediate shaft right through to the screwshaft is joined together at the flanges by the coupling bolts. The propeller is keyed on to the screwshaft on its cone-shaped end, and held in place by the propeller nut which is screwed on to the threaded outboard end of the screwshaft. The torque is transmitted through the coupling bolts to the conical end of the screwshaft and through the key into the propeller boss.

It must not be forgotten that due to the tightening of the coupling bolts and the method of fitting the propeller on to the screwshaft an enormous amount of friction exists between adjacent coupling flange faces, and also between conical end of the screwshaft and the propeller. This friction plays a large part in the transmission of the torque from the engine to the propeller.

Note In some ships, fitted with machinery aft, it may be found that the intermediate shaft is larger in diameter than both the engine crankshaft and the screwshaft. The screwshaft is commonly called the tailshaft or propeller shaft. In some ships the flywheel may be fitted in 'sandwich' fashion between the engine shaft and the first section of the intermediate shaft. In similar manner a turning-wheel may be fitted between the main gearwheel shaft and the first section of the intermediate shaft, if the screw and intermediate shaft turning gear is not incorporated within the reduction gearing.

■ **13.2   What form do coupling bolts take and how are they fitted?**

There are three types of coupling bolt in use. The most common has a parallel shank; the others have either a very small taper with a normal bolt head, or a larger amount of taper and no bolt head. The parallel-shank bolts have a normal head, cylindrical in shape, and the parallel section is reduced in diameter where it joins the head with a well-radiused stretching length. The threaded section is smaller in diameter than the shank and is joined to it with a stretching length well radiused and slightly smaller in diameter than the bottom of the thread. The end of the threaded section is turned parallel and drilled to take a split pin. The coupling-bolt nut is of the normal hexagonal type with a depth of between 0.75 and 1.0 of the thread diameter. The bolts have a taper over a short length on the shank end to facilitate entry of the bolt into the bolt hole in the coupling flange. The coupling-bolt hole is bored or reamed to size and the parallel shank is made to fit into the hole with a light interference fit.

Some engineering works produce a coupling bolt with a very small taper amounting to 1 mm in 100 mm. The coupling-bolt hole is made to the same taper and the bolt is made to such a size that when tapped home in the bolt hole there is a small clearance between the bolt head and its landing face on the flange. The nut tightens the bolt-head landing on to the spot facing on the

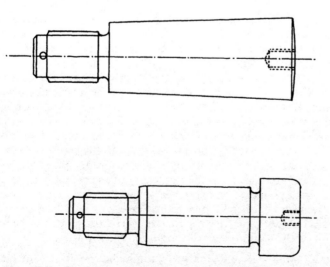

Fig. 13.1   Shaft coupling bolts – taper and parallel bolt types.

coupling. The advantage with this type of bolt is that removal is comparatively easy with drawing gear, and the fitted section of the bolt does not spoil so easily during removal. The hoop stress across the section between the side of the hole and the edge of the coupling flange is controlled by the amount of draw used to pull the bolt head hard home.

Tapered coupling bolts have a taper on the diameter of about 6 mm per 100 mm of bolt length; they do not have heads. The tapered bolts are fitted to accurately reamed holes, and pulled tight with hammer and spanner. Tapered coupling bolts are not favoured so much in British practice but are widely used in the United States and the continent of Europe (Fig. 13.1).

■ **13.3 What other types of coupling may be used on intermediate shafts and screwshafts in place of the solid forged flange type?**

The other types are those which are semi-permanently fitted to the shaft, and those where the shafts must be dismantled, such as the coupling between the screwshaft and the aft section of the intermediate shaft, as is necessary with some variable- or controllable-pitch propellers.

The semi-permanently fitted type consists of a flange and boss forged as one piece. The inside of the flange and boss is bored out to the limits required for shrink fits, and circumferential oil grooves are machined in the flange bore. The flange is fitted to the shaft by normal shrinking. It can be removed if necessary by connecting a high-pressure oil pump to the oil grooves and forcing oil into the space between the shaft and coupling flange bore. This expands the flange and boss and allows it to be easily removed from the shaft. This type of coupling is used with shafts fitted with normal roller bearings (i.e. solid outer and inner rings). It is removed only to fit new bearings. The flange is replaced again by shrinking. As the flanges are removed only to fit new roller bearings they are referred to as semi-permanent.

The other type of coupling is of the muff type. The muff is bored with a smooth fine taper. A thin sleeve is machined conical on the outside to match the fine taper in the muff. The inside of the sleeve is bored to suit the outside diameter of the shafts being connected. The taper bore in the muff has circumferential oil grooves machined into it. Each end of the thin sleeve has a circumferential groove in its outer surface. When the coupling is fitted the thin sleeve is mounted on one shaft and slid along so that it is half-way over each of the shafts being coupled together. The muff, previously mounted on the opposite side to the sleeve, is slid along and on to the sleeve until it is tight on the taper. The oil grooves in the muff are connected with a high-pressure oil pump. The hydraulic tightening tool is a split ring in which are fitted a series of small hydraulic rams. The tightening tool is fitted on the small end of the thin sleeve and engages in a circumferential groove. The rams in the tightening tool come on to the muff. The oil pump connected to the grooves in the muff is operated so that the muff is expanded by the oil pressure. The hydraulic tightening tool is also pumped up and the rams pull the sleeve into the muff. The operation is continued until the muff is expanded out to the correct circumference. The pressure on the oil between the muff and the sleeve is released, the oil is discharged out as the muff tightens itself on to the sleeve which in turn is tightened on to the two ends of the shafts being coupled. To dismantle the coupling the tightening tool is fitted on the large end of the sleeve. When the muff is put under internal oil pressure, pumping of the tightening tool causes the sleeve to be withdrawn from the muff.

Couplings of this type are often used in twin-screw ships where the screw-shafts are withdrawn outboard for survey. They are also used on some single-screw ships where the propeller is mounted on a flange on the outboard end of the screwshaft instead of a normal cone end. This type of screwshaft is withdrawn outboard for survey whereas cone-ended screwshafts are withdrawn inboard.

■ **13.4   Why are intermediate shafts smaller in diameter than screwshafts? What factors govern the size of these shafts?**

The sizes of engine- and screwshafting are governed by the formulae adopted by the various ship classification societies. The factors used take into account the strength of the steel used; the quality of the steel is controlled by the upper and lower specified limits of the tensile strength. The horsepower being transmitted together with the rotational speed of the shaft, the maximum torque and the respective mass moments of inertia of the propeller and the flywheel are used in the formula to evaluate the intermediate shaft diameter. The mass moment of inertia of the propeller includes the entrained water.

Once the diameter of the intermediate shafting has been found, the diameter of the screwshaft is found by increasing the diameter of the intermediate shaft by some percentage, to which is added a further amount which takes into account the propeller diameter and the method by which the shaft is protected from the corrosive action of sea-water. In practice the screwshaft is approximately 15% to 17% larger in diameter than the intermediate shaft. The larger percentage is associated with the larger-diameter propellers.

The reason for intermediate shafting being smaller in diameter than the screwshaft is because the intermediate shafting is not subjected to such high stresses as the screwshaft. The stresses to which an intermediate shaft is subject are the stresses due to the transmitted torque, the compressive stress (running ahead) from the propeller thrust and the stresses induced by its own weight which are often of little consequence. The screwshaft is subjected to similar stresses but the forward side of the conical end is heavily loaded due to the manner in which the weight of the propeller is supported by the screwshaft. This end of the screwshaft is like a cantilever, the support being the screwshaft bearing and the load being the propeller. The weight of the propeller creates a bending moment which in the static condition causes a tensile stress in the upper part of the shaft having a maximum value at the top. This tensile stress diminishes to zero at the sides of the conical end, and at this point the stress starts to become compressive in nature and increases to a maximum value at the bottom. When the screwshaft is rotating a point or small portion of the shaft will in turn be subjected to a maximum tensile stress, when it is in its uppermost position, which changes to zero stress at the sides and a maximum compressive stress at its lowest point.

From this we can see that the large part of the conical end of the shaft undergoes cyclic stress reversal so is therefore subject to fatigue failure. The chances of fatigue failure are minimized if the stress range is reduced, and this is done by increasing the shaft diameter. This accounts for the propeller shaft being larger in diameter than the intermediate shaft.

■ **13.5 It is known that when screwshafts develop fractures the fracture often starts in the end of the keyway. Why do cracks start at this point?**

It is an established fact that most screwshaft fractures result from a crack which has begun from the region on the forward end of the keyway. It was explained in the previous answer that the forward end of the screwshaft cone was subjected to cyclic stress reversal; it is in this region that the keyway extends. The keyway, due to its form acts as a stress raiser. If there is some small ingress of sea-water into the large end of the cone it creates a corrosive environment which will hasten fatigue failure. This starts in the highly stressed region at the end of the key. There are other factors which create problems in this region; they are the hard spot at the end of the key and the region at the forward end of the propeller boss where there is a very small change of shaft diameter due to the compressive effect of the propeller boss on the shaft. At the forward end of the conical fit the propeller boss compresses the steel of the shaft, and just forward of the fit, the steel, in effect, 'bulges' out where it is not constrained; this constitutes a stress raiser. Sometimes fractures start in this locality; often a series of pitting marks is found running round the shaft circumferentially. These pitting marks also act as stress raisers.

■ **13.6 How is the aft end of a screwshaft protected from sea-water? Where is the likely point of ingress if leakage occurs?**

Screwshafts are protected from contact with sea-water by the brass liner which covers the greater part of the screwshaft. The liner extends from the largest part of the conical end over the length of the stern tube, through the gland and stuffing box and some distance beyond, so that the liner extends forward of the gland when it is just entering the stuffing box. The large end of the conical bore in the propeller boss is bored with a counterbore into which a square cross-section rubber sealing ring is fitted. The end of the screwshaft liner is made so that when the propeller is hard on the cone the rubber ring is compressed between the end face of the liner and the counterbore in the forward end of the propeller boss. Some clearance is allowed in the counter bore: this allows the sealing ring to squeeze out and prevents it becoming volume bound.

A common point of ingress for sea-water is at the sealing ring. Leakage may come about if the sealing ring is allowed to twist when the propeller is being fitted, or if the ring has insufficient compression. If the sealing ring is too large and becomes volume bound the ring will prevent the propeller coming hard home on its cone. In service the propeller will move and chafe the rubber ring causing it to leak, and further damage will occur as the movement damages the conical end of the propeller shaft and the conical bore in the propeller boss.

If leakage of sea-water should occur at the sealing ring it usually results in a fractured screwshaft; every care must therefore be exercised when propellers are being fitted on the screwshaft during assembly on the building berth or following survey in dry dock.

■ **13.7 Some screwshafts do not have liners fitted. In such cases how is the shaft end protected from sea-water? Where is the most likely point of ingress of sea-water?**

Where screwshafts are not fitted with liners the shaft is oil-lubricated and the stern tube bearings are white-metal-lined cast iron bushes. In order to retain oil in the stern tube the inboard end of the shaft is fitted with a mechanical seal which prevents oil running out. At the aft end of the stern tube another mechanical seal is fitted so that the space between the stern frame and the forward end of the propeller boss is sealed off. This seal, often called the outer seal, prevents egress of oil when the pressure of the oil in the stern tube is greater than that of the sea-water pressure outside, and it also prevents ingress of sea-water when the pressures are reversed. This is the most likely point of ingress of sea-water to the screwshaft.

■ **13.8 What kind of lubricant can be used in stern tubes with white-metal-lined bearings?**

The lubricant used in stern tube systems must have the ability to maintain a lubrication film in the presence of water so that it is not washed away. The lubricant must also have an affinity for metal surfaces so that it affords good protection of the metal in the stern tube and shaft areas against sea-water. This affinity for metal surfaces is also necessary when the screwshaft starts to revolve, as boundary lubrication conditions are present at this time.

Compounded oils have these properties; they are blends of mineral oils and fatty oils. The fatty constituent causes water physically to combine with the oil and form an emulsion. The fatty constituent may be lanolin or a synthetic fatty oil having similar properties.

The lubricants used have a specific gravity at 15.5°C within the range of 0.92 to 0.95, with a viscosity Redwood 1 at 60° of about 300. The viscosity index of many brands of stern-tube lubricant have wide ranges which is unfortunate, as the conditions under which they operate make a high-viscosity index desirable.

See also questions in Chapter 15 dealing with stern-tube lubrication systems.

■ **13.9 If leakage of sea-water into an oil-lubricated screwshaft occurs, what indications will there be? When is sea-water leakage into the system most likely to occur?**

If the stern-tube system is pressurised by a pump and the oil pressure within the outer seal is greater than the sea-water pressure outside the seal, leakage will always be outboard. This is due to the head differential between the pressures on either side of the seal. If the pump maintaining the pressure inside the stern tube should fail, the pressure inside the outer seal may be lower than that outside, and leakage of sea-water into the system may occur if the outer seal is defective.

Leakage of sea-water into the stern tube is indicated by emulsification of the lubricant. The best time to carry out checks on leakage is when the vessel is in port.

As the ship loads or discharges cargo, the aft draught of the vessel changes and the static head of sea-water above the screwshaft outer seal will change in a similar manner. If the pressure gauge showing the oil pressure in the stern tube system changes with the change of draught it may under many conditions indicate a defective outer seal.

Prior to accepting stand-by when sailing, the stern-tube-system drain cock should be opened and any water present drained off. If a large amount of water is drained off it indicates a seal leak.

If the ship has a light aft draught and there is no, or only a small, current round the ship, the stern-tube system can be pressurised to some pressure greater than the head of water outside the outer seal. If the seal is defective oil globules will be seen coming to the surface from the seal.

Sea-water leakage into the system can occur only when the pressure of water outside the seal is greater than the oil pressure behind the seal. When this condition exists, coupled with heavy weather and cold conditions adversely affecting the lubricant viscosity, leakage will occur if the seal is defective. The chances of seal defects coming about are greatest when the stern-tube bearing is approaching its maximum allowable wear.

■ **13.10 Why is it important with oil-lubricated screwshaft bearings that the stern tube drain cock is operated prior to starting the main engine?**

If the stern tube should contain water and the engine is started, the rate of wear on the stern tube bearing will be very great until an oil emulsion is created in the bearing. Even then the emulsion may not effectively lubricate the bearing. If the amount of water is large it would be possible for the bearings and screwshaft to be damaged and scored to the point of ruin. For this reason it is imperative that all water in the stern-tube system be removed by draining and that the stern-tube space be completely filled with oil prior to starting the main engine.

Note It is unfortunate that there is not enough attention given to the access arrangements to get to this drain. It is often placed on the aft peak bulkhead right under the shaft. It is then difficult to get at and difficult to observe when draining. The outlet drain-off point within the stern tube is also badly located in some designs.

■ **13.11 How can outer-seal leakage be found when at sea? What can be done to keep the system safe without dry-docking the vessel?**

Outer-seal leakage has often been found by an increase in the consumption of the stern tube lubricant. It is more likely to be noticed when the ship is proceeding into tropical waters from cold-water areas. Leakage inboard is shown by oil emulsification and reduced oil consumption.

Note No stern-tube gland remains a hundred percent tight over a long period and usually the oil shows some small amounts of emulsification, indicating the presence of small amounts of water.

Seepage from the inboard seal often has the appearance of bad emulsification. This comes about due to condensation on the water-cooled wear ring on the forward end of the stern tube. The wear ring is the junction of the static and moving points of the shaft seal.

■ **13.12   How much is the screwshaft bearing allowed to wear down before correction is made?**

The amount a screwshaft is allowed to wear down is governed by the rules of some ship classification societies. In other cases it will be decided by the experience of the people concerned. Figures sometimes quoted suggest that the screwshaft bearing in the stern tube should be re-wooded when the wear-down is about 8.5 mm for shafts of diameter of 300 mm (diameter of shaft under liner). For shafts of 200 mm diameter the allowable wear-down would be 7 mm. In other cases from Class Rules the allowable wear-down is of the order of 5 mm for 300 mm diameter and 4.75 mm for 200 mm diameter shafts. The figures given are related to stern-tube screwshaft bearings of lignum vitae, rubber, or plastic.

White-metal-lined bearings with oil lubrication often have a lining which is considerably thinner than the wear-down figures given. With this type of bearing the allowable wear-down is dictated by the effectiveness of the oil seal when the stern-tube bearing becomes slack. In practice it is often found that oil seals will begin to give trouble, particularly in heavy weather, when the bearing clearance is approaching 2 mm. Some seals give trouble with less clearance, others work well with considerably more.

In all cases the seal manufacturer's recommendations covering maximum allowable bearing clearance should be followed, but clearances must be viewed together with the thickness of the white-metal lining where applicable.

Note   In this question we have used the terms wear-down and clearance because with oil seals clearance becomes the critical factor.

Wear-down of bearing = newly measured clearance − original clearance.

■ **13.13   The screwshaft bearing fitted in the aft end of the stern tube has its wear-down checked when the vessel is in dry-dock. How is this check carried out? Is it necessary to dismantle the oil gland or outer seal in oil-lubricated stern-tube bearings?**

When a vessel enters a dry-dock or floating dock the screwshaft wear-down is checked as soon as possible after the dock is dry. This work is usually done from a ladder placed against the stern frame boss surrounding the stern tube. The method of checking varies with the type of stern-tube bearing. With lignum vitae, plastic, or rubber bearings the measurement of the bearing clearance may be carried out with a wedge gauge or feelers.

If a wedge gauge is used, the side of the wedge contacting the bearing is chalked and inserted into the clearance space between the top of the screwshaft and the bearing. The gauge is pressed home and withdrawn. The clearance of the bearing is measured on the wedge at the point where the chalk marking has been scraped off by the bearing.

If the clearance is measured with feelers it is done in the conventional manner.

If the screwshaft liner is worn so that a ridge has formed on the liner at the end of the bearing, care must be taken as the step formed by the ridge can lead to false clearance readings. Some difficulty arises in taking these measurements as they are usually taken through a hand hole in the top of the rope guard.

When oil lubrication is used for the screwshaft bearing it is not necessary to remove the oil gland to find the screwshaft bearing wear-down. With oil-lubricated stern-tube systems a hole is drilled right through the stern-tube nut, or stern frame boss, the stern tube and the bush, at the time of building. The upper end of the hole is drilled to a larger size and tapped out. It is closed with a screwed plug. When the ship is in dry-dock the stern-tube lubrication system is shut down and the pressure on the system is relieved. The plug on the end of the stern tube is removed. Another plug with a central hole is then inserted. The top of this plug is the datum from which all measurements are made, and a depth micrometer is used to measure the distance from the datum to the top of the screwshaft. The difference between the measurement just taken and the original measurement gives the bearing wear-down. The bearing wear-down and the original bearing clearance added together gives the new clearance.

A third method is to fit a dial indicator on the rope guard or stern-tube nut so that the indicator spindle is vertical and touching the propeller boss. A hydraulic jack is placed on the stern frame skeg at some point over a keel block so that the skeg is supported. A wood shore is placed between the jack and the propeller boss. The jack is then used to lift the propeller until the screwshaft contacts the upper part of its stern bearing. The lift recorded on the dial indicator gives the bearing clearance. This method is used in many dry-docks as it gives a direct positive reading of bearing clearance and is suitable for all types of stern bearings.

■ 13.14 List the design improvements made in the method of sealing the space between a screwshaft liner and a propeller, the key and keyway, and the forward end of the conical bore in the propeller boss.

The answer in Question 13.5 was based on the older design forms which are still present in ships operating now and will be for some years to come.

A design improvement made in the sealing method between the screwshaft liner and propeller boss is to machine the liner and propeller boss so that the liner extends into the boss. A counterbore in the propeller boss allows three sides of the sealing ring to make good contacts; the inner circumference of the ring bears on the liner; the outer circumference and one face bears on the propeller. Complete control is possible when the sealing ring is slid along the liner into the sealing-ring space in the propeller boss. A twisted ring is immediately noticed. A split gland ring is used to compress the sealing ring. When the gland ring is tightened by the nuts and studs on the forward face of the propeller boss the sealing ring is compressed until it is volume bound. The clearance between the gland-ring flange and the propeller boss is then measured and a flat split spacer ring of correct thickness is made and fitted into the space, and the gland is pulled up tight on both the sealing ring and the spacer.

The advantage of this design is that the propeller is fastened on to its conical landing on the screwshaft before the sealing ring is fitted. The sealing ring can then be kept under observation while being fitted. Further, there is no fear of a volume-bound sealing ring preventing the propeller fitting properly on the conical end of the shaft. The sealing-ring space, being of larger diameter, is less influenced by the keyway in the propeller boss. A much better landing is provided for the seal ring. In the older design the keyway extended nearly all the way across the sealing-ring landing.

The keyway in the conical end of the screwshaft was at one time cut with an end milling cutter which left a sharp corner profile at the bottom of the keyway and a semi-circular end. Some improvement was made when a radius was put on the bottom of the keyway to remove the sharp corner, but the semi-circular end remained.

The sled runner end keyway form was next used, this being cut with a horizontal milling cutter and the end of the keyway followed the radius of the cutter. This gave an easier transition between the bottom of the keyway and the conical shaft surface. The end of the keyway was then less of a stress raiser.

Finally, the keyway end is now usually made to what is referred to as spoon-ended form. In effect the end of the keyway is made with an easy transition into a shallow hollow space which resembles the shape of a spoon. There are no sharp corners with this form and it has a greater resistance to fatigue than the other forms.

With normal keyway forms ship classification societies allowed a three-year period between screwshaft surveys. With the new forms of keyway end, four years is allowed between screwshaft surveys.

An improvement in key form is to make the forward end of the key with a V-shaped end. The bottom of V is radiused. This key form does not create a 'hard spot' as others have towards the forward end of the keyway.

In older propellers the forward end of the conical bore in the boss had only the sharp edge removed with a scraper. When the propeller was hardened home on the screwshaft the metal of the screwshaft was compressed whereas just forward of the boss the screwshaft metal is unrestrained. The junction of the uncompressed and compressed shaft material forms a stress raiser as the transition from the compressed state to the uncompressed condition takes place over a very short distance. This effect is now minimized by putting a radius on the corner formed by the forward side of the propeller boss face and the conical bore. The radius is best formed as a quarter ellipse, or as a curve with large radius starting on the inside of the bore which gradually decreases to a sharper radius merging with the end face of the boss.

The ends of shrunk-on shaft liners also create stress raisers in a similar manner to the end of the propeller where it fits on the cone. To obviate this the ends of the liners are machined down to a smaller diameter, or a semi-circular or semi-elliptical groove may be machined into the end faces.

■ **13.15** If a propeller were being removed from a screwshaft, would you allow heat to be used on the boss to make removal easier?

A flame which heats a small localized area in a short space of time should never be used to heat a propeller boss. This prohibition applies to all forms of flame heating whether it be a gas burning metal cutting torch or a vaporizing paraffin burner. The drawback to using flame (even if the flame is kept moving around the boss) is that distortion occurs and the accurate fit of the propeller bore on the screwshaft cone is spoiled. If heat must be used it is better to use a low-pressure steam lance.

■ **13.16** If you had to supervise the fitting of a propeller on to a screwshaft in extremely cold conditions, how would you ensure a proper fit?

When a screwshaft survey is carried out it is usual to pull the shaft into the engine room while the propeller remains on the dock. When the propeller is to be refitted it will have cooled off to the low air temperatures prevailing at the time. The screwshaft will, however, be at near engine-room temperature. Due to the differences in the coefficient of expansion of each material the difference in their relative size is such that the warm screwshaft will prevent the cold propeller from returning to its original position. If no special action is taken and the propeller is fitted to the shaft it will only appear to be tight and there is every possibility that it will go slack in tropical waters.

In such temperature conditions the propeller boss is carefully steam-heated prior to fitting; it can then be brought back to its original position axially on the conical end of the screwshaft.

After the propeller is fitted and hardened home it should be kept relatively warm by the heat from open coke braziers until the ship is put down in the water. Cases have been known where the propeller boss has split when it has not been kept warm after refitting. In such conditions it must be remembered that although large amounts of ice may be present on the water, the temperature of the water is usually much higher than that of the air.

■ **13.17** What is a keyless propeller and what is its advantage? How is it possible to drive the propeller if it has no key?

As its name implies a keyless propeller has no key. Technically the keyless propeller has an advantage over normal arrangements in that the stress raisers (even though minimized in modern design) associated with the forward end of the shaft keyway do not exist. The shaft is therefore stronger and has a greater resistance to fatigue failure. When the propeller is fitted on the conical end of the shaft and tightened up, the friction between the surfaces of the propeller and shaft is sufficient to prevent slip between each part, so the propeller is driven in much the same way as torque is transmitted across a friction clutch. The tightening of the propeller on to the conical end of the shaft must be especially carefully controlled. Calculations based on Lamé's Theorem are made so that the necessary friction is obtained without overstressing the material near the bore or even bursting the propeller boss by forcing it too far on to the cone.

The control is effected by the use of a hydraulically operated device in the propeller nut (Pilgrim nut) and measurements are taken of the amount the propeller is forced on the cone.

■ **13.18   What is a Pilgrim nut and where is it used?**

The term Pilgrim is a trade-name for a patented type of nut often used for fastening a propeller on to the screwshaft. The landing face of the nut is bored out with a circumferential slot into which a steel ring like a short piece of tube is fitted. Grooves are machined on the outside of this ring to accommodate an O-ring seal. A similar groove is turned into the inner circumferential side face of the slot, and an O-ring seal is fitted in to this groove so that is bears on the inside bore of the ring. This then seals the steel ring and when connections are drilled for an air-release plug, and for coupling up a hydraulic pump, the assembly forms a hydraulic jack.

When the propeller is mounted and fitted-up on the propeller shaft, the propeller nut is run down the shaft thread. The steel ring in the propeller nut is then loaded with the hydraulic pump to a predetermined pressure, and this forces the propeller hard on to its cone. The pressure is taken off the jack and the air-release plug is opened. The nut is then hardened up with the nut spanner and locked in the normal way. The plugs are replaced in the pump connection and air release and the Pilgrim propeller nut is locked in the normal manner.

An advantage of this type of nut is that it can be used to remove the propeller from the shaft. When the propeller is to be removed the nut is taken off the end of the shaft, reversed so that the steel jacking ring is facing outwards, and screwed back on to the shaft, leaving some clearance between it and the propeller. Studs are screwed into the aft face of the propeller boss and a 'strong back' plate is fitted over the studs. Nuts are fitted on the studs so that the plate contacts the propeller nut and steel ring. When hydraulic pressure is put on the steel jacking ring the propeller is pulled off the conical end of the screwshaft.

The Pilgrim nut is the best means of propeller fastening and the fastest means possible for releasing a propeller from the end of a shaft. It is the type of nut used with keyless propellers.

When this type of nut is used, either for keyless or keyed propellers, extreme pressure lubricants, molybdenum disulphide lubricants, or similar materials must not be put on the screwshaft cone or in the propeller boss bore. If these materials are used the elasticity of the material in the propeller causes it to slide back down the screw-shaft cone when the hydraulic pressure on the nut jacking ring is released, making it impossible to tighten the propeller on the shaft.

Nuts of the Pilgrim type are shown in Fig. 13.2. They may also be used in many places where accurate stress values must be known. The product of the pump pressure and the effective area of the tightening ring under hydraulic pressure gives the force acting on the bolt. The maximum stress in the bolt is then obtained by dividing the tightening force by the cross sectional area of bolt in way of the minimum diameter subjected to the tightening force.

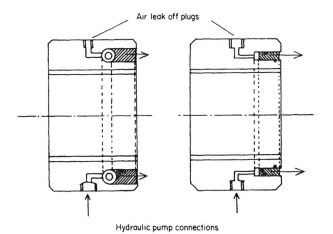

Air leak off plugs

Hydraulic pump connections

Fig. 13.2 Pilgrim-type nuts tightened with a hydraulic pump. The product of the pump pressure and the area of the ring subjected to pressure gives the load on the bolt. When this product is divided by the sectional area of the bolt the result gives the stress on the bolt.

■ **13.19** How is a tight propeller removed from the conical end of a screwshaft? Would you allow a propeller to be wedged off the shaft using the thrust block to take the reaction?

Although the thrust bearing is designed to take the propulsive thrust from the rotating propeller, which is quite a high load, it should on no account be used to resist the reaction from wedges used to free a propeller from a screwshaft without a Pilgrim nut.

The usual practice for removing a propeller is first to remove the rope guard and the boss fairing cone, or 'top hat', to expose the propeller nut. After removing the propeller nut locking device the propeller nut is loosened. While the work is going on outside in the dry-dock, the coupling bolts in the screwshaft and the last section of the intermediate shaft are also being removed. The distance between the aft side of the screwshaft flange and the forward side of the stern tube is measured, and steel packing pieces are made up to fit in this space. Box wedges or folding wedges are then fitted between the forward side of the propeller boss and the aft side of the stern-tube nut.

When the wedges are driven up and hardened a force is exerted on the propeller tending to loosen it on the screwshaft cone. The load from the wedges puts the screwshaft in tension, and this tensile loading goes right back to the screwshaft coupling. The load is resisted by the steel packing pieces fitted between the flange and the forward end of the stern tube. The packing pieces are therefore subjected to compressive loading which is transferred to and through the stern tube right back to the stern-tube nut. The stern tube being of cast-iron is well able to resist this compressive load.

Note   If the stern gland is not of the split type it must remain on the screwshaft. In this case steel packing is fitted between the gland flange and the stern tube. The packing behind the screwshaft flange is then fitted on to the gland flange. Care must be taken when doing this or the gland may fracture. If enough space is available, and the coupling flange diameter is large enough, the steel packing can be put on each side of the gland flange and on to the stern tube. This packing must be supported by the stern tube and not by the aft peak bulkhead.

■ **13.20   Where would you expect to find defects (if any are present) in a screwshaft which has been in service? How would you carry out the examination and what equipment would you use?**

The first place to examine is around the forward end of the keyway to see if any fractures have started; usually they are found to run in a circumferential direction from the forward end of the keyway. If the keyway is of the older type the lower corners of the keyway must be carefully examined from the forward end on one side round the radius of the end of the keyway and along the forward end of the other side. To assist with this examination it is normal to use magnetic particle crack detection equipment. In order to find cracks at the corners of the bottom of the keyway dye-penetrants can be used as they usually show up cracks better in this location. If the shaft is of older design it will not be 'eased' at the ends of the liner. Cracks have been found running in a circumferential direction in older shafts just under the end of the liner; in these circumstances two or three millimetres is machined off the ends to expose the steel shaft underneath. When this operation is carried out it is usual also to machine a semi-circular groove in each end of the liner. This relieves the hard edge which comes from the shrink fit and prevents it acting as a stress raiser.

The parts of the shaft exposed after machining the ends of the liner are checked for cracks using the magnetic particle method or dye-penetrants. The forward end of the shaft, where it is reduced from the large diameter down to the radius which is swept into the back of the flange, is also a location which must be examined. The critical region is at the smallest diameter. Old shafts, which may have had the coupling-bolt holes enlarged, should also have the thinnest section of material examined for cracks running radially outwards from the hole to the outer circumference of the flange. The aft end of the shaft at the reduced section from the end of the thread to the face meeting the smaller end of the cone should also be examined – though this area rarely gives trouble. If the shaft liner is made in two pieces the joint between the two sections should be carefully examined. Dye-penetrant tests would be carried out on the joint. (Non-destructive testing is dealt with in Question 6.50).

■ **13.21   What are the advantages and disadvantages of controllable-pitch propellers?**

The only real disadvantage of the controllable-pitch propeller is the first cost which is considerably higher than that of a solid cast propeller. Some of these first cost-differentials are considerably reduced when savings from engine reversing gear, starting air reservoirs, smaller-size starting air compressors, and reduction of similar equipment are taken into account. The advantages to be gained are simpler methods of bridge control and immediate astern reponse under an emergency when the vessel is at sea. Normally with motorships and turbo-charged engines hull fouling reacts unfavourably on the engine and it is normal to design the propeller with a finer, compromise pitch so that when the hull is fouled the engine characteristics of power and speed are still matched. With controllable-pitch propellers the engine speed can be matched to the condition of the hull – and to weather conditions – both of which lead to savings in fuel and ability to maintain higher service speeds.

■ **13.22 Name and describe the various parts which make up a controllable-pitch propeller.**

The centre part of the propeller is referred to as the boss or the hub, the term hub being more commonly used. The hub houses the mechanism which controls the pitch of the propeller blades. The blades, which are individually cast, have a flange which merges into the root of the blade at its least radius. The blade is fastened to the blade carrier which in appearance is similar to a blank flange with a rebate around one face and its circumference. The blade carrier is called the crank ring on some propellers; it fits into a recess or a circular hole in the hub. In some propellers it is assembled from inside the hub and in others from outside. When the blade carrier is assembled from inside the hub, the rebate fits into the circular hole which is smaller in diameter than the outside diameter of the carrier. The flange on the base of the propeller blade is mounted on the blade carrier and fastened to it with set bolts which fit into tapped holes in the plate. When the blade carrier is fitted into a recess in the hub it is kept in place with a threaded ring which fits over the rebate and is screwed into the hub. The blades are fastened to the blade carrier as mentioned before. The blade carrier, being circular, can rotate, and rotation of the carrier alters the helix angle of the blade and therefore the propeller pitch.

Within the hub is a yoke which moves in a fore and aft direction within guides. On each circumferential face on the yoke there is a hole or slot. A crankpin fitted on the inside face of the blade carrier engages in the hole or slot in the yoke. Fore and aft movement of the yoke then moves this crankpin and causes angular movement of the blade carrier and blade, which varies the blade pitch. The yoke is actuated by an oil-operated servo motor which may be fitted in the hub cone or fairing and is then directly connected to the yoke. Sometimes the servo motor is fitted within the propeller shafting, in which case it is connected to the yoke with a push-pull rod.

Controllable-pitch propellers are flange mounted on to the screwshaft.

Access to the propeller mechanism and its internal parts is made, either through the hole in the centre of the hub flange, which necessitates removal of the propeller from the shaft, or, in other designs, by removal of the hub cap or fairing.

The important parts of a controllable pitch propeller used when altering pitch are shown in Fig. 13.3, the yoke actuating mechanism and the sealing rings are not shown. These omissions have been made for the sake of clarity.

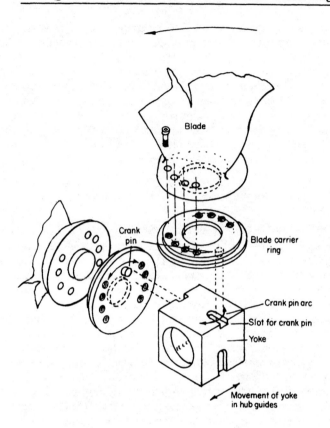

Fig. 13.3   Sketch showing the mechanism of a controllable pitch propeller.

Note   In single-screw ships a controllable pitch propeller is usually arranged for left-hand rotation.

■ 13.23   In what manner are the internal parts of a controllable-pitch propeller lubricated?

The internal parts are lubricated by keeping the hub space full of lubricating oil under pressure. The oil under pressure prevents sea-water entering the hub if any of the seals should leak. The pressure is maintained by the servo-motor pressure pump, and oil enters into the hub space either through balance holes from the pressure side or through small clearances arranged in sealing bushes. If the hub pressure becomes too great the pressure is relieved through a relief valve which lets the excess oil go into the return part of the servo-motor oil system.

■ **13.24 How is the oil delivered from the servo-motor pressure pump into the servo motor in controllable-pitch propellers?**

The oil delivered by the servo-motor pressure pump is led through high-pressure flexible hoses (or solid pipes with some flexibility) into a sleeve which surrounds the shaft but does not turn with it. Annular grooves which line up with radial holes in the shaft are machined in the sleeve. The radial holes connect up with an axial bore and an internal pipe within the axial bore. Oil from the servo-motor pressure pump passes through a control unit and into the sleeve through the flexible hoses, then into one of the annular grooves, and on into one of the radial holes leading to the servo motor through an internal tube. It returns from the servo motor on the outside of the tube back to the other radial hole, and out through the sleeve. The ends of the sleeve surrounding the intermediate shaft are fitted with seals with leak-off pipes. In some units the sleeve is incorporated with the propeller servo-motor control unit.

The weight of this heavy unit is sometimes relieved by placing springs under the unit to reduce the loads on the bearing bushes fitted in the sleeve.

■ **13.25 Where are seals fitted between the moving parts of the blade assembly and the hub in a controllable-pitch propeller?**

The seals must be located in some place where they are reasonably accessible; they are therefore fitted just below the blade flange and make contact with the flange and the hub which become the sealing surfaces. In some cases two O-rings are fitted into a seal supporting ring. One O-ring bears against the flange on the foot of the propeller blade, the other contacts the propeller hub. The seals may be in the form of O-rings or may be specially moulded to suit the ring housing and landing face.

Renewal of seals involves dismantling the propeller blades and lifting them off the blade carrier plate or crank ring.

■ **13.26 What types of lubricating oils are used for the servo-motor systems in controllable-pitch propellers?**

The same oil is used for the servo-motor pressure system and for internal lubrication of the parts within the hub, and is similar to the lubricating oils used in marine steam turbines. They will have the following characteristics.

Specific gravity 0.875–0.90
Viscosity Redwood 1 at 60°C = 100 sec.
Viscosity index 104.

The pour point must obviously be very low to suit very cold sea-water and atmospheric conditions; its value will be within the range −10°C to −30°C depending on the service of the ship. These grades of oil have additives with properties to resist oxidation, corrosion and foaming. One of the features of lubricating oils within this grading is that water separates itself out from the oil very easily.

■ **13.27   What attention do controllable-pitch propellers require at sea and in port?**

The condition of the oil must be constantly kept under observation for indications of water or condensation entering the servo-motor oil system. The quantity of oil in the servo-motor system must also be regularly checked by taking soundings of drain tanks and other tanks within the servo-motor oil system. A record of the oil quantities used for topping-up the system should be kept so that any changes in the amounts used are quickly noticed. It must be remembered that changes in the oil temperature will cause corresponding changes in the drain-tank soundings.

Filters on the oil system and control air system require regular attention and cleaning, and particular care must be given to this if pressure gauges indicating the pressure before and after the filters are not fitted or, if fitted, are inoperative. Particular attention must be paid to filters in new vessels, and following repair work, drain tank cleaning, and the like.

During each watch the various parts of the piping system should be checked; slack pipe clips must be hardened up. The temperature and electrical loads of servo oil pump motors must also be checked out.

On long sea passages when the engine room is put on stand-by and the engine is at slow for boat and fire drill, it is the practice in some Companies to move the propeller to the reverse or negative pitch position and back before or at the end of the stand-by, before going to full ahead operation. The same check should be put on the propeller pitch control at the end-of-passage prior to propeller manoeuvres for picking up the pilot. It is also good practice to change over to clean filters and clean the shut-down filters some few hours before the end-of-passage.

■ **13.28   What attention do controllable-pitch propellers require when in dry-dock?**

The parts of the propeller external to the hull must be checked out for seal leakages. Oil within the hub is drained off and its quality checked for water content. In some cases it is changed, and the oil drained off is purified and returned to the system for make-up.

Programmed blade seal renewal is usually carried out on one or two propeller blades at the time the vessel is in dock. This prevents breakdown of seals in service as they all are renewed over a reasonable time period depending on the number changed and the period that the vessel is out of dry-dock, which in many cases may be up to two years. The remaining propeller blade fastenings and locking devices (or spot welds used for locking blade set bolts or nuts) must be carefully examined. After changing blade seals the propeller hub is pressure tested.

The shaft outer seal (to stern-tube oil system) must be inspected for leakage and wear of parts (see Question 13.13).

After completing renewal of blade seals and pressure-testing the hub and other parts, or completing other work, the propeller should be tested from full positive to full negative pitch positions before the dry-dock is flooded. This test should be carried out from the bridge, engine room, and local control stations. The emergency operation procedures should also be checked out at this time.

■ **13.29 What do you understand by the term propeller pitch? What is slip?**

We know that if a thread has a pitch of two millimetres the screw will move a distance of two millimetres if it is turned one revolution when screwed into a nut. A ship's propeller may be likened to a multiple-start thread, a three-bladed propeller being similar to a short length of a three-start thread, a four bladed propeller being similar to a four-start thread and so on.

The pitch of a propeller is the distance it would advance in one revolution if it were turning within a nut. In practice the propeller pitch is usually of some value between 2.5 metres for a small coaster with a high propeller rev/min and 6 metres for large fast ships. For motorships with speeds of 16 knots, common pitch values are about 3.8 to 4.2 metres depending on various factors such as propeller rev/min and coefficients of hull fineness. When a propeller is revolving in water it does not advance through the water in one revolution as much as the value of its pitch; it is said to slip.

The slip value, which ships' engineer officers record at noon each day, is the average or mean apparent slip for the previous 24-hours period.

The mean apparent slip per day is given by:

$$\frac{\text{distance run propeller-distance run by ship}}{\text{distance run propeller}}$$

It is usually expressed as a percentage so that the distance in nautical miles run by the propeller (the DRP) is

$$\frac{\text{pitch metres} \times \text{total engine revolutions for day}}{1852}$$

Then the apparent slip per day is given by

$$\frac{\text{DRP} - \text{distance run by ship}}{\text{DRP}} \times 100\%$$

Note There is another slip value referred to as true slip which takes into account wake speed. This matter is dealt with in books on naval architecture and ship propulsion.

■ **13.30 Is the propeller pitch the same at the various radii of the propeller blades, in solid cast or monobloc propellers?**

With modern propellers the pitch gradually increases as the blade radius increases. There are various mathematical ways in which the mean pitch is found for purposes of the slip calculations made on board ship.

A British method uses a moments system. Other methods use an average obtained from the sines of the helix angles; or the pitch at 0.7 of the blade radius is sometimes used. The pitch to be used for apparent slip calculations is stated on the propeller drawings.

■ **13.31 Intermediate shafting fitted between the engine and the screwshaft is supported on bearings. What types of bearing are used for this purpose?**

There are four different types of bearing used for supporting the intermediate shafting.

Normal plummer or pedestal. The housing of the bearing is made of a good quality shock-resistant cast-iron. The bearing surface is white-metal, cast directly on to the parent cast-iron of the bearing. The bearing is made to support the weight of the shaft only and no white-metal-lined keep is fitted. A cover housing the oil box and tubes for wool syphon feeders is fitted. The cover also keeps out dirt and foreign matter. This type of bearing has limitations on its load-carrying capacity and rubbing speeds. A pocket is sometimes made under the bearing which is circulated with cooling water.

Tilting-pad. This bearing is also made to support the weight of the shaft only. The weight of the shaft is supported on three or four bronze white-metal-lined shells each of which extends about 30° to 40° round the journal circumference. The back of each shell has a ridge or projection running axially (along its length) which becomes the tilting fulcrum for the bearing shells. The white-metal lining is well radiused along its length for oil lead-in. The ridge is supported in the body of the bearing which also forms on an oil sump. A carrier ring is fitted on the shaft. It is large enough in diameter to extend down into the oil in the sump. When the shaft is revolving, oil is picked up by the oil carrier ring and carried up to a scraper which skims it off the ring; the oil then runs down a channel and is led into the bearing shells. The oil sump is fitted with a cooling coil or a series of cooling tubes through which sea-water is circulated to cool the oil so that it retains its vicosity. This type of bearing can carry very heavy loads at high rubbing speeds.

Solid-ring ball or roller bearings may be found in older ships. Provision must be made for the fore and aft movement which takes place when the shafting runs in the ahead or the astern direction.

Split-ring roller bearings are commonly used for intermediate shafts. They are held together in a housing which has a spherical outer form. The spherical portion fits into the bearing body and can swivel within it. The body is fastened down on chocks with holding-down bolts. Ball and roller type bearings are grease lubricated and do not require cooling water.

■ **13.32 What attention must be given to intermediate screwshaft bearings?**

Ball and roller intermediate shaft bearings require periodic greasing and inspection of the fastenings and chocks. Attention should also be given to noise levels coming from the bearings and the temperatures at which they are running. Changes in these levels may indicate chipped rollers or 'brinelled' inner or outer rings. Oil-lubricated shaft bearings with drip-feed wool syphons require the oil boxes filling at regular intervals. If a cooling space is fitted on the underside of the bearing it will require periodic cleaning to remove sand and mud. The bearing fastenings or holding-down bolts and chocks also require periodic checking.

Tilting-pad bearings require regular checks to be made on the oil level in the sump. The oil carrier ring requires regular inspection to see that it is lifting the oil correctly and that enough is passing from the oil carrier ring to the bearing pads. The cooling water supply also needs attention because as the temperature of the oil rises the lowering of the viscosity causes less oil to be picked up by the carrier ring, also when the oil viscosity is reduced the ability of tilting-pad bearings to carry load is reduced.

The aft bearing on the intermediate shafting is most vulnerable from water leakage from the stern gland, and requires close attention. If the oil goes creamy or milky in appearance it will require changing, and the water leakage from the stern gland must be reduced. Leakage of water from the cooling coils or tubes in the bearing oil sump will also cause the oil to go creamy or milky.

■ 13.33 What is a stern tube and what is its purpose?

The stern tube houses the screwshaft, and its purpose is to carry the screwshaft bearing bushes. Due to the shape of the aft end of a ship where it is streamlined down to allow an easy flow of water to the propeller, little space remains in the narrow section to fit normal bearings. This narrow space within the hull structure at the screwshaft level accommodates the aft peak. The stern tube passes through the aft peak and is supported at its forward end by the aft peak bulkhead. Along its length the stern tube is supported by narrow floors or brackets to which the after framing is attached. The after end of the stern tube is supported by the stern frame.

The stern-tube bearings support the (overhanging) weight of the propeller and screwshaft. The load on these bearings is transferred on to the stern tube which is supported by the stern frame and the internal parts of the ship's structure around the aft peak. The forward end of the stern tube houses the stern gland which prevents sea-water passing through the stern tube and into ship, and also oil leakage from the stern tubes.

■ 13.34 Where are the stern-tube bearings fitted? How are they lubricated?

The stern-tube bearings are fitted just behind the stern-gland neck bush and at the aft end of the stern tube. The bearing at the aft end of the stern tube has to support a large weight and usually has a length of about four times the shaft diameter. The bearings may be built up from staves of lignum vitae fitted into a bronze bush. Phenolic resins and rubber are also used for stern-tube screwshaft bearings. These materials are lubricated by sea-water, and consequently a bronze or brass liner is shrunk on to the steel screwshaft to protect it from the corrosive action of sea-water.

White-metal-lined bushes may also be used, and these are oil lubricated. To prevent sea-water coming in contact with the oil, a seal is fitted on the outside of the screwshaft between the propeller and the stern tube. Brass liners are not fitted on oil-lubricated screwshafts.

■   **13.35   Where are thrust bearings fitted and what is their purpose?**

If a force of any sort acts axially along a shaft, some form of thrust bearing must be used to prevent the shaft sliding along in its bearings as it revolves. The magnitude of the force may be small, as might be experienced in electrical generators. A simple type of thrust bearing in then used. This consists of nothing more than the provision of some bearing surface on both ends of one of the engine crankshaft main bearings. The main bearing nearest the generator is the one used; one end of the main bearing contacts the side of the crankweb, and the other may contact a collar on the shaft or the side of a camshaft drive wheel. Lubrication for the end surfaces is provided from the bearing oil supply. This type of thrust bearing is often called a location bearing as it also locates the crankshaft in its correct position relative to the cylinders.

The thrust from the propeller is taken up by the main thrust bearing which transmits the thrust to the ship's hull and causes the ship to be propelled in the direction of the thrust. The main thrust bearing is always fitted at the aft end of the main engine crankshaft. The correct location of the crankpins relative to the centre of the cylinders is controlled by the main thrust bearing.

Turbo-blowers have a small thrust bearing on the shaft to hold the rotor in its correct axial position. The thrust bearing also balances thrust which comes from the action of the exhaust gases on the turbine blades, to which must be added the thrust from the axial-flow section of the blower.

Many centrifugal pumps of the single-entry type have a thrust which is set up in the pump rotor shaft from the action of the liquid flow. This thrust is usually balanced within a ball bearing of the motor. The ball bearing is fitted nearest to the pump. The housing for this bearing does not have any axial clearance. With large, heavy, low-pressure vertical centrifugal pumps the weight of the motor armature and rotor is usually greater than the end thrust from the flow, so the weight becomes the thrust factor. All shafts and rotating components of machinery fitted vertically have a thrust set up in the shafting from the action of the weight of the parts.

■   **13.36   What type of bearing is used to take the thrust set up by the propeller in propulsion machinery?**

In ship's lifeboats the thrust bearing commonly used is a normal ball bearing fitted in the reduction gearbox. The thrust acting along the propeller shaft is transferred from the inner ring into the ball race and then from the race into the outer ring which is held in the gear case. The allowable limit for the transmission of thrust through a ball bearing is relatively low. Small craft commonly use a double conical roller bearing for transmission of propeller thrust. The double conical roller bearing is fitted on the main shaft of the gearbox in which it is held. One bearing takes ahead thrust and the other the astern thrust. The holding down bolts of the engine and gearbox transmit the thrust to the hull.

In propulsion machinery on larger craft, through the range from coasters to the largest ships afloat, the thrust bearing is most commonly of the tilting-pad type. In this type of bearing a thrust collar is forged integrally with the thrust shaft. On the forward and aft sides of the thrust collar, the thrust pads are fitted. The thrust pads are lined with white-metal and face on to the finely machined and polished surface of the thrust collar. The back of the pad has a radial ridge which forms a fulcrum on which the pad can tilt. The tilting fulcrum on the back of the pad comes in contact with a solidly constructed housing. The housing is rigidly held in the thrust-bearing casing. In slow-speed direct-coupled engines the casing is usually integral with the engine bedplate. In geared diesel-propelled ships the thrust bearing may be separate from other parts, in which case it will have its own casing, cooling service and foundation.

This type of bearing builds up an oil pressure between the white-metal face of the thrust pad and the thrust collar when the shaft revolves. The oil pressure is due to the formation of an oil wedge which can build up only when the thrust collar is supplied with oil and is revolving. As the pad is able to tilt it becomes self-adjusting to the shape of the wedge. An essential for the build-up of the oil wedge is a good radius on the leading edge of the thrust pad; without this radius the leading edge would act as an oil scraper and the bearing could not function. The radial ridge on the back of the pad which becomes the fulcrum for the tilting action is often made off centre. If the thrust pads are viewed from the top, the tilting point is away from the centre moving in the direction of rotation of the thrust collar. When designed and constructed in this manner the pad tilts more easily to form the oil wedge.

■ **13.37   What are the indications of faulty operation of the main thrust bearing?**

The temperature at which a thrust bearing is operating is a good indication of the condition of the bearing. If the running temperature shows an increase (above what would be expected for climatic and sea-water temperature increases) it will indicate either that the oil flow to the bearing has been reduced, or that alignment has changed. When the main engines are manoeuvring from ahead to astern it is good practice to sight the intermediate shaft where it protrudes from out of the intermediate shaft bearings. If the shaft moves fore and aft when the engine goes from ahead to astern and vice versa, it is a sure sign that something is amiss. Even small amounts of movement can be detected with the eye.

Note   There are questions on thrust lubrication systems in Chapter 15.

■ **13.38   Why is the thrust bearing fitted forward of the main gear wheel in some gearing arrangements? What particular attention does this arrangement require?**

When the thrust bearing is fitted forward of the main gearwheel the cost of the heavy forging for a normal thrust shaft with integrally forged collar can be reduced to that of a plain shaft. The forward end of the main gearwheel shaft is machined to a taper with a thread at the end. The thrust collar is machined from

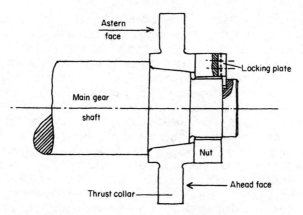

Fig. 13.4   Sketch showing a detachable thrust collar as fitted on many geared diesel propulsion systems (see fig. 12.1).

a flat forged disc, with a tapered hole which fits on the taper of the shaft end and is held in place with a heavy nut capable of taking the astern thrust. A key is fitted within the taper. The thrust housing is arranged so that the thrust collar can be removed without lifting the shaft, thus allowing the thrust collar to be removed for re-machining without a lot of disturbance to other equipment.

Periodic checks must be made on the thrust collar nut and locking arrangements to ascertain that nothing is coming loose. To make a check on the thrust collar nut only involves the removal of the forward end cover in way of the thrust bearing housing.

A detachable thrust collar is shown in Fig. 13.4. This type of thrust collar is used in the reduction gear arrangement shown in Fig. 12.1.

■ **13.39   Describe with the aid of sketches how in one type of thrust bearing the load on each thrust pad is made to balance the load on adjacent thrust pads.**

In this type of bearing the pivot point or pivot edge of the thrust pad is supported on another pad or levelling plate. The levelling plate is supported by two other tilting plates. If the load on one shoe is increased or decreased from any cause, the load on the levelling plate is increased or decreased. This acts on the tilting plates supporting the levelling plate and causes an adjustment to take place to decrease or increase the load on the adjacent thrust pads.

This type of bearing is known as a Kingsbury-type thrust bearing. (Fig. 13.5)

■ **13.40   How is the thrust-bearing clearance checked? What precautions must be taken when checking thrust-bearing clearance and what would you expect the main thrust-bearing clearance to be?**

There are various ways of checking the thrust-bearing clearance.

The most common is when the thrust bearing is opened up for examination,

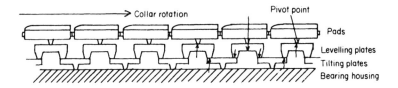

Fig. 13.5 Arrangement of pads in a Kingsbury thrust bearing.

and checked with a feeler gauge. If it is possible to move the shafting in its bearings by a jack or wedges so that the thrust collar is hard on one of the sets of the thrust pads, it facilitates the taking of the feeler readings. They then have to be taken only on the side of the thrust collar which is open. The feelers should be long enough to extend from one corner of the pad to the corner diagonally opposite. They must be inserted at one corner and eased diagonally across to the other. When the feelers are inserted in this manner they measure the clearance correctly without allowing the pad to tilt. If the collar cannot be hardened on to either the ahead or astern side pads, two sets of readings must be taken on both faces of the thrust collar. The addition of the two readings gives the clearance.

It is possible to take thrust-bearing clearances without dismantling the casing and the locking devices used to prevent the pads sliding round in their housing. A clock gauge with a magnetic base is used to make the measurement. The stern gland is then slackened off so that the compression on the packing is minimal. A small hydraulic jack is inserted on some part of the hull structure such as the top of the stern-tube flange and the coupling. The shafting is eased forward with the hydraulic jack. The dial gauge is fastened to a coupling guard and set to zero on the side of the coupling or bolt head. The jack is then moved to another position and operated to move the shafting so that the thrust collar comes on the astern pads. The dial gauge reading gives the thrust-bearing clearance.

The shafting can be similarly moved by using wedges and inserting them between the sides of the after crank webs and the bedplate cross-members. When moving the shafting backwards and forwards with wedges a dial gauge should be set between the two webs of the aftermost crank and observation kept on the deflection.

If the after end of the screwshaft liner is worn with a step or ridge, and the forward end of the liner is similarly ridged, some difficulty may be experienced in moving the shafting. If the shafting does not move easily in response to the jack or wedge action, care must be taken to avoid overloading. If this method is to be used it is the best carried out just after the engine is shut down and before the oil has drained from the main and running bearings.

An alternative method can be used when the engines are being manoeuvred. This method uses the small step at the radius of the smooth-turned journal portion of the intermediate shaft where it joins the rough or turned portion which is slightly less in diameter.

When the engines are stopped one of these radiused fillets should be cleaned off smooth. A magnetic dial gauge is then set up on the bearing so that the follower is contacting the edge of the radius. By setting up the gauge and noting the readings when running ahead and astern the clearance can be obtained. Care must be taken when this method is used to obtain *clearance* readings. If there is a 'throw' of the needle during a revolution of the shaft the readings could be misleading.

The thrust-bearing clearance (tilting pad bearings) when cold is between 0.35 mm for small shafts and up to 1.00 mm for large shafts.

Note   When setting up dial gauges care must be exercised with the position of the follower which should always be perpendicular or near to perpendicular to the direction of the 'throw' or difference to be measured.

■   **13.41   How can excess thrust-bearing clearance be corrected?**

Normally the amount of wear on thrust bearings is very small and provided the lubricating oil is kept clean and free of water they will operate for many years without adjustment. The small amount of wear is due to the fact that there is no metallic contact between the pad and the thrust collar during operation. If adjustment is necessary it is carried out on the forward side of the bearing. The crankshaft or intermediate shaft is pushed aft so that the thrust collar is hard on the astern pads; this leaves the ahead pads free (see Question 13.39 for details). Assuming that the cover is off the thrust-bearing casing, the oil spreader and other loose-fitting parts are lifted out. The stopper piece on the top of the bearing is dismantled, leaving the forward pads free. If it is a small-size bearing the pads can be lifted out of the housing by hand. In larger bearings a long eye bolt is screwed into the contact piece on the side of the pad and it is lifted out with tackle. The pads remaining are then pushed round in their housing to the lifting-out position and lifted out. Care must be exercised in removing the thrust pads to avoid damage to the fine finish on the thrust collar. The pads must be checked for marked numbers as they are removed, and they must be reassembled in their original locations. The pad housing is turned out of the casing and lifted out. The screws on the back plate are removed, shims and liners are made up in the amount that the clearance is to be reduced, and fitted between the back plate and the housing.

The pad housing is then replaced. If any marks or blemishes have been made on the collar they must be cleaned back flush with an oil stone. The pads are reassembled and the new clearance confirmed with feeler gauge readings. The rest of the parts are then replaced. Before the casing cover is replaced the lubricating oil supply pipes must be carefully checked for position, if they are open ended. After adjustment of clearance the oil flow to the bearing must be checked when oil is put on the engine or gearing.

The thrust bearing, after being adjusted, must be kept under observation for the first few hours of running.

■   **13.42   In service a propeller develops defects which roughen the smooth surface of the blades. What are the factors that cause this roughness?**

The factors are corrosion, cavitation, bubble impingement from badly placed anodes, and build-up of deposits from anodes.

Cavitation may occur in various forms due to the action of the water flowing across the front and back of a propeller. Changes of velocity occur when water flows across a hydrofoil. From 'Bernoulli' we know that as the velocity increases the static pressure diminishes. If the static pressure falls too low, bubbles form. When the velocity of flow is reduced, and the pressure therefore rises, the bubbles collapse and cause loss of material from the blade surface. This form of damage is usually seen near the blade tips; it may show itself as a brighter area or, in bad cases, as rough-bottomed craters.

Bubble impingment from badly placed anodes shows itself as a roughened area on the back face (forward side) of the blade. The roughened area extends circumferentially at a constant radius across the blade.

In ships such as bulk carriers and tankers, where a large part of the ship's service time is in ballast, the rough marks often show up at two or three radii which correspond with the height of the anodes above the centre of the propeller shaft.

Anodes fitted in the stern area adjacent to the upper portion of the propeller aperture require locating very carefully to avoid this trouble.

The build-up of deposits from anodes on the surface of the propeller blades appears as a relatively rough grey-white to grey-brown deposit.

If the edge of a propeller blade suffers damage and is part distorted, the damaged locality interferes with water flow and may be the cause of surface damage similar in appearance to cavitation damage.

■ **13.43 Intermediate shafting is sometimes fitted with an electrical slip ring or slip rings and carbon brushes. What is the purpose of this? How are the electrical connections made and what attention must be given?**

This device is fitted to intermediate shafting to create a short circuit between the propeller and the hull of the ship; in effect it reduces or stops propeller corrosion which comes about due to the action of dissimilar metals in sea-water. The brushes and brush-holders are connected by a heavy electrical cable directly with the ship's hull. Attention must be given to the cleanliness of the brushes, brush-holders and slip ring surfaces to ensure that their resistance to current flow is not increased by oil or dirt.

14

ENGINE AND SHAFTING ALIGNMENT

■ **14.1 When main bearings and crankshaft journals wear, what effect does it have on the operation of the engine?**

Normally the wear on crankshaft journals is slight, even over very long periods, and journal wear normally has little effect on engine operation. Wear on the main bearings, however, is considerably greater. The effect of this wear is to lower the compression ratio on the engine cylinders, which reduces the thermal efficiency of the engine and increases fuel consumption. In opposed piston engines the wear on main bearings lowers the running position of *both* the upper and lower pistons, and therefore the compression ratio remains unchanged.

If the wear on the main bearings and crankshaft journals were similar in amount throughout the engine the alignment of the crankshaft would not be affected. In practice, however, main bearing wear is never the same throughout the engine, and the alignment of the crankshaft is therefore adversely affected as the unequal wear progresses.

■ **14.2 How is the wear on journals and main bearings measured?**

The amount of wear that has taken place on a crankshaft journal can only be found by measuring the journal with a micrometer. The load on the journal varies around its circumference. This is due to the load on the journal being the algebraic addition of the effects of gas and inertia loads and the weight of the various parts. In consequence the wear is uneven so the measurements must be taken at various locations along the journal, and at each location the diameters must be measured at several angular positions. The measurements are plotted in tabular form and the minimum diameter together with its location can then be found. This method applies to all types of engines; in some cases the measurements can be properly taken only when the crankshaft is lifted or lowered out of its main bearings. In other cases the thickness of the main bearing shell is such that it is possible to measure the diameters with a narrow frame micrometer when the bearing is removed. In the remaining cases special equipment is used to take the measurements.

Main bearing wear is measured with a bridge gauge when the bearing keep and upper shell are removed. The bridge gauge is fitted over the journal, and the clearance between the gauge and the journal is measured with a feeler gauge. If the original gauge reading is subtracted from the new reading the amount of wear on the bearing together with the wear on the journal is given. As the wear on the journal is usually very small, the wear-down reading given by the bridge gauge readings is normally accepted as the bearing wear-down. When dealing with very old engines, however, it is important to remember that use of the bridge gauge gives both the wear on the journal *and* the bearing; this is particularly important when dealing with alignment problems.

If a record was made of the crown thickness of the lower halves of the main bearings when they were new, the wear on the bearings can be found by measuring the thickness. The difference between the two measurements gives the amount of wear. Some engine builders record these measurements on a brass plate which is fastened on the engine and kept as permanent record.

Note In taking measurements of bearing thickness a micrometer with a ball-faced anvil should be used. If this type of micrometer is not available an ordinary micrometer can be used in conjunction with a short piece of standard-size round rod or a steel ball.

■ **14.3 Using the bridge gauge and feelers, how would you estimate the maximum wear on a crankshaft journal?**

The maximum wear and location of maximum wear can be determined with a bridge gauge and feelers when a main bearing keep and the upper half bearing shell are removed. Bridge-gauge readings are taken with the journal in various positions, some point such as an adjacent crankweb being used as a base position. The readings are taken at 30 degree intervals over a crankshaft rotation of 180°, and then plotted on a graph, with the gauge readings as ordinates. The difference between the maximum and minimum readings gives an *estimate* of the amount of wear which has taken place on the crankshaft journal; the actual amount of the maximum wear will be slightly higher than this.

In old engines, particularly when difficulties arise with maintaining lubricating oil pressure, the figures can be used when adjusting main bearings. The figures can also be used in extreme cases when making crankshaft alignment checks by optical or taut-wire methods.

■ **14.4 Give some details of how and where errors may occur when taking bridge-gauge readings.**

Errors may arise because the feet of the bridge gauge were placed on dirt, or from small burrs on or under the feet. These faults will give high readings. Before taking bridge-gauge readings the location around the bearing, and the bridge gauge itself, must be thoroughly cleaned. Errors can also arise if the crankshaft is stiff and the span between adjacent main bearings is small – leading to a 'bridged' bearing. The wear-down shown in a 'bridged' bearing may appear to be normal, though in actual fact a dangerous situation could be present. When the engine is working, the loads imposed on the crankshaft force the journal down into the bearing which, if 'bridged', could cause large amounts of strain from which high stresses result. 'Bridged' bearings may also occur in locations near to overhung weights such as flywheels, armatures, couplings, and some types of detuners. (There is further information on 'bridged' bearings in Question 14.10.)

The location where the bridge gauge is placed to take the measurements is also important, as in some engines a special location is scraped up so that bridge-gauge readings are the same throughout the engine when it is new. If the bridge gauge is moved from this specially scraped area, errors result. In opposed piston engines with main bearings fitted in spherical housings the bridge gauge is placed mid-length on the journal. This takes care of any errors which may result if the spherical bearing allows one end of the journal to be up or down (see also Questions 10.20 and 14.13).

■ **14.5   How would you check a dial micrometer gauge of the type used for taking crankweb deflections?**

Note The dial micrometer gauge is often termed a dial gauge, a deflection gauge, or a clock gauge.

The first essential is to check that the zeroing and return spring in the gauge is adequate for its job of holding the instrument in place between the crankwebs and following the change which occurs in the distance between the crankwebs as the engine is turned (some gauges have an external spring system to hold the gauge between the webs).

If any dirt adheres to, or damage occurs to, the spring-loaded contact point spindle it may cause sluggish action of the gauge needle so that, though it will read correctly moving inwards, sluggish action when moving outwards may give false readings. This action can be checked by carefully loading the contact point through the full extent of its travel with the finger and then letting it return by slowly unloading the contact point. Any sluggishness together with its location is then found. Some designs of gauge suffer damage if they are set up in a crank location where the deflection is greater than the allowable amount of contact-point travel. This strains the delicate gear teeth within the instrument and, although the instrument may appear to be working satisfactorily, the readings at some points within a part revolution of the indicator needle may have large errors.

This error can be checked in a lathe which has a micrometer scale on the hand feed screw. The dial gauge is set up with the extension pieces corresponding to the distance between the crankwebs. It is then placed between the lathe face plate or chuck and the side of the table holding the tool post. The table is fed inwards carefully towards the headstock with the hand feed screw. The amount of feed on the micrometer hand feed screw should correspond with the amount shown on the dial gauge – at all points. If a lathe with this type of hand feed is not available, the dial gauge can be checked with feelers when it is fitted between the cranks on which it is to be used.

These two error factors, while most important for medium-speed engines where only minimal deflections are allowable, are slightly less important in large engines where larger deflections are permitted. However, this fact must not be used as an excuse for not checking dial gauges, because *correct* readings are essential if deflections are approaching or are at a limiting value.

Note Though robust in appearance, dial micrometer gauges are delicate instruments. They consist of a contact point spindle held in guides. The spindle is connected to a rack which is geared to a set of step-up gearing. A spring is used on the contact spindle to push it out to its point of maximum travel. A spiral spring is fitted to the needle axle so the load on the rack teeth is always in the same direction, thus cancelling out the backlash.

■ 14.6 Imagine that you have a section of an engine crankshaft consisting of two crankwebs, a crankpin and two journals supported between two main bearings. Describe how the crankwebs behave when they are loaded with the weight of the piston, piston rod, crosshead, connecting-rod and bearings, (a) when the crankpin is on top centre and (b) when the crankpin is on bottom centre.

In order to analyse what is happening to the crankwebs we can first consider what would happen to a straight length of shafting supported in the main bearings and similarly loaded. When the straight length of shafting is loaded with a static weight of the same amount as the piston, connecting-rod, etc. it will sag downwards under the load. From what we know of beams, we know that the deflection caused by the load will compress the material in the upper half of the shaft, reducing the length of the material at the top. Similarly the lower half of the shaft is in tension and its length will be increased.

(a)   When the crankshaft is considered, a similar action takes place. With the crankshaft on top centre the load is on the crankpin. The journals and crankpin deflect downwards in the centre as before. This causes the crankwebs (at the opposite side to the crankpin) to open up. The distance between the webs increases just as the length of the bottom part of the shaft did.

(b)   With the crankshaft on bottom centre, the load is still on the crankpin, the journals deflect downwards in the centre, but now the position of the crankwebs corresponds to the upper part of the shafting. This is in the position where the shaft is in compression and shortened, so when the crankpin is on the bottom centre position the distance between the crankwebs is reduced under the load of the piston.

Summing up, we can see that the action of the weight of the running gear on the crank assembly is to cause the crankwebs to open outwards when the crankpin is on top centre, and to close inwards when the crankpin is on bottom centre.

Obviously the length of the crankpin does not change so the distance between the crankwebs in way of the crankpin remains the same. The action of the

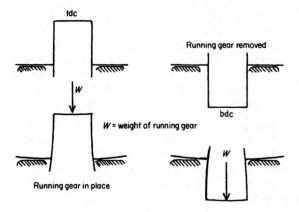

Fig. 14.1 Sketches showing a crankshaft with running gear removed and running gear in place. The weight of the running gear causes a deflection in the crankwebs, pins and journal as shown.

change in the distance between the crankwebs comes about from the elasticity of all the parts making up the crankshaft, but a large part of the movement comes from spring in the crankwebs. (Fig. 14.1)

■ **14.7 If a dial micrometer is fixed between the crankwebs at the opposite end to the crankpin in the arrangement described in Question 14.6 when the crankpin is near bottom centre, describe what happens when the crankpin is turned to the top centre position and rotation is continued round to the near-bottom centre position.**

Normally when checking crankwebs' deflections, the first reading is taken at a point when the bottom-end bearing has just passed bottom centre position so that the side of the connecting-rod is just beyond the centre of the uppermost part of the journal. To get to this position from the absolute bottom dead centre position, the engine is turned in the ahead-running direction if a main engine, or in its normal direction of operation if it is an auxiliary. The dial micrometer is fastened between the crankwebs usually at a point just above the uppermost part of the journal. The gauge is set to a zero reading.

Referring to the last Question we can see that the crankwebs near bottom dead centre will be in the position where they had sprung inwards. As the engine is turned in the ahead direction the crankwebs will begin to open and the dial micrometer will give a small plus reading. When the crankpin is on the port side the dial gauge will be on the starboard side. As the gauge passes this position the amount of deflection of the crankwebs will have further increased. The reading should be noted and recorded. With continued rotation the crankpin reaches its uppermost position and the gauge its lowest position. In this position the gauge reading will be noted and recorded again. Referring to Question 14.6 we can see that the webs will be deflected outwards so the gauge reading will be plus and

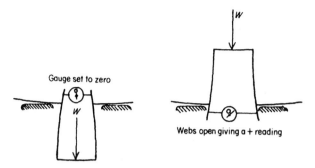

Fig. 14.2 Sketch showing a dial gauge micrometer set to zero between a pair of crankwebs near bottom centre and the 'plus' reading obtained when the cranks are turned up to top dead centre.

twice the amount it indicated when it was in the starboard position. Continuing rotation of the crank, it will be noticed that as the crankpin goes over top centre position the reading on the dial gauge is still shown as a plus value but begins diminishing in amount. When the gauge is in the port position it will still show as a plus value and should be the same in amount as when on the starboard side. When the crankpin approaches bottom centre the gauge will be in its uppermost position and will read zero as the side of the connecting-rod approaches the gauge. Now the gauge will be on the port side of the connecting-rod.

It will be seen from this and the previous question that the weight of the piston, piston rod, and running gear gives a plus deflection reading on the dial micrometer when it is set to zero in its uppermost postion. This is often referred to as the natural deflection due to the weight of the running gear. (Fig. 14.2)

■ **14.8** Referring to Question 14.7, describe how the dial micrometer readings will be affected (a) if the dial gauge is fitted between the crankweb at a position midway between the crankpin and journal, and (b) if the indicator cock is left closed when turning the engine.

(a) Due to the rigidity and strength of the crankpin the distance between the crankwebs adjacent to the crankpin does not change in amount. When the crankpin is on top centre, the weight of the running gear causes the crankwebs to move outwards (so that plus readings are given). As the distance between the crankwebs cannot change in the region of the crankpin it follows that the distance between the crankwebs will increase as the distance from the crankpin increases and will be a maximum at the edge of the crankweb furthest from the crankpin. So it can be seen that the dial micrometer must be set in the same position on the webs each time deflection readings are taken. The position where the dial micrometer should be fitted is always given in the engine instruction book. In many engines the builder puts centre punch marks in this position, and for easy identification a letter 'O' or a small square is cut around each centre-punch mark.

(b)   If the indicator cock were inadvertently left closed when turning the crank to record crankweb deflections, the build-up of gas pressure inside the cylinder would have the same effect as an increase in weight of the running gear – and would increase the deflections.

■ **14.9   In a single crank arrangement consisting of a pair of crankwebs, a crankpin and two journals, describe what happens to the distance between the crankwebs if one journal is rigidly held in its bearing and the other bearing is lowered from its original position.**

If a dial gauge is placed between the cranks when the crankpin is in the bottom centre position and the engine is turned to its top centre position the crankwebs

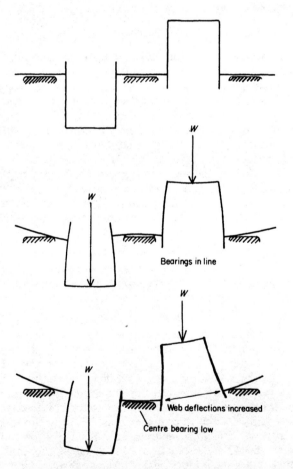

Fig. 14.3   Sketch showing the effect on deflections caused by a low centre bearing in a crankshaft for a two-cylinder engine.

will be opened an amount corresponding to the deflection produced by the weight of the running gear. With one of the main bearings tightened hard down on one of the journals (so that it cannot cant), it can be seen that if the other bearing is lowered, the weight of the running gear will cause the crankwebs and crankpin to bend in such a manner that the distance between the webs decreases and will continue to decrease until the bearing no longer supports the journal. At this point the distance between the crankwebs will not change, no matter how much more the bearing is lowered. From this we can see that when the crankpin is on top centre (micrometer dial gauge at lowest position) the effect of a low bearing is to cause the crankwebs to close inwards to one another.

The action shown on the dial micrometer gauge as the bearing is lowered would be to reduce the natural plus reading to zero and then to show an increasing negative reading until the crank assembly was supporting the weight of the running gear without assistance from the lowered main bearing. At this point the deflection would reach its maximum negative value and no further change would occur no matter how much further the bearing was lowered. The effect then of a low main bearing is to reduce the natural plus deflection of the crankwebs (found when bearings are level). If the wear down of one bearing is large in amount relative to another, the deflection will be reduced to the point where negative values will be recorded when deflection readings are taken.

Figure 14.3 shows line sketches of the deflections in a two cylinder engine crankshaft with a centre bearing of proper height; and a low centre bearing causing misalignment.

■ 14.10 How can bridged bearings be found?

If it is suspected that a crankshaft journal is not seating itself into its bearing when the engine is stopped it can be found by the following means.

1   Feeler gauges can be used to check if there is a clearance on the underside of a journal between the journal and the bearing. The engine is turned into the top centre position and from the bottom of the crankcase underneath the web an attempt is made to insert a thin narrow-width feeler between the journal and the bearing. The job is made easier if telescopic feelers are available. If the space between the side of the web and the main bearing is small, it is virtually impossible to use this method with any certainty of a definite result.

2   If method 1 cannot be used the journal must be loaded downwards with a light jack. The crank adjacent to the suspected bridged bearing is put to a position near to bottom centre and a dial gauge is inserted between the webs. The journal is then loaded with the jack acting downwards. If the bearing is bridged a deflection will be shown on the gauge. In some cases it is possible to do this check without removing the main bearing keep. This is done by removing the oil pipe and using a brass bar passed through the oil hole on to the journal. Extreme care must be taken with regard to cleanliness and to prevent damage to the journal. The jack is fitted between the top of the brass bar and the underside of the A-frame above the main bearing.

If the oil hole through to the bearing keep and shell is small it will be necessary to remove the upper half bearing and keep so that access to the journal is easily made.

3   See Question 10.20. End of answer deals with method of finding natural sag of crankshaft in way of main bearings.

■   **14.11 In multi-cylinder engines what is the effect of the weight of the crankshaft and running gear on either side of a crank where deflections are being taken?**

The effect of the weight of the crankwebs and running gear in crank units adjacent to the crankweb at which deflection readings are being taken is to reduce the deflection being recorded.

Referring to Question 14.6 it can be seen that the weight of the adjacent parts of the crankshaft itself will to some extent balance out the weight of the parts acting on the crank unit where deflections are being taken.

■   **14.12 Would you expect deflections to be shown on a micrometer dial gauge when the alignment of the crankshaft is correct?**

With the dial gauge set up between the crankwebs, yes. The amount of deflection of the crankwebs will be dependent on the weight of the running gear, the flexibility of the crankshaft and the distance between adjacent main bearings. In large slow-speed engines the deflection would be of the order of 0.05 mm whereas in medium-speed engines it would be about 0.01 mm, or so small that it would be difficult to measure. In older, opposed piston engines having three cranks to a cylinder unit and main bearings with spherically shaped housings the deflection could be as high as 0.4 mm. These deflections might be referred to as the natural deflection of the crankshaft, which comes about from the load of the running gear and the weight of the crankshaft itself.

■   **14.13 What is the purpose of taking deflections? How and where may errors arise which could lead to misleading results?**

It has been seen in earlier questions that when uneven wear takes place along a line of main bearings, the effect is to cause changes in the deflections measured between the crankwebs. Any change in deflection from the natural deflection can then be related to misalignment in the main bearings, as the change of deflection is proportional to the differences in height of the main bearings. Deflection measurements are taken to ascertain quickly whether the alignment of a crankshaft is within acceptable limits or whether re-alignment is necessary.

Errors may arise in engines with rigid crankshafts and relatively light running gear. In such engines it is possible for the crankshaft to support the weight of the running gear without being properly seated in its main bearings. The deflection readings recorded could then appear to be satisfactory, but, when such an engine is in operation, the gas pressure in the cylinder causes the journal to seat in the low or 'bridged' bearing. The flexing of the crankshaft which then takes place causes high stresses to be set up which will lead to early failure. The location of the failure will be dependent on the proportions of the various parts of the crankshaft and the effects of localized stress raisers.

If part of the running gear of an engine such as a connecting-rod, piston and piston rod has been removed for survey or repair, the amount of deflection in way of removed gear will change and not be the same as when these parts are present.

■ **14.14 Describe how crankshaft deflections are taken, how they are recorded and how it is ascertained that alignment is acceptable?**

The deflection readings are taken at each crank in five different angular positions. Two readings, the first and last, are taken with the crankpin near bottom centre, the other three readings are taken with the crankpin on the port side, top centre and starboard side. It must be remembered that the dial micrometer is fitted in a position between the webs on the opposite side to the crankpin. A table of measurements is drawn up relative to dial gauge position. To set the dial gauge between the crankwebs the engine is turned *ahead* until the crankpin is approximately 10° to 15° past bottom centre (see Note towards end of answer). The dial gauge is inserted between the webs in the location fixed by the engine builder and set to zero.

The turning gear is used to turn the engine in the ahead direction and readings are taken when the gauge is passing starboard position, bottom position, port position, and as it approaches the top position again. The turning gear is stopped just before the connecting-rod touches the dial gauge. This is repeated at each cylinder unit.

The readings are recorded so that they cover the alignment of the crankshaft in the vertical plane, and in the horizontal plane. The readings taken when the dial gauge is in the bottom and top positions cover the crankshaft alignment in the vertical plane and those taken in the starboard and port positions cover the alignment in the horizontal plane.

The table for recording the readings is made up as shown. Some examples of how the figures are treated are also given, and it must be remembered that in dealing with the deflection figures taken when the crankwebs are horizontal the sign of the lower line of figures is changed and the figures are added algebraically.

The figures taken in the engine room are recorded in tabular form with six horizontal lines and a vertical column for each cylinder unit, as shown in the table.

Engine room record sheet. Example: six cylinder engine
(Deflections in mm)

Gauge position	Cyl. No. 6	Cyl. No. 5	Cyl. No. 4	Cyl. No. 3	Cyl. No. 2	Cyl. No. 1
Top	0.00	0.00	0.00	0.00	0.00	0.00
Starboard	−0.01	0.00	+0.02	−0.01	+0.02	+0.03
Bottom	+0.01	+0.04	+0.04	−0.02	+0.03	+0.06
Port	+0.02	+0.04	+0.02	−0.01	+0.01	+0.04
Top	0.00	+0.01	0.00	0.00	0.00	0.00

The table drawn up is for a direct-coupled main engine with a right-handed propeller. Ships with controllable-pitch propellers usually run with left-hand turning. The first vertical column would then have the port and starboard gauge positions reversed to that given above. The turning direction for taking clock gaugings should be the ahead direction for main engines, and direction of rotation for diesel generators. The units of the measurements must be stated on the record sheet.

The analysis of the figures taken is drawn up as shown in the table for vertical plane alignment. If the variations in the port and starboard gauge readings show a tendency to misalignment in the horizontal plane or some plane near to the horizontal plane, the second table is drawn up.

Vertical plane alignment (mm)

Gauge position	Cyl. No. 6	Cyl. No. 5	Cyl. No. 4	Cyl. No. 3	Cyl. No. 2	Cyl. No.1
Bottom A	+0.01	+0.04	+0.04	−0.02	+0.03	+0.06
Top B	0.00	+0.005	0.00	0.00	0.00	0.00
Difference A − B	+0.01	+0.035	+0.04	−0.02	+0.03	+0.06

Horizontal plane alignment (mm)

Gauge position	Cyl. No. 6	Cyl. No. 5	Cyl. No. 4	Cyl. No. 3	Cyl. No. 2	Cyl. No. 1
Starboard C	−0.01	0.00	+0.02	−0.01	+0.02	+0.03
Port D	+0.02	+0.04	+0.02	−0.01	+0.01	+0.04
Difference C − D	−0.03	−0.04	0.00	0.00	+0.01	−0.01

Maximum vertical plane deflection recorded	+0.06 mm
Maximum vertical plane deflection allowed	(As stated in engine instruction book)
Maximum horizontal plane deflection recorded	−0.04 mm
Maximum horizontal plane deflection allowed	(As stated in engine instruction book)

Alignment: satisfactory/unsatisfactory *(Delete as appropriate)*

In order to check whether the alignment is acceptable, the maximum allowable deflection figures are obtained from the engine instruction book. The values recorded in the vertical and horizontal plane of the table are checked and compared with the maximum allowable figure. In many engines the allowable deflection in the horizontal plane is the same as that allowed in the vertical plane, while in others the allowances are different.

Note Once a position of the cranks is found for the initial setting of the dial, each crank can be set to its correct position to commence readings by using the degree markings on the flywheel.

Time can be saved in taking deflection readings by starting the readings on the crank nearest to the bottom centre position and after completion going to the next crank nearest the bottom centre position. It is not necessary to deal with each crank in its numerical order in the engine.

It will be noticed that the vertical columns in the tables have No. 1 cylinder unit placed on the right-hand side. This is for convenience as in many cases the 'front' of the engine is on the starboard side, and this will be the side on which the crankcase is entered. Some engine builders number the cylinder and main bearing numbers from aft to forward in the same way that the shipbuilder numbers the ship's frames. In such cases No. 1 cylinder would be marked-up on the left-hand side column.

■ **14.15** When taking crankshaft deflection readings on large engines having a worm- or gear-operated turning gear, certain precautions must be taken when taking the deflection readings on the cranks nearest to the turning gear. What are these precautions and why are they necessary?

If the main engine turning gear is fitted on the port side and the engine turns right-handed when running ahead there will be a tendency to lift the turning wheel or flywheel when the engine is moved by the turning gear. A similar action takes place with a left-hand turning main engine when the turning gear is fitted on the starboard side of the engine. In such cases the lifting action exerted by the worm or pinion when turning the flywheel may cause the journal nearest the flywheel to lift off the bottom of its bearing. This will then be reflected in the reading given by the micrometer dial gauge, and may influence the two after-most cranks on the engine. In order to avoid false readings on these cranks it is good practice to take the load off the worm or pinion by running the gear back by hand at the motor coupling, in the same manner as is done when the turning gear is to be taken out of engagement with the flywheel.

In geared propulsion systems and twin-screw ships the turning direction of the engines when running ahead and the location of engine turning gear would have to be examined individually to ensure that lifting did not occur when operating the turning gear to take deflection readings. (There is more information on this in Questions 12.29 and 12.30.)

Note The importance of an error in a false reading increases as the true deflection approaches a limiting value. It may also be of great importance in medium-speed engines where allowable deflections are a lot less than in slow-speed engines.

■ **14.16** In a diesel generator with one pedestal bearing what is the effect on the crankweb deflection (in the unit nearest to the generator) when the pedestal bearing is low?

If the camshaft drive is between the crank and the flywheel there are two journal bearings between the crank and the flywheel. When the camshaft drive is at the

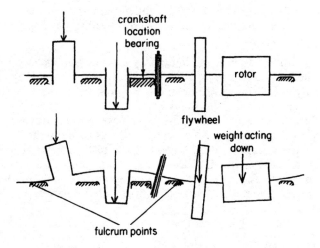

Fig. 14.4   Effect of overhung weight on the crankshaft of a diesel generator set.

other end of the engine (opposite end to flywheel and generator) there may be only one journal or main bearing between the crank, flywheel and generator. The effect of a low pedestal bearing will be similar in each case, but it may be less apparent when taking deflection readings if there are two journal bearings between the crank and the flywheel. On the crank nearest the generator, starting with the crankpin near bottom centre and the gauge in the uppermost position and set to zero, the crankwebs will begin to close inward and show a negative reading. This will be continued until the crankpin is on top centre and the gauge is in its lowest position when it will show the greatest negative value.

In engines with two bearings between the crank and the flywheel, the weight of the armature or rotor and the flywheel causes the bearing nearest the flywheel to become a fulcrum point and the journal nearest the crank may lift off its bearing. With relatively lightweight running gear, as found on engines operating at 720 or 750 rev/min and above, it is possible to lift the two journals on either side of the crank off their bearings. In this case acceptable deflection readings may be given whereas a dangerous condition of misalignment could exist due to the bridged bearings. (Fig. 14.4)

The manner in which the crankshaft deflects is shown in Fig. 14.4. The fulcrum points where the crank journals are supported are indicated in the line sketch. The amount of deflection shown is greatly exaggerated to show readily how dangerously false deflection readings could be obtained when taking the deflection readings of a diesel generator crankshaft.

Note   When the engine is in operation the cylinder pressure causes the journals to seat themselves on the main bearings, but a low bearing if undetected could lead to early crankshaft failure.

■ **14.17** Which is stiffer, the hull structure or the engine structure including crankshaft and intermediate shafting when considered relative to their resistance to bending?

The ship's hull can be likened to a beam. For the purposes of this question it can be considered as uniformly supported in calm weather and still water conditions. When the weather is bad the hull will be subjected to large upward-acting forces from the buoyancy given by the crests of waves. As the waves move along the length of the ship these upward-acting forces move along with them.

The stiffness of a beam is related to its moment of inertia or *I* value. If the ship's hull is likened to a hollow rectangular girder, it is readily seen that the stiffness of the hull will be much greater than the stiffness of the engine crankcase, crankshaft and intermediate shafting.

When the ship's hull is subjected to forces from heavy weather, abnormal loading and the like, it will deflect and small changes in the shape of the hull will occur. As the engine crankcase, crankshaft and intermediate shafting is less stiff, i.e. more flexible than the hull, it follows that the engine and shafting will also deflect and will follow the changes in shape of the hull which supports it.

■ **14.18** If the shape of the hull changes and the main engine crankshaft follows the shape of the hull, will the changed shape or line of the crankshaft be indicated when taking deflections with a micrometer dial gauge?

A line passing through the centres of the main bearings will follow the changes in hull deflection as bending occurs and the line of the crankshaft will change in an identical manner. The changes that will occur in the shaft alignment will be shown when crankweb deflections are taken with the dial gauge. It is thus possible to make a quick crankshaft alignment check if the cargo distribution within the hull follows an unusual pattern or if the ship should go aground.

■ **14.19** What do you understand by the term 'true line' of a crankshaft and intermediate shafting?

If some 'dead' straight line is taken and the vertical distances of the axial centre-line of each crankshaft and intermediate bearing is measured, it would be possible to plot these distances in graphical form and obtain the true line of the crankshaft and shafting in the vertical plane. In a similar manner it is possible to measure the distance of the axial centre-line of each main bearing in a horizontal direction from an imaginary straight line at the side of the engine. The true line of the crankshaft in a horizontal plane can then be obtained.

The true line of a crankshaft is plotted in graphical form. The X or horizontal axis shows the fore and aft location of each bearing, to some suitable scale which will be dependent on the length of shafting being checked, and the length of the sheet used for plotting the true shaft line. The axial centre-lines of the first and last main bearing in the engine are made datum points having zero height and

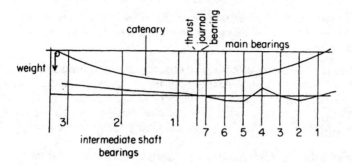

Fig. 14.5   Sketch showing the true line of a crankshaft and intermediate shafting obtained by using a stretched piano wire or a micrometer telescope.

the X axis of the graph passes through these two datum points. To obtain the true shaft line in the vertical plane the height of each of the other bearings above or below this datum is then plotted as an ordinate at its correct fore and aft location on the X axis line. As these heights are only small measurements amounting to a few hundredths of a millimetre, the scale adopted makes the actual heights shown on the graph 25 or more times larger. A height of 0.02 of a millimetre would then be drawn on the graph as a half millimetre or more depending on the scale chosen. When the true shaft line is drawn in this manner, with the fore and aft distances scaled down and the vertical distances scaled up, the true line of the shafting is readily seen and understood. The line of the crankshaft in the horizontal plane is also plotted in a similar manner.

Normally the true line of the shafting should follow a smooth curve, and interferences to the smooth curve from high or low bearings are readily seen when the curve is plotted with the scales as described. (Fig. 14.5)

Note   The straight line referred to as a 'dead' straight line is one which is mathematically straight and the shortest distance between two points.

The term 'true alignment' of shafting is often used instead of the term 'true line'.

■ **14.20**   Name the items of equipment used to establish the true line of a crankshaft and intermediate shafting.

There are two methods. One method uses a taut piano wire, and an internal micrometer of suitable length. The head of the micrometer is fitted with an insulated contact arrangement. A low voltage is applied to the crankshaft and taut wire, so that the micrometer operates as a switch. When the measuring head of the micrometer contacts the taut wire it allows current to flow which can be used to light a small lamp, or to give an audible noise in a pair of ear phones. This equipment is essential for obtaining accurate and repeatable measurements. The weight of the wire and its flexibility prevents the taking of a measurement on 'feel' as is normally done with micrometers. A length of piano wire, ball bearing pulley, anchoring or support clamp, and weight are the other items of equipment required to make a check by the taut-wire method.

The other method uses optical means for establishment of the true line of the crankshaft and intermediate shafting. The equipment required is a special telescope, intermediate targets, and a base target. The special telescope has built into it the capability of measuring small vertical and horizontal distances from which the true line of the shafting is established. This capability is obtained by a combination of mechanical and optical means. The method by which the heights of the bearings are calculated is similar; it is only the methods of measurement which are different.

■ 14.21 The diameters of crankshaft and intermediate shaft bearings are different, and in old engines there may also be differences in the crankshaft journals. How is the axial centre-line of the various bearings found when taking measurements to establish the true shafting line?

Irrespective of whether taut-wire or optical methods are used, all measurements are taken from the very top of each journal. The axial centre-line of the bearing will then be one half of the diameter of the journal below its top. Addition of the semi-diameter of a journal to any measurement taken from its top will give a result which is related to the axial centre-line of the bearing. By using the semi-diameter of each journal in this way differences in bearing or journal diameter are cancelled out as we are then relating measurements taken from the top of a journal to its axial centre-line. It is therefore usual when checking the alignment of a shaft to measure accurately each journal diameter.

■ 14.22 How is the mathematically straight line obtained from which the height of the axial centre-lines of the bearings are measured when establishing a true shaft line?

In any method used for establishing the true line of a system of shafting, it is necessary to have a real or imaginary straight line to which the height of the axial centre-line of each bearing is related.

In the optical method of obtaining height measurements the straight line is the line of sight observed through the micrometer telescope on to some base target. Displacements in the heights of the tops of journals are related to this line of sight, and when the semi-diameters of the journals are added to the displacement found it is related to the axial centre-line of the bearing.

When a taut wire is used to obtain the relative heights of the bearings, the wire is set up vertically above the top of the journals. It is passed right through the crankcase from forward to aft; one end is fixed and the other passes over a pulley and is tensioned with a heavy weight. The wire naturally sags to some curvature. Two points at the ends of the wire vertically above the bearings at the extreme ends of the shafting are taken, and the straight line is considered to be between these two points. It is an imaginary straight line because the taut wire will be sagging below it. The measurements taken from the tops of the journals are related to this imaginary straight line.

■ **14.23   What form does the curvature of a taut wire take?**

Any flexible wire or chain having uniform weight over its length takes up curvature when its ends are supported. The curve formed is a catenary which has certain mathematical properties. When the sag at the centre of the curve is less than one-tenth of the distance between the supports the curvature approximates to a parabola which also has certain mathematical properties. These properties are related to conic sections, and with a knowledge of the tensile load in the taut wire and the mathematical formula related to the parabola, it is possible to calculate the maximum sag at mid span and the sag at various points over the span.

Note   The mathematics required to formulate the relationship of these curves is covered in mathematics books under sections dealing with the geometry of conic sections, series and hyperbolic functions.

■ **14.24   Show how the maximum sag in a taut wire can be found by calculation. After finding the maximum sag show how the amount of sag can be found at any point between the supports. Assume that the wire is fastened at one end, the other end being passed over a pulley mounted on a ball bearing and supporting a weight which keeps the wire taut.**

It can be proved that the relationship between the sag in the wire (S), the weight of a unit length of wire (w), the distance between the supports (L) and the weight (W) suspended on the free end of the wire is given by the formula

$$S = \frac{wL^2}{8W}$$

Since

$$w = A\rho$$

where A is the sectional area of the wire and ρ the density of the wire, the formula can be written

$$S = \frac{A\rho L^2}{8W}$$

As $W/A =$ stress in wire

$$\frac{A}{W} = \frac{1}{f}$$

where $f =$ initial stress. Substituting

$$S = \frac{\rho L^2}{8f} \qquad \text{(Eq. 1)}$$

The units chosen must cancel to leave a linear measurement.

It can also be shown that the sag in the wire at a distance x from a support is given by the formula below, where S_m is the maximum sag.

$$S_x = S_m \left[1 - \left(\frac{L - 2x}{L} \right)^2 \right] \qquad x < \frac{L}{2}$$

Let $L - 2x = L_1$. Therefore

$$S_x = S_m \left[1 - \left(\frac{L_1}{L} \right)^2 \right] \qquad \text{(Eq. 2)}$$

In order to make the necessary calculations the maximum sag is obtained from (Eq. 1), and used in (Eq. 2). The distance values used for the various values of x can be obtained from the crankshaft and shafting drawings.

In making the calculations the work can be facilitated by setting out the values in columnar form. A column would be assigned for each main and intermediate shaft bearing under consideration.

The maximum sag occurs at the centre of the span if the pulley and support are horizontal. If one of the supports is slightly lower than the other the maximum sag will be a small amount off-centre and towards the low side.

■ **14.25** Using the formulae in Question 14.24, calculate the sag of a taut wire at each main bearing in an engine 12 000 mm long. The spacing of the cylinder centres is 1150 mm. The engine has six cylinders and the main bearings are equidistant from one another. The piano wire used is 18 s.w.g. and the weight supported by the wire and keeping it taut is 20 kg.

The first thing that must be fixed is L, the distance between the extreme measuring points. In the case shown, this distance is the distance between the first and last main bearings and would normally be read off the print of the crankshaft drawing. Assume that the camshaft drive is taken from a point midlength on the crankshaft; this then requires seven spaces between adjacent main bearings – six for the cylinders and one for the camshaft drive. The number of main bearings will equal the number of spaces plus one which gives 8 main bearings in all.

Note   Many slow-speed engines have another journal bearing just aft of the thrust block which is built up within the main engine bedplate structure. This is named the thrust journal bearing and is not included.

The distance, L, = seven spaces × distance between main bearings, i.e.

$$L = 7 \times 1150 = 8050 \text{ mm}$$

The density, ρ, of the wire is 7849.86 kg/m³. (Note this unit must be divided by 1000^3 to give an answer in mm.) From tables the sectional area, A, of the wire is

$$0.7854 \times 1.219^2 = 1.1671 \text{ mm}^2$$

Therefore stress is given by

$$f = \frac{W}{A} = \frac{20}{1.1671} = 17.1365 \text{ kg/mm}^2$$

and sag (max) by

$$S_m = \frac{\rho L^2}{8f} = \frac{7849.86 \times 8050^2}{8 \times 17.1365 \times 1000^3} = 3.71 \text{ mm}$$

Sag at no. 1 main bearing $= 0$

Sag at no. 2 main bearing

$$S_x = S_m \left[1 - \left(\frac{L_1}{L} \right)^2 \right] \qquad x = 1150$$

Given that

$$L_1 = L - 2x = 8050 - 2 \times 1150 = 5750$$

it follows that

$$\left(\frac{L_1}{L} \right)^2 = \left(\frac{5750}{8050} \right)^2 = 0.5102$$

and

$$\left[1 - \left(\frac{L_1}{L} \right)^2 \right] = 1 - 0.5102 = 0.4898$$

Sag in wire above no. 2 bearing is therefore

$$S_x = 3.71 \times 0.4898 = 1.82 \text{ mm}$$

Sag at no. 3 main bearing

In this case

$$x = 2 \times 1150 = 2300 \text{ and } L_1 = 8050 - 2 \times 2300 = 3450$$

It follows that

$$\left[1 - \left(\frac{L_1}{L} \right)^2 \right] = 1 - \left(\frac{3450}{8050} \right)^2 = 1 - 0.1837 = 0.8163$$

Sag in wire above no. 3 bearing is therefore

$$S_x = 3.71 \times 0.8163 = 3.03 \text{ mm}$$

Sag at no. 4 main bearing

In this case

$$x = 3 \times 1150 = 3450 \text{ and } L_1 = 8050 - 2 \times 3450 = 1150$$

It follows that

$$\left[1 - \left(\frac{L_1}{L} \right)^2 \right] = 1 - \left(\frac{1150}{8050} \right)^2 = 1 - 0.0204 = 0.9796$$

Sag in line above no. 4 bearing is therefore

$S_x = 3.71 \times 0.9796 = 3.63$ mm

As the bearings are equidistant to one another:

the sag in the wire above no. 8 bearing will be 0

the sag in the wire above no. 7 bearing will be similar to no. 2

the sag in the wire above no. 6 bearing will be similar to no. 3

the sag in the wire above no. 5 bearing will be similar to no. 4.

The sag in the wire at the point where a measurement is taken between the wire and the top of the journal is then set out in tabular form.

Main bearing no.	8	7	6	5	4	3	2	1
Sag in wire (mm)	0	1.82	3.03	3.63	3.63	3.03	1.82	0

■ **14.26** How are measurements taken and utilized to establish the true line of a crankshaft when a taut wire is used?

The taut wire must be passed right through the crankcase of the engine vertically above the centre-line of all the main bearings. Holes covered with blank flanges or screwed plugs are often provided in the ends of the crankcase for this purpose or for using a micrometer telescope. After the flanges or plugs have been removed a light is placed adjacent to one hole, and a sight is taken from the other hole while the engine is being turned with the turning gear. The engine is turned until the light is seen; at this point all the connecting-rods are clear and an uninterrupted path is made for the wire to be passed through without making contact with any of the connecting-rods or other parts of the running gear (see Note). The oil supply pipes to the main bearings are removed to enable the micrometer to be passed down on to the crankshaft journals. The wire is then passed through the crankcase and supported at one end and passed over the pulley at the other. The weight is suspended on the wire to keep it taut.

The micrometer measurements are then taken between the top of the crankshaft journals and the taut wire. The measurements are set down in tabular form with a vertical column assigned to each main bearing. There are horizontal columns for micrometer measurement, calculated wire sag, and journal semi-diameters; the readings are recorded in the appropriate 'boxes'. Various corrections are applied to get the true shaft line relative to the first and last main bearings as the datum line. The corrections are drawn up in further horizontal columns.

After the corrections have been applied the true shaft line is plotted on suitably-sized graph paper.

Note Engines with seven or more cylinders do not usually have a satisfactory angular position of the crankshaft that will leave a clear path for the wire to be put through the crankcase. In such cases the connecting-rod or rods must be dismantled from the bottom-end bearing and swung clear. In some cases an oil hole in the journal will be in line with the oil hole in the bearing keep so measurements cannot be taken. The keep must be removed and the micrometer measurement between the top of the journal and the wire is taken just forward or aft of the oil hole.

■ **14.27   State the methods used to ascertain the amount of weight that can be put on the free end of a wire to hold it taut for obtaining a true shaft line.**

The amount of weight used must be such that the sag in the wire is minimal and less than one-tenth of the span between the supports. In practice it is aimed to keep the wire as near to a straight line as possible. This limits the number of changes that have to be made with the extension pieces on the micrometer. The limit to the allowable weight which can be put on the wire is obviously related to the ultimate tensile strength of the wire and the safety factor used. Generally the ultimate tensile strength of steel piano wire increases with increase in the s.w.g. number, or as the diameter of the wire gets smaller. For example and for guidance purposes, the ultimate tensile strength of 19 s.w.g. steel piano wire will be of the order of 2060 N/mm² (210 kg/mm², 133 tons/in²) and wire of 38 s.w.g. size may have a ultimate tensile strength of twice the previous figure.

The safety factor used should be such that it gives a stress in the wire of not more than one-third of its ultimate tensile strength.

■ **14.28   What effect does the trim of the ship have on a taut wire when it is set up in an engine?**

It was shown earlier that the support points for a taut wire must be horizontal for the maximum sag to occur at the centre of the span. If one end support is higher than the other the point of maximum sag is displaced slightly from the mid-span position towards the low end. If the support points are at the same height above the shafting at each end, it can be seen that the ship's trim will cause the position of the maximum sag to be displaced. Generally if the amount of trim is small the effect can be neglected as no appreciable error will arise. In many cases the effect of trim can be cancelled out by adjustments to the height of the wire (see Question 14.30).

■ **14.29   Why are true shafting lines established in engines which have been in service for some time?**

Generally the wear which takes place on main bearings is such that at some time in the ship's life corrective action must be taken with main engine shaft alignment. A true shafting line is established, and the line plotted shows high and low bearings as peaks and dips in the shaft line. The allowable curvature of the crankshaft is plotted on this line from which it is readily seen which bearings need attention. In some cases it is possible to machine out the high bearings to obtain a satisfactory shaft line. The amount to be machined out is read off the plotted shaft line. In other cases it may be necessary to re-metal some bearings and machine others to get the required line.

When bearings have to be re-metalled the amount of lift required is also obtained from the plotted line, and at the time the bearing is removed from the engine its lower-half crown thickness is measured. The amount that the shaft must be lifted is added to this measurement and this gives the new lower-half crown thickness to which the bearing must be machined.

14.30 A set of micrometer measurement readings taken between a taut wire and the top surface of the main engine crankshaft journals are given in the following table. All the journals are 550 mm diameter with the exception of nos 2 and 5 which were ground *in situ*. These journals are 4 mm less in diameter. Assume that the wire and weight used is as given in Question 14.25 and use the sag values calculated in the question. Show how the true line of the crankshaft is obtained.

The main shafting line is parallel to the keel, the ship is 185 m long and is trimming 0.50 m by the stern.

Bearing	8	7	6	5	4	3	2	1
Micrometer reading	621.87	617.37	613.26	611.18	605.17	603.50	603.70	600.10

Although the possible errors are small when the support points or extreme measuring points on the wire are not truly horizontal, it is incumbent on the engineer to reduce errors to a minimum if the cost is not too great. In the case considered the time taken to reduce error is worthwhile, as only a small calculation is involved.

As the ship is trimming by the stern the line of the shaft will be low at the aft end. Some correction is possible by raising the aft end of the wire at the support pulley. The amount it should be lifted is found as follows.

The ratio of ship length to crankshaft length between no. 1 and no. 8 bearings is

$$185:7 \times 1.150, \text{ or } 185:8.050 \approx 185:8$$

The ship is trimming 0.5 m by the stern so

$$\frac{\text{ship length}}{\text{trim by stern}} = \frac{\text{crankshaft length}}{y}$$

where y should be the difference in micrometer measurements between no. 1 and no. 8 main bearing. Then

$$y = \frac{\text{crankshaft length} \times \text{trim}}{\text{ship length}}$$

$$= \frac{8 \times 0.5}{185} = \frac{4}{185} = 0.022 \text{ m or } 22 \text{ mm}$$

The pulley supporting the wire should be raised so that the micrometer measurement at no. 8 main bearing is approximately 22 mm more than the measurement at no. 1 main bearing. In this case no. 1 main bearing is at the forward end of the engine and nos 1 and 8 journals are similar in diameter. This is seen in the measurements given where the pulley height was adjusted and gave a difference of 21.77 mm.

A table is first drawn up with 10 vertical columns and 10 horizontal spaces. The top line is marked up with the bearing numbers 1 to 8. The micrometer reading, wire sag, and journal semi-diameters are recorded in the horizontal spaces under the bearing number to which they apply and each horizontal line is marked A, B and C respectively. The summation of these amounts is then recorded in the fifth line, reference letter D. A + B + C = D. The figures in line D are the amounts that the axial centre of each main bearing is below the straight line passing through the wire at the measuring points of no. 1 and no. 8 main bearings. It is seen by inspection of the table that the forward end of the straight line passes through a point 875.10 mm above no. 1 main bearing, and the aft end of the line passes through a point 896.87 mm above no. 8 main bearing. If 875.10 mm is subtracted from the measurements in line D, the effect is to lower the straight line by the amount subtracted; the forward end of the line then passes through the centre of no. 1 main bearing.

If we take a straight line, which is the datum, through no. 1 and no. 8 main bearing, and the straight line passing through no. 1 and the point 21.77 mm above no. 8 main bearing, it is seen that a triangle is formed. Similar triangles are also formed between the vertical passing through each main bearing. The vertical height of the perpendicular of each of these triangles is calculated and recorded in line G.

The base of the largest triangle is 7×1150 mm, and as each bearing is equidistant, the base can be considered as 7 units. The perpendicular of the triangle at bearing no. 7 is

$$\frac{21.77}{7} \times 6 = 18.66 \text{ mm}$$

Similarly at bearing no. 6 the perpendicular is

$$\frac{21.77}{7} \times 5 = 15.55 \text{ mm}$$

At bearing no. 5 the perpendicular is

$$\frac{21.77}{7} \times 4 = 12.44 \text{ mm}$$

The perpendicular for each of the triangles is calculated in a similar way and entered in line G.

If the distance shown in line F is greater than the calculated value of the corresponding perpendicular given in line G it follows that the axial centre-line of the bearing is below datum. If the distance is less, the axial centre-line of the bearing is above datum. The values given in line G are subtracted from the values given in line F.

Table for alignment of main bearings when ship is trimming by the stern

Bearing number	8	7	6	5	4	3	2	1
A Micrometer reading (mm)	621.87	617.37	613.26	611.18	605.17	603.50	603.70	600.10
B Sag in wire (mm)	0.00	1.82	3.03	3.63	3.63	3.03	1.82	0.00
C Journal semi-diam. mm	275	275	275	273	275	275	273	275
D A + B + C = D	896.87	894.19	891.29	887.81	883.80	881.53	878.52	875.10
E Subtract D − E	875.10	875.10	875.10	875.10	875.10	875.10	875.10	875.10
F Corrected line D − E	21.77	19.09	16.19	12.71	8.70	6.43	3.42	0.00
G Triangle correction	21.77	18.66	15.55	12.44	9.33	6.22	3.11	0.00
H F − G	00.0	0.43	0.64	0.27	−0.63	0.21	0.31	0.00
I Shaft line journal position	Datum	−0.43	−0.64	−0.27	+0.63	−0.21	−0.31	Datum

Journals 2, 3, 5, 6, and 7 are below datum.
Journal 4 is above datum.
All measurements in mm. Datum line passes through centres of journals 1 and 8.
Students should plot the true shaft line and its construction on graph paper.

When the axial centre-line of the bearing is below datum, the amount is marked with a negative sign, when it is above datum with a positive sign.

Figure 14.6 shows how the various measurements given in the table are applied to obtain the position of the centre axis of the bearing above or below the datum line.

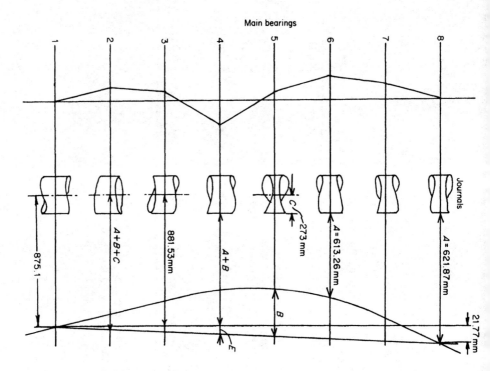

Fig. 14.6   Sketch showing the method of obtaining a crankshaft line when using a stretched piano wire passed through the engine.

■ 14.31 When an engine is being erected in a new ship or parts are being assembled after a major repair, which method is used to carry out crankshaft alignment?

When new engines are erected on the test bed in the engine works, the lower half of the crankcase is set up on blocks on the foundation of the test bed. The height of these blocks is adjusted so that the machined surface on the top of the lower half of the crankcase is dead flat. The micrometer telescope is set up at some point outside the crankcase, or on one corner, and radial sweeps are made with the telescope on to fixed-height targets. The blocks are adjusted to bring the bedplate flat which will be shown when the micrometer is sighted on the fixed-height targets. The amount that the targets are up or down indicates the amount the blocks must be lifted or lowered. The line of the crankshaft is corrected with the micrometer telescope in a similar manner.

When the engine is erected in the ship the crankshaft flange is adjusted to its correct position relative to the intermediate shaft flange and the bedplate is brought to the same line that it had on the test bed. The micrometer telescope is then used in a similar manner during erection of the engine in the ship.

When a crankshaft is being re-aligned after major repair work such as grinding crankshaft journals and re-metalling main bearings, taut-wire or optical alignment methods can be used.

■ 14.32 What is the maximum amount that a bearing can be raised or lowered when aligning intermediate shafting? How would you find the amount of support being given by a bearing?

The maximum theoretical amount that an intermediate shaft bearing can be lifted will be such that the support given by the adjacent bearings is reduced to zero. In actual practice a bearing lifted this amount would run hot in service, so the limiting value is related to the amount of weight the bearing can carry without becoming overheated. In a similar manner the amount that a bearing could be lowered would be such that the support it gave to the shaft would be zero. Again, in practice, it is most likely that adjacent bearings would run hot in service.

In order to find the weight that an intermediate shaft bearing is carrying, it is necessary to set up a hydraulic jack under the shaft on the bearing pedestal or from the tank top. A micrometer dial gauge is set up to bear on the top of the shaft journal and indicate any lift that occurs in the shaft. The area of the ram in the jack will be known. The jack is then gradually loaded with the hydraulic pump, keeping a watch on the pressure gauge. At some point during the gradual loading of the jack the dial micrometer gauge will move indicating that the shaft has started to lift. The pressure at which the shaft lifts must be noted. The pressure is applied continuously and the shaft is lifted a small amount. The amount of lift and the corresponding pressure to obtain this is noted as the shaft is lifted. The shaft is then gradually lowered, and the pressure and micrometer dial gauge readings are noted as the shaft is allowed to come down.

The various pressures are multiplied by the area of the ram to derive a series of values of the upward force exerted by the jack.

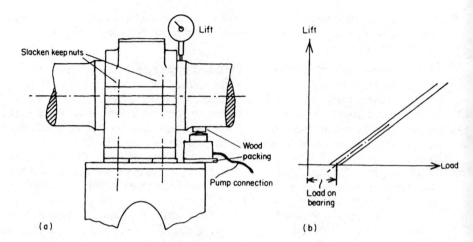

Fig. 14.7   (a) Sketch showing the set-up for finding the load acting on an intermediate shaft bearing.
(b) Method of plotting values obtained when jacking up the shaft.
Note. The product of the pump pressure and ram area gives the force required to lift the shaft.

Axes are then drawn with the abscissa scaled to suit the upward forces exerted by the jack. The ordinates are scaled to give a good slope when the lift recorded by the dial gauge is plotted.

A graph is then drawn of upward force plotted against the shaft lift. The graph when completed will give the force at which the shaft started to lift and will be a straight line for the various points of force and lift. A similar line is plotted for the shaft lowering period. The two lines will be parallel to each other with the lowering line nearer to the *y* axis. A line is drawn centrally between the plotted line indicating 'raising' and 'lowering'. The intersection of this line through the *x* or force axis gives the weight that the intermediate shaft bearing is carrying.

This method of checking the load that a shaft bearing is carrying is sometimes used for aligning intermediate shafting. When bearings give trouble overheating and the problem is not associated with faulty lubrication or cooling services, this method of checking bearing load is useful for correcting alignment without breaking shaft couplings or taking sights of shafting lines. (Fig. 14.7)

■ ˙ **14.33 If a single section of intermediate shaft is supported on two bearings, what angles do the coupling faces take?**

The angle that the coupling faces make relative to one another will be dependent on the distance between the bearings which support the shaft.

If the bearings are close together near the mid-length of the shaft, the ends of the shaft will sag or deflect downwards. Vertical projections off the faces of the

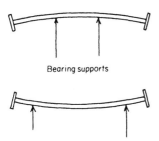

Fig. 14.8 Sketch showing how an intermediate shaft hogs or sags due to the location of the shaft bearings.

coupling flanges would then no longer be parallel lines; they would be oblique to one another and opening upwards.

Now imagine the bearings are slid away from one another towards the coupling flanges; the weight of the shaft will cause its centre portion to deflect downwards, and the projected lines will again be non-parallel – and closing upwards.

In the first case the polar axis of the shaft is curved convex upwards; in the second case the curvature is reversed and is concave upwards. (Fig. 14.8)

■ **14.34** When adjacent couplings of two sections of intermediate shafting are correctly lined up so that their faces are parallel and their circumferences level, what height will the bearings supporting the shaft be relative to one another?

With normal distances between the bearings supporting the two sections of shafting, the mid-lengths of the shafts will sag or deflect downwards as shown in Question 14.33. If the height of the bearings is the same right through it

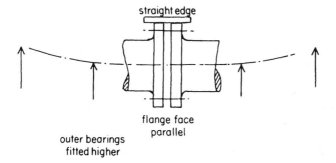

Fig. 14.9 Sketch showing the height of normally spaced bearings when coupling shaft flanges are brought into line. **Note.** Care must be taken to ensure that the flanges are of the same diameter when using a straight-edge in the manner shown. If the flanges are of different diameters a correction must be made.

follows that the coupling flange faces will be open at the top. By raising the height of the outer bearings the coupling face flanges will be brought parallel. In practice the true line of shafting follows a curve in this manner. (Fig. 14.9)

Note   When using a straight edge in the manner shown to align shaft flanges care must be taken to ascertain that each flange is of the same diameter. If the flanges are of different diameters a correction of half of the difference must be applied to the flange with the smaller diameter. A feeler gauge set to half the difference is used to make the correction.

■ **14.35   If a ship goes aground, what effect is it likely to have on crankshaft alignment? What precautions must be taken after the vessel is afloat prior to operating main engines?**

If a ship goes aground it is always possible that the bottom of the ship will be damaged; also the hull may be strained if part of the ship is in clear water and tidal changes occur before the ship is re-floated. In such cases it is quite likely that the crankshaft alignment will have suffered, and it is therefore necessary to make a check on alignment before the engine is operated for any length of time. Provided weather conditions permit, some check should be made into the bearing operation within the engine to see that no bearings are overheating. This check should naturally be carried out as soon as practicable after the engine is operating.

No particular set of rules can be drawn up but prudence must be exercised until it is established that the engine is safe to operate for lengthy periods; at the same time it must be remembered that it is useless to save the engine if the safety of the ship is at stake.

■ **14.36   How is the alignment of a cylinder relative to the crosshead guides and crankshaft established? Which parts of the running gear must be removed to carry out this check?**

The cylinder cover must first be removed, followed by the piston, leaving the crosshead clear. The crosshead is hung in the guides. The crosshead bearing keeps are also removed and the connecting-rod is swung clear of the crosshead, and the crank is set to the position of half stroke, approximately.

Piano wire is passed through the cylinder bore, down through the hole in the crosshead which takes the piston-rod fastening, then through to the bottom of the crankcase. The top end of the wire is fastened to a bar supported on wood blocks clear of the top of the cylinder bore. The lower end of the wire is attached to a stretching screw. The stretching screw is in turn fastened to a bar passed under the drain holes between the crank pits. The wire is centred at the top relative to an unworn part of the bore. The wire is then centred in the lower part of the cylinder liner by adjustments to the position of the bar passed under the drain holes between the crank pits. After the adjustments are made the wire is tensioned further and its position rechecked.

The crosshead is raised to the upper position by chain blocks and then lightly jacked or wedged over on to the ahead guide faces. The position of the wire

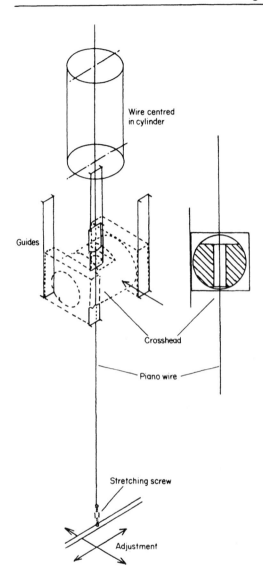

Fig. 14.10 Sketch showing a method of checking cylinder and crosshead guide alignment with a wire centred in the cylinder liner. When alignment is correct the wire is central in the piston rod stud hole in the crosshead.

relative to the sides of the hole is then measured in the fore and aft and thwartships positions. The crosshead is then placed at mid and bottom stroke and the position of the wire checked in a similar manner. The position of the wire between the crankwebs in the fore and aft plane is also measured and recorded. (Fig. 14.10)

Note   The instructions given in the engine instruction book must always be followed when checking alignment.

■   **14.37   Why are the clearances between the outsides of the crankwebs and the sides of the adjacent main bearings comparatively large? Are they the same throughout the engine?**

The relatively large clearances between the sides of main bearings and crankwebs must in the first place be sufficient to accommodate the float in the crankshaft which comes about when going from ahead to astern, and also to accommodate the change in the length of the crankshaft relative to the length of the bedplate during engine operation.

Further, during manufacture of fully built and semi-built crankshafts, difficulties arise during the assembly and shrinkage of the various parts. In effect this can cause differences to arise in the axial centre-line distances between adjacent journals. The large clearances avoid difficulties that might arise from these causes.

■   **14.38   When an engine is started the temperature of the crankshaft and crankcase will be approximately the same. After the engine has been operating for some time what will be the relationship between the temperatures of the crankshaft and the crankcase? Would you expect differences in the temperature of each and what effect will any differences have?**

Generally the temperature of the material forming the crankcase will be considerably lower than the temperature of the crankshaft when the running temperatures have stabilized. The temperature of the crankshaft will approximate to the 'splash-off' temperature of the bearing lubricating oil.

The lower half of the crankcase is rigidly fastened to the tank top. The effect of the temperature differences between the crankshaft and crankcase, together with the fact that the lower half of the crankcase is rigidly fastened down, is to cause the crankshaft to be longer relative to the crankcase during periods of engine operation.

Put in another way, the increase in length of the crankshaft is greater than the increase in length of the crankcase during the period in which the engine is coming up to running temperatures.

■   **14.39   In examining the overall width (axial length) of a bottom-end bearing it is noticed that it is considerably less than the distance between the crankwebs. Why is it made in this way?**

The axial length of bottom-end bearings is made considerably less than the length of the crankpin to accommodate expansion of the crankshaft in coming up to working temperature, and movement of the crankshaft in going from ahead to astern. In medium-speed engines with solid forged crankshafts the clearances, while relatively large, are considerably smaller than in slow-speed engines with fully or semi-built crankshafts.

This clearance, together with the similar journal clearances, permits considerable movement of the crankshaft in the fore and aft direction without the sides of the bottom-end bearings contacting the crankshaft.

If the clearance were insufficient and the crankshaft moved forward, because of a worn thrust bearing for example, the side pressure on the bottom-end bearings would change the loading pattern on crankpin and crosshead bearings.

15

HEAT EXCHANGERS, COOLING SYSTEMS, LUBRICATING SYSTEMS

■ **15.1 What is a heat exchanger?**

A heat exchanger is a piece of equipment in which two fluids are separately circulated in separate adjacent spaces so that some of the heat in the fluid at the higher temperature is transferred into the fluid at the lower temperature. The fluid having the higher temperature is therefore cooled and the fluid having the lower temperature is heated.

Note The term heat exchanger covers fired and unfired equipment. It also covers heat exchangers in which two fluids or gases are circulated, and those in which evaporation or condensation takes place. In diesel machinery installations most of the heat exchangers are associated with circulation of two liquids, though ancillary steam plant involves heat exchangers in which evaporation and condensation of the steam takes place. Similarly refrigeration plant involves heat exchangers in which condensation and evaporation of the refrigerant occurs.

■ **15.2 Name the types of heat exchanger used in marine diesel machinery installations.**

Heat exchangers used in marine diesel machinery installations are either of the multitubular type or plate types. Generally the multitubular type heat exchanger has been favoured in the past but the plate type is now finding increasing usage particularly in high-powered installations. There are various types of multitubular heat exchanger, classified by the flow arrangements, tube arrangements, and the like. Plate type heat exchangers consist of a number of thin metal sheets (often made of titanium) which are sandwiched between cast header plates. By arranging the cast headers the flow of the liquids is such that the liquid to be cooled is passed between the one pair of sheets and the coolant between an adjacent pair of plates. Figure 15.1 shows various types of heat exchangers.

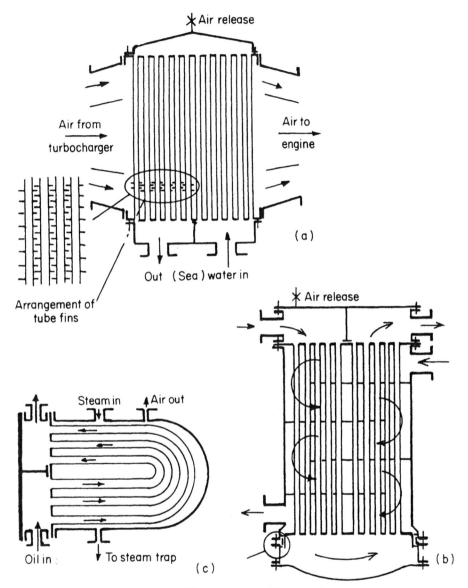

Fig. 15.1 Sketches showing different types of heat exchangers.

A multitubular heat exchanger with fins surrounding the tubes is shown in Fig. 15.1 (a), this type of heat exchanger is used in conjunction with a turbocharger to cool the charge air before it passes through to the engine cylinders. The coolant flow makes a double pass, the air flow makes a single pass.

Figure 15.1 (b) shows another type of multitubular heat exchanger often used for removing heat from main engine cooling water and lubricating oil. In this case the coolant makes a double pass and the jacket water makes a multiple pass.

Figure 15.1 (c) shows a U-tube heat exchanger commonly used for heating fuel oil.

■ **15.3   Where and for what purposes are heat exchangers used in association with main propulsion engines? Describe the type of heat exchanger used for each purpose and the coolant used.**

Jacket cooling water system. A heat exchanger of the multitubular type, or plate type, is used to remove heat from the jacket cooling water. The coolant to which the heat from the jacket cooling water is transferred is sea-water, which is circulated through the heat exchanger by the sea-water circulating pump.

Piston cooling water system. A heat exchanger is used to remove heat from the piston cooling water; it may be of the multitubular or plate type and will be circulated with sea water as the coolant. (When pistons are oil cooled, the oil used is lubricating oil.)

Lubricating oil system. The lubricating oil is cooled in a heat exchanger of the multitubular type or plate type. The coolant is sea-water. In engines with oil-cooled pistons it is usual for the lubricating oil heat exchanger to serve both the lubricating oil and the piston cooling oil which mix together in the lubricating oil drain tank.

A heating coil is sometimes fitted in L.O. drain tanks.

Scavenge air system. A heat exchanger is used to cool the scavenge air after it leaves the turbo-blower and before it enters the cylinders. Scavenge air heat exchangers are shaped to fit the air ducting which carries the scavenge air. Headers are fitted on both sides of the cooler, gilled or finned tubes connect the two headers, and sea-water is circulated through the inside of the tubes. The scavenge air passes over the outsides of the tubes and between the fins. In some engine systems heat exchangers are used to heat the cooling water and the lubricating oil up to running temperature prior to starting the engine, or to hold the cooling water and lubricating oil at running temperatures during stand-by periods. Steam is usually used as the heating medium but in some cases the cooling water from diesel auxiliary sets may be used.

Reduction gearing oil cooler. The heat generated in the reduction gearing and bearings in geared diesel installations must be removed from the lubricant to hold it at its correct temperature. The reduction gearing lubricating oil system is separate from the engine lubricating oil system and therefore requires its own heat exchanger for cooling purposes. This is often of the multitubular type, but in more modern installations it may be of the plate type.

Fuel system. A heat exchanger is used within the main engine fuel system to heat the fuel prior to entry into the main engine fuel injection pumps. The purpose is to heat the fuel and lower its viscosity so that it will atomize correctly when injected into the combustion space. Fuel heaters normally use steam as the heating medium, but in some smaller ships electrical heaters are used.

■ **15.4 Name and give the purpose for which heat exchangers are used in the engine room of a motor ship excluding those used for the main engine. Give the heating or cooling media.**

Fuel service. Heat exchangers are used for heating the fuel before it is treated in the centrifugal purifiers prior to use in the main engine. These heat exchangers are usually termed fuel purifier heaters, and use steam as the heating medium. In small ships electrical heaters may be used. Heat exchangers are also used within the fuel system for boiler firing; boiler fuel-oil heaters are usually steam heated. The heating coils in fuel storage tanks, settling tanks and the like are in effect a form of heat exchanger. Steam is used in fuel tank heating coils for the heating medium. Fuel purifier sludge tanks are also fitted with a heating coil to make sludges more fluid for pumping and removal purposes.

Lubricating oil services. A heat exchanger or heater is used for heating lubricating oil prior to its being treated in lubricating oil centrifugal purifiers. This is usually termed the L.O. purifier heater. Lubricating oil tanks used for batch treatment of the main engine lubricating oil are fitted with heating coils. L.O. purifier heaters and heating coils in separator tanks use steam as the heating medium.

Steam and feed water systems. A heat exchanger referred to as the condenser is used to condense exhaust steam by removing its latent heat so that it is converted to water. The cooling medium is sea-water which is circulated through the condenser to remove the latent heat from the exhaust steam. A feed heater is used to heat the condensed water when it is pumped back into the boiler as feed water. Feed heaters take exhaust steam as the heating medium. Within the boiler firing system steam heaters are used to preheat the combustion air before entering the burner front or register. In other cases combustion gases from the boiler may be used in a multitubular air heater to heat the combustion air.

Fresh water production. Heat exchangers in the form of multitubular or coiled heaters are used to heat sea-water in evaporators or fresh water generators. The heating media may be either cooling water from the main engine fresh water cooling system, or steam. The vapour produced in the evaporator of a fresh water generator is condensed back to fresh water in a heat exchanger called a condenser or distiller. Condensers and distillers use sea-water as the cooling medium.

Cargo tank cleaning. In oil tankers heat exchangers are used to heat the sea-water used in cargo tank cleaning sprayer heads. Steam is used as the heating medium. Another heat exchanger is used to preheat the sea-water before it goes into the main heater. This heat exchanger is referred to as a drain cooler. Apart from pre-heating the sea-water it also cools the condensed steam and prevents it 'flashing' back to vapour after it passes out of the drain cooler into low-pressure drain lines.

Heat exchangers fitted on auxiliary diesel engines (see Question 15.3) are similar to those fitted on main engines, but are naturally smaller and mounted on the auxiliary engine.

■ **15.5 What is a heat balance sheet or diagram?**

Heat balances are drawn up to investigate how the heat in the fuel burned is used around the various parts of the machinery installations, and to evaluate the capacity requirements of the various items of machinery, heat exchangers, and ancillary equipment, which make up the whole installation. Use of the heat balance diagram and statements enables the designer to optimize the overall performance of the installation, and the operating engineer to make comparisons of the machinery operation with an acceptable norm. The use of heat balance diagrams and calculations is a prerequisite in the early stages of the design of a steam turbine installation; the diagrams are also used, but to a lesser extent, in the design of diesel machinery installations.

The heat balance diagram in Fig. 15.2 shows a thermal efficiency of 53%, 15% of the heat in the fuel is shown going to heat exchangers. Some of this heat may be recovered in a flash type evaporator when engine cooling water is used to evaporate sea water.

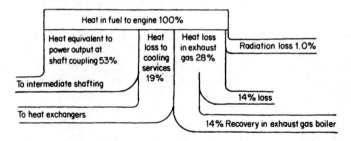

Fig. 15.2 Heat balance diagram for modern two-stroke cycle, slow-speed, long-stroke engines used for ship propulstion.

In some modern machinery installations boiler feed water is used as the coolant for the charge air cooler; this heat exchanger becomes a water preheater for the boiler feed water.

Approximately half of the heat in the exhaust gases may be recovered in the exhaust gas boiler.

The heat balance diagram shown is sometimes referred to as a Sankey diagram.

■ **15.6 How are the flow paths defined in a multitubular heat exchanger?**

The flow paths of the fluids are fixed by the division plates in the heat exchanger heads and the internal baffles or tube support sheets within the body of the heat exchanger. The division plates in the heads or boxes fix the position of the inlet and outlet branches for the fluid passing through the tubes. The internal baffles fix the position of the inlet and outlet branches for the fluid passing through the body of the heater on the outside of the tubes.

If the heat exchanger heads do not contain any division plates the fluid passing through the tubes enters at one end of the heat exchanger and leaves at the other. This arrangement is referred to as a single-pass heat exchanger. If the heat exchanger is a double-pass type a division plate is fitted in one head. The inlet and outlet connections for the fluid passing through the tubes are fitted on this head. The division plate prevents the fluid bypassing the tubes and causes it to pass through half the tubes in the heat exchanger which is referred to as the inlet bank. After the fluid passes through the inlet bank it enters the other head which is just a 'bobbin' piece and a cover. The direction of the fluid flow is reversed in this head or box and it passes back through the outlet bank of tubes and leaves at the outlet branch. The fluid has passed through the tubes in two separate paths from which it gets the name two-pass, or double-pass type.

By dividing the total number of tubes into three banks, fitting a division plate in each head, and fitting the inlet branch on one head and the outlet branch on the other, a three- or triple-pass arrangement can be formed.

■ 15.7 When a cold fluid is passing through the tubes of a heat exchanger and a warm fluid is circulating outside the tubes, the body expands more than the tubes. How are these differences of expansion catered for?

In some types of heat exchanger the body is made of mild steel plates, and in order to cater for the differential expansion between the body and the tubes a bellows ring is welded circumferentially round the body of the heat exchanger (as shown in Fig. 15.3(a)). Expansion and contraction of the tubes are then catered for by the bellows ring which deforms slightly to accomodate the changing length of the tubes. This type of expansion arrangement can be used for any number of fluid passes through the tubes. The tube plates are bolted directly on to the flanges of the body and the tubes can be roller-expanded at both ends.

In heat exchangers with cast bodies, one tube plate is fastened to the flange on one end of the body. The other tube plate is made to slide within the end of the body. Sealing is effected by fitting O-rings in circumferentially cut grooves. The O-rings contact the circular bored end in the body and the tube plate to make a seal. (Fig. 15.3(a))

When there is a difference in expansion between the tubes and the body the differences are accommodated by the tube plate which slides relative to the body. With this arrangement the tubes can be arranged only for single- or double-pass flow. This arrangement is more commonly used than any other for engine cooling water and lubricating oil coolers. It also has the advantage of allowing very easy removal of the tube stack from the exchanger body for chemical cleaning or repair. This is further facilitated if the cooler is mounted vertically and the sliding tube plate is at the lower end.

A third arrangement is to fasten the tube plates to the body of the exchanger and roller-expand and bell-mouth the inlet ends of the tubes. The outlet end of the tubes is fitted with a cotton cord packing tightened by a threaded gland

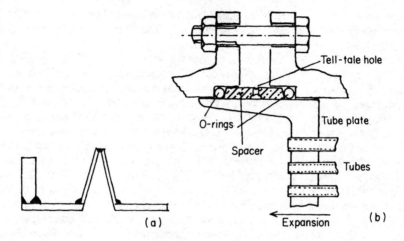

Fig. 15.3 (a) Arrangement of sliding tube plate used in many multitubular-type heat exchangers. The spacer is held in place between the exchanger shell and the water box and compresses the rings. Tell-tale holes are drilled through the spacer in various locations around the circumference to indicate leakage of either primary or secondary liquids. (b) Arrangement of a bellows piece in the welded shell of a heat exchanger.

ferrule screwed into the threaded end of the stuffing boxes. In other cases sealing of the tube is made by caulking fibre and white-meal foil rings into the stuffing box. Differences of expansion between the tubes and exchanger body is covered by the tube ends sliding through the packing in the tube plate gland. This is a common arrangement for exhaust steam condensers.

■ **15.8   Describe what happens to the material of a tube end when it is rolled or expanded into a tube plate.**

When a tube end is to be expanded into a tube plate it must first be annealed to prevent splitting which might arise due to work hardening of the material as it is expanded. At the commencement of rolling, the end of the tube is stretched circumferentially, so hoop tension is present until the outside of the tube contacts the hole in the tube plate. After contact is made between the tube end and the tube plate, continuation of the rolling or expanding process further stretches the tube and, due to the restricting effect of the hole in the tube plate, causes the hoop stress to reverse and become compressive in nature. At this stage an axial compressive stress is also set up because there will be some restriction on the free movement of the tube surface contacting the tube plate. This axial compressive stress is increased if serrations or circumferential grooves are machined in the tube plate holes.

■ **15.9 Why must tubes be expanded very carefully when being fitted into the tube plate of a heat exchanger?**

When tube ends are being expanded into a tube plate with a roller expander, great care must be exercised to prevent excessive thinning and hardening which comes about from over-rolling or expanding. When the expander roller cage is fitted into the tube end, the tapered mandrel must not be hammered into the rollers. The mandrel should preferably by pushed into the rollers by hand loading or at most only with extremely light tapping blows with a very light hammer. In some cases electrically operated expanders may be available, and it is equally important to exercise care in their use so as to prevent over-expanding. Excessive rolling materially affects the expected life of a set of exchanger tubes as the thinning of the tube-end reduces the effective life of the tube.

■ **15.10 What is a contra- or counter-flow heat exchanger?**

The two terms are synonymous. A contra-flow, or counter-flow heat exchanger is one where the two fluids flow in opposite directions. For example in a contra-flow heat exchanger the inlet branch for the fluid passing through the tubes will be adjacent to the outlet branch on the body for the fluid circulating outside the tubes. Similarly the outlet branch for the fluid passing through the tubes will be adjacent to the inlet branch for the other fluid.

Generally contra-flow type heat exchangers have a greater heat-transfer capability based on unit tube area than other types.

■ **15.11 When a tubular-type heat exchanger is mounted on a foundation or bulkhead, how is the heat exchanger fastened and how is contraction and expansion allowed for?**

Generally the mounting feet or bossings are cast integral with or welded on to the heat exchanger body shell. One pair of the feet is made with circular holes and the other pair has holes elongated in the axial direction of the body shell. The feet with the circular holes are bolted up hard to the foundation. The other feet are fitted with shouldered bolts which give a small clearance under the bolt head. This allows the shell of the heat exchanger to expand or contract freely in an axial direction without inducing stresses as would occur if expansion could not take place freely.

The pipework connected to the inlet and outlet branches or nozzles on the heat exchangers is also designed and fastened in such a manner that the forces and moments set up in the branches from pipe expansion and contraction are kept within safe limits.

Some heat exchangers have the feet rigidly bolted at each end, and in such cases one mounting foundation is designed rigid while the other is able to flex in an axial direction relative to the heat exchanger body. The only load on the shell is that required to flex the mounting. With this type of fastening the same care must be given to the connecting pipes to prevent high forces and moments on the inlet and outlet branches.

■ **15.12   Name the usual causes of a lowering in performance of heat exchangers.**

1 Air locking of fluid spaces.
2 Fouling of heat transfer surfaces.
3 In two- or multi-pass heat exchangers, fluid by-passing the tubes because of division plate joint failure or holed division plates.

Air locking in the fluid spaces of a heat exchanger can occur in many heat exchanger and pipe work configurations, particularly if the heat exchanger and adjacent piping form inverted letter U shapes.

The rate at which air builds up in the sea-water circuits is influenced by the cleanliness of the pump glands or seals; the build-up of marine growth on water box suction grids which if large in amount, reduces the pump suction pressure. The position of the water box on the ship's shell also has an influence; for example, in light or ballast condition with the ship pitching, air is carried in with the water under the bow and is picked up in the suction water boxes on the ship's shell.

Fouling occurs mainly on the sea-water side of oil and fresh-water coolers and on the oil side of oil coolers.

■ **15.13   What does the phrase 'terminal temperature differences' mean?**

'Terminal temperature differences' refers to the inlet and outlet temperatures of each of the fluids passing through a heat exchanger. By noting the temperature rise of the coolant and the temperature drop of the fluid being cooled the terminal temperature differences can be obtained. A knowledge of these figures can be useful in evaluating the performance of a heat exchanger when it is in service.

Note   In design work, logarithmic mean overall temperature differences are used.

■ **15.14   How does air locking occur in the heaters and coolers in a marine diesel machinery installation?**

Any place or point at which ingress of air into a system can occur can be the cause of an air lock. A heat exchanger and connecting pipework that has the configuration of an inverted letter U is more prone to air locking than systems without this feature.

In sea-water circuits air ingress may come about at pump glands, and the chances of air ingress are increased if the grids on the water boxes or sea-water suction valves become badly fouled. In such cases the suction pressure in the pump suction line is reduced and may even become sub-atmospheric. If the gland or glands on the suction side of the pump shaft leak, air will be drawn in.

Note   Some pump glands are fitted with a lantern ring which is connected through a small pipe to the pump discharge. The water in the lantern ring is then under pressure and acts very effectively as a seal. If the pipe becomes damaged or blocked the sealing action is lost and air leakage into the sea-water system can occur very easily.

When a ship is light, in ballast or pitching in heavy weather, air is drawn under the bow of the ship and passes along the bottom just under the bilge keels and leaves at the aft end of the bilge keel. Sometimes this air finds its way into the water boxes or sea suction chests, passes through the pumps and causes air locks on the sea-water side of coolers. It is usual to fit an air leak-off pipe on the top of shipside water boxes or chests. The pipe leads up to a goose-neck at the top of the engine room. After a vessel comes out of dry-dock the valve to this pipe on the top of the chest is usually closed by repairers. Before machinery start-up it should be checked and left open. If air locking in coolers occurs the valve should again be checked to ascertain that it is open.

Joints on sea-water strainer or filter boxes are also a common point of air ingress. If the sea chests are fitted with compressed air de-icing and weed-clearing connections, leakage of air from the compressed air line through the valve into the sea chest is another cause of air locking problems.

Air leakage into lubricating oil systems on engines may occur at glands on pumps, and on suction filter, or strainer cover, joints and the like. Another cause of air leakage into lubricating systems may be due to the sucking in of air at the tail end of the suction pipe in the drain tank. This condition can easily come about if foreign matter such as rag, old jointing or similar material is left in the crankcase. This material finds its way into the crankcase strainers at the rundown point into the sump tank and causes an accumulation of oil to occur in the bottom of the crankcase, this in turn causes a low level in the drain tank and allows the pump to pull in air. A similar condition occurs if the drain holes within the tank structure become blocked. The flow of oil to the pump suction becomes reduced and air is drawn into the pump.

Cooling systems are of two different types; there is one type of system which is enclosed and where the cooling water is under a pressure head from an expansion tank, and the other type is an open system where all the cooling water is returned to a collecting tank from which the cooling water pump suction line is connected. In the former type, where the system is enclosed and under pressure, problems with air locking do not normally arise; however, in ships with a limited height in the machinery space the height of the expansion tank may be less than the static head in the cooling water at the engine cylinder tops. In such cases a circulation of water may come about from the de-aerating cocks on the uppermost part of the engine cooling system. When the cooling water circulates in this manner it goes up to the gooseneck at the top of the de-aerating pipe and goes into the expansion tank via the 'tundish' or 'save all' funnel. The water in passing into the expansion tank pulls in air with it. If the expansion tank is not fitted with air separation plates or if they become corroded, air finds its way into the cooling system and can cause air locking in pipes or coolers.

In the second type of cooling-water system the sources of air leakage are mainly from pump glands and stuffing boxes. If air passes into the cooling water at cooling-water-return sight glasses this sometimes causes trouble.

Normally the coolers in the fresh-water and lubricating-oil system are fitted with air leak-off cocks which should be opened at regular intervals and also during start-up of lubrication and cooling water services.

■ **15.15   Which parts of a diesel machinery installation use steam as the heating medium? Which types of heater are used for such service?**

Steam is used as the heating medium in fuel systems and lubrication systems. The type of heater used is multitubular; the tubes are bent into a U shape and fitted with expanded ends into a single tube plate. The liquid head or box is fitted with a division plate in its middle so that the oil flow takes a two-pass flow. The oil enters the inlet branch of the liquid head, passes through one leg of the U, round the bend and leaves by the other leg. The division plate separates the inlet and outlet legs of the U-shaped heater tubes. The inlet branch is always on the lower side.

This type of heater is used to prepare the fuel (by raising its temperature which in turn lowers its viscosity and specific gravity) for treatment in the fuel purifiers, and also in other heaters to make the viscosity of the fuel correct for use in the engines. The fuel used for boiler firing is also heated in this type of heater, which forms part of the boiler firing units.

Steam-heated lubricating oil heaters of this type are also used to heat the engine lubricating oil prior to treatment in centrifugal oil purifiers.

Bunker tanks, settling tanks and daily service fuel tanks and in some cases main engine lubricating oil drain tanks are fitted with heating coils which use steam as the heating media.

Ships built for service in very cold climates often have steam pipes led around the bottom of the engine room. These pipes are used to maintain a satisfactory temperature in the engine during service and to prevent freezing during periods when the main engine is shut down.

■ **15.16   What is the cause of air locks in the steam side of steam-heated heaters and heating coils?**

The boiler feed system used in most diesel-propelled ships is of the open hotwell type. In this type of system all the condensate returns to the hotwell tank to which the suction side of the feed pump is connected. If problems arise with control of the feed pumps they usually pull in air from the atmosphere which mixes with the feed water and gets pumped into the boiler. Some air from the atmosphere also mixes with the feed water even when the feed pump controls are is good condition. In the boiler the air separates from the boiler water and passes into the steam lines where it can find its way into the steam space of heaters and heating coils. The air locked in the steam system gives up its sensible heat and as it cannot condense and drain out like steam it remains locked in the system and causes a fall away of the heater or heating-coil performance. The air leak-off cocks on heaters and bleed-off points on tank heating systems should be regularly used to bleed off air and prevent performance reduction.

Note   If air locking occurs frequently and rapidly the boiler feed system should be checked; the indications are that a fault exists which will most likely be in the feed-pump controls. If the fault is not corrected serious pitting and corrosion may occur in the boilers, condensate lines and other parts of the feed system.

■ **15.17 The term 'fouling factor' is used in heat exchanger design. What is the meaning of this term?**

When the capacity of a heat exchanger is calculated, the calculations are based on the various conditions applicable. When the heat exchangers become fouled or dirty their capacity becomes reduced and in such conditions it would be necessary to reduce power on the engine to suit the reduced capacity of the heater. By introducing a fouling factor the effective capacity of the heat exchanger is increased (in effect tube surface area is increased) and some fouling can take place before its effect becomes critical. By increasing the fouling factor a greater degree of fouling can be accommodated before it becomes necessary to reduce engine power or clean the heat exchangers.

■ **15.18 How does fouling affect performance of a heat exchanger?**

The tube of a heat exchanger transmits heat from the hot fluid across the thickness of the tube wall and into the cold fluid. When the surfaces of the tubes become dirty their ability to transmit heat is impaired so that the rate of heat transmission is reduced. When this state of affairs is reached and the reserve capacity given by the fouling factor is exceeded, difficulties arise in removing all the heat put into the jacket cooling water from the cylinder liners and cylinder heads. Similarly the heat put into the piston cooling water on its passage through the pistons is not wholly removed. As the heat put into the jacket and piston cooling water during its passage through the engine is only partly removed in the coolers, it follows that the outlet temperature of the cooling water will begin to rise above what is required. After the cooling water leaves the coolers it passes through the connecting pipe lines to the engine and enters the cylinder jacket at a slightly higher temperature than that required. The engine then slowly begins to overheat and the power must be reduced by shutting in the fuel lever (reducing the fuel) to bring about a balance again.

Note The effect of dirty lubricating-oil and cooling-water coolers is readily seen when a vessel is running from a temperate zone to a tropical area. When the sea-water is cold the heat put into the main engine cooling water and lubricating oil is easily removed in the coolers by the cold sea-water circulating through them, and to maintain the correct temperature the by-pass control valves will be·partly open. As the sea-water temperature increases the by-pass valves must be gradually closed to maintain the correct temperatures. When the by-pass valves are closed the cooler is working at its maximum heat removal capacity; as the sea-water temperature rises still further the engine cooling-water outlet temperature from the cooler and into the main engine rises. To maintain the correct cooling-water temperatures the heat exchangers must be cleaned or the power on the engine must be reduced until the ship enters a location where the sea-water is lower in temperature.

■   **15.19   What form does the fouling take in the sea-water circuits, fresh-water circuits, lubricating-oil circuits, and the fuel-oil circuits in heat exchangers?**

In the sea-water circuits fouling takes place within the tubes in the form of build-up of scale and sedimentation from mud and earthy matter picked up in coastal and river water. The scale found is small in amount and approximates to eggshell thickness; further incrustation, however, takes place from the sediment build-up which hardens in the course of time from its exposure to the heat passed from the liquid on the outside of the tubes.

When distilled water is used in the engine cooling circuits the amount of fouling that takes place is negligible. If contamination from sea-water occurs or raw hard fresh water is used in the cooling systems small amounts of incrustation may occur.

In the lubricating oil sides of coolers, waxy and greasy deposits settle out from the lubricating oil and sometimes from its additives.

In all cases of fouling there is a considerable reduction in the rate of heat transfer through the walls of the tubes or cooler plates; this comes about from the insulating properties of the materials causing the fouling.

■   **15.20   How are heat exchangers cleaned?**

The sea-water circulated through the various coolers is usually arranged to pass through the bore of the cooler tubes. To clean the insides of the tubes the water box covers are removed and the tubes are cleaned by mechanical means; if the deposits are soft mud and earthy matter, they can be cleaned with compressed air. The mechanical means of cleaning tubes is usually with wire brushes made to fit into the bore of the tube. The ends of the brushes are threaded and screwed into rods or flexible wire handles which are used to push the brush through the tubes. Some ships are provided with rotary cleaning gear.

The fresh water and lubricating oil sides of coolers are cleaned with solvents which are sold under various trade names. The solvents are usually weak acids containing an inhibitor to prevent metal corrosion. In lubricating oil coolers alkaline-base cleaning media are sometimes used to remove waxy materials and sludges. When solvents are used for cleaning, the tubes are cleaned *in situ* within the casing of the cooler.

If there is any chance of solid or semi-solid matter settling in the lower part of the heat exchanger casing there is risk that the flow of fluid through the casing and round the tube support plates and division plates will be impaired. In such cases the tube nest or stack should be removed and the various spaces cleared with compressed air.

When certain solvents and weak acids are used for cleaning, the space being cleaned must be vented so that any gases formed from the cleaning action are allowed to escape. It should be noted that the gases vented off are often inflammable so care must be exercised to prevent explosion. Some chemicals and solvents used for cleaning require neutralising agents after the cleaning medium has been removed from the casing.

After cleaning has been completed it is a good policy to put a hydraulic test on the casing. If the stack has not been removed the test will indicate whether deterioration of the sealing rings has taken place. It is better to find a leaking tube nest immediately after cleaning than a short time before stand-by. If the tube nest has been removed the same provisions apply.

■ **15.21 When coolers leak what indications will be given that leakage is taking place?**

When machinery is in operation, there are difficulties in establishing that a cooler is leaking. This comes about from the pressures in the various systems. Usually the working pressures in the jacket cooling system, the piston cooling system, and the engine lubrication system are higher than on the sea-water side of the coolers. If leakage takes place, flow from any point of leakage is into the sea-water side of the cooler, from where it is passed overboard. The only indications of cooler leakage given in the engine room when machinery is operating are losses of cooling water or lubricating oil. If the leakage is small in amount an indication of leakage by system loss is not easily noted or found.

If coolers and pumps are properly shut down in port, difficulties still arise with knowing if a cooler is leaking. There is only one positive way to ascertain whether a cooler is leaking or tight and that is to carry out a test on it. A good time to do this test is after cleaning the sea-water side of the cooler.

When coolers are shut down for cleaning, the end covers to the sea-water boxes are removed. After the tubes have been cleaned and dried out it is a simple matter to start the fresh-water cooling pump or the lubricating-oil pump and circulate through the cooler spaces on the outside of the tubes. Leakage within the tubes or at the junction of the tube plate and tube is then readily seen on the dry surfaces. When cleaning of the sea-water sides of the coolers is scheduled, the other work in the engine room should be programmed so that it does not interfere with the testing of the coolers.

Note Lubricating-oil cooler leakage is sometimes noticed by the presence of oil in the overboard discharge, but usually when oil is visible the leakage is apparent from the losses noted when sump soundings are taken.

Some coolers need to have clamp rings fitted in place to secure the sealing rings and prevent them being blown out when the shell of the heat exchanger is put under pressure.

■ **15.22 If you found leakage occurring in a multitubular heat exchanger how would you deal with it?**

If the leakage is at the junction between the tube-end and the tube plate, the leakage can be stopped by expanding the tube-end with a roller expander. Should the leakage be within the tube wall the ends of the tube must be plugged.

Leakage from tube-ends, which are made tight by packing fitted in small stuffing boxes round the tube-ends (as is often found in exhaust steam condensers and distillers), can be made tight by tightening up the screwed ferrule on to the packing. If the thread on the outside of the screwed ferrule is covered by scale, the ferrule should be carefully removed and the thread cleaned off with a thread chaser before tightening is attempted.

■ **15.23   Describe the types of plugs used to plug the ends of leaking heat exchanger tubes.**

Plugs used as a purely temporary measure may be made of soft wood or hardwood; often, in the absence of properly-sized plugs, pieces of a broom handle are turned down in the ship's lathe to the size required. In other cases metal conical tapered plugs are used. The large end of the pluds is usually drilled with a blind hole and tapped so that it can be easily withdrawn when it is desired to remove the plug. The material used to manufacture the plugs is usually a non-ferrous metal or alloy, the same as the tube material. It is common practice for the manufacturer of the heat exchanger to supply some plugs for this purpose. They are usually located in the heat exchanger tool box which also includes the correct-size cleaning brushes, the tube roller expanders, plug-drawing gear and the like.

■ **15.24   What type of equipment can be used on water coolers, distillers and condensers to find small leaks, particularly if time precludes properly drying out the sea-water cooling spaces? How is the equipment used?**

Small leaks can be extremely difficult to find, particularly if time does not allow for the proper drying out of the sea-water cooling spaces. The damp tube plates unfortunately mask out small amounts of leakage. In such cases a small amount of fluorescent sodium crystals can be dissolved in the water within the space surrounding the tubes. The tube plates are then viewed under a source of ultra-violet light. Where leakage is present it is sharply delineated as a fluorescent area at the point of leakage. The equipment required is an ultra-violet lamp and a supply of fluorescent sodium crystals or some 'Fluorescene' solution.

Ultra-sonic testing equipment can also be used to detect the location of a leakage point in a cooler. This equipment consists of a detector or ultra-sonic microphone which can be used with a contact probe or a sound concentrator. The pick-up from the microphone is amplified and gives a visual warning of leakage by read-out on a meter, or by audible warning through a pair of headphones. For this type of equipment to be used it is necessary to drain the space that surrounds the tubes and then pressurize the space with compressed air to some suitable pressure. After this has been done the detector is used with the sound concentrator to find the general location of the leakage. The probe can then be used to find the individual tube or tubes leaking.

Note   Fluorescence takes place in fluorescent materials through their ability to absorb light of one colour or wavelength and emit light of another colour or wavelength. After using fluorescent solutions the cooling space must be washed out prior to circulating the cooling water. If the heat exchanger is used for the production of potable water, extreme care must be taken in washing out and removing all traces of the chemicals used.

Ultra-sonic equipment of the type mentioned is also used to find leaking valves.

■ **15.25 How can leaks be found in lubricating-oil coolers?**

As there are usually mud and sandy-type sludges left by the sea-water in the tubes of lubricating-oil coolers, leakage of oil discolours the sludges. The presence of lubricating-oil globules may also be noticed together with the discoloured sludges within the tubes that are leaking. The discolouration is often seen as darker patchy areas in the sludge after the water has dried out. Care must be taken in the examination, however. If there is a carry-over of oil globules with the flowing sea-water, staining of sludge and oil globules may be found in the second bank of tubes and lead one to think more tubes are leaking than is actually the case. If doubt exists after the inside of the tubes have been cleaned and sponged out, the cooler can be pressure-tested with the lubricating oil pump. Ultrasonic testing equipment can also be used as mentioned in Question 15.24 (see Note at end of Question 15.21 regarding clamp rings).

■ **15.26 Describe any equipment used for checking the tightness of lubricating-oil coolers while they are in operation.**

On the cooling-water outlet lines from the lubricating-oil coolers, at suitably positioned take-off points, small outlet connections are fitted. These connections are led to a glass-windowed fitting which is similar in appearance to a flow indicator. The upper part of the windowed space forms a reservoir, and small quantities of the cooling water are passed continuously through the indicator. Any lubricating oil leaking from the cooler floats to the top of the indicator and is retained in the reservoir where its presence can be observed. Some indicators are fitted with a lamp and a photo-cell. If oil particles are present in the sampled cooling water an alarm is given when the light path between the lamp and the photo-cell is obstructed.

■ **15.27 How does corrosion occur in heat exchangers?**

Corrosion within heat exchangers may take various forms according to the behaviour of the materials under their environmental conditions. Generally the corrosion is electro-chemical in nature and may occur as general wastage, which takes place over large areas, or as localized wastage which shows as pitting. These forms of attack can occur only where there are dissimilar metals in an electrolyte. The two dissimilar metals will form an anode and a cathode. Wastage takes place from the anode or anodic material. Sometimes this form of corrosion is referred to as wet corrosion or metallic corrosion. The cathode will be lower in the electromotive series than the anode. When there is a large difference in their positions in the table the rate of corrosion will be greater. The metals that are low in the electromotive series are referred to as noble metals.
  Corrosion of cast-iron water boxes, covers and the like is usually in the form of general wastage in which the iron wastes away leaving the graphite or carbon

behind in a soft spongy form. This form of corrosion is referred to as graphitization. When pitting occurs it is localized in nature and often takes place at a location where a small breakdown has occurred in a protective coating or in the natural passive protecting film. The effect is similar to a large cathode in the presence of a small anode and attack proceeds rapidly.

Alloys of copper and zinc forming brass sometimes break down due to the zinc being attacked. When the zinc is attacked a spongy copper structure is left which has little mechanical strength, and leakages soon occur when the spongy copper structure breaks away.

Corrosion may also occur under joints and porous jointing materials in regions where scale has exfoliated, and in similar locations where a very narrow space or crevice is formed and stagnation takes place within the liquid contained in the crevice. This form of attack is referred to as crevice corrosion.

Note The subject of corrosion is a highly specialized field, covering in-depth studies in physics, chemistry, and metallurgy. As marine engineers we must be cognizant of the various forms of corrosion we meet in our work, and have a good knowledge of material selection for difficult environments and of the means whereby the ravages of corrosion can be met and prevented.

The electromotive series of metals is sometimes referred to as the galvanic series. In some tables the cathodic metals are placed at the bottom while in other tables they are placed at the top. It must be remembered that the noble or cathodic metals have a positive potential while the anodic metals have a negative potential.

Electromotive series
Anodic metals: magnesium (Mg), aluminium (Al), zinc (Zn), chromium (Cr), iron (Fe), nickel (Ni).
Cathodic metals: copper (Cu), silver (Ag), platinum (Pt), gold (Au).
Silver, platinum and gold are the noble metals.

Chromium is sometimes given two positions in the table due to its ability to create a highly corrosion-resistant passive film on its surface. Stainless steels have the ability to build up this film naturally, but in certain cases it is built up by washing the stainless steel surfaces with acid (nitric) solutions to accelerate the production of the passive film.

■ **15.28 How are heat exchangers protected from the effects of corrosion?**

There are two ways to protect the sea-water cooling side of heat exchangers: by coatings and by cathodic protection systems. The fresh-water side of the heat exchanger is protected by the addition of chemical additives to the cooling water.

Coatings, varying from bitumen-based paints through to epoxy coatings of various forms, can be used to protect water boxes and water-box covers. In other cases rubber sheet materials may be bonded to the mild-steel or cast-iron surfaces so that sea-water cannot make contact with the steel or cast-iron. The steel or cast-iron is then, in effect, electrically insulated from the sea-water which is in contact with the other metals of the heat exchangers.

Cathodic protection systems using sacrificial anodes are also used within the sea-water spaces of heat exchangers. The metal used for the sacrificial anodes must be such that they are higher in the electromotive or galvanic series of metals than the metals which they have to protect. These metals or alloys are referred to as being active or anodic. In descending order in the galvanic series they are cadmium, commercially pure aluminium, zinc, and magnesium alloys (see note in Question 15.27). In the presence of sea-water and with continued electrical continuity the anodes waste away and so protect the other metals within the heat exchanger. In some cases a protective or passive film is deposited on the metals being protected.

Chemical additives placed in the fresh-water side of the cooler are referred to as inhibitors. Their action is to create protective films of passive material on the metallic surfaces which they protect. They not only protect the heat exchanger but also the other parts of the cooling system in which they circulate.

■ **15.29** How would you know whether anodes are effective or non-effective? What causes anodes to be ineffective and how can an ineffective condition be rectified?

Ineffective anodes are readily seen when the sea-water spaces of a heat exchanger are opened up; obviously if the condition of the anodes is the same as when they were fitted, they are ineffective. This is not uncommon and comes about due to either incorrect selection of anode material, bad design of anode fastening, or badly fitted anodes. Some anode materials build up a passive layer between the metal of the anode and its fastenings. This passive layer acts as an insulator and the anode is not sacrificed, and cannot therefore protect the cathodic material. This condition can be rectified by excluding sea-water from the anode fastening. In many cases anode material such as magnesium alloy or zinc is cast around a mild-steel strip. The mild-steel strip is then drilled and held in place with studs and nuts or set screws.

The contraction of the molten anode material when it freezes round the mild-steel strip precludes any ingress of sea-water between the anode and the steel strip. Electrical continuity is then always maintained.

■ **15.30** How may badly located anodes cause damage to the tubes in water and lubricating-oil coolers and similar types of equipment?

If the anodes are badly located within the flow pattern of the sea-water passing through the water boxes, turbulence and cavitation may be set up.

This allows the release of dissolved air in the sea-water, which leaves the point of flow disturbance as a series of bubbles. The bubbles may then impinge on some area of the tubes and cause leakage at the point of impingement. This form of attack is referred to as bubble impingement. It may also occur due to a piece of sea shell or other debris lodging at the entrance to a tube and causing cavitation (see Question 13.42).

■ **15.31** If coolers are examined, one finds small metallic strips, held in place by round-head screws, fitted into the heater casing flange, tube plate, and water-box flange. What is the purpose of these little strips and what attention do they require?

Their purpose is to provide electrical continuity between the various parts of the heat exchanger and so complete the circuit for the cathodic protection anodes. The strips, usually made of brass, require periodical removal and cleaning. The landing surfaces for the strips on the heat exchanger parts also require cleaning so that good conductivity is given when the strips are replaced. This job should be done at least after the heat exchangers are painted but preferably every time heat exchangers are cleaned.

■ **15.32** Name the types of pump used for cooling and lubricating systems on propulsion and auxiliary diesel engines.

Centrifugal-type pumps are used for the circulation of sea-water and fresh water. Centrifugal-type pumps are also sometimes used in the lubricating oil system, but they are then mostly of the two-stage type. The most common type used for lubrication systems is the rotary positive-displacement type of pump.

■ **15.33** Why are centrifugal pumps used for the sea- and cooling-water services with marine diesel engines? What are their drawbacks?

In the first place they are small and light for the volumes of water they can handle. The speed of the rotating element is such that it is very suitable for either a.c or d.c electric motor drive. Except for the pump bearings and seals or gland packings there is no mechanical contact between the fixed and moving parts of the pump, so wear is limited to the parts mentioned. In consequence the maintenance necessary is small provided that the lubrication requirements of the bearings are properly attended to.

The only drawback to these pumps is the fact that if they are empty they cannot produce a vaccum and so prime themselves. In cases where the pumps are always submerged, as in sea-water systems and many cooling systems, they are always full of water and so do not require priming. In other cases, such as the cooling systems of some engines, where the cooling-water returns to a collecting tank, it is necessary to have a rotary wet-air pump fitted to the centrifugal pump so that a vacuum is formed in the volute casing. This draws liquids into the centrifugal pump rotor and so allows it to start pumping.

In other cases where it is necessary to prime the pump it may be connected to a ring priming system.

■ **15.34** Why are rotary positive-displacement pumps preferred for lubricating-oil services?

These pumps are generally preferred because their characteristics are more suitable than those of centrifugal pumps and, being of positive displacement, they are inherently self-priming. This preference also applies if the pump has to supply lubricating oil for cooling the pistons.

Sometimes two-stage centrifugal pumps are used for lubricating oil services, the second stage of the pump being necessary for the pressure requirements. To facilitate priming, non-return or foot valves must be fitted in the bottom of the suction pipes, together with other arrangements for positive supply at start-up or automatic change-over in case of pump failure. In some cases where centrifugal pumps are used the pump is actually fitted at the bottom of the drain tank and driven by a motor outside the tank. A vertical extension of the motor shaft is connected to the pump rotor.

■ **15.35 What do you understand by the term 'pump characteristics'? What is net positive suction head?**

The characteristics of a pump show in a graphical manner its performance capabilities. After the pump is tested for performance the discharge head H, the pump efficiency η, and the power absorbed in pumping are plotted as ordinates against the volume of fluid pumped which forms the abscissa.

Net Positive Suction Head (N.P.S.H.) is the term used to indicate the ability of a centrifugal pump to deal with fluids which may vaporize in the inlet branch and entrance to the rotor. Although we should be conversant with this characteristic of a centrifugal pump and its significance, the conditions under which centrifugal pumps work in the engine rooms of motor ships are such that vaporization of the fluid being pumped does not commonly occur.

Note N.P.S.H. of pumps and piping systems is of importance when dealing with cargo pumps in oil tankers. It is measured as a head in metres absolute.

■ **15.36 Which types of rotary pumps are used for lubrication and piston-cooling services on main propulsion and auxiliary diesel engines?**

There are many different types of rotary pumps ranging from simple pumps having flexible plastic vanes to multiple-screw rotary pumps whose rotating elements are machined to the closest possible tolerances.

The basic types used in smaller engines for lubricating services are flexible-vane pumps, sliding-vane pumps, gear pumps and screw-and-wheel pumps. In larger engines, where lubricating oil is also used for cooling pistons, multiple-screw pumps having two or three screws or helices are commonly used. Flexible-vane pumps, sliding-vane pumps, gear pumps and screw-and-wheel pumps are often fitted on the engine and connected to the engine crankshaft through some drive train. The pump is then driven by the engine. The single-screw type of pump has a rotor which is threaded. This operates within a stator which is made of oil-resisting rubber or suitable plastic material. The oil flows axially along the thread which is eccentric to the axis of the rotor. This type of pump is independently driven. Multiple-screw type pumps have their own electric motor and operate independently of the main engine. Two- and three-screw pumps have various arrangements of the individual screws. In some cases the oil goes into each end of the screw and is discharged at the centre; in others the flow goes in at the middle part of the screw with discharge at the ends. Some two-screw pumps have pinions fitted on the ends of each screw. The pinion on the driven screw drives the follower screw. With such an arrangement the only mechanical contact is between the pinions, giving the rotating screws a long service life.

■ **15.37   Give details of the flow path, and of the parts associated with it, of the lubricating oil through the lubricating system of a slow-speed, direct-coupled propulsion engine. What changes would you expect to find in the system if the pistons are oil cooled?**

Irrespective of the fluid used to cool the pistons, the lubricating oil within the system progressively drains down into the lubricating-oil drain tank from where it is pumped by one of the lubricating-oil pumps. The lubricating-oil drain tank is sometimes called the double bottom lubricating-oil tank or crankcase drain tank. The tank forms part of the hull structure so that the sides of the tank running fore and aft form intercostals and the thwartship members connect up with the 'floors' in the double bottom structure. The structure is robust because it forms part of the engine foundations. Within the oil drain tank is a suction well (see Note) into which the tail of the suction pipe is fitted. The tail is prevented from moving by welded fins which also keep the flow line of the oil entering the pipe from rotating and forming a vortex.

The oil, after passing into the suction pipe, will flow through valves, a strainer, and then another valve where it enters the lubricating oil pump. From the lubricating-oil pump it will pass through a pump discharge valve of the non-return type and then through a strainer or strainers which filter out quite small particles of foreign matter. From the strainer the oil passes up to the oil coolers and then to the lubricating oil manifold on the engine.

The various bearings on the engine are fed with oil under pressure through pipes connected to the manifold. After lubricating and cooling the bearings the oil runs down to the bottom of the crankcase from where it drains down through 'rose' plates or grids into the drain tank in the double bottom. Various valves are fitted on the lubricating oil system pipe lines to isolate various sections for testing purposes. If the pistons are oil cooled it is usual for the lubricating oil line from the cooler to be led to a branch piece so that one side of the branch goes to the bearings and the other side to the piston-cooling system. The oil to the piston cooling passes through a valve which bleeds it to the piston-cooling system and so maintains adequate pressure on the lubrication system.

When pistons are oil cooled the lubricating-oil pumps are naturally larger as they have to pump more oil to supply both the bearings and the piston-cooling system. The lubricating-oil coolers are also larger as they have to remove not only the heat from friction in the bearings but also the heat passed into the piston cooling oil from the piston crown.

Note   In some ships the lubricating-oil drain tank extends down to the keel plate and garboard strakes so that the bottom of the tank forms part of the outer bottom plating. Obviously with such an arrangement there is no suction well. The tail end of the suction pipe is fitted with a conical-shaped or bell-mouth suction fitting.

■ **15.38 What attention do the various parts of the lubricating oil system up to the suction side of the pump require?**

Suction lines. The flange bolts on the suction lines must be checked periodically and tightened as required, particularly near bends and where the pipework is more rigidly fastened. This prevents suction loss in the lubricating-oil pump which is noted in some ships during heavy weather. In some cases suction loss can be associated with a low level of oil in the drain tank, in other cases it is associated with pipework flexing and joints opening slightly when flange bolts go slack.

Valves on suction lines. The glands on suction line valves require regular tightening down to prevent air ingress into the suction line. It is the practice in many well-run ships to slacken the gland prior to opening the suction line valves and 'nip' the gland after the valve is open. At the same time a few drops of lubricating oil may be put on the valve spindle and gland so that parts are kept working freely.

Suction-line strainers. The suction strainers in the suction line to the pump require periodic cleaning, particularly after work in the engine crankcase or cleaning out the lubricating-oil drain tank. The cover to the suction strainer casing requires particular attention when it is replaced. Doubtful joints should be renewed and care must be taken to ensure that the cover is pulled down squarely on the casing flange. Air ingress through suction strainer covers and glands on valves in the suction line are the most common causes of loss of suction or similar difficulties when starting lubricating oil pumps. A common cause of trouble with glands and stuffing boxes is bent valve spindles.

Pressure gauges on the suction line or suction inlet to the lubricating-oil pump should be kept in working order. The read-out on these gauges gives the best indication of causes of problems on the suction side of the pump. High vacuum readings indicate choked strainers, blocked lines, partially opened valves and the like. Low vacuum readings indicate air ingress into the suction line.

■ **15.39 Give some details of the attention that must be paid to the various parts and fittings on the discharge side of the lubricating-oil pumps.**

In most ships a non-return valve of some type is fitted on the discharge lines from the lubricating-oil pumps. Its purpose is to prevent a back flow of lubricating oil through the stand-by lubricating-oil pump when the pump suction line and discharge line valves are kept open. They must be left open in lubricating oil systems where automatic start-up of the stand-by pump occurs with low lubricating oil pressures. The non-return valve must be kept in good working order so that it can open freely. This keeps the head or friction loss in the valve to a minimum.

The pressure filters, which are usually of the self-cleaning type, either hand or automatically operated, require the sludge and dirt reservoir at the bottom of the filter to be cleaned out regularly. Air-locking within the filter casings would reduce the effective strainer area, and it is therefore essential that the air bleed or leak-off cocks be kept in proper working order and used regularly. It is essential to use them when the lubricating-oil pumps are started to prepare the engines for sea.

The pressure gauges, inlet and outlet, on either side of the pressure filters need to be kept in working order; a large difference in the readings of the two gauges indicates a large head loss across the filters which comes about when a filter is dirty or air-locked.

The requirements of the oil coolers have been covered in earlier questions, and the air leak-off cocks on coolers and the upper parts of piping systems require the same attention as the air cocks on the discharge filters.

The shut-down valves require the normal attention given to valves such as gland packing renewal, spindle lubrication, together with regular opening and closing to keep the valves working freely.

■ **15.40   Describe a common method of lubricating the main thrust bearing in the main propulsion machinery. What attention does the system require?**

The thrust collar of the main thrust bearing revolves in an oil bath. The revolving collar picks up lubricating oil from the bath which is scraped off the top of the collar by a diamond- or rhombus-shaped scraper or spreader. The oil is then diverted into two channels which keep the thrust faces of the collar adequately supplied with oil. The lubricating oil is supplied to the thrust bearing oil bath from the lubricating oil manifold or rail within the crankcase. The supply pipe passes through the end of the crankcase and then leads through filters down into the thrust bearing and the lubricating oil leaves from an open-ended pipe.

The take-off point for the supply pipe on the main manifold is often the location of sludge and solid matter build-up which can throttle the oil supply to the thrust bearing. The end of the manifold requires regular cleaning to prevent this build-up and so prevent choking of the thrust oil supply. The end of the tail pipe is sometimes fitted with a restriction plate: this also requires regular cleaning.

In some ships a thrust lubricating oil pressure alarm is fitted, and it can be seen that if the end of the tail pipe becomes blocked there will be no pressure fall and therefore no alarm call-out. The purpose of the low-pressure alarm is to call attention to the condition of the filter.

In other ships, particularly those classed for unmanned machinery spaces, a temperature sensor is fitted in the oil bath. This sensor requires a check-out at regular periods.

If the precautions listed are regularly taken, troubles with thrust bearings should be almost entirely eliminated.

Note   Generally the same methods of thrust bearing lubrication are used in geared installations where the thrust bearing is integral with the gear casing. The same precautions must be taken with the oil supply lines, pressure alarms and temperature sensors.

■ 15.41 When sea-water circulating pumps are started it is noted that the pressure on the suction line falls after the pump is set in motion. Why does this occur?

When the pump is opened up to the sea-water suction and discharge lines the pressure registered on the suction gauge is the static head pressure of the sea-water outside the ship. When the pump is started, water begins to flow through the sea-water grids on the ship's side, into the water boxes and through the ship-side suction valve. It continues through the sea-water strainer, through the various suction lines, bends, T-pieces, valves, and then through a conical reducer into the suction eye of the pump rotor. During its passage from outside the ship the water meets with resistance to its flow as it passes throught the pipe work and fittings. This resistance to flow reduces the effect of the static head of water outside the ship and causes the lower reading shown on the suction line (pressure) gauge.

It sometimes happens that suction line pressure gauges are fitted on the pump inlet branch and on the outlet side of the filter. When the pump is stopped both gauges give the same reading but when the pump is operating each gauge gives a different reading. The different values recorded on each suction gauge are accounted for by changes in water velocity along the line and in the pump suction branch.

Note It is important that a record be kept of the pressure drop on the suction line gauge to the sea-water, as increasing head loss for any one pump speed indicates fouling of strainers, ship side grids, etc.

A clear understanding of the different values recorded on pressure gauges is greatly helped by an understanding of Bernoulli's Law.

■ 15.42 What is the reason for starting centrifugal pumps with the discharge valve closed?

It is common practice to start large-capacity centrifugal pumps (such as sea-water circulating pumps and fresh-water circulating pumps) with the discharge valve closed. If the characteristic curves for a centrifugal pump are examined it will be seen that when the quantity of water discharged is zero the power required by the pump is zero or a very small amount. By starting the pump with the discharge valve closed the power demand made by the pump on the pump motor is kept to the very minimum.

After the pump has started and the momentary high motor current demand has stabilized, the discharge valve is opened.

■ 15.43 What is the purpose of the automatic and hand-controlled by-pass valves on positive-displacement lubricating-oil pumps?

The automatic by-pass valve is used to control the discharge pressure of a positive-displacement pump. The automatic by-pass is a necessity with pumps driven by alternating current motor which have no speed control.

The hand-controlled by-pass valve is opened before the pump is started so that the oil discharged by the pump is returned to the sump or drain tank. This limits the load placed on the pump motor while the oil is comparatively cold; as the temperature of the oil rises its viscosity decreases and the pumping load decreases. The hand-controlled by-pass valve is gradually closed as the motor load falls. When the lubricating oil is warm or its viscosity is not too high it is often possible to start the lubricating-oil pump without over-loading the motor even though the hand-controlled by-pass valve is not used.

In pumps with variable-speed motors the hand-controlled by-pass valve is not always fitted.

■ **15.44 What attention does the fresh-water cooling system require in service?**

The first requirement is to keep the cooling water in proper condition. This is accomplished by regular testing of the water and the addition of the correct quantities of chemical additives so that there is always a reserve of chemical in the water. No specific instructions can be given in this answer beyond stating that the instructions of the treatment supplier should be closely followed.

The parts making up the system are the cooling-water pumps, valves, piping system, expansion tank, de-aerating leak-off lines and reserve cooling-water tank. The pump is connected through the pipework system to a manifold at the bottom of the engine cylinder cooling spaces. Separate connections are made to each cylinder jacket space, and each cylinder jacket is in turn connected with a cylinder cover. The outlets from the cylinder covers lead into a manifold which is connected with the suction side of the cooling pump.

As the pipework is commonly made of steel the external surfaces require painting to protect the steel. This is most important particularly on the unseen lower sides of any piping that may be led under the floor plates of the main platform. The valves within the system should be opened and closed at regular intervals to keep valve spindle threads free. Similar attention is required by the air leak-off cocks and similar fittings. The fresh-water circulating pump needs to have the motor and pump bearings lubricated at the recommended intervals. Attention must also be given to pump glands and seals so that they are kept in good order. Pump glands should not normally be topped-up with turns of packing; they should be completely repacked, to prevent bad wear ridges and grooves forming. When hardened packing is squeezed down on to the rotor shaft packing sleeve, damage is caused by the hard inflexible packing. If packing seals are fitted on the pump any lubrication requirements must be attended to. Thermometers fitted into the various parts of the system should be regularly cleaned to facilitate reading. The bulb portion of a thermometer fitted into a thermometer pocket should be cleaned periodically and the pocket topped-up with oil. The expansion tank and its internal division plates should be checked for corrosion and wastage. The make-up or reserve cooling water tank also requires checking internally for pitting and corrosion. These checks must obviously be made when the engines are shut down and the tanks drained down.

Some chemical treatments form sludges, particularly if used with raw, town-supply water. In such cases when the engine is shut down the engine cooling spaces should be drained, and any sludges which have formed and settled should be washed out.

■ **15.45 If you have opened up a main sea-water circulating pump (centrifugal type) for survey or examination, describe the checks you would make, and state your conclusions regarding any faults found.**

The rotor and rotor shaft are first examined, paying particular attention to the rotor sealing surfaces, gland packing locations, the rotor blades, and the journal surfaces of the bearings.

The internal sealing surfaces on this type of pump are at the sides of the rotor and have a small radial clearance. The surface should be examined for contact marks and wear. Contact marks and wear are indicative of slack bearings, particularly in the bottom bearing if it is fitted internally in the lower part of the pump casing. If the clearance is greater than that recommended by the pump maker, new sealing rings must be fitted.

The gland packing locations must also be examined and if there is considerable wear new packing sleeves must be fitted on the rotor shaft. If the wear has come about over a short period it may indicate that incorrect packings have been used, or if it is a water-sealed and lubricated gland it may indicate incorrect location of the lantern ring, or choked or damaged water pipes.

The blades of the rotor should be examined for erosion and cavitation. A general thinning of the blade extremities indicates erosion while a pitted appearance on the back tip area of the blade indicates cavitation. Cavitation damage usually indicates a restriction of inflow into the pump which may come about from improperly opened valves (not fully open), dirty strainers, or shipside grids choked with weed or shell.

The journal surfaces of the bearings (if not of the ball or roller type) must also be examined for condition and clearance. Particular attention must be paid to the bearing clearances if wear and contact marks are present on the sealing rings.

Slack bearings (if not from normal wear and tear over a lengthy period) indicate deficiencies in lubrication to the bearings. In vertical pumps it is fairly common practice to locate the bottom bearing within the pump casing. This bearing is then either water lubricated, by a small flow of water through the bearing, or grease lubricated. If faults are found in this bearing they are often associated with blocked or damaged water supply pipes or lack of attention to regular greasing of the bearing. Excessive wear on bearings may also be associated with misalignment between the electric motor and the rotor shaft: misalignment will be indicated at the couplings.

The pump casing must also be examined for general deterioration, particularly if it is made of cast-iron. The spaces in which the rotor sealing rings are located must also have attention. If any deterioration has occurred remedial work is often difficult. Temporary, and even permanent, repair work may sometimes be possible with epoxy fillers and similar materials. Glass fibres can be used as a reinforcing material to strengthen the filler.

■ **15.46   Give some details of the checks you would make on a positive-displacement lubricating-oil pump when it is dismantled and opened up for examination and survey.**

The ability of the pump to maintain correct lubricating-oil pressures with correct engine bearing clearances would be known before the pump was dismantled. If the pump was maintaining good pressures when the lubricating oil was up to its working temperature it is unlikely that there will be anything amiss with the clearances of the rotating parts within the pump. The first examination of the internal parts after opening the pump should be carefully directed towards their cleanliness. If any hard sediment or lacquer from the lubricating oil is present, careful consideration must be given as to whether it should be cleaned off. It is likely that if the deposits were removed the effects of wear would be found and the pump would not maintain the correct pressure when it was put back in service. This effect has often been noted in the past after thorough cleaning of rotor parts, and also when solvent-type flushing compounds have been used. It is noticed more in older ships than in, say, ships during their first survey.

Once a decision on the amount of cleaning has been made, the important things to consider are the end and radial clearances of the rotors within their housings. In pumps with two helical-form rotors the driving gears should be carefully examined together with the surfaces of the helices on the rotors. This is to establish and confirm that no contact is being made between the rotor surfaces. In pumps with three helical rotors the contact pattern between the rotor helices should be examined to see that it is uniform over the length of the rotor. Non-uniformity of contact indicates unequal wear in the rotor bearings, which will require correction. The general conditions regarding packing contact surfaces on rotor shafts as set out in Question 15.45 also apply.

■ **15.47   How are engine cylinders and pistons lubricated? What do you understand by the term, 'timed injection' or 'timed lubrication'?**

Pistons, piston rings, and cylinder liners are lubricated either by oil splashed up from the crankcase as in trunk piston engines, or by special lubricants pumped into the cylinders as in crosshead-type engines. The oil is pumped into the cylinders through a quill fitting which may or may not pass through the jacket cooling space. A small single-acting pump is used to pump the cylinder oil through the quill, one pump being used for each quill point. A number of the small pumps are mounted together on the front side of a metal box which forms the oil reservoir from which the small pumps take their oil supply. The pump pistons are driven by eccentrics mounted on a shaft which passes through the length of the lubricator box. The lubricator shaft is driven by the engine camshaft, either by a roller chain or gearing.

Each small pump discharges through a sight glass. In some cases the sight glass is made like a venturi tube, and a small ball is fitted in the upper part. The height to which the ball rises in the sight glass indicates the amount of lubricant flow. In other cases the glass is a small parallel tube which is filled with distilled water or a clear fluid which does not affect the additives in the lubricant. With such glasses the flow is indicated by the droplets of lubricant passing up through the sight glass.

There is usually a small heater in the oil reservoir or lubricator box to hold the oil at some steady temperature and viscosity. Once the lubricators have been set, the flow rate to the cylinders is unaffected by engine-room temperature changes as the oil viscosity in the lubricator remains constant.

Timed lubrication or timed cylinder lubricant injection is arranged for on some engines. With such systems the oil is pumped into the cylinder at the instant when the piston rings pass the quill point during the piston stroke. This requires the cylinder lubricator to be timed relative to the piston position in much the same way as the fuel injection is timed to occur at the correct instant. It must be remembered that the cylinder lubricant is pumped into the cylinder at a slow rate and that it flows down the cylinder lubricating grooves or cylinder walls rather than being expelled from the quill into the cylinder space in the way that fuel is injected. The end of the oil quill is usually drilled in such a manner that the lubricant flow is directed into the cylinder liner oil grooves or down the cylinder wall. The cylinder lubricant is then spread over the cylinder wall working surfaces. Some of the oil is picked up on the horizontal faces of the

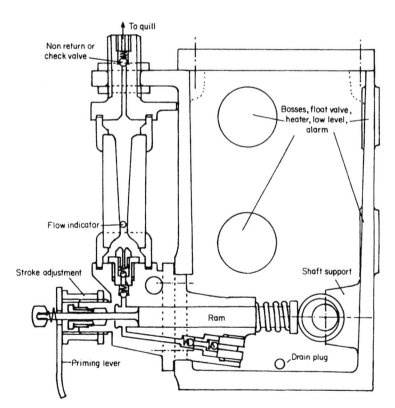

Fig. 15.4 Common type of cylinder lubricator.

piston rings where it lubricates the contact faces between piston ring and groove.

Figure 15.4 shows a common type of cylinder lubricator. A small set screw (not shown) is fitted below the sight glass. This connects with the pump chamber and is used to release air so that the pump can be primed. The material of the various parts must be compatible with the additives used in the cylinder lubricating oil.

Note Some builders of medium-speed trunk piston engines lubricate the pistons with lubricators in the same manner as a crosshead engine. When the engine is operating on heavy fuel oil a lubricant with a higher TBN (total base number) than the crankcase oil is used to lubricate the pistons and cylinders in addition to the oil splashed up from the crankcase. The higher TBN oil used for cylinder lubrication does not lose all its alkalinity; it drains into the crankcase and assists in the crankcase oil make up requirements and in maintaining the TBN of the crankcase oil at a satisfactory level.

■ **15.48 Name the various parts which make up the lubrication system for white-metal-lined bearings as fitted in stern tubes.**

These bearings are commonly lubricated by gravity feed systems, though there are exceptions in which an electrically driven pump, or a pump driven by a cam on the intermediate shaft, is used. Whether a pump is used or not, the essential parts of the lubrication system are similar. If the tank is in a gravity system it is naturally fitted at such a height that sufficient head is available to force lubricant into the bearings at all times. A gauge glass is fitted on the tank to give an indication of the lubricant level. A stop valve is fitted on the bottom of the tank at the inlet to the supply pipe to the stern tube. A pipe leads from the tank through the aft peak bulkhead into the top of the stern tube. Another separate pipe leads from the top of the stern tube and runs parallel with the supply pipe back towards the gravity tank, ending with an open-ended goose neck situated above a tundish or funnel leading back into the tank. The other fittings in the system are the drain line and drain cock. The drain cock is fitted on the aft peak bulkhead below the stern gland or shaft seal.

Temperature sensors are sometimes fitted to indicate the working temperature of the stern tube or screw shaft bearing. When a pump forms part of the system it is fitted into the feed line from the supply tank and pressurizes the stern tube and bearing spaces.

■ **15.49 What particular attention must be given to stern tube bearing lubrication systems when low temperatures are experienced in the region of the gravity tank and connecting pipework?**

A problem for the machinery designer is the question of the diameter of the supply pipe fitted between the gravity tank and the stern tube. There is virtually nil flow between the tank and the stern tube, and on this basis one might think that a small-diameter pipe should be used. In practice the pipe fitted is often too small, and with low temperatures the stern tube lubricant can thicken up so that it in effect 'plugs' the supply pipe. In such circumstances the lubricant cannot pass into the bearing even though the amounts required are very small. This condition is easily checked by slackening a pipe flange or union coupling and noting whether the lubricant comes through easily or just oozes out. If the latter is the case some heat must be applied to the supply pipe from the gravity tank. This is best done by rigging a temporary steam tracer line against the supply pipe to the stern tube.

If it is the usual practice to operate the ship with some water in the aft peak to transmit heat away from the stern tube, consideration should be given to discontinuing this practice in cold weather. Better lubrication and oil entry into the tube might be obtained if some temperature rise were permitted.

16

AIR COMPRESSORS, AIR STORAGE TANKS

■ **16.1 For what purposes is compressed air used in motor ships?**

Compressed air is used for starting both main and auxiliary engines. It is also used for control equipment and instrumentation purposes. Many of the portable tools such as drilling machines, impact wrenches, torque wrenches, hand grinders and lifting gear also use compressed air. Auxiliary boilers and economizers are often fitted out with soot blowers which use compressed air. It is also used for chipping and scaling machines, paint-spraying equipment and similar services both in the engine room and on deck.

■ **16.2 Give the working pressures for the compressed air or pneumatic services used on board motor ships.**

The starting air pressure (gauge) for the various types of diesel engine, both propulsion and auxiliary, varies between

24 bar (350 lbf/in², 25 kgf/cm²) and
41 bar (600 lbf/in², 42 kgf/cm²).

For most of the other pneumatic and compressed air services the working pressure is in the region of

4 bar (60 lbf/in², 4 kgf/cm²) to
7 bar (100 lbf/in², 7 kgf/cm²).

For control engineering services using compressed air a wide variety of air pressures may be required; it is normal practice to use a small reducing valve to supply air at the correct pressure for each individual fitting.

■ **16.3 Why are air compressors constructed to compress air in two or three stages in preference to compressing it in a single stage?**

The number of stages used is governed by the required final pressure of the compressed air. As the pressure increases more stages are required. When the

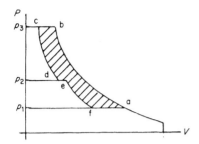

Fig. 16.1 PV diagram showing compression of air in one or three stages. The reduction in work compared with compression in a single stage is shown by the area bounded by the letters a-b-c-d-e-f. The re-expansion of air from the clearance volume spaces is not shown on the diagram.

air is compressed in stages it is easier to control the temperature of the air and to hold it at lower temperatures during its passage through the air compressor.

This is accomplished by water jacketing the air compressor cylinders and passing the air through heat exchangers, or intercoolers as they are often named on air compressors. As the air leaves each stage of the compressor it is cooled in the intercooler. This lowers the work done in compressing the air and prevents a lot of the mechanical problems which could arise if the air temperature were uncontrolled. By keeping the air temperatures low, less difficulty is experienced with the lubrication of the pistons and cylinders, and the suction and delivery valves remain in a cleaner condition without becoming fouled with carbonized oil.

Apart from practical considerations, a three stage air compressor is more desirable as it requires less energy or work input than a single stage air compressor when compressing air over the same pressure range.

This is illustrated in Fig. 16.1 where the reduction in work is shown by the shaded area bounded by the letters a-b-c-d-e-f. The pressure at the end of each stage is denoted by p_1, p_2 and p_3. The discharge pressure p is the same for each compressor. The extra energy required to compress the air in the single stage compressor will be converted into heat and the air discharged will be at a much higher temperature, although the discharge pressure is p.

■ **16.4 Trace the path and name the parts through which the air passes when it is compressed in a three-stage reciprocating air compressor.**

The air first passes through an air filter as it enters the compressor. After passing through the air filter it enters the low pressure (L.P.) stage of the compressor and goes through the L.P. suction valves during the suction stroke of the L.P. piston. During the discharge stroke of the L.P. piston the air passes through the L.P. discharge valves.

The air then passes through the medium or intermediate pressure (M.P. or I.P.) stage and on through the high pressure (H.P.) stage in a similar manner.

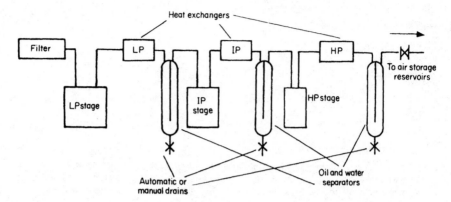

Fig. 16.2 Arrangement of intercoolers and oil and water separators in a three-stage air compressor. The arrangement of these parts in a two-stage compressor is similar but has one stage less.

The arrangement shown in Fig. 16.2 shows each stage and includes the relative locations of the inlet air filter, compressor cylinders, intercoolers and moisture-oil separators. The valve at the outlet end of the compressor is shown as a screw down stop valve, in many cases a screw down non-return valve is fitted.

The suction and delivery valves for each stage are not shown.

■ **16.5   Give the reasons for the reduction in the work done when air is compressed in stages and cooled between each stage of compression.**

When air is compressed and no heat is removed during compression the temperature of the air rises and the air is compressed adiabatically. If heat is removed during compression so that there is no change in the temperature of the air, the air is compressed isothermally. For isothermal compression, the compression curve on the PV diagram will be such that: $PV = $ a constant.

In actual compressor practice it is not possible to compress the air isothermally, but by jacketing and cooling the cylinders, and cooling the air between each stage, we can approach isothermal compression. For compression between adiabatic and isothermal, the compression curve on the PV diagram will be such that $PV^n = C$.

The value of the index n will be less than the value of gamma (see Note); in consequence as the value of n is reduced the steepness or slope of the compression curve on the PV diagram is reduced. The area under the curve represents the work done and this will be a lesser amount under the curve with least slope.

In practice the actual value of n will depend on the number of stages and the amount of intercooling between stages, the cleanliness of the intercoolers, and the temperature of the cooling water.

It is normally between 1.35 and 1.25.

Note   The value of gamma γ is obtained by dividing the specific heat of air at constant pressure by its specific heat at constant volume.

■ **16.6   Define the term 'volumetric efficiency' in reference to reciprocating air compressors.**

The volumetric efficiency of an air compressor is the relationship between the quantity of air discharged when brought to standard atmospheric conditions and the swept volume of the low pressure piston.

$$\text{Volumetric efficiency} = \frac{\text{volume of air discharged as 'free' air}}{\text{swept volume of L.P. piston}}$$

Free air is air at atmospheric pressure and a temperature of 15°C. The time value is the same for both the numerator and denominator in the efficiency equation.

In the case of two- or three-stage compressors the volumetric efficiency always refers to the low-pressure piston.

■ **16.7   How is the capacity of an air compressor specified?**

The capacity of an air compressor is stated in terms of cubic metre per hour (m³/h), this being the volume that the air actually discharged in one hour would occupy if it were expanded down to atmospheric pressure and cooled to atmospheric temperature. Air at atmospheric pressure and temperature is referred to as normal air. Normal air has an absolute pressure of 1.013 bar and a temperature of 15°C.

The details of an air compressor would therefore be specified as:

Capacity	– Nm³/h or m³/h F.A.D. (15°C, 1.013 bar)
Pressure	– discharge pressure in bar
Temperature	– discharge temperature in °C
Power input	– kW

Note   The letter N preceeding m³ stands for normal and not newton; F.A.D. denotes free air delivery. Sometimes the figure of 0°C is used for the temperature value when referring to normal air. This value is often used on the continent of Europe.

The volumetric efficiency is approximately equal to the discharge capacity F.A.D. in m³/h divided by the displacement of the low pressure piston measured in m³/h.

■ **16.8   Mention the factors that have an adverse effect upon the volumetric efficiency of an air compressor.**

The factors that affect the volumetric efficiency of an air compressor are as follows.

1 The clearance between the cylinder cover and the end of the piston when the piston is at the end of its discharge stroke. The larger the clearance the less air is discharged per stroke.
2 Sluggish opening and closing of suction and delivery valves.
3 Leakage past compressor piston rings.
4 Insufficient cooling water or the cooling-water inlet temperature too high. With air-cooled compressors an insufficient number of air changes into the cooling space or the cooling air temperature too high.
5 Inlet temperature of the air to the first or low-pressure stage of the compressor too high.
6 Throttling of air supply to L.P. suction – example, dirty air inlet strainers.

■ **16.9 Explain how excessive piston end-clearance reduces the volumetric efficiency of an air compressor.**

Excessive piston end-clearance gives a large clearance volume which will be full of air at high pressure when the piston reaches the end of its stroke. On the outward stroke of the piston the air within the clearance volume expands until the pressure within the cylinder is low enough to allow the suction valves to open. With a large clearance volume the outward travel of the piston will be greater before the pressure is low enough within the cylinder to allow the suction valves to open. In consequence a large part of the suction stroke is made ineffective and the amount of air taken into the cylinder during each suction stroke is reduced. This lowers the volumetric efficiency.

■ **16.10 How is a check made on the clearance volume of an air compressor? What would you expect the clearance volume to be?**

It is not normal practice to make a check on the clearance volume of an air compressor as this is something which is fixed by the designer. We do, however, make a careful check on the mechanical clearance between the piston head and the cylinder cover. This check is made by making up a small, loosely woven ball of lead wire. This is placed on the top of the piston which will have been moved a little off the end of the stroke. The cylinder cover is replaced on the cylinder with a joint in place and tightened down. The compressor is then barred slowly over top centre so that the ball of lead wire is compressed. After it is removed from the top of the piston it can be measured with a micrometer. The mechanical clearance measured is compared with the compressor manufacturer's recommendations. Adjustments are made by altering the cover joint thickness or by fitting or removing shims between the foot of the connecting rod and the bottom-end bearing.

The clearance volume in air compressors varies widely according to cylinder diameter, running speeds and final discharge pressures. The mechanical clearance in the H.P. stages of high-pressure air compressors may be only a fraction of a millimetre when the machine is cold and less at running temperatures.

Note In making up the ball of lead wire, care must be taken to prevent an excess of wire being used. It is better to make one or two attempts to get a reading than to use an excess of wire and strain the cover studs or other parts. The ball of wire must be placed centrally on the top of the piston; if the wire is on one side of the piston there is a possibility that the piston rod could be bent when the piston is barred over the top centre position.

Similar care must also be exercised with regard to the suction and delivery valve port locations in the cylinder cover.

■ 16.11   In most air compressors it will be found that the L.P. cylinder absorbs the most power and that the H.P. cylinder absorbs the least: in other words, the major part of the work of compressing the air is done in the L.P. cylinder. Why is not an equal amount of work done in each stage?

The reason for doing the smallest part of the work in the H.P. cylinder is because the comparatively small diameter affords much less surface for the dissipation of heat to the cooling water. Furthermore, the high pressure in this stage necessitates a thicker cylinder wall, which is not conductive to the rapid transfer of heat. These considerations make it advisable for the greater proportion of the work to be done in the L.P. or largest cylinder.

■ 16.12   Explain how sluggish action of the suction and delivery valves reduces the volumetric efficiency of an air compressor.

If a suction valve does not reseat promptly at the end of the suction stroke, due to a weak spring or to carbon deposit, part of the air drawn into the cylinder will be returned through the defective suction valve during the first part of the delivery stroke. If a delivery valve is slow in reseating, for the same reason, part of the air compressed and delivered during the delivery stroke will return to the cylinder during the first part of the suction stroke.

■ 16.13   Explain how the temperature of the air drawn into the low-pressure cylinder affects the efficiency of an air compressor.

The higher the temperature of the air drawn into the low-pressure cylinder, the less will be the weight of air discharged by the compressor in a given time, since the higher the temperature of the air the greater will be the volume occupied by a given weight.

■ 16.14   Explain how a restricted inlet passage to the low-pressure cylinder reduces the volumetric efficiency of a multiple-stage air compressor.

If the inlet passage to the low-pressure cylinder is restricted, the air will be prevented from following up the piston at the correct rate during the suction stroke, with the result that when the piston begins the delivery stroke, there will be a partial vacuum in the cylinder and less air will be discharged during each delivery stroke.

Note   The air inlet passage to the low-pressure cylinder may be restricted owing to insufficient suction-valve lift, suction-valve springs being too strong, or to a dirty and partly chocked air strainer.

■ **16.15 What would be the effect of the suction valves of an air compressor having too much lift?**

The valves would be late in closing and this would reduce the volumetric efficiency of the machine. Valves with too much lift reach the end of their travel with greater force and therefore are more liable to break.

■ **16.16 Briefly describe the most common forms of air compressor intercoolers, and state how they are arranged.**

An air compressor intercooler generally consists of a number of small-diameter copper tubes contained in a cast-iron chamber forming a water jacket. The air passes through the tubes and the cooling water circulates around them. Sometimes a single long copper coil takes the place of the nest of straight tubes. Each intercooler is provided with what are commonly called 'purgepots,' the purpose of which is to collect and drain off water and oil which finds its way into all intercoolers. Intercoolers are fitted after each stage of compression and arranged as close to the cylinder as conveniently possible, to reduce the length of hot delivery pipes. The chamber containing the nest of tubes or coil, as the case may be, is fitted with a large diaphragm of rubber, lead, or very thin brass, to release the pressure in the event of a tube bursting.

■ **16.17 What are the advantages and disadvantages of the single-coil type and the straight-tube type of intercooler for air compressors? Why are the tubes and coils generally made of copper?**

With the single-coil type there is less likelihood of leakages, but the coil is more difficult to clean and wears more rapidly than the straight tubes. When a straight tube leaks it can be plugged and remain in service until it is convenient to renew it. Tubes and coils are generally made of copper because this material is a good conductor of heat.

Note A good deal of wear sometimes takes place inside cooling coils. One theory is that this is due to corrosion as a result of oxidation of the lubricating oil to organic acids. Another probable cause is that the air, travelling at a high velocity, carries with it small particles of carbonised oil and abrasive dust, which have a scouring effect inside the coil. This latter theory is supported by the fact that the coils wear most on the outer part of the bore. Copper coils become hard and more likely to break if allowed to vibrate; they must therefore be well secured by lead-lined clips.

■ **16.18 Why is it necessary for air compressors to have efficient drains fitted between each stage of compression?**

The air that is drawn into compressors contains moisture held in suspension. The amount of moisture contained in the air depends on the density (the lower the barometer reading the greater amount of moisture the air will contain). The moisture is deposited in the form of water when the air is compressed and afterwards cooled. Unless the water is drawn off at each stage it will go with the air through the various stages and wash the oil film off the cylinder wall. Proper attention to the draining of compressors is necessary if excessive cylinder liner and piston ring wear is to be prevented. The drains are also a means of drawing off any surplus lubricating oil which has been injected into the cylinders.

■ **16.19 What do you understand by the term 'valve loss' relative to reciprocating air compressors?**

The suction and delivery valves of an air compressor allow the air to flow in only one direction. The suction valves allow air flow into the cylinder and prevent back flow out. The delivery valve similarly allows air to flow out of the cylinder and prevents flow back. To make the action of the valves more positive the valves are spring loaded. When the piston is making a suction stroke, the pressure within the cylinder space is lowered to the point where the air pressure behind the suction valve overcomes the spring load and causes the valve to open. When the suction valve opens, the external air flows into the cylinder by virtue of the greater air pressure outside the cylinder. During the suction stroke the air within the cylinder clearance volume expands to some value lower than outside. Once the air commences to flow through the suction valve the pressure in the cylinder rises slightly but remains slightly lower than the pressure outside. This pressure difference is necessary to get air to flow into the cylinder. The difference in pressure is the suction valve loss.

Similarly on the discharge stroke of the piston the pressure in the cylinder rises to some point at which the discharge valve opens. After the valve opens the pressure in the cylinder falls slightly, then remains approximately constant, and falls again just before the discharge valve closes. This approximately constant pressure would be slightly higher than the pressure into which the compressor is discharging. This pressure difference comes about due to flow resistance set up within the valve. These valve losses lower the mechanical efficiency of the air compressor. From this it can be seen that careful attention must be given to suction and delivery valves when they are overhauled and that it is essential that the correct type and weight of springs be fitted.

■ **16.20 How are air compressor pistons, piston rings, and cylinders lubricated?**

Some air compressor pistons are connected to the connecting-rod through a gudgeon pin and are similar to trunk pistons. With such an arrangement the cylinders are lubricated by oil splashed from the crankcase. Condition of scraper rings is of the utmost importance in this type of compressor.

When air compressor pistons are connected to the running gear through piston rods and crossheads, the piston rings, pistons, and cylinders are lubricated with lubricator pumps similar to those fitted on crosshead engines for cylinder lubrication. The oil is passed through a quill fitting to the cylinder liner. There is further information in Question 15.47.

■ **16.21   Can sea-water be used to cool air compressors and air-compressor intercoolers?**

Sea-water, fresh water, or distilled water can be used for cooling compressors. Sea-water has the advantage of always having an inlet temperature to the compressors lower than that available from fresh water or distilled water. It also has simplicity in that no heat exchangers are required within the cooling-water system, therefore there are fewer cooling-water pumps. The main disadvantages with sea-water are its tendency to form scale, the passage of mud and sandy matter through the cooling spaces of the compressor and intercoolers, and the greater possibility of active corrosion within the cooling spaces.

   Fresh or distilled water has the advantage of being less corrosive than sea-water and it is on this count that it is mainly used, irrespective of the fact that an extra heat exchanger and pump are required. As the fresh water used for cooling must pass through heat exchangers to give up the heat it receives, it will always be at some temperature higher than sea-water.

■ **16.22   How is the running gear of reciprocating air compressors lubricated?**

Small air-cooled compressors having the running gear enclosed within the crankcase may be lubricated by splash. Larger compressors, both air-cooled and water-cooled, usually have the running gear lubricated by force lubrication. A pressure pump draws oil from the crankcase sump and discharges it through the main bearings to holes drilled within the crankshaft journals. The oil passes through the crankshaft passages and lubricates the bottom-end bearings. The oil for the top-end bearings and guides is passed from the bottom-end up through the connecting rods. The pump used to supply oil to the bearings is sometimes operated from an eccentric which is mounted on the compressor crankshaft; in other cases a small rotary gear-type pump is used. When gear-type pumps are used they may be fitted within the crankcase and be driven by a roller chain, or external to the crankcase and driven by a V-belt. In many compressors the lubricant discharged from the lubricating oil pump is passed through a full-flow self-cleaning oil filter before it enters the main bearings.

■ **16.23   What are oil-free air compressors, and for what purposes are they used?**

When it is necessary to supply compressed air that is completely free of any contaminating oil, oil-free air compressors are used. Oil-free reciprocating air compressors do not have any normal lubricant supplied to the pistons. Some oil-free compressors have a solid piston with labyrinth grooves and a very small clearance in the cylinder. They operate with high piston speeds to reduce the effects and losses from piston leakage. In other cases piston rings may be used. Piston rings in oil-free compressors may be made from bronze, phenolic resin plastics laminated with linen or cotton fibres, or PTFE reinforced with bronze wires. In other cases, rings may be made from carbon or graphite materials. To reduce friction between the piston rings and the cylinder, molybdenum disulphide compounds are sometimes bonded in with the ring materials. Double-acting compressors have the piston-rod packing rings made of materials similar to the piston rings.

In marine service oil-free compressors are used for the supply of compressed air for instrumentation purposes. The location of the inlet to the air compressor is important; it must be in a position where the air drawn in will be uncontaminated. In some cases the air suction is ducted to some point external to the engine room.

Some ships are fitted with rotary compressors of the positive displacement type for supplying oil-free air to the instrument air system.

■ 16.24 What attention would you give to the running gear contained within the enclosed crankcase of an air compressor?

The crankcase should be opened up periodically to examine the running gear of the compressor. At the time the crankcase is opened up for examination it is common practice to drain off the lubricating oil and pass it through a very fine filter to take out any foreign matter. Attention is given to all the nuts and nut locking devices on the main bearings, bottom-end bearings and the crosshead bearings. The guide-bar fastening nuts must also be checked for tightness.

The lubricating oil pipe union nuts, oil pipe clips and oil pipe fittings are also checked to see that none of the parts are loose. Attention should also be given to the lubricating oil suction filter or strainer to ensure that it is clean before the lubricating oil is put into the crankcase. The fine filter fitted on the discharge side of the lubricating-oil pump should also be cleaned.

■ 16.25 How many starting air compressors are usually fitted on motorships? Is any limit put on the temperature of the air discharged from the compressors?

It is usual to have at least two starting air compressors and sometimes there are more than two. The compressors may be independently driven by an electric motor or a steam engine. In other cases they may be driven by a diesel generator set; the compressor is driven off the free end of the electrical generator through a friction clutch. This enables the compressor to be used as required, and allows the generator to be operated without driving the compressor. It is usual to design the starting air compressors and the inter-coolers so that the temperature of the air discharged from the final stage is not greater than 93°C.

■ 16.26 If, during the course of an overhaul or as a result of some mishap, it becomes necessary to fit new bearings to a reciprocating compressor, where must special precautions be taken?

When new bearings are fitted to either main bearing journals, crankpins or crosshead pins, the effect will be to lift the working position of the pistons within the cylinders. This lifting of the working position reduces the mechanical clearance of pistons at the upper part of the stroke, and in double-acting compressors it also increases the mechanical clearance at the lower part of the stroke. In consequence when new bearings are to be fitted the crown thickness of the new bearings must be compared with the old and the mechanical clearances of the pistons must be carefully checked. The bore of each cylinder must be carefully checked for ridges at the ends of the stroke and any ridges found must be carefully ground down. If new bearings have been fitted as a result of a mishap and the white-metal in the old bearings has wiped or run, care must be taken to ensure that no white-metal has entered and blocked the drilled holes forming the oil channels within the crankshaft. If white-metal has plugged these holes it must be drilled out or otherwise removed.

■ **16.27 Describe the method of construction of large and small air reservoirs. What material is used?**

The material used for the construction of large or small reservoirs is good-quality, mild-steel plate having similar specifications to boiler plate material. The steel will have an ultimate tensile strength within the range of 360 MN/m² to 500 MN/m² with an elongation of not less than 23% to 25%. Large starting air storage reservoirs have dished ends; one end has an opening formed with a lip to take an elliptical manhole door. The form of the dished end may be either torispherical or semi-ellipsoidal. The cylindrical part of the air vessel may be rolled from one or two steel plates and will have either one or two longitudinal welded seams. The dished ends are usually made by the spinning process. The edge left by the spinning is machined and tapered down to the thickness of the cylindrical shell. The ends are welded to the cylindrical shell by full-penetration welds. The longitudinal seams are machine welded. The circumferential seams where the dished ends join the cylindrical shell may be either machine or hand welded.

Smaller air storage reservoirs often have hemispherical ends, and the inspection holes may be fitted in the cylindrical shell or in the hemispherical ends. The ends may also be fastened to the cylindrical shell as a lap joint with double fillet welds.

■ **16.28 Are welded air reservoirs X-rayed and heat treated after completion of the welding operations?**

There are various rules covering the manufacture of welded pressure vessels. These rules are issued by Ship Classification Societies, Government Authorities, Technical Societies and Institutions. The necessity for X-ray testing and heat treatment or otherwise is set out in these rules and construction codes, and as is to be expected there are some variations.

Generally one may say that as the thickness and working pressure of the air reservoir increases, the rules governing welding quality and inspection become more stringent. As the welding rules become more stringent, so the necessity for X-ray examination of the welding and subsequent heat treatment increases. In many cases air reservoirs are designed by carefully selecting the dimensions in conjunction with the required working pressures so that they come within the least stringent rules. This allows them to be manufactured at the lowest cost. Air reservoirs built to the least stringent rules require only to be hydraulically tested.

■ **16.29 What are compensation rings and why are they necessary?**

When a hole is cut or machined in the shell of a compressed air reservoir the material around the hole will be subjected to a higher stress when the reservoir is under pressure. In order to reduce the stress in the material around the hole a compensating ring is fitted. The material in the compensating ring in effect replaces the material removed when the hole is cut. When the diameter of a hole exceeds a specified amount relative to the thickness of the shell it becomes essential to fit a compensating ring to reduce the shell stresses to a safe figure. It is common practice to utilize the compensating ring as a flange on which to mount the valves or fittings for which the hole was originally cut.

■ **16.30 How are compensating rings fastened to the parent material of an air vessel?**

Compensating rings are made from thick plate of suitable material specification. In some instances they are made of material considerably thicker than the shell to which they will be fitted, in which case a hole is cut in the shell similar to the outside diameter of the compensating ring. The ring is then fitted in the hole and fastened to the shell by a full-penetration weld. The side of the hole in the shell is prepared to a suitable profile before the welding.

   In other cases one flat side of the ring is machined to the curvature of the outside diameter of the shell. The opposite side is left flat. The side with the

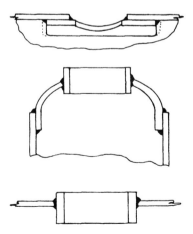

Fig. 16.3 Compensation for the material removed in compressed air receiver shells and dished ends.

machined curvature is then bedded to the shell and the ring is fastened with two large fillet welds, one between the outside circumference of the ring and the shell, and the other between the sides of the hole in the ring and the surface of the shell. For this weld procedure to be practicable the hole in the compensating ring must be larger than the hole in the reservoir shell. There is further relevant information on welding in Questions 6.33 and 6.34. (Fig. 16.3)

■ **16.31　When flat plate compensating rings are to be fitted to pressure vessels by fusion welding, it is usual to drill a small hole through the plate thickness of the compensating ring prior to welding. What is the purpose of this hole?**

The hole drilled through the thickness of the plate compensating ring is kept less than 9.5 mm in diameter. The hole allows release of air during welding and heat treatment so that no pressure rise can occur as it could if air were entrapped between the compensating ring and the shell. If any failure in service should occur with the welding within the hole in the compensating ring and the shell it will be indicated by air leakage from the small hole in the compensating ring. The hole is also sometimes used for testing purposes, in which case a thread is tapped in the hole. A threaded connection piece is screwed into the hole and connected to an air supply with suitable pressure. When the welding is under pressure it is tested for leakage and porosity with a soapy-water solution brushed on with a paint brush.

■ **16.32　Name the fittings and mountings likely to be found on a main compressed air reservoir for starting main engines.**

There will always be a large outlet valve of a size suitable to supply compressed air at the required rate of the main engine starting air system, and a valve of much smaller size which is used for the air inlet to the reservoir from the main compressor discharge. Many air reservoirs are also fitted with other outlet valves which have connections with other air systems such as auxiliary engine starting systems, instrument air, workshop air services, ship's whistle, filter cleaning systems, emergency air supply to boiler feed pump, boiler light up sets and the like.

Excess air pressure is prevented by spring-loaded relief valves on the air reservoir. In some cases the relief valves are fitted on the common discharge line from the compressors and fusible plugs are fitted on the air reservoirs. The fusible plugs prevent serious pressure rise if a fire should break out near the air reservoir. Drain valves or cocks are fitted to the bottom of the reservoir to drain off oil or moisture carried over with the air from the compressor. An independent connection is fitted for the pressure gauges used to indicate the air pressure within the reservoir.

■ **16.33　Is there any rule governing the size of main engine starting air reservoirs?**

There are no rules governing the actual physical dimensions, but there is a rule used by government bodies and classification societies which states that the total capacity of the reservoirs or receivers must be sufficient to make 12 or more starts without refilling the reservoirs. If the ship has twin screws the size of the reservoirs must allow not less than 12 starts on each engine. When the ship has controllable-pitch propellers which are used for supplying astern power, it is usual to have the air reservoir capacity such that six starts can be made.

In practice, ships are usually supplied with air reservoir capacity well in excess of these requirements. The actual number of starts that can be made without refilling the air reservoirs is dependent on engine temperatures, lubricating oil temperature, the skill of the engineer manoeuvring the engine and the minimum air pressure that will start the engine. Usually most engines will still start when the air pressure has fallen to half or even less than half of the designed maximum starting air pressure.

■ **16.34   How does the stress vary in the shell material of a compressed air receiver in the longitudinal and circumferential direction?**

The shell thickness of a compressed air receiver or reservoir is thin in relation to the diameter. It cån therefore be considered as a thin-walled vessel where the stress is uniform across the thickness of the material.

Let

D = internal diameter of the receiver
t = shell thickness
P = working pressure

and

stress = load/area

If a section of unit axial length is taken the circumferential or hoop stress is

$$f_c = \frac{DP}{2t}$$

and the longitudinal stress is

$$f_1 = \frac{\dfrac{D^2 P}{4}}{Dt} = \frac{DP}{4t}$$

from which it can be seen that the stress in the material of the cylindrical shell in the axial or longitudinal direction is only fifty per cent of the hoop or circumferential stress, or the hoop stress is twice the axial stress.

■ **16.35   Give an estimate of the increase in circumference of an air reservoir when it is pumped up from zero gauge pressure to 41.4 bar gauge pressure. The internal diameter of the reservoir is 1000 mm and the thickness of the shell is 13 mm.**

The load of a semicircular section 1 cm long is equal to

$D \times 1$ cm × pressure

Therefore

$$\text{Load} = 1 \times 0.01 \times 41.4 \text{ bar} = 0.0414 \text{ MN}$$

and the area of material resisting the load is equal to

$$0.01 \times 0.013 \times 2 = 0.00026 \text{ m}^2$$

Young's modulus E is stress/strain and is equal to 210 GN/m². Stress is given by load/area which is

$$\frac{0.0414 \text{ MN}}{0.00026 \text{ m}^2} = 159.23 \text{ MN/m}^2$$

Stress/Strain $= E = 210 \text{ GN/m}^2$
Strain $=$ Stress/E

$$\frac{159.23}{210 \times 10^3} = 0.000\,758$$

The increase in circumference in mm is given by $\pi \times 1000 \times$ strain which is

$$3.142 \times 1000 \times 0.000\,758 = 2.38 \text{ mm}$$

■ **16.36 Why are the outlets from relief valves and fusible plugs fitted on compressed air receivers arranged with lines from the fitting to some point outside the engine room?**

In the event of a fire in the engine room the function of the fusible plug and relief valve is to discharge air and hold the pressure in the receiver at some safe level. If the air is discharged into the engine room the air will aid combustion of any burning material and cause any CO_2 discharged in the engine room to be displaced or weakened by mixing with air from the receiver. When the air discharged is piped out of the engine room there is no risk of air from the receivers supporting the fire.

17

BALANCING AND VIBRATION

■ **17.1 When a ship's propeller is driven by some power source a torque-resisting rotation is created. Is this resisting torque of constant or varying value?**

When the ship is in calm water at some particular draught and proceeding at uniform speed the torque-resisting rotation can be considered to remain at some constant value. When the ship is in seas that cause the vessel to pitch, the propeller immersion changes as the stern rises and falls. The torque-resisting rotation then changes with the change of immersion of the propeller. As the immersion increases the torque increases and the engine driving the propeller slows down. In a like manner the engine speed increases when the propeller comes towards the surface of the water.

If the resisting torque is plotted against the angle turned through by the propeller the result is a straight line of constant value.

Note Due to the pattern of the wake at the stern of a ship some variation of torque will exist; for our purposes, however, we can assume that the torque is constant in smooth water.

■ **17.2 When an electrical generator is being driven a torque-resisting rotation is set up. How does this torque vary?**

While the electrical load remains constant the torque-resisting rotation remains constant. When the electrical load is increased on the generator the resisting torque increases, and when it is reduced the resisting torque is reduced. In order to suit the voltage and frequency requirements of the various types of electrical machine the rotational speed of a generator or alternator should remain constant (See Question 17.4).

■ **17.3   The gas pressure on the piston of a diesel engine changes as the piston reciprocates within the cylinder. How do these gas pressure changes affect the output torque of the engine?**

Starting with the piston on top centre at the beginning of the firing stroke the torque created by the gas pressure is zero, because the crank angle is zero (piston, connecting-rod and crank form a straight line). As the crank moves over the top centre position the gas pressure forces the piston downwards. Some of the pressure is used to accelerate the piston, and the remainder creates a turning moment or torque which overcomes the resisting torque from the propeller, generator or machine being driven by the engine. When the crank is approximately 90° from the top centre position the piston speed is at a maximum. From this position the piston speed begins to fall until the piston reaches bottom centre where its speed is zero. While the piston is being retarded the retardation force is added to the gas pressure on the piston, thus increasing the torque.

While the expansion stroke is taking place the forces acting on the piston create a torque which is considerably in excess of that required by the machine being driven. During the other strokes required to complete the working cycle of the engine (induction, compression, exhaust) energy must be obtained from some source other than the fuel, and the engine torque is therefore negative. In particular, during compression a considerable amount of energy is required to compress the air.

It can be seen then that there are considerable fluctuations in the torque output from a single cylinder at the crankshaft coupling and that there will be larger fluctuations from four-stroke cycle engines than from two-stroke cycle engines. As the number of cylinders on an engine are increased the torque fluctuation is reduced.

Figure 17.1 shows the forces coming on to a piston during engine operation. The gas pressure within the cylinder creates a downward acting force on the piston. Initially the primary inertia force acts in opposition to the gas pressure.

Soon after passing the mid position of its stroke the primary inertia forces aid the movement of the piston.

During the upward stroke of the piston the inertia forces act on the piston in the opposite manner. The gas pressure under the piston (the charge air pressure in crosshead engines and the crankcase pressure in trunk piston engines) continually act upward in opposition to the gas pressure above the piston. Piston and piston ring friction also act continuously in a direction contrary to the piston movement.

■ **17.4   What is the purpose of the flywheel on a diesel engine?**

The turning moment from an engine crankshaft is subject to considerable variation during its rotation, and the variation can be shown on a torque diagram for the engine. If the average torque required to drive the load is plotted on a graph as a straight line with a constant value it will be seen that the actual torque at any instant will be greater or less than the mean torque. When the torque from the engine is greater than the mean torque requirement the excess torque causes some increase in the rotational speed of the crankshaft. The excess energy is thus stored in the flywheel by virtue of its increased speed. During the periods of the cycle when the torque from the engine crankshaft is less than the mean torque requirement of the load, the extra energy stored in the flywheel is returned to the shafting system, and helps to maintain the rotational speed of the crankshaft. The flywheel is designed to hold the fluctuation in the rotational speed of the engine within the limits required by the load.

The flywheel also assists in holding the speed of an engine nearer to the desired value while the engine governor is altering the fuel pump setting to suit some changed condition in the load on the engine. It also stores energy from the starting system during the starting operation and keeps the engine turning at sufficient speed to initiate combustion.

The speed fluctuations discussed above do not alter the speed of the engine when measured as revolutions per minute, but cause small changes in angular velocity within the period of one or two revolutions depending on whether the engine is of the two- or four-stroke cycle.

Note The flywheel is also used in conjunction with the turning gear for turning the engine. In smaller engines teeth are cut in the flywheel rim which are engaged by the pinion of a starting motor when the engine is started.

■ **17.5 The terms 'fluctuation in energy' and 'fluctuation in speed' are used in reference to reciprocating engines. What do you understand by these terms?**

The term fluctuation in energy refers to the changes in the energy output that occur during a complete cycle. It is calculated from the twisting moment or torque diagram and the positions at which the angular velocities of the crankshaft are at a maximum and minimum value. The term fluctuation in speed refers to the changes that occur in the angular velocity of the crankshaft during a complete cycle.

After the fluctuation in energy and speed have been obtained, coefficients of these values are calculated and used for establishing the required mass moment of inertia for the flywheel.

■ **17.6 How would the requirements of a flywheel for a main propulsion engine be different from the requirements for a diesel engine driving an alternator?**

The purpose of the flywheel on each type of engine is the same. The allowable limits of speed fluctuation for a propeller are generally higher than for an alternator. The flywheel for an alternator must therefore control the speed fluctuation to much narrower limits. The necessity for controlling the speed fluctuation to narrow limits becomes particularly important for engines driving alternators which have to be run in parallel; further, the allowable tolerances in electrical frequency are very small for much of the electronic equipment in use on a ship.

■ **17.7   Does the speed of a piston change or remain the same throughout the stroke?**

When we speak of piston speed we are usually referring to the mean piston speed which (in m/s) is given by

$$\frac{\text{stroke (mm)} \times 2 \times \text{rev/min}}{1000 \times 60} = \frac{\text{stroke (mm)} \times \text{rev/min}}{30\,000}$$

Example

What is the mean piston speed of an engine of 1600 mm stroke running at 115 rev/min?
The mean piston speed in m/s is

$$\frac{1600 \times 115}{30\,000} = 6.13 \text{ m/s}$$

The mean piston speed in ft/min is

$$\frac{1600}{25.4 \times 12} \times 2 \times \text{rev/min} = 5.2493 \times 2 \times 115 = 1207.35 \text{ ft/min}$$

Actually, during the stroke of the piston, the piston velocity changes from zero at the beginning of the stroke to some maximum value at approximately mid stroke. The velocity then decreases to zero again at the end of the stroke. From our studies in mechanics we know that any change in velocity is associated with acceleration, therefore as the velocity of a piston varies it follows that acceleration is occurring. When the piston velocity is zero the piston acceleration is at a maximum.

If a connecting-rod were of infinite length, the piston velocity and acceleration would be such that it would move with simple harmonic motion.

■ **17.8   Define the term 'simple harmonic motion'.**

Simple harmonic motion is a form of movement that is quite often met with in nature: the strings of a musical instrument, the simple pendulum, and the particles in an elastic body which is vibrating all move with simple harmonic motion. It is often denoted by the letters s.h.m. We know from the motions mentioned that it is a form of oscillating movement of a point or body which takes place between defined limits or points. If a body or point moves with s.h.m. it has an acceleration which varies as its distance from the central point of the movement. The distance that the point moves from the central position is termed the displacement.

Mathematically, if the velocity of a point moving with s.h.m. is plotted on a time or angular displacement base, a sine curve is formed. The acceleration is given by a cosine curve when plotted on a time or angular displacement base. Zero time commences with movement from the centre point of the travel or from when the displacement is zero.

■ **17.9 What do you understand by the terms 'displacement', 'oscillation', 'periodic time and frequency' when related to s.h.m.?**

If a body or point is moving with s.h.m. it will oscillate between two fixed points; if the centre between the two fixed points is taken, the maximum displacement is the distance from this centre to the extreme point of travel. The displacement is in effect half the distance between the extreme points of travel.

An oscillation is the movement of a point or body from the centre to one extreme point, over to the opposite extreme point and back to the centre point again. More simply it could be described as the movement starting from one extreme point, over to the other extreme point and back to the original extreme point. An oscillation is sometimes referred to as a cycle.

The periodic time is the time taken to make one oscillation or cycle. The frequency is the number of cycles or oscillations made in unit time. It is the reciprocal of the periodic time.

■ **17.10 Give a statement or formulae covering the velocity and acceleration of a piston when the connecting-rod is 'n' times the length of the crank. Say whether the formulae are applicable for both slow- and high-speed engines.**

Let

V = piston velocity
f = piston acceleration
l = connecting rod length
r = crank length = half stroke
θ = angle turned through by crank
ω = angular velocity of crank

$$n = \frac{l}{r}$$

Then

$$V = \omega r \left(\sin \theta + \frac{\sin 2\theta}{2n} \right)$$

and

$$f = \omega^2 r \left(\cos \theta + \frac{\cos 2\theta}{n} \right)$$

The formulae given are sufficiently accurate for slow- and medium-speed engines; in balancing calculations for high-speed engines it might be necessary to take the part of the formulae within the brackets out to a larger number of terms, which can be done by placing the appropriate input into a Fourier series. The mathematical work involved in this is beyond the scope of this book, but the method can be found in books on mathematics.

■ **17.11   When a piston is moving through its stroke it experiences acceleration and retardation. Where do the forces causing acceleration and retardation originate?**

When we consider the balance of the moving parts of an engine the gas pressures are not taken into account (for reasons which are explained in the next Question). The piston is connected with the other moving parts of the engine through the connecting-rod. The force causing acceleration of the piston from the top centre position can be considered as coming from a pull exerted by the connecting-rod. After the piston reaches its maximum velocity, retardation occurs, the retarding force coming from a push exerted by the connecting-rod. This push from the connecting-rod continues while the piston goes through the bottom centre position and continues so that the piston is accelerated in the vertical direction upwards. After the piston reaches its maximum velocity it is again retarded as it continues towards the top centre. The retardation is caused by a pull exerted by the connectng-rod.

The reactions to the forces in the connection-rod can be traced back to the main bearings and the engine bedplate.

■ **17.12   Do the forces associated with piston acceleration have any effect on the balance of an engine? What effect does the gas load on the piston have on balance?**

When we speak about the balance of an engine we refer to the balancing of forces associated with the acceleration of reciprocating parts, and the balancing of centrifugal forces arising out of the revolving parts. The forces arising out of piston acceleration do have an effect on balance. The forces associated with each piston do not act in the same plane so couples are created which must also be taken into consideration when the complete balance of the engine is being studied. This form of balancing is associated with the dynamics of the various moving parts; in consequence the gas load on the piston is not considered in the calculations covering the balance of these forces and couples. However, the effects of the gas load on the piston must obviously be considered when stresses are being calculated.

When calculations are made to find the actual loads on crosshead bearings, crosshead slippers, piston pins and on the sides of piston skirts, a correction must be applied to the gas load on the piston for the inertia loads arising out of the reciprocating parts.

■ **17.13   How is it possible for the forces arising out of the acceleration of the piston to affect the balance of an engine?**

It is known that a force is required to accelerate a mass (Newton's Second Law of Motion)

accelerating force = mass × acceleration

The acceleration of the reciprocating parts is denoted by f and

$$F = M \times f$$

where M = reciprocating mass. As

$$f = \omega^2 r \left(\cos \theta + \frac{\cos 2\theta}{n} \right)$$

the original equation can be rewritten

$$F = M\omega^2 r \left(\cos \theta + \frac{\cos 2\theta}{n} \right)$$

If the parentheses are removed we have

$$F = M\omega^2 r \cos \theta + M\omega^2 r \, \frac{\cos 2\theta}{n}$$

The component $M\omega^2 r \cos \theta$ is referred to as the primary disturbing force and is comparatively easy to balance because it has an action similar to centrifugal force.

The second component $M\omega^2 r \cos 2\theta/n$ is referred to as the secondary disturbing force and is more difficult to balance.

When the piston is on top centre the piston is in effect pulled down by the connecting-rod to give the acceleration, and the reaction to this pull tends to lift the bedplate. When the piston approaches bottom centre it is retarded, and the reaction from the retarding force tends to push the bedplate down.

When the piston moves upwards from the bottom centre position the reaction to the accelerating force continues to push the bedplate down until maximum piston velocity is reached. Beyond this point and as the piston approaches top centre moving upwards, the retarding force reaction tends to lift the bedplate up.

This can be summarized by saying that the primary inertia forces from the piston and the reciprocating masses cause the bedplate to press hard down on the chocks when the piston approaches, goes through and leaves the bottom centre position. Similarly the bedplate tends to lift when the piston approaches, goes through and leaves the top centre position.

■ **17.14 Give details of how primary disturbing forces can be balanced on single-cylinder engines.**

Primary disturbing forces from the reciprocating parts can be balanced by placing balance weights on the crank webs at 180° to the crankpin. The centrifugal force exerted by the revolving balance weights works in opposition to the primary inertia forces associated with the acceleration of the reciprocating parts. Vertical forces coming on to the bedplate (vertical engines) from the reciprocating parts are then balanced by the centrifugal forces coming from the balance weights. Unfortunately when the cranks are at 90° to the line of the piston stroke, the centrifugal force from the balance weights tends to rock the engine transversely. A compromise is therefore necessary and it is customary to balance out the vertical primary forces only partially, by using balance weights lighter than would be required for full balance, and so keep the transverse disturbing forces down to reasonable proportions.

The primary inertia force at any point of piston travel is given by

$$M_1 \omega^2 r_1 \cos \theta$$

where M_1 = the mass of the reciprocating parts.
  ω = the rotational speed of the crank radian/second.
  r_1 = the crank radius.
  θ = crank angle.

The centrifugal force caused by the rotation of the balance weights is given by

$$M_2 \omega^2 r_2$$

where M_2 = the sum of the mass of each balance weight.
  ω = the rotational speed of the crank radian/second.
  r_2 = the radius of action of the balance weights.

Figure 17.1 shows in diagramatic form a piston and crank with a balance weight attached. A vector triangle shows the centrifugal force *CF* from the balance weights acting at an angle of θ. From this diagram it can be seen that the vertical component *VC* of this force is equal to *CF* cos θ.

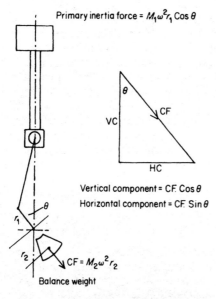

Primary inertia force = $M_1 \omega^2 r_1 \cos \theta$

Vertical component = CF Cos θ
Horizontal component = CF Sin θ

CF = $M_2 \omega^2 r_2$

Balance weight

Fig. 17.1 Balance weights fitted to crank webs to balance or partially balance the primary inertia forces arising out of the reciproacting parts of an engine. The primary inertia forces are governed by the mass of the reciprocating parts and the speed of rotation. The mass of the balance weights may be varied to suit the degree of balance required. The speed of rotation given by ω is the same for the reciprocating parts and the balance weights.
In order to reduce the value of the horizontal component of the centrifugal forces coming from the balance weights it is usual only to partially balance the primary inertia forces. The reduction of the horizontal component value reduces the side forces acting on the engine and minimises their effect.

By substitution, the vertical component of the centrifugal force is shown to be:

$$VC = M_2 \, \omega^2 \, r_2$$

The value of ω is the same in both the centrifugal force equation and the primary inertia force equation. If the percentage amount of balance required is given by x then:

$$M_2 \, r_2 = xM_1 \, r_1 \,/100 \qquad \text{as } M = W/g$$
$$W_2 \, r_2 = xW_1 \, r_1 \,/100$$

■ **17.15** Is it possible to balance the secondary disturbing forces? Give some details of how it is done.

For an engine running at a constant speed the variables in the formulae for the primary and secondary disturbing forces are the values of cos θ and cos 2θ, where θ represents the crank angle movement from the top centre position. The table below gives the values of the cosine for θ and 2θ.

Crank angle θ	Cosθ	2θ	Cos 2θ
0	+ 1.0000	0	+ 1.0000
45	+ 0.7071	90	0.0000
90	+ 0.0000	180	− 1.000
135	− 0.7071	270	0.0000
180	− 1.0000	360	+ 1.0000
225	− 0.7071	450	0.0000
270	0.0000	540	− 1.0000
315	+ 0.7071	630	0.0000
360	+ 1.0000	720	+ 1.0000

Examination of the table shows that the secondary disturbing force reaches a maximum and minimum value twice in one revolution of the crank, while the primary disturbing force reaches a maximum and minimum value once, i.e. the secondary disturbing force goes through two cycles while the primary disturbing force goes through one. It is possible to balance the secondary disturbing forces but the balancing masses must revolve at twice the crankshaft speed to accommodate the two cycles that occur in one revolution of the engine crankshaft.

The way this is done in practice is to fit two auxiliary balance weight shafts on either side of the engine bedplate, geared to run at twice the engine speed, and in opposite directions. The balance weights on these shafts are in the top centre position when the piston is on top centre. When the piston is at approximately half stroke the balance weights on the auxiliary shafts are in the bottom position, and so on. A set of balance weights is required for each piston, and they are fitted in the transverse plane of the piston. Because the auxiliary shafts run in opposite directions the centrifugal forces from the weights on them balance out when they are in the horizontal plane: they are either both acting outwards at the same time when the weights are outboard or both acting inwards when the weights are inboard.

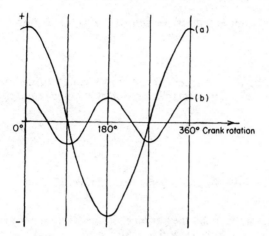

Fig. 17.2 The primary and secondary inertia forces plotted as a function of crank rotation degrees.
(a) Plot of primary inertia forces.
(b) Plot of secondary inertia forces.

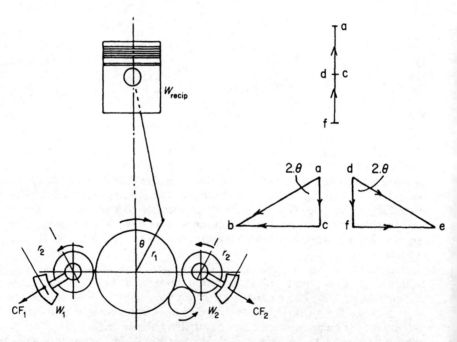

Fig. 17.3 Secondary balance weights and their horizontal components. By arranging the balance weights in the manner shown, the horizontal component of each balance weight is balanced. As the horizontal components are balanced the secondary inertia forces can be completely balanced.
A pair of balance weights is fitted at each crank position.

The values of cos θ and cos 2θ and the crank angle θ as set out in the table are shown plotted, with different peak amplitudes in Fig. 17.2. The value of cos 2θ has twice the number of peak values as cos θ in the same amount of crank displacement.

Figure. 17.3 shows a diagramatic arrangement of secondary balance weights as described earlier. The number of teeth on the crankshaft gear wheel is twice that on the balancer shaft pinions. The number of teeth on the idler pinion has no effect on the speed of the right-hand balancer shaft.

The centrifugal force created by each balance weight is shown in the vector diagrams as a–b and d–e, their vertical components are a–c and d–f respectively. The mass of each balance weight and its centre of gravity is arranged so that the sum of the two vertical components is equal to the secondary inertia force arising out of the motion of the reciprocating parts.

The horizontal components b–c and f–e are equal in amount but act in opposite directions, therefore, they balance one another when the engine is in operation.

Note The mathematics related to this form of balancing can be found in text books dealing with theory of machines and the Lanchester methods of balancing.

Some engines in the higher-speed range of both in-line and V-types have auxiliary balancing shafts of this kind.

■ **17.16 In engines with auxiliary shafts and weights for balancing secondary inertia forces, where must special attention be given during crankcase examination?**

When secondary-force balancer shafts are fitted to an engine the loads coming on to the bedplate and holding-down bolts are reduced. The centrifugal forces from the revolving balance weights on each shaft, although balancing the vertical secondary forces and in opposition to one another in the horizontal plane and therefore balancing out so far as the engine structure is concerned, do imposed heavy loads on the individual auxiliary shaft bearings. The stress on the bearing studs changes during one revolution of the auxiliary shaft from the tensile stress due to tightening of the stud and nut, to a maximum value made up of the tightening stress plus the maximum stress induced from the centrifugal force exerted by the revolving balance weight. The induced stress in the studs from centrifugal force is variable. The load on these studs is cyclic in nature, so they are prone to slackening off while in operation.

During crankcase inspection special care must be given to the auxiliary shaft bearings, the bearing cap fastenings and locking devices. If these studs slacken, particularly at the driven ends of the shafts, considerable damage may come about, often involving costly renewals and repairs. If the studs slacken at the driving end, the driving gears may come out of mesh, jam and cause the auxiliary shaft or crankshaft to bend.

When these bearing studs are tightened they must be carefully pulled up to the torque recommended in the instruction book. Split pins or split cotters should not be used if they do not fit properly. The fit should be such that the split pin can be lightly tapped home with a light hammer. If special locking nuts with fibre inserts are used, they should not be re-used without the most careful consideration. Their cost is so little in relation to the potential damage if they slacken, that they should be considered as expendable items during overhaul – unless the instruction book specifically states otherwise.

■ **17.17** **In vertical diesel engines each unit, so far as the masses of the reciprocating parts are concerned, is identical. The cranks for each unit are at some angle to one another and at some distance from one another. What effect does this have on the disturbing forces created by the reciprocating parts?**

As the cranks are at an angle to one another there is a difference in the value (at any one instant) of the disturbing forces between two units. The difference in the instantaneous values of the disturbing forces and the fact that they are acting at some distance from one another causes a couple or moment to be created. The product of the difference in the value of the disturbing forces and the distance between the cylinder centres produces the moment. These disturbing moments tend to rock the engine bedplate in a fore and aft plane. The action tends to force one end of the engine bedplate down and the opposite end up.

Balance weights are sometimes fitted at one end of a propulsion engine to balance out moments. When a primary moment is being balanced the weights revolve at engine speed; they are timed or phased so that the moment from the action of the weights balances the moment produced by the engine.

When a secondary moment is being balanced the weights revolve at twice the engine speed; they are also so phased that they balance out the secondary moments. If the weights are timed or phased incorrectly they will increase the out-of-balance moments.

The fitting of these balance weights assists in reducing any hull vibration that may arise out of any of the moments coming from the engine. In some cases balance weights become necessary if the engine is located at a node point in the hull and the frequency of the moment matches or nearly matches the natural frequency of any hull vibration and creates a resonant condition. (Fig. 17.4)

■ **17.18** **In vertical opposed piston diesel engines where the upper and lower pistons operate through one lower crankshaft, the upper piston stroke is less than the lower piston stroke. Why is the engine made in this way?**

In this type of engine the pistons always move in opposite directions so it is possible to balance the primary disturbing forces from the upper and lower pistons and their attendant reciprocating parts. Taking the formula for the primary disturbing force we have

$$F = M\omega^2 r \cos \theta$$

and in order to balance the disturbing forces from the upper piston they must be equal and opposite to the disturbing force from the lower piston.

As ω and cos θ are identical for both upper and lower pistons they cancel out and we can say

top piston mass × r upper piston = bottom piston mass × r bottom piston

It follows that the weight of the upper piston reciprocating parts consisting of the piston, transverse beam, side rods, etc. is naturally greater than the reciprocating parts of the lower piston. As the mass of the upper piston reciprocating parts is greater, the value of r, the length of the crank, must be made smaller to make the primary disturbing forces equal and opposite.

It should be noted that as the primary forces are equal and opposite for the upper and lower pistons they equate to zero in any one cylinder unit. As they equate to zero no couples are set up.

Note This applies only to engines where the lower piston is connected to a centre crank and the upper piston is connected through two side rods and two connecting-rods to a crank on either side of the centre crank.

■ 17.19 If one examines the lower half of the crankcase or the bedplate of opposed piston engines it is noted that the scantlings are lighter than the bedplate of a normal engine. Explain why.

As opposed piston engines of the type discussed have no external primary forces or couples and the firing loads are contained between the upper and lower pistons and not transmitted down to the bedplate and main bearings, it follows that loads coming on to the bedplate are much less than in ordinary engines. In consequence they are often made lighter than the bedplates of other types of engine.

It should be noted that the gas loads and inertia forces from the individual upper and lower pistons and running gear are transmitted through to the crankshaft from the centre and side rod bottom-end bearings.

■ 17.20 If one examines a crank and bottom-end bearing assembly it is seen that quite a large mass is on one side of the centre-line of the crankshaft journals. How does this off-centre mass affect the operation of the engine?

The mass of material which forms the crankwebs, crankpin, bottom-end bearing assembly, and the lower part of the connecting-rod, constitutes a large mass which rotates in a circular path having its axis on the polar axis of the journals. When the engine is running at operational speed these parts set up centrifugal forces. The summation of these forces is such that they add up to a large force which acts radially outwards from a point midway between the crankwebs passing outwards through the centre of the crankpin. The value of this force is naturally dependent on the weight of the parts and their angular velocity. The effect of this force on the operation of the engine is to induce loads in the main bearings which in turn tend to move the bedplate in the direction in which the force is acting. For example, when the crankpin is passing over the top centre the force is acting upwards and is tending to lift the bedplate. When the crankwebs are at 90° to the line of piston stroke the force tends to slide the bedplate to one side in the direction of the crankweb. When the crankpin is on bottom centre the force is acting downwards and tends to push the bedplate hard down on the chocks and engine foundations.

■ **17.21   How do the primary disturbing forces from the inertia of the reciprocating parts compare with the forces caused by the revolving parts?**

The primary inertia forces from the reciprocating parts vary in amount but always have the same line of action; the centrifugal force created from the revolving parts on the crankshaft is always the same in amount but is changing in direction.

The mass of the bottom part of the connecting-rod, the bottom-end bearing, the crankpin and part of the crankweb sets up a centrifugal force, the vertical component of which is algebraically additive to the inertia force arising out of the mass of the reciprocating parts.

The horizontal component tends to slide the engine bedplate from side to side (see Fig. 17.1).

■ **17.22   What do you understand by the terms 'static balance' and 'dynamic balance'?**

A crankshaft or turbo-blower rotor is in balance statically if it has its centre of gravity on the polar axis of its journals. If the rotor or crankshaft were set up on knife edges it would remain in the position it was left in and would not roll. If it is out of balance statically it will have its centre of gravity off the polar axis of the journals. When placed on knife edges and moved slightly it will then always roll to the same position where its centre of gravity is directly under the polar axis. In any other postion a turning moment exists due to gravity which causes the rotor or crankshaft to roll until its C.G. is vertically below its polar axis.

When a turbo-rotor or crankshaft is in dynamic balance the conditions for static balance must be present, but in addition, when the shaft or rotor is revolved in the bearings or on two sets of rollers the load on each bearing or set of rollers must remain constant throughout 360° of rotation or during the time the shaft or rotor is turning. If the load on a bearing or set of rollers does not remain constant throughout a revolution it indicates that some mass in one plane is out of balance with another mass in another plane and a couple is set up which causes a variation in the load on each of the bearings or rollers.

If a turbo-blower rotor with two blade discs is considered it can be seen that if a blade were removed from one rotor disc and another blade were removed from the other rotor disc but diametrically opposite in angular position, the rotor would be in good static balance if the moment (mass × distance from axis) of each blade were the same. If this rotor were run up to speed in a balancing machine it would exhibit bad dynamic balance due to the moment created by each disc with a missing blade. The out-of-balance forces not being in the same plane and therefore some distance apart, create a couple or moment.

■ **17.23 What do you understand by the statement that an engine is in good balance so far as moments and forces are concerned?**

If an engine is in good balance it is understood that the forces and couples coming on to the bedplate are within acceptable limits and are not likely to cause trouble with the foundation chocks or holding-down bolts, or set up unacceptable vibration in the engine foundation and hull structure. For an engine to be in good balance all the forces caused by the reciprocating and revolving parts must be preferably fully balanced out, or at least partially balanced out. Any moments or couples caused by these forces must also be fully or partially balanced out. In many cases it is not possible to completely eradicate all the forces and moments but they must be balanced to satisfactory and acceptable values.

Generally it can be said that six-cylinder and twelve-cylinder engines are the easiest to balance completely, followed by eight-cylinder engines – in which out-of-balance couples may remain. Usually engines with four or more cylinders

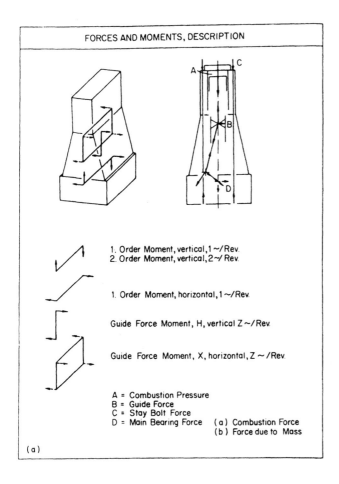

FORCES AND MOMENTS, DESCRIPTION

1. Order Moment, vertical, 1 ~/ Rev.
2. Order Moment, vertical, 2 ~/ Rev.

1. Order Moment, horizontal, 1 ~/Rev.

Guide Force Moment, H, vertical Z ~/ Rev.

Guide Force Moment, X, horizontal, Z ~ / Rev.

A = Combustion Pressure
B = Guide Force
C = Stay Bolt Force
D = Main Bearing Force (a) Combustion Force
(b) Force due to Mass

(a)

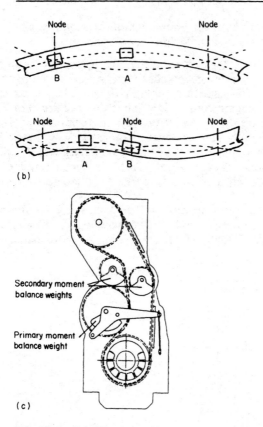

(b)

(c)

Fig. 17.4 (a) Forces and moments occurring in two-stroke crosshead engines. These forces and moments occur in any make of engine and are dependent on engine speed, the mass of parts, etc.

The first and second vertical order moments may cause a forced vibration in the hull. If the engine speed is such that the forced vibration frequency is near a natural frequency in the hull, resonant conditions may occur and lead to very high hull stresses and extreme discomfort to those on board. Although electronic navigation equipment is tested on vibrating stands to ascertain the life of components, experience has shown that breakdown of electronic navigation equipment occurs early in life when any hull vibration that may appear acceptable is present. Some failure of electronic equipment is also associated with local vibration caused by propeller excitation. (MAN B&W drawing)

(b) Two- and three-node hull vibration. Engines with large out of balance vertical moments may cause hull vibration when fitted at points marked 'B'.

Note. Some main propulsion engines with a small number of cylinders and large out of balance primary forces have caused hull vibration to occur when fitted at points marked 'A'. When hull vibration occurs while operating under light ship condition the vibration may be reduced by increasing the amount or changing the location of ballast.

(c) Primary and secondary moment balance weights fitted at one end of an engine. These balance weights can be fitted after the shop trials if a serious hull vibration not found in the hull design stage is experienced. (MAN B&W)

have the cranks arranged so that the primary reciprocating forces and centrifugal forces are balanced.

Note The only perfectly balanced reciprocating engines were those built to the designs of the late Dr Lanchester and used in automotive practice many years ago.

■ **17.24 List the forces and moments which may come about in a vertical multicylinder engine. What effects will these forces and moments have?**

1 Primary and secondary disturbing forces arising out of reciprocating parts, together with their associated moments.
2 Horizontal and vertical forces and moments arising out of the centrifugal forces from revolving parts.
3 Forces and moments arising out of guide forces.
4 Reaction of the engine structure due to torque from the crankshaft.

The forces and moments are dependent on the number of cylinders, crank angles, masses and speed of the moving parts. In some cases the forces and moments are balanced internally and have no external effect, in other cases the unbalance may manifest itself in vibration of the hull structure. Where there is no external effect arising out of the balance moments there may be an increase in the loads on main bearings around the mid length of the crankshaft.

Figure 17.4 (a) shows the out of balance forces and moments associated with a two stroke cycle crosshead type diesel engine. The effects of gas pressure are also shown.

Figure 17.4 (b) shows how an engine having out of balance forces and moments can cause hull vibration.

An engine having out of balance forces could cause a two node, three node or multi node form of vibration if it is located between nodes. This is shown when an engine is located at points marked 'A'. The upward and downward acting forces coming out of the engine create a push and pull action on the hull structure and lead to vibration.

Engines having out of balance moments may cause vibration if located at points marked 'B'. The out of balance moments create a couple; when this couple acts in way of any node point the rocking action coming out of the engine acts on the hull and could lead to vibration.

If the hull structure is not strong enough to resist or dampen these actions vibration will be experienced. If the engine is operating at a speed which coincides with a natural hull frequency a resonant condition will arise and the vibration amplitudes may become excessive or even dangerous. A change of engine speed, up or down, usually gives the desired effect.

If service demands make it impracticable to change engine speed, changing fuel location or moving ballast may give the required degree of damping. If the vibration remains excessive then a speed change becomes necessary to prevent the start of hull fractures.

Figure 17.4 (c) shows a type of balancer gear capable of balancing moments that lead to hull vibration.

The engine designer aims to eliminate or reduce to acceptable levels of the various forces and moments which come from an engine when it is in operation. The reactions to these forces and moments come down to the engine foundation which forms part of the ship's hull structure and in consequence may cause vibration of the hull or parts of the hull.

■ **17.25   Apart from ease of manufacture, why is the semi-built crankshaft more popular in slow-speed direct-coupled engines?**

Semi-built crankshafts may be cast or forged in steel, and the two webs and crankpin form one piece. By enlarging the size of the crankpin and boring it out, a very stiff section can be formed. This, together with the fact that the web does not encircle the crankpin as in fully built shafts, reduces the out-of-balance rotation weight at the crankpin and makes it less difficult to achieve the balance required. Some crankshafts are balanced by varying the size of the holes in the crankpins and casting a balance weight section within the engine flywheel.

■ **17.26   How can the action of vibration be described?**

If a constrained piece of elastic material is loaded in some manner, deformation occurs. When the load is removed the deformed material returns to its original condition. In returning to its original condition the elastic strain energy stored in the deformed material is converted to kinetic energy which causes the material to continue moving beyond the point of zero strain. The movement continues until all the kinetic energy is again converted to elastic strain energy and the material is in a deformed condition. The action is again repeated as the deformed materials revert to the position of zero strain. The action continues and the material is said to be vibrating.

The motion is described as free or natural harmonic vibration.

■ **17.27   Imagine you have a square-sectioned length of rubber constrained by holding the ends in the right and left hands. Describe how you could stress the rubber and name the types of vibration that would be associated with the stresses described.**

The piece of rubber could be stretched by moving the hands apart thereby increasing its length, axially. If vibration occurred so that axial stresses are set up it would be described as axial vibration.

If the wrists are turned in opposite directions the square section would be twisted and the edges of the square corners would change from straight lines to helices. The rubber would then be subjected to torsional or shear stress; when vibration occurs associated with torsional stress it is described as torsional vibration. In the unstressed condition the corners of the square-section rubber lie along straight lines; when torsional vibration occurs the straight lines formed by the square corners move to form right-handed and left-handed helices. When the shaft is twisted in this manner it is slightly shortened in an axial direction. When under the action of a varying torque changing in direction as in a crankshaft, the periodic decrease and increase in length will bring about an axial movement at the end of the shaft and may lead to an unwanted axial vibration.

By moving the hands, one clockwise and the other anti-clockwise, the length of rubber can be bent to form the arc of a circle. The stresses will be similar to those in a loaded beam and if a vibration occurs from this form of loading it is referred to as transverse vibration.

The three forms of vibration are: axial vibration in which the particles move backwards and forwards in lines parallel to the axis of the bar or shaft; torsional vibration in which the particles move in small circular arcs around the centre of the axis of the shaft; transverse vibration in which the particles move in lines perpendicular to the axis of the bar.

■ **17.28 Is it possible for a shaft to experience torsional vibration while it is revolving and transmitting torque?**

Yes. Axial and transverse vibrations can also occur in revolving shafts. In making studies of the vibration characteristics of shafting, engine components and the like, the part being studied is considered as static except for the motion of the vibration which is occurring.

■ **17.29 Sometimes a piece of material can be seen vibrating but the vibration dies away and the vibrating part becomes static. Why does this occur?**

All elastic materials when vibrating experience some resistance which dissipates the energy causing the vibration, and so the vibration may die away and the vibrating parts become static. This resistance and its action is referred to as *damping*. It is made up of internal and external resistance. The internal resistance is caused by the resistance of the particles to motion or by internal friction. The external resistance is due to the resistance set up by the medium in which the part is vibrating. It should be pointed out that as the amplitude increases the damping action also increases.

■ **17.30 If a vibration is damped describe what occurs.**

When a free vibration is damped the energy changes at each oscillation are gradually reduced. The amplitude is also gradually reduced until it becomes zero. During this period of decay the periodic time is slightly increased and the frequency is slightly reduced.

■ **17.31 The term 'forced vibration' is sometimes used. What do you understand by this term?**

Forced vibrations are of fairly common occurrence, and therefore of considerable importance in vibration studies. For forced vibration to occur the vibrating body must be subjected to a rhythmically applied force or load. The initial motion of vibration is complicated in nature and consists of both free or natural vibrations and the forced vibration. The natural vibration rapidly dies away (see Note) and the forced vibration continues under the action of the rhythmically applied force. The frequency of a forced vibration is the same as the frequency of the rhythmically applied force.

Note   The rapid decay of the natural vibration is sometimes referred to as the transient vibration or transient condition.

■   **17.32   The term 'resonance' sometimes appears in articles on vibration. What is resonance and why is a resonant condition of great importance?**

It is known that any elastic system has a natural vibration which will have some frequency. If the system is subjected to forced vibration and the frequency of the rhythmically applied forces is very near to or the same as the natural frequency of the elastic system, resonance occurs. When the frequency of the applied force approaches the natural frequency of the elastic system the amplitude of the vibration increases at an alarming rate. The amplitude of the vibration is equivalent to strain. As stress is proportional to strain, stress also increases at an alarming rate. Operation of machinery where a resonant condition occurs is extremely dangerous and early failure must be expected of the part operating under resonant conditions. It is for this reason that resonant conditions are of great importance.

■   **17.33   What is a harmonic?**

If we have a vibration with a fundamental frequency, the harmonics would be vibrations having frequencies which are whole number multiples of the fundamental frequency. For example, if the fundamental frequency were 4 cycles per second, the second harmonic would have a frequency of 2×4 cycles per second (8 c/s), similarly the fourth harmonic would have a frequency of 4×4 cycles per second (16 c/s) and so on.

■   **17.34   When we speak of constrained elastic systems vibrating, what would be the constraint in the shafting system of a direct-coupled slow-speed engine?**

The crankshaft and shafting system of a slow-speed direct-coupled diesel engine is made up of a series of large masses connected by relatively slender and flexible shafts. If we start at the aft end of the system the first large mass is the propeller. This is connected to the engine flywheel by a long shaft; being long, relative to its diameter, it is naturally flexible. The two constraints are the masses of the propeller and the flywheel.

Between the flywheel and the engine is the thrust shaft and the aftermost crankwebs. The thrust shaft, while relatively slender, is much shorter and therefore its elasticity is less. Each crank can also be considered as a constraint connected by a relatively flexible journal.

The large masses are therefore the constraints and the elastic parts of the system are the connecting sections of the shafting.

■ 17.35 If we have two masses of different values fastened at each end of a flexible shaft supported on frictionless bearings, how would the system behave when the shaft is vibrating torsionally and in a natural manner?

Imagine that a straight line has been scribed along the shaft longitudinally between the two masses. Vibration would be started by giving the masses some angular displacement in opposite directions. If the masses and shaft dimensions were chosen so that the vibratory movement were slow enough to be easily visible, the scribed line would be seen to go through a swinging movement so that right-handed and left-handed helices were formed. It would also be noticed that the movement of the scribed line was a radial or angular movement which took place about some point which remained fixed while the rest of the shaft was torsionally vibrating. On one side of the fixed point the shaft would swing in one direction while on the other side it would be swinging in the opposite direction. The fixed point is referred to as the *node* or *nodal point*.

The periodic time would be the same for the sections of shaft on each side of the node, and the node would be nearer to the larger mass; the amplitude of the larger mass would be less than the amplitude of the smaller mass.

In this example the amplitudes would be angular and would normally be measured in radians. The line scribed on the shaft when in a position of extreme displacement is referred to as the elastic line.

■ 17.36 Show how the pressure ordinates on an indicator diagram can be redrawn on a uniform time base in the form of a periodic pressure or force acting on the piston.

An indicator diagram taken from an engine where the indicator drive is taken from the crosshead or from a lever and eccentric off the camshaft, will incorporate the effect of connecting-rod angularity.

The connecting-rod angularity and its effect on piston position must be taken into account when re-drawing the diagram on a time base or as a polar diagram.

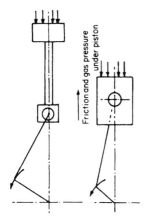

Fig. 17.5 Sketch showing the forces acting on an engine piston.

The method of making this correction is shown in Fig. 17.8. The gas pressure on the top of the piston causes motion downwards. The forces resisting piston motion are the gas pressure under the piston, piston ring and piston friction, and piston skirt or crosshead slipper friction. The inertia forces resist the piston motion during the time the piston is accelerating from the top centre position and reaching its maximum velocity; during the lower part of the piston stroke when the piston is being retarded the retardation assists the movement of the piston. When the piston is moving upwards from the bottom centre position the inertia forces retard piston movement until the piston reaches its maximum velocity. When the piston is being retarded the inertia forces assist the piston motion. The length of the connecting-rod must be known. This is drawn to the same scale as the base of the indicator diagram. The positions of the ordinates can then be corrected to obtain their relative crank angles. When this has been done the ordinates from the diagram can be plotted on the time or crank-angle base (see later and Fig. 17.8 which shows how this is carried out).

If the gas pressure at any point is multiplied by the area of the piston the diagram can be drawn with the ordinates as the force acting on the piston. This diagram will be identical in form with the plot of the gas pressure because the area of the piston has a constant value.

■ **17.37   Show how the effect of two harmonic motions having the same frequencies but not occurring at the same time can be combined.**

The two harmonics are first plotted as separate harmonics on the same time base, but with the time difference correctly positioned. The vectors of each harmonic motion are then added geometrically to obtain a third periodic motion. This is shown in Fig. 17.6. The time difference between the two harmonics is referred to as the phase angle or phase difference.

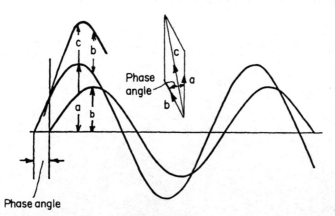

Fig. 17.6   Combination of two harmonics of the same frequency. **Note.** The term 'phase angle' applies only to two harmonics with the same frequency value. Harmonics with different frequencies do not have a phase angle.

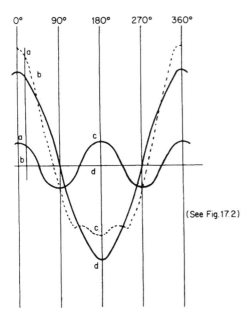

Fig. 17.7 Addition of primary and secondary inertia forces.

■ **17.38** Show how two periodic motions can be combined when one periodic motion has a frequency of twice the other.

This case can be likened to the case of the primary and secondary inertia forces in an engine.

The two periodic motions are plotted on the same time base and the ordinates are added algebraically to obtain the combined effect of the two periodic motions. (Fig. 17.7)

■ **17.39** If a periodic motion is the same as or similar to the combined motions shown in the previous Questions, could the periodic motion be broken down to the separate motions from which the combined motion was originally obtained?

It is possible to break down a combined form of a set of periodic motions into the basic forms from which it was obtained. Sometimes the basic forms are referred to as harmonics. This is difficult to carry out and requires a good knowledge of mathematics.

One method is to use a mathematical process known as Fourier Analysis. The method is beyond the scope of this book, but the student should be familiar with it in order to achieve a better understanding of the cause and effect of vibration.

■ **17.40** Show how the diagram of the forces acting on a piston (shown in Fig. 17.5) can be converted to a diagram showing the tangential forces acting on the crankpin.

The force acting on the piston at any point in the stroke is obtained from the product of the pressure at that point in the stroke and the area of the piston.

If the effect of inertia force is ignored, the force acting on the piston due to gas pressure is the same as the force acting on the crosshead. This force can be drawn as a vector in a triangle of forces from which the force acting on the connecting-rod may be obtained. The force acting on the connecting-rod can then be drawn as a vector in another triangle of forces to obtain the tangential force acting on the crankpin. The values obtained can then be plotted as ordinates on a time base. (Fig. 17.8)

The method for carrying out this procedure is as follows.

Figure 17.8 (a) The atmospheric line of the indicator card or diagram is projected to the right to accommodate the scale length of the connecting rod and the crankpin circle. The length of the indicator card and the diameter of the crankpin circle will be the same. In the example shown the connecting rod length is three times the crank throw. The crankpin circle has been divided up into 10 degree intervals. Using a compass set to the scale length of the connecting rod, arcs are swung from the various points on the crankpin circle to pass through the atmospheric line of the indicator card.

The height of the ordinate at each point is then lifted with dividers and carefully measured. The product of ordinate height and spring number gives the cylinder pressure at that point in the piston stroke. The calculated value for the cylinder pressure at each ordinate is recorded.

Figure 17.8 (b) shows a base line drawn up to represent 360 degrees of crank rotation divided up into 10 degree intervals.

The cylinder pressure recorded for each ordinate is set up to some scale on the corresponding ordinates in (b). The curve showing the gas pressure in the cylinder may then be drawn in as shown.

Figure 17.8 (c) shows the line of the piston stroke set up vertically and the crankpin circle drawn to the same scale as in (a). An arc corresponding to the scale length of the connecting rod is swung from a point on the crank pin circle, corresponding to the point on the crankpin circle shown in Figs 17.8 (a) and (b), to pass through the line of stroke.

In Fig. 17.8 (d) the gas pressure is set up in a vector diagram as a vertical line a–c. The angle of the connecting rod to the line of the piston stroke is set up at the correct angle as shown in a–b, this will run parallel to the line of the connecting rod shown in Fig. 17.8 (c). The length of the vector (a–b) represents the force acting down the connecting rod per unit area of piston.

The vector b–d passing through point (b) is drawn parallel to the normal on the crankpin circle. The line of the crank is drawn parallel to the line of the crank in Fig. 17.8 (c) to pass through point (a) in Fig. 17.8 (d). The vector b–d represents the force acting tangentially to the crankpin per unit area of piston. The product of this force, the area of the piston and the length of the crankthrow gives the torque acting on the crankshaft at this point in the piston

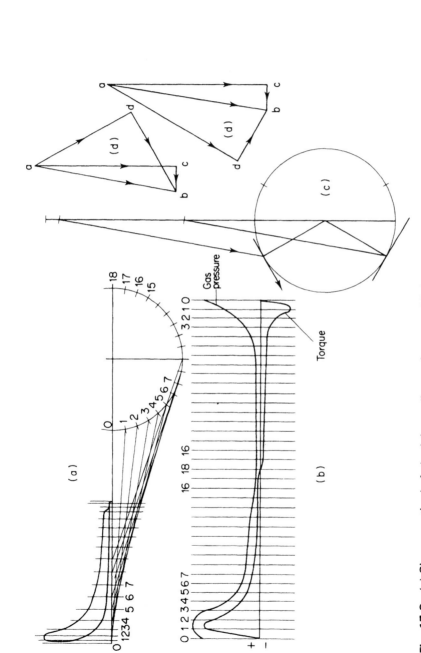

Fig. 17.8 (a) Shows method of obtaining ordinates for every 10 degrees of crank travel to make the correction for connecting rod angularity. (b) Gas pressure ordinates from indicator card set up at 10 degree intervals on time base. The product of the tangential force on the crank pin obtained from (d) and the crank-throw gives the torque. The torque values obtained are set up as ordinates on (b). (c) Shows the free body or space diagram for the connecting rod and crank-throw in two positions. (d) The vector diagrams show the method of obtaining the tangential force on the crankpin.

stroke. The value of the torque is plotted to some scale on Fig. 17.8 (b). This is repeated for each of the crank positions drawn up in (a).

After marking in each torque value the curve of torque values plotted against crank angle may be drawn in.

When dealing with four stroke cycle engines the base for the cycle must extend over 720 degrees of crank rotation.

Note If a greater degree of accuracy is required, the inertia forces arising out of the motion of the reciprocating parts will have to be used as a correction factor and added algebraically to the force on the piston arising out of the gas pressure.

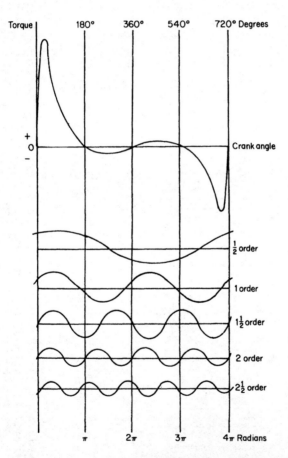

Fig. 17.9 The torque plotted against the crank angle for a single-cylinder, four-stroke cycle, pressure-charged engine. The torque, crank angle diagram has been broken down into a set of curves representing the half order, first order, one and a half order, second order and two and one half order harmonics. **Note.** There are no half order harmonics in two-stroke cycle engines.

■ 17.41 Show the form of the diagrams you would expect if the curve obtained from plotting the tangential force on the crankpin shown in Fig. 17.8 were to be broken down into its various harmonics.

The various curves obtained by Fourier Analysis are shown in Fig. 17.9.

The harmonics shown sketched in below the torque curve were at one time obtained mechanically by using a special drawing and plotting instrument. In design office practice tables and sets of curves may be used to find the harmonics and calculate the corresponding stresses.

Today, this design work is carried out with a computer and a plotter using the various programs now available. It should be noted that two stroke cycle engines do not have half order harmonics.

■ 17.42 If three masses were attached together by two relatively flexible shafts how would the system vibrate torsionally? What is the relationship between the number of masses and the number of nodes?

Let one mass be located between the two flexible shaft sections and the other masses be fitted at opposite ends on each side of the central mass. If the system described is supported on a pair of frictionless bearings it can be shown that there are two *modes* of natural torsional vibration. The system is sometimes referred to as a three-mass system.

One *mode* of vibration would be similar to that described in 17.34 and have one *node*. In this mode, two of the masses would move in one direction while the other mass moved in the opposite direction, and vice versa. The nodal point would naturally be on that section of shafting where the masses moved in opposite directions. In the second mode of torsional vibration the central mass would move in one direction while the two outer masses moved in the opposite direction and vice versa. This mode of vibration has two nodes – one on each section of shafting.

Generally it can be assumed that the number of nodes will be one less than the number of masses in the system, or equal to the number of lengths of elastic connecting-shaft sections.

■ 17.43 In a system in which a six-cylinder, direct-drive, slow-speed engine is connected to a propeller, how many nodes would there be? How many modes of natural vibration could such a system have; are some more important than others?

If each crank is considered as a mass there will be six masses within the engine, the mass of the flywheel, and the mass of the propeller. This is equal to a total of eight masses. The possible number of nodes could be up to seven. From this it may be assumed that there could be seven different modes of vibration. The first mode of vibration would have one node, the second two nodes, the third three nodes and so on up to seven. Generally, the lower modes of vibration are investigated and analysed as these are normally of more importance; the higher modes of vibration are often outside the operating speed range of the engine.

■ **17.44   What do you understand by the terms 'critical speed' and 'barred speed-range'?**

The graph of the turning moment from a single-cylinder engine is made up from a combination of the forces acting on the pistons and the inertia forces from the running gear (Question 17.3). The turning moment reaches a peak each 720° of crank rotation in a four-stroke cycle engine and each 360° for a two-stroke cycle engine. In multi-cylinder engines there will be a peak for each cylinder. In a four-stroke cycle engine the number of peaks will therefore equal the number of cylinders firing in 720° of crank rotation, or half the number of cylinders for 360° of rotation. In two-stroke cycle engines the number of peaks per revolution will be equal to the number of cylinders. If the peaks per 360° of crank rotation are multiplied by the engine rev/min we can get a frequency per min. If this frequency is similar to any one of the various natural frequencies of the engine shafting system we have a resonant condition (Question 17.31).

The speed of the engine in rev/min at which the resonant condition occurs is referred to as the critical speed. The high stresses associated with resonant conditions start to build up as the engine speed approaches the critical speed, and do not come back to some safe value until the engine speed is beyond the critical speed. The unsafe stresses either side of the critical speed are referred to as the flank stresses. The barred speed-range of an engine is the range of speed from the beginning of build-up of unsafe flank stresses to the dying-away of these stresses at some higher speed. Obviously the engine must not be continuously operated at speeds within the barred range.

■ **17.45   How are ship's officers made aware of the barred speed-range of an engine where a barred speed exists?**

The barred speed-range or ranges are often marked on the engine-room and bridge propeller-shaft speed tachometers. The marking consists of a red-coloured sector on the tachometer. The width of the sector is such that it corresponds to the upper and lower speeds of the barred speed-range. In other cases an engraved plate giving the lower and upper speeds of the barred range is fitted on the bridge and engine-room main-engine control stations. The barred speeds are also recorded in the main engine instruction book, usually on the page giving the engine's characteristics.

■ **17.46   What is the effect on the natural frequency of a two-mass system if any of the following changes are made?**
(a)   An increase or reduction in the distance between the masses.
(b)   An increase or reduction in one of the masses.
(c)   An increase or reduction in the shaft diameter.

(a)   An increase in the distance between the masses reduces the natural frequency of the system; conversely a reduction in the distance increases the natural frequency.
(b)   Increasing the size of one of the masses so that there is an increase in its mass moment of inertia will cause a reduction in the natural frequency of the system; decreasing the mass moment of inertia increases the natural frequency.

(c) An increase in the shaft diameter will make the shaft stiffer; consequently the natural frequency of the system will be increased.

■ 17.47 If the spare propeller has to be fitted in place of the normal working propeller does the barred speed change, or is a barred speed likely to be introduced?

If the spare propeller and working propeller are made of similar materials and have similar characteristics the barred speed will not alter. But if, for example, the working propeller is made of one of the lighter high-tensile bronzes and the spare propeller is made of cast-iron there will be a considerable difference in their weights. Consequently there will be a difference in their mass moments of inertia (WK^2) and when the spare propeller is fitted the barred speed will change. It can generally be assumed that if a heavier propeller is fitted the natural frequency of the shafting system will be reduced. The actual amount of change will depend on various factors. Where there is a large difference between the weights of the spare propeller and the working propeller, details are given at the time the torsional vibration characteristics are submitted by the engine builders to the classification society for their approval.

The actual change in the barred speed-range is usually small and is often covered by making the red sector on the tachometer sufficiently large to include both the working and spare propeller barred ranges.

■ 17.48 If in the preliminary design of a shafting system it is found that the torsional vibration characteristics are not satisfactory what can be done to improve the situation?

If the system is to run at constant speed (for example, a diesel generator) there is normally no great difficulty in bringing the critical speeds outside of the operating speed of the engine. This can usually be done by increasing the torsional stiffness of the shafting system.

In the shafting systems associated with propulsion machinery it becomes more difficult to modify the natural frequency of the system, as usually more than one mode of vibration must be covered. If alterations to the torsional stiffness of shafting, or changing the values or location of the masses does not give the desired results, additional masses must be brought into the system. Sometimes the additional mass is in the form of a damper or detuner. In geared propulsion systems, couplings giving some damping effect can be fitted between the engines and the gearbox (described in Question 12.6).

■ 17.49 What do you understand by the term 'anti-node'?

In an earlier question it was shown that within a vibrating system certain points did not move or had zero amplitudes. These points having zero amplitudes were referred to as nodes. The anti-nodes within a vibrating system are those points where the amplitudes of vibratory movement have maximum values.

■ **17.50   Give some details of the commoner forms of vibration, detuning and damping devices used in conjunction with marine diesel engines.**

The commonest detuning and damping devices used in marine practice are viscous-fluid dampers and spring-loaded detuners. In each of these both damping and detuning occurs. The viscous-fluid damper consists of a flanged circular casing which is rigidly fixed to the shafting usually at the forward end of the crankshaft. Within the circular casing is a ring which has a large mass and completely fills the casing except for small clearances around it. The ring is not fastened to the casing in any way and is completely free to move angularly within it. The clearance space is filled with a fluid which has the property of retaining its viscosity over a wide temperature range. The ring fitted within the casing is sometimes referred to as the seismic mass.

Spring-loaded detuners have been made in many forms but a common form used in merchant ships consists of two parts. One part is rigidly fixed to the crankshaft and has a centrally located shaft projecting from it. The other part,

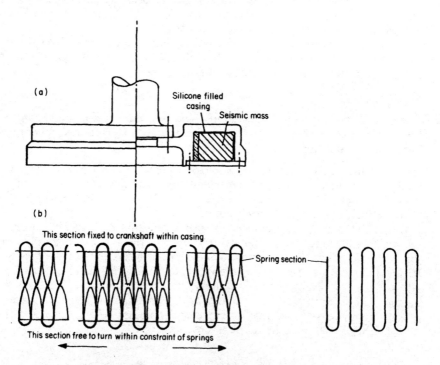

Fig. 17.10   Silicone filled and spring-loaded detuners.
(a) Silicone filled detuner. The small clearance space between the seismic mass and the casing is filled with a silicone liquid.
(b) Spring-loaded detuner shown operating with maximum damping. As increased damping becomes necessary when approaching a critical speed the deflection of the springs increases to give the required degree of damping.

consisting of a large mass (sometimes called the seismic mass), is fitted on the first part in such a manner that it can move angularly. The centre part of this mass is bored out to fit on the projecting shaft. The two parts have a series of slots cut around their circumferences in an axial direction. The two parts are connected by a series of flat springs, the inside ends of the slots being cut with a curve which accommodates the curvature of the springs when they are deflected under load. The moving part of the detuner is enclosed within a casing which is attached to the part rigidly fixed on the crankshaft. The inside of the casing is connected with the bearing lubricating oil supply and drain holes allow a small circulation of lubricating oil through the casing. (Fig. 17.10)

■ **17.51   How does a viscous damper operate?**

The effectiveness of any mass on the natural frequency of a shafting system vibrating torsionally is dependent on the mass moment of inertia of the mass. A viscous damper is made up of two masses: the comparatively light casing and the heavy inner ring which has a high mass moment of inertia. When the engine is started the inner ring lags behind the outer casing, which moves with the crankshaft. The viscous fluid drags the heavy ring round and it quickly comes up to the same speed as the casing.

When the inner ring is moving with the speed of the casing the effective value of the mass moment of inertia of the damper is the summation of the value for the casing and the inner ring. As the speed of the engine is increased and a resonant condition is approached, the crankshaft and outer casing commence to vibrate torsionally. When this occurs the viscous fluid separating the casing and the ring allows some slip to take place and the effective mass of the damper is reduced to that of the casing only. The reduction in the mass alters the frequency of the system by increasing it. When slip occurs the energy within the vibrating system is dissipated by the movement of shearing action within the viscous fluid, and the amplitude of the vibration is kept within safe bounds. As the amplitudes are kept low the stresses remain low. This will occur at any critical speed.

■ **17.52   How does a spring-loaded detuner work?**

The connection between the fixed and moving elements of a detuner commonly takes the form of flat strip springs, or laminated springs made up from a series of flat strips. These springs fit into slots which have a curved profile. The curved profile is such that when the detuner operates with very small vibration amplitudes the distance between the spring supports in the slots is nearly a maximum. When the vibration amplitudes are large the distance between the spring supports is reduced. If we liken the flat strip spring to a beam we can see that as the distance between the supports is reduced the stiffness of the beam or the spring increases. This change in stiffness of the spring in effect alters the mass moment of inertia of the moving part of the detuner and so varies the natural frequency of the shafting system.

When the engine is in operation and the speed is altered so that it coincides with or approaches a critical speed, vibration of the shafting system commences. The element of the detuner fixed to the crankshaft vibrates with the crankshaft. The moving part of the detuner then lags or leads the vibratory movement of the crankshaft and the natural frequency of the shafting system is changed or detuned and the resonant condition vanishes. While the moving part of the detuner is lagging or leading the part attached to the crankshaft, the energy used to reverse or change the loading on the flat springs dissipates a large amount of the energy of the vibration, and damping occurs. The damping effect reduces the amplitude of vibration of the crankshaft and keeps the stresses within safe bounds.

■ **17.53   Is it possible to operate an engine continuously at a speed which causes the damper or detuner to be in full operation?**

Detuners and dampers designed for use with main engines will usually operate continuously. The heat generated from the damping action of the viscous fluid or the load reversals on the flat strip springs must be continuously removed so that no serious temperature rise takes place in the damper or detuner.

Viscous-fluid dampers are sometimes fitted external to the engine crankcase in such a manner that sufficient air circulates around them to remove the heat generated. The lubricating oil allowed to circulate through a detuner is usually of sufficient quantity for adequate heat removal.

■ **17.54   Are detuners and dampers fitted at the nodal points or anti-nodal points within a shafting system associated with a diesel engine?**

At the nodal points in a shafting system subjected to torsional vibration there is no torque variation or amplitude arising out of the vibration. As there is no torque variation or amplitude nothing is available to activate the movement of the seismic mass in the detuner or damper. No detuning or damping can therefore take place.

Detuners and dampers are most effective if they are fitted at the anti-nodes. In practice this may not be possible so they are often fitted at the forward end of the engine. There is nearly always enough vibratory movement at the forward end of the crankshaft to activate fully detuning and damping devices. In some special cases detuners have been fitted at the aft end of the engine and on the intermediate shafting.

■ **17.55   Why is it desirable to have the drive for the camshaft and fuel pumps fitted at or near a nodal point on the crankshaft?**

If the drive is taken from a nodal point the turning moment is even; this is desirable as it prevents any crankshaft torsional vibration from being transmitted to the fuel pump and camshaft drives. If the drive is taken from some point on the crankshaft where there is appreciable torsional amplitude, considerable variations in torque will be transmitted through to the camshaft and fuel pump shafts.

The torque variation can cause heavy loads to come on the drive chains or gearing. Chains may dampen the vibration to some extent, but when the drive is through gearing it may impose heavy loads and cause fatigue failures of gear teeth. This is more likely to occur in idler wheels.

If drives are also taken to the governor a speed increase is involved through gearing; the torque variation may then cause the governor drive gearing to wear excessively.

■ **17.56 How would you know whether the drive to the fuel pumps and camshaft was, or was not at a nodal point?**

The elastic curve line drawn for each mode of vibration crosses the base line or line of zero amplitude at the nodal points. An examination of the nodal points and their position relative to the camshaft drive will show if the drive is at a nodal point and indicate the amplitude of the various modes.

On an actual engine on a deep-sea ship, where the number of hours running ahead is very large relative to running astern, the wear patterns on the crankshaft gear and the camshaft gear teeth will show wear on both sides of the teeth if any appreciable amplitude or torque variation exists. If wear is present on both sides of the teeth, the teeth on the idler gears should be inspected periodically for fatigue failure starting at the root fillets, more particularly if the root fillet radius is small.

Note Dye-penetrant tests are very effective for this form of examination (see Question 6.50).

■ **17.57 Why are dampers fitted in the camshaft drives of some medium-speed engines?**

Dampers may be fitted for either of the following reasons.

1 They may be fitted to smooth out the torque variations produced by the engine at the point where the camshaft drive gear is fitted on the crankshaft.
2 In some engines, particularly those where the fuel pumps are driven by the main camshaft, the variation in torque driving the camshaft through two revolutions can be very wide and in many cases (particularly with small cylinder numbers) the torque may have negative values for part of the time. When the camshaft has negative torque values it is actually assisting in driving the crankshaft and in so doing reverses the loading on the driven and driving gears. The function of the damper in this case is to smooth out the back driving periods, reduce tooth loads and noise, as back lash is taken up.

Note Dampers used on camshaft drives are fitted with restriction devices to limit the relative motion between the two parts of the damper.

■ **17.58 If you read in the instruction book for the main engine that a third-order, second-degree vibration existed at some engine speed what would you understand by this?**

A vibration of the second degree is the same as the second mode of vibration. It indicates that when the system is vibrating two nodes are present. The third order indicates that the engine speed is such that the number of impulses per revolution coincides with the third harmonic of the natural frequency of the system, in this case the two-node natural frequency (see Question 17.41).

Note   In four-stroke cycle engines $\frac{1}{2}$-order harmonics are possible. For example a $1\frac{1}{2}$-order harmonic would have a frequency of $1\frac{1}{2}$ times the fundamental frequency.

■   **17.59   If the natural frequency of a shafting system is N and the speed of the engine is such that a second-order critical speed exists, what is the speed of the crankshaft?**

It can be shown that the natural frequency divided by the critical speed gives the order of the critical speed.

$$\frac{\text{natural frequency}}{\text{critical speed}} = \text{order number}$$

$$\text{critical speed} = \frac{\text{natural frequency}}{\text{order number}}$$

In the case given

$$\text{critical speed} = \frac{N}{2}$$

$$= \tfrac{1}{2} \text{ natural frequency}$$

If the frequency is measured in cycles per minute the second-order critical speed in rev/min will be half the frequency (see Question 17.41).

■   **17.60   It is known that the friction between the molecules in a material is responsible for the ability of that material to dampen out vibratory movement. Is the ability to dampen out vibratory movement the same for all the constructional materials used in diesel engines?**

The ability of a material to dampen out vibratory movement is sometimes referred to as its damping capacity or dynamic ductility. Usually the materials with high tensile strength have low damping capacity and those with lower tensile strength have higher damping capacities. For example, high-tensile steels have less ability to dampen than low-tensile steels. Generally cast-iron has a better damping ability than steel, which is one reason it is sometimes preferred for crankcase construction.

In other cases (apart from providing easier production to any shape) cast-iron (special alloy type) is preferred to steel for the crankshafts of smaller engines.

Note   Generally the ability to withstand fatigue or the resistance to alternating stresses is better with the higher tensile materials.

■   **17.61   When axial vibration occurs in the shafting system of a main propulsion engine or diesel generator, where is the node located?**

The node is at the fixed point within a vibrating system. In a main engine (direct drive) the propeller when running ahead forces the shafting system to take up a position where the ahead face of the thrust meets the thrust pads. This becomes the fixed point or the node in the system. Similarly the node will be coincident with the crankshaft location bearing in diesel generator sets. Any vibratory movement in an axial direction will therefore take place away from the thrust bearing or location bearing.

■ 17.62 Where does axial vibration usually occur in marine diesel engine shafting systems?

Axial vibration is usually found within the crankshaft section of the system and makes itself manifest by movement of the forward section of the crankshaft; in multi-cylinder, in-line, slow-speed engines the movement may reach an appreciable amount. The shortening and lengthening of the crankshaft journals and crankpins due to the shear stresses arising out of the torque placed on them may set up axial vibration. In older engines with more flexible crankshafts this may be exacerbated by bending of the crankwebs, crank pins and journals in a similar manner to the vibration in a tuning-fork (see Question 17.27).

Axial vibration can be controlled by a damper on the forward end of the crankshaft. The damper usually consists of a piston moving in a cylinder. The movement of the piston causes oil to be forced through small openings so that resistance to movement is set up. The resistance damps out the vibration.

■ 17.63 Why are air inlet and exhaust valve springs arranged in such a manner that more than one spring is used per valve?

If the frequency of the natural vibration of the air inlet or exhaust valve springs is a harmonic of the camshaft speed, the springs may vibrate axially and are said to surge.

Surge can be avoided by modifying the sizes of the springs. By arranging the springs in pairs and fitting one inside the other the avoidance of surge is made less difficult. Another advantage of having pairs of springs is that if one spring fails the valve is held up and is not damaged by striking the piston.

Note A stopper or safety ring is usually fitted around valve stems to prevent a valve falling into the cylinder, but in the lower position it is still possible for the valve to be struck by the piston.

■ 17.64 When a shaft is vibrating torsionally how can the stresses caused by the vibration be ascertained? If the shaft is driving some external load does the vibration increase the stress?

Referring to the two-mass system in Question 17.34, it was found that the line scribed on the shaft in the unstressed condition would swing between two positions and that the line would swing radially with the nodal point as the centre. The elastic line gives the position of extreme displacement. The angle subtended between the elastic line and the scribed line in the zero-stress position at the nodal point gives the angular strain along the surface of the shaft.

Let this angle be denoted by α radian. Let the distance from the node to one of the masses be l, and the angular displacement of this mass be θ, and the diameter of the shaft be D. It can be shown that

$$l\alpha = \frac{\theta D}{2}$$

from which

$$\theta = \frac{2l\alpha}{D}$$

Let q be the stress from the torsional vibration (shear stress) and C be the modulus of rigidity (shear modulus). From the torsion formula it is known that

$$\frac{q}{D/2} = \frac{C\theta}{l}$$

Therefore

$$q = \frac{CD}{2l}\theta \quad \text{or} \quad q = \frac{CD}{2l} \times \frac{2l\alpha}{D}$$

and

$$q = C\alpha$$

The stress caused by the torsional vibration is equal to the product of the modulus of rigidity of the shaft material and the angular strain α on the surface of the shaft. This stress, from the torsional vibration, is additive to the stress from the torque set up by the external load. When torsional vibration is taking place, therefore, the maximum stress is equal to the addition of stress from the external load, the stress from the vibration, and the induced stress from the bending action which also takes place when the engine is in operation.

Note   In practice, obtaining the various frequencies either by calculation or tabular methods is reasonably straightforward but beyond the scope of this book. The calculation of stresses from vibration, and under resonant conditions, is more complex.

18

INSTRUMENTATION AND CONTROLS

■ **18.1   How is the speed of a main engine controlled? What types of device can be used to control speed?**

The speed of main engines is controlled primarily by the fuel-lever or fuel-wheel setting. The fuel lever or wheel controls the fuel pump settings which in turn control the amount of fuel injected per working cycle into each cylinder. Provided the load on an engine did not change the speed of the engine would remain constant for any fuel-lever setting. Unfortunately this condition occurs only in very smooth water; as soon as a ship starts to pitch the propeller rises and falls and the load on the engine changes. If the speed of the engine were controlled only by the fuel-lever setting, the speed would rise and fall with the pitching of the vessel and the corresponding load changes.

Small changes in engine speed can be tolerated. But in bad weather when the ship is pitching heavily it is possible for the propeller to come clear of, or nearly clear of, the water, and in such circumstances the speed of the engine could rise dangerously. A similar situation could arise if the propeller shaft fractured and/or the propeller was lost.

In order to keep the engine speed within reasonable bounds in heavy weather, or in the event of shaft failure, a governor is fitted to the main engine. When the engine speed rises the governor overrides the fuel lever and reduces the fuel injected into the cylinders, so preventing the engine speed from rising further.

There are four types of governor used for this purpose: the inertia-type governor; the mechanical-type governor with spring-loaded sleeves and flyweights; the mechanical hydraulic-type governor; and the electronic governor. The inertia-type governor was at one time used only on main engines, to limit the maximum speed when the engine was racing in heavy weather (see Question 18.16).

■ **18.2   How do inertia-type governors control the engine speed? What are their disadvantages?**

Inertia-type governors are fitted on to a swinging arm or beam which receives its motion by a link connection with some reciprocating part of the engine, such as the crosshead pin. The inertia governor then moves up and down through the arc of a circle and an angle of approximately 45°. The governor consists of a weight held down by a spring. The weight is normally held by the spring in the lower position, and when the weight moves from its normal position, one pawl (the upper) is retracted and another pawl (the lower) is extended. When the speed of the engine rises the inertia of the weight is such that it overcomes the spring holding it. This retracts the upper pawl and extends outwards the lower pawl. The lower pawl engages with a lever and lifts it. The lever is connected with the fuel pumps or fuel pump suction valves. Movement of this lever reduces the quantity of fuel injected, or may allow injection to occur only in one cylinder. When the speed of the engine returns to normal the weight returns to its normal position and reverses the pawls. The upper pawl then pushes the lever downwards and restores the fuel pumps to the setting given by the fuel lever. The lever connected with the fuel pumps has its fulcrum pin in the same centre-line position as the axis of the swinging arm.

Although this governor is very simple it requires an engine speed increase of the order of 5% or more to make it operate. In some cases this speed increase will bring the speed of the engine into or near to a critical speed. During heavy weather the engine speed is continuously rising and falling and the reduction in average rev/min may be considerable. This type of governor is fitted only on slow-speed direct-coupled engines and is found mainly on older engines. It has been superseded by mechanical hydraulic-type governors and electronic governors.

■ **18.3 What are the maintenance requirements for inertia-type governors?**

The only regular maintenance required for inertia-type governors is lubrication of the inertia-weight pin, the pawl pins, and connecting lever and linkage pins.

After some lengthy period it becomes necessary to dismantle the governor and clean the various parts, pins and bearings. The spring also needs periodic greasing to prevent wastage from rusting. The pin holding the inertia weight tends to wear on opposite sides and eventually requires renewal. The holes in which the pin fits are usually reamed round again when a new pin is fitted. After reaming, an oversize pin must be used.

■ **18.4 On what principles does the mechanical centrifugal governor operate?**

Mechanical centrifugal governors used in marine practice are of the Hartnell type or modifications of it. The governor consists of a vertical shaft on which a yoke is fitted. The opposite ends of the yoke are fitted with pins, and these pins hold flyweights which are attached to the ends of bell cranks. The other ends of the bell cranks are fitted into a circumferential slot at the top of a sleeve which surrounds the vertical shaft. As the radius of the path of the flyweights is increased the sleeve is raised up the vertical shaft. In some governors springs are fitted to the weights so that the centrifugal force on the weights is balanced by the pull on the springs. In others a helical compression spring is also fitted on the top of the sleeve, in which case the springs between the balance weights may sometimes be omitted.

The action of the governor is such that when the flyweights are revolving the load on the springs is balanced by the centrifugal force of the flyweights. The sleeve then takes up some position on the vertical shaft. An increase in the engine speed increases the centrifugal force of the flyweights and they move outward to a new radius. This causes the sleeve to rise and a new equilibrium position is taken up when the increased centrifugal force is balanced by the springs. The movement of the sleeve is transferred by levers and linkage to the fuel pumps which then reduce the fuel quantity injected and slow the engine down to its proper running speed.

The governor operates on the principle that the centrifugal force exerted by the flyweights is balanced by the load on the spring or springs. Changes in engine speed result in changes in the radius at which the flyweights operate, and this in turn alters the height of the sleeve which controls, through levers and linkage, the fuel pump settings.

Note The spring that is used to load the governor sleeve is also used to make slight adjustments to the engine speed. This is accomplished by increasing or decreasing the load on the spring by a screw mechanism.

■ **18.5** What do you understand by the following terms when referred to speed control governors: 'sensitivity', 'stability', 'effort', 'droop', 'hunting'?

Sensitivity. The sensitivity of a governor is a measure of its ability to control the speed of the engine within narrow limits. It can be understood that a sensitive governor will give a large movement of the control sleeve for only a small change in the radius in which the flyweights are revolving.

Stability. Normally with a governor the no-load position of the flyweights will be such that at some speed N_2 the weights will be at a maximum radius R_2. As the load on the engine increases a slight drop in speed occurs and the flyweights move to a new equilibrium position slightly less in radius than R_2; further increase in load causes a further slight decrease in speed and a further reduction in the radius at which the flyweights are operating. A governor is stable when there is only one radius of rotation of the flyweights for each speed at which the governor operates, within the speed range of the governor.

Note As a mechanical centrifugal-type governor becomes more sensitive it becomes less stable.

Effort is the force exerted by the governor sleeve on the mechanism controlling the fuel pumps (and the fuel pump delivery) when a change of engine load and in turn a change of engine speed occurs. If there is no change of load there is no change of speed and the effort is zero.

Droop. When the load on an engine increases from no-load to full load the governor flyweights move to an equilibrium position with a smaller radius and a slight reduction in speed occurs. The reduction or change in speed which occurs from no-load to full load is referred to as the droop or governor droop.

Hunting. When the load on an engine changes the governor tends to over-control and under-control. This over- and under-control causes a fluctuation in rotational speed which is referred to as hunting. For example, if load is removed from an engine the speed increases some amount above normal. The governor then comes into operation and reduces the fuel supplied by the fuel pump. Due to friction, time lag, etc., the governor causes the fuel reduction to be in excess of that required and the fall in speed is too much. This causes the governor to increase fuel again and the engine speed goes slightly above normal. This swing in speeds above and below the mean operating speed for that load continues until equilibrium is reached and hunting ceases. It can be seen that the more sensitive a governor is, the greater will be the tendency to hunt.

■ **18.6   How is the operation of a generator governor affected by tight sleeves, slack flyweight and bell-crank bearings, and slack cod pieces in sleeve groove?**

If the sleeve becomes tight on its spindle, or other causes affect its free up and down movement, the effect on the governor will be that larger changes in the rotational speed of the engine must occur before the governor can act. When the load on the engine is being increased the spring assisting the downward movement of the sleeve will reduce the effects of the increased friction but the engine speed will drop more than normal before the engine takes up the increase of load at the correct engine speed. With decreasing load the sleeve spring load is additive to the increased friction and a considerable increase in engine speed must occur before the governor can operate and reduce the fuel supply. The governor droop will be increased and the speed at which the engine operates will become erratic.

Slackness in flyweight bearings must be taken up when engine load increases and engine speed falls. This has the effect of increasing the lag before the governor starts to increase the fuel supply and so increase the engine speed to normal. With reduction in load the slack does not have to be taken up and the speed change will be approximately normal.

Slackness in the sleeve cod piece will make for erratic speed changes when the load on the engine changes; the effect of load change on speed will depend on the system of levers and linkage between the governor and the fuel pumps.

■ **18.7   What types of engine use the mechanical centrifugal-type governor?**

All types of engine used for most purposes may use the mechanical centrifu-gal-type governor. Some main engines have a governor of this type fitted to enable the engine to be run at near-constant speeds. In some medium-speed two-stroke cycle engines the governor is fitted horizontally direct on to the fuel pump camshaft. On slow-speed engines the governor is fitted separately and geared to revolve at a speed considerably higher than the engine speed.

The governor is connected with the fuel pump suction valves or the fuel pump rack controlling the circumferential position of the ram, and is arranged so that the engine can run at near-constant speed. This type of governor has been used with diesel generators, diesel-driven emergency compressors, diesel-driven fire pumps and the like.

■ **18.8   When the effort given by a mechanical centrifugal-type governor is insufficient to operate the fuel pump control gear what can be done?**

Formerly if the governor effort were insufficient to give the desired control it would have meant re-designing the governor. Steam turbine designers had to face up to this problem many years ago and they used a Hartnell-type mechanical centrifugal governor to operate a pilot or control valve to supply oil under pressure to a relay piston. The relay piston was connected either through linkage or directly with the throttle valve. The advantages to be gained were that the governor flyweights could be reduced in size with consequent reduction in inertia effects; this increased the sensitivity of the governor. By using oil under pressure as a hydraulic medium, large efforts can be obtained for operating the control equipment.

The basis of this type of governor is used by companies who specialize in the design of centrifugal hydraulic-type governors. They produce a 'packaged' unit in which is contained a centrifugal-type governor, an oil pump and the necessary control valves, linkage, and hydraulic piston which is connected to the external controls on the fuel pumps.

■ **18.9   Name the various parts of a centrifugal hydraulic-type governor. What are the advantages of this type of governor over a normal centrifugal mechanical governor?**

The main part of a centrifugal hydraulic-type governor is the speed sensor which is similar to that of a Hartnell-type governor. The flyweights operate through bell cranks on to a spring-loaded sleeve. The springs on the sleeve consist of a conically wound helical spring and a normal helical spring. The sleeve controls, through a lever, the control valve on the hydraulic system.

Within the system is incorporated a reversible gear pump driven off the main shaft, which pumps oil into two spring-loaded hydraulic accumulators. The accumulators when pumped up also act as relief by-pass valves to assist in maintaining constant pressure in the hydraulic system when large load changes occur.

A direct hydraulic connection is made to the control valve and the hydraulic servo piston. The hydraulic servo piston is of the differential type. The connection with the servo piston from the accumulators goes to the top of the differential piston which has the smaller effective area. The oil pressure in the accumulator is always tending to push the servo piston down which will reduce fuel supply and speed. If load comes on to the engine the momentary or transient speed reduction causes the control valve to open to the oil supply from the accumulators. This puts oil to the underside of the differential servo piston and lifts the piston, thereby increasing the fuel supply to the engine. If load comes off the engine the control valve opens to the underside of the differential servo control piston and allows oil to bleed out and drain to the governor sump. The oil pressure on the top of the piston then forces it down and reduces fuel supply to the engine.

The levers and linkages are arranged so that there is feedback and compensation, and the governor may be termed truly isochronous except for the transient speed changes when load is increased or reduced. Incorporated within the lever and linkage arrangement, controls may be fitted to control load limit; synchronizing control (to raise or lower speed of engine operation), and speed droop control.

This type of mechanical hydraulic governor is supplied by specialist firms which manufacture them as complete units for mounting on the engine. The casing for the unit also forms the oil reservoir or sump of the hydraulic system. The whole of the lever and linkage system is contained within the casing so it always receives adequate lubrication.

The advantages of this type of governor over the mechanical centrifugal type are that, due to the lightness of the parts, inertia effects are at a minimum and the governor is very sensitive without losing stability. The controls for adjustment of governor operation make it particularly suitable for diesel engines generating alternating current and running in parallel, or geared engine installations where two or more engines are connected through gearing to a single propeller.

■ **18.10   What do you understand by the term 'feedback' and how is it arranged for in a mechanical hydraulic governor?**

When there is no change in load on an engine controlled by a mechanical hydraulic governor the control valve in the governor remains in one position. There is sufficient bleeding of oil to the underside of the differential servo piston to ensure that the total pressure on both sides of the piston remains the same and the piston therefore maintains a set position. In consequence the discharge of fuel from the fuel pumps remains constant. A three-point lever is introduced with pivot points at the governor sleeve, the top of the control valve, and the top of the differential servo piston, or as shown in Fig. 18.1. Assume that the load on the engine is increased. The first effect from this is a speed reduction of the engine, followed by the flyweights moving inwards and the governor sleeve moving downwards. The pivot at the top of the differential piston is at this stage a fixed pivot point; the downward action of the governor sleeve moves the lever downwards around the fixed pivot point and also causes the control piston to move downwards. This allows flow of oil to take place from the hydraulic accumulator into the underside of the differential servo piston and causes it to move upwards. During this stage of operation the sleeve is the fixed pivot point. As the differential servo point moves upwards the control valve moves upwards from the action of the three-point lever being moved around the fixed pivot point at the sleeve. The control valve is then moved back to its original position so that the total pressure on each side of the servo differential position is the same. While the control valve is being moved back to the mid position the increase in engine speed also assists in moving the

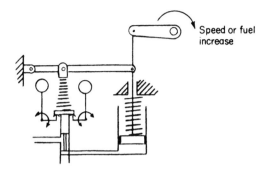

Speed or fuel increase

Fig. 18.1 Woodward hydraulic/mechanical governor with feedback to the speed sensor spring.

control valve upwards. The engine speed is returned to its original value at the new load. The hunting action of the three-point lever moving from the action of the differential servo piston is referred to as feedback.

Figure 18.1 shows a governor arranged with feedback. It is of a type manufactured by the Woodward Governor Company.

Note This action is included in the governor discussed in Question 18.9.

18.11 What do you understand by compensation action within a mechanical hydraulic governor? Why is compensation introduced?

The mechanism for giving compensation action to a mechanical hydraulic governor consists of two hydraulic cylinders and pistons, one being considerably larger than the other. The bottoms of the cylinders are connected with each other. An adjustable needle valve has an opening into the connection between the cylinders and allows oil to bleed into or out of the cylinders from or to the oil sump formed by the governor casing. The piston in the larger cylinder is referred to as the transmitter piston while that in the smaller cylinder is called the receiver piston. The receiver piston is spring loaded in such a manner that it acts in opposition to the governor sleeve spring.

The compensation mechanism is connected into the governor in the following manner: the larger piston is connected to the differential servo piston through a lever. The lever is pivoted in the middle and has a fixed point, and the larger piston and differential servo piston are connected to the opposite ends of the lever. The smaller piston, control valve and governor sleeve are connected together through a three-point floating lever. The control valve is connected to the mid point of the lever. The springs on the governor sleeve and the smaller piston, which act in opposition to one another, always hunt back to the same equilibrium position at any constant load and the engine speed remains exactly the same for any load.

The compensation mechanism is fitted to a mechanical hydraulic governor to control an engine to run at the same speed irrespective of load. Except for the transient speed reduction which must occur when load changes, the governor might be described as isochronous.

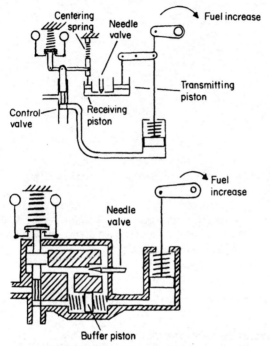

Fig. 18.2   Isochronous governors: speed droop occurs in this type of governor but is a transient condition.
(a) Woodward governor with compensation as arranged in a 'UG'-type governor.
(b) Woodward governor with compensation as arranged in a 'PG'-type governor.

Figure 18.2 (a) shows a diagramatic arrangement of an isochronous governor of the UG type manufactured by the Woodward Governor Company. Figure 18.2 (b) shows a diagramatic arrangement of an isochronous governor of the PG type manufactured by the same company.

Note   The definition of the phrase 'isochronous governors' is given differently in older text books on the Theory of Machines. Today it is usually accepted that an isochronous governor will control an engine with zero speed droop under an increasing load with the exception of transients or momentary speed changes following a load change. After an engine speed change the engine almost immediately comes back to its original speed. The words iso and chronous come from ancient Greek and mean of equal time or uniform in time.

■   **18.12   If a generator fitted with a mechanical hydraulic-type governor shows evidence of erratic speed control how would you find the fault?**

Normally mechanical hydraulic governors give very little trouble, and the units are usually sealed to prevent interference with their internal parts. Virtually the only servicing they require in operation is to check that the system is full of oil by observing the height of the oil in the sight glass, or dip. In the event of operational difficulties the engine should be stopped. The connection between the governor and the control linkage connecting with the fuel pump racks should be dismantled and checked for stiffness, which will almost certainly be the cause of the trouble. Once the stiff area is located it can be freed up so that the governor can operate properly.

Most generators with Bosch, C.A.V. or similar-type fuel pumps have an arrangement on the shaft arm connected to the fuel pump rack which consists of a loose clamp held in place by a spring. With this arrangement, if one of the fuel pump racks sticks or seizes in the full-load position, it does not prevent the governor from operating the fuel pump racks on the other fuel pumps. The fuel pumps on all cylinders except the one with the seized rack cut back the fuel. When this occurs the claw opens on the seized fuel pump rack and allows the governor to continue to control engine speed. A runaway or over-speeding engine is thus prevented. Sometimes a defective fuel pump rack can be located by observation of the claws when wide load changes occur. If no fault can be found within the linkage and shafting system between the governor and the fuel pump racks the fault will be in the governor.

This fault may be an imbalance in the governor controls in the needle bleed valve in the compensation mechanism. If the governor does not operate correctly after checking out controls it is the usual practice to change it.

If it becomes necessary to break the seals and open up a mechanical hydraulic governor the dismantling must be carried out with extreme care and the various parts must be handled with a delicate touch. Assembly of the dismantled parts must be carried out on a perfectly clean bench.

■ 18.13 Develop a statement covering the relationship between power output, engine speed and torque.

Power is defined as the rate of doing work, i.e.

> work done ÷ time

Torque is given by

> tangential force × radius of action

Work done when force moves through 360° is given by

> tangential force × radius × 2π = torque × 2π

and work done per second by

> torque × $2\pi \dfrac{\text{rpm}}{60}$

Power is equal to

> torque × rpm × a constant

(See Note for value of constants).

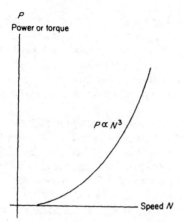

Fig. 18.3 Characteristics of power or torque and speed of a diesel engine used to propel a ship. The characteristics of an engine driving a centrifugal pump or fan are similar. The cubic law is usually used, but in many cases the exponent has some value other than three. For accuracy in calculations the correct value for the exponent should be used.

Note

$$1 \text{ watt} = 1 \text{ joule/second} = 1 \text{ Nm/s} \qquad 1 \text{ hp} = 33\,000 \text{ ft lbf/minute}$$
$$1 \text{ joule} = 1 \text{ newton metre} = 1 \text{ Nm} \qquad \text{torque} = \text{force} \times \text{radius}$$

If torque is in Nm then power in kW is equal to

$$\frac{\text{torque} \times 2\pi \times \text{rpm}}{1000 \times 60} = \frac{\text{torque Nm} \times \text{rpm}}{9549.3}$$

If torque is in kgf m the power in kW is equal to

$$\frac{\text{torque} \times 9.807 \times 2\pi \times \text{rpm}}{1000 \times 60} = \frac{\text{torque kgf m} \times \text{rpm}}{979.7}$$

If torque is in lbf ft then horsepower is equal to

$$\frac{\text{torque} \times 2\pi \times \text{rpm}}{33\,000} = \frac{\text{torque lbf ft} \times \text{rpm}}{5252.1}$$

■ **18.14** Why is the speed of some forms of prime mover more easily controlled than that of others? Give examples.

When the torque of a prime mover decreases with an increase in speed the prime mover is easy to control. When the torque increases with an increase in speed the prime mover becomes difficult to control.

A steam-driven engine operated with the steam control valve or throttle held in a fixed position is an example of a prime mover where the torque decreases with an increase in speed. A machine or engine such as this has to some extent a self-regulating characteristic and the governing of such machinery is relatively easy to accomplish.

A practical example of this form of self-regulation is the older propulsion steam turbine. Their speed was controlled by the number of nozzles opened or by fixing the position of the steam manoeuvring valve or throttle. These older turbines had no speed control governor other than a safety device known as an overspeed trip. This came into play if the load were entirely removed from the engine for any length of time such as when the vessel was pitching in heavy weather.

An unsupercharged four-stroke diesel engine will behave to some extent in a similar manner to the steam turbine. If the fuel lever is put into some fixed position an increase in speed will increase the pumping losses, the windage losses, and the friction in the engine. These increased losses compensate to some extent for the decrease in load, and the engine approaches the condition of being self-regulating.

If an engine-driven supercharger were fitted to the engine and the engine were operated with a fixed fuel lever position, the increased pumping and other losses could be over-compensated by the supercharger and result in an increase in torque. Such an engine would not be self-regulating as the increase in torque would produce a further increase in speed. The engine would then be difficult to govern. Figure 18.4 shows two different speed and torque characteristics. (a) shows the characteristic of increasing torque with increasing speed. (b) shows the characteristic of decreasing torque with increasing speed.

Note If an engine is operated with a fixed fuel lever position it is sometimes referred to as steady state operation.

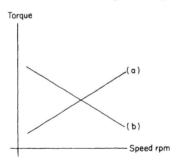

Fig. 18.4 Torque and speed characteristics.
(a) Increasing torque with increasing speed.
(b) Decreasing torque with increasing spped.

■ **18.15 How do the load characteristics vary with speed for an ungoverned, single-screw propulsion engine and a single diesel engine driving an electrical generator.**

A propulsion engine directly driving (no clutch between engine and propeller) a screw propeller will always have a load on the engine. As the speed of the engine is increased the power demand on the engine follows the cube law. That is the power given out by the engine varies as the cube of the speed.

Within load changes where the engine does not over-speed, an engine driving a generator will have a slight decrease in speed as the load on the engine increases, provided the voltage is kept at a constant value. The change will be linear or follow the form of a graph where $y = mx + c$.

Figure 18.3 shows the torque speed relationship of an engine propelling a ship. The relationship is similar when an engine is driving a centrifugal pump or a centrifugal fan.

The cubic law is commonly used for this relationship but in many cases the exponent will have some value other than three. For accuracy the correct value for the exponent should be used when making calculations.

■ **18.16   What is an overspeed trip?**

Overspeed trips are fitted on engines where the governor does not fail safe. Their function is to shut off the fuel supply to the cylinders in the event of the speed of the engine rising to a dangerous level. They are always fitted on steam turbo alternators or generators. The overspeed trip usually consists of a 'bolt' with a relatively heavy head. The bolt is fitted at the forward end of the engine shaft where no torque is transmitted. It is fitted in a space bored out across the diameter of the shaft. The bolt is held in place by a nut and supported by a spring. When the engine overspeeds the centrifugal force exerted by the bolt head overcomes the support of the spring and flies outwards until restrained by the nut and compression of the spring. The bolt head in the 'thrown' position strikes a trip lever which is also thrown and shuts off the engine fuel eventually bringing the engine to rest.

The means whereby the fuel is shut off varies with the make of engine. As some motor ships have a steam turbo alternator it is worth mentioning that in many turbo sets the overspeed trip often works through the lubricating oil low pressure cut-out and the steam stop valve. In other cases the overspeed trip operates through a system of levers and links to the turbine steam stop valve. When the overspeed trip operates, the pawl holding the steam stop valve open is released and a spring closes the stop valve. The tripping speed can be adjusted by adding or removing thin spacing washers under the spring so that its compression is altered.

A small permanent magnet alternator driven from the engine camshaft in conjuction with a rectifier can also be used with an electronic governor to prevent the engine racing. When the engine speeds up the voltage increases. At some level it will operate the governor to reduce the fuel supply to the engine and prevent a dangerous increase in engine speed.

■ **18.17   Why does the speed of an engine change when the load changes.**

When an engine is in operation under the control of a governor it can be considered as acting in a steady state condition. The governor is controlling the flow of fuel to the engine. When the load on the engine changes the power output remains constant for a short delay period because the flow of fuel to the engine remains constant during this period.

After a load reduction there is an excess of torque while the power remains constant during the delay period and the speed of the engine then increases.

After a load increase there is insufficient torque while the power remains constant during the delay period and the speed then decreases. The speed changes remain until the governor acts and adjusts the fuel supply to match the new load.

It must be remembered that during the delay period the energy input to the engine in the form of fuel remains constant and during this period the product of torque and speed remain constant (see Question 18.13).

■ **18.18 What do you understand by the term 'electronic governor'? Name the various parts of an electronic governor.**

Electronic governors are manufactured in various forms. They contain parts which act in a similar manner to the parts in other governors. These parts are a speed-sensing device, some means of comparing the signal from the speed-sensing device to some selected norm, and a means of processing the signal from the speed-sensing device to control some form of actuator and give a mechanical output to control the fuel pump and the flow of fuel from the injection pumps to the individual cylinders of the engine.

The speed-sensing device may be a set of flyweights controlled by a spring and the engine speed. Speed change causes flyweight movement; this movement is transferred to an output rod when the speed of the engine changes. A number of electronic governors use a set of spring-controlled flyweights as the speed sensor. The spring forms the speed reference device.

In other cases the field from a revolving permanent magnet driven by the engine is used to generate an alternating current in a set of stator windings. The alternating current output may be rectified to give a direct current output having a voltage proportional to the engine speed. If the output is not rectified the frequency changes occuring with changes of engine speed give a form of speed sensing.

Proximity sensors may also be used as speed-sensing devices on engines (see Question 18.52). The proximity sensor is located by the engine flywheel-turning gear teeth. In some engines the voltage impulses generated by the proximity sensor may be sufficient without amplification for an input signal giving the engine speed.

In cases where the proximity sensor cannot be located near enough to the flywheel or the form of the gear teeth is not satisfactory, the proximity sensor is used as a switching device to control an input voltage in an on/off manner. The impulses from the switch may then be used to act as a speed-sensing device.

Diesel engine-driven alternators may use the frequency of the current generated as the speed-sensing device. In this case the alternating current is passed through a frequency converter, which outputs a direct current having a voltage proportional to the engine speed in the same manner as the output frequency was treated in the case of the small permanent magnet alternator.

The direct current voltage output proportional to engine speed is then compared with the output from a Zener diode voltage regulator. If there is a difference between the voltages it is amplified to control an output to a solenoid-controlled hydraulic actuator or to supply current to a special form of electric motor and screw. The motor becomes the actuator used to control the fuel pump setting. The armature of the motor must have a very low moment of inertia. (Fig. 18.5)

■ **18.19   What are the advantages and disadvantages of electronic governors?**

The advantages of electronic governors are: few mechanical parts and the ability to build up complex engine-speed control systems incorporating input from various sources such as engine speed, engine load, electrical load; overspeed control; load sharing between engines in geared diesel propulsion installations; and load sharing in engines generating electrical power when connected in parallel to a common set of bus bars either for isochronous control or with speed droop.

The disadvantage of an electronic governor is the loss of governing that can occur if input current to the governor from any source fails.

It is for this reason that some governor manufacturers combine an electronic governor with a mechanical- hydraulic governor. If the electronic portion of the governor supplying current to the electrically controlled actuator valve fails, the mechanical portion of the governor acts as an emergency system and takes over the operation of the actuator control valve. Figure 18.5 shows a diagrammatic arrangement of an electronic governor as manufactured by the Woodward Governor Company.

■ **18.20   Torsionmeters and torsiographs are instruments used in diesel engine operation. What is the purpose of each instrument, and where and how are they used?**

Torsionmeters are used to obtain the shaft horsepower output of a diesel engine. In most cases for practical purposes the shaft horsepower may be considered the same as the brake horsepower (see Note). Torsiographs are used to obtain a record of the vibratory movement that a shaft undergoes when it is vibrating torsionally.

Torsionmeters are fitted on the intermediate shafting between the engine and the screw shaft. The torsionmeter actually measures the angle of twist over a measured length of shaft. When the angle of twist is obtained it is possible to obtain the torque on the shaft by using the torsion formula for shafts, provided that the modulus of rigidity of the shaft material is known.

$$\frac{T}{J} = \frac{C\theta}{l}$$

and so

$$T = \frac{C\theta J}{l}$$

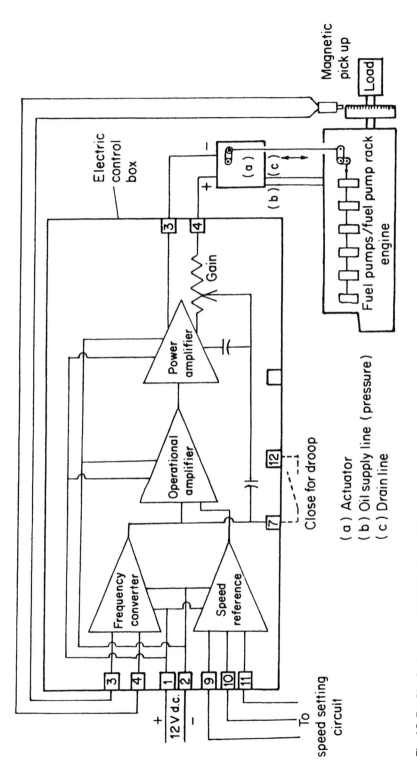

Fig 18.5 Block diagram of a Woodward electronic governor. The actuator is controlled electrically but the servomotor is of the hydraulic type. For the sake of clarity the actuator is shown separately from the engine. It is usual to mount the actuator directly on the engine.

Magnetic pick up

Load

Electric control box

Gain

Power amplifier

Operational amplifier

Frequency converter

Speed reference

Close for droop

3
4
12
7

+
−

(a)
(b)
(c)

Fuel pumps/fuel pump rack engine

(a) Actuator
(b) Oil supply line (pressure)
(c) Drain line

3
4
1
2
9
10
11

12 V d.c.
+
−

To
speed setting
circuit

where

T = torque
J = polar moment of inertia of shaft
C = modulus of rigidity
θ = angle of twist
l = length of shaft

After the value of T is found and the speed (rev/min) of the shaft is obtained it is possible to find the power that is being transmitted by the shaft.

Note The brake horsepower measured with a water brake on the engine test bed is not the same as the shaft horsepower when the engine is in service. This is due to the friction lost in the thrust bearing and any intermediate shaft bearings fitted between the torsionmeter and the engine. Normally they are considered as being approximately the same except in cases where the trial performance of a ship is being very carefully studied or in design calculations involving the power requirements for a new hull design.

Torsiographs used in marine practice are portable and are used at the time a ship is undergoing builder's trials to check the torsional vibration characteristics of the main engine crankshaft and shafting. They are also used on prototype diesel generating sets to make checks on the torsional vibration characteristics of the crankshaft and generator shaft. Torsiographs of various types are available; they are driven by a belt from a temporary pulley fitted at the forward end of the crankshaft. When a torsiograph is being used the vibratory movement of the shaft is recorded on specially prepared paper. The record shown on the paper ribbon requires considerable expertise in its analysis.

■ **18.21 Briefly describe the component parts of a torsiograph which can be used on a diesel engine. Show how the torsiograph is used.**

The main parts of a commonly used torsiograph consist of a relatively heavy circular mass fitted concentrically within a lightweight aluminium pulley which enshrouds the circular mass.

This mass is sometimes referred to as a seismic mass. A spring connects the pulley and the seismic mass so that the pulley drives the mass through the spring. The pulley on the torsiograph is driven by a portable pulley of similar size fitted on the forward end of the engine. A non-stretchable belt connects the two pulleys.

When the torsiograph is in operation the lightweight pulley movement is identical with the actual movements of the engine shaft. The inertia of the seismic mass causes it to revolve at a uniform speed similar to the average speed of the crankshaft. The relative movement between the lightweight pulley and the seismic mass is the same as the vibratory movement of the crankshaft.

A lever and bell crank are fitted between the seismic mass and the lightweight pulley and revolve with them. The relative motion between the pulley and the seismic mass is transferred by the bell crank lever and linkage to a pin which passes right through the centre of the pulley axle. The reciprocating motion of this pin is transferred to a pen and recorded on a specially prepared ribbon and

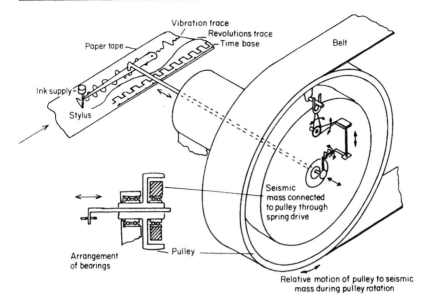

Vibration trace
Revolutions trace
Time base
Belt
Paper tape
Ink supply
Stylus
Seismic
mass connected
to pulley through
spring drive
Pulley
Arrangement
of bearings
Relative motion of pulley to seismic
mass during pulley rotation

Fig. 18.6 Working parts of an older type mechanical torsiograph.

reproduces the vibratory amplitudes of the engine shaft on an enlarged scale. The ribbon moving at constant velocity causes a sinusoidal wave to be drawn on the paper. Clockwork or electrical means are used to drive the paper ribbon. Also traced on the ribbon is a time base line which is used to identify and find the vibration frequency. In other torsiographs a third line may be drawn, showing a small impulse for every revolution of the engine.

Of the three lines traced on the ribbon, the first gives the vibration double amplitudes multiplied to some scale, the second line is a sine curve which has a known frequency and forms the time base from which the torsional vibration frequency is found. The third line, showing the impulses of every engine revolution, is used to obtain the true engine speed (rev/min). The record traced on the ribbon is analysed and used to verify the torsional vibration characteristics found by calculation and also as a check for the stresses found by calculation (see also Question 17.63).

Note The torsiograph described is the one most commonly used in the marine industry. There are, however, other types of torsiograph; the optical and electrical types. (Fig. 18.6)

■ **18.22 What are engine pressure indicators; for what purposes are they used? Are there any limitations on their use?**

Engine pressure indicators used on marine diesel engines are of two types. The first type is used to record the pressure within the cylinder at any part of the engine cycle. A graph of these pressures on a stroke base (to some scale) is then recorded on an indicator paper or shown as a trace on the cathode-ray tube of an oscilloscope. The model usually used on board ship is of the mechanical type and uses small, specially prepared paper sheets called indicator papers. The limitations of a mechanical indicator are due to the inertia effects of the moving parts. Their mass is therefore kept to a minimum, but even so they are not considered reliable when used on engines with speeds above 150 rev/min. Very small lightweight indicators are also made and some models are considered to be suitable for use on engines up to 350 rev/min approximately.

The graph of the pressures recorded during the cycle and shown traced on the indicator paper is commonly called an indicator diagram or card. The average pressure throughout the cycle can be obtained from the indicator card and is referred to as the mean indicated pressure (m.i.p). From the m.i.p. value the indicated power of the engine is obtained.

The second type of engine pressure indicator measures maximum pressures and is referred to as a maximum-pressure indicator or recorder. These indicators are used to obtain the maximum combustion pressure. If the fuel is shut off a cylinder the maximum-pressure indicator may also be used to find the compression pressure. There are no engine speed limitations restricting the use of maximum-pressure indicators. They are often used on medium speed engines.

Note The value for the mean indicated pressure must not be confused with mean effective pressure (m.e.p.); there is a relation between the two values given by

$$\frac{\text{m.e.p.}}{\text{m.i.p.}} = \text{mechanical efficiency of engine}$$

There are no speed limitations in the use of electronic indicators.

■ **18.23 How is a record of cylinder pressures and other data obtained with electronic equipment? What are the precautions that must be observed with the use of this equipment?**

A piezoelectric transducer is used to measure cylinder pressures. When a piezoelectric crystal is subjected to a pressure on its surface it generates a voltage related to the applied pressure. The voltage generated is low, so it is usually amplified; the amplified voltage can be used in several ways to obtain the cylinder pressure values.

The speed of the engine is also obtained by various methods together with a record of the piston position for one cylinder; this is used to obtain the piston position for the other cylinders. The voltage changes representing pressure changes can be supplied to an oscilloscope or an analog computer.

The engine speed and crank position can also be processed for supply to an oscilloscope for giving a time base. The pressure over the cycle can then be shown as a normal indicator card or as an out-of-phase or draw card depending on where the piston position is fixed on the time base.

The varying voltage representing pressure can be processed in a computer to give a direct digital readout of cylinder pressure at any crank angle. This data can be stored and printed as a table of pressures at corresponding crank angles. It can also be processed to give a digital readout of the mean indicated pressure in each cylinder. The availability of this data expedites the balancing of power among the cylinders on the engine.

If a torsion meter is incorporated to give input to the computer, a brake mean effective pressure can also be given together with a figure for mechanical efficiency.

The precautions to be taken in the use of this equipment are to ensure that the pressure transducer is not damaged during handling, and that the cooling system is properly used to prevent false outputs caused by high temperature.

There are no limitations on engine speed when electronic equipment is used to obtain cylinder pressures (see Question 18.49).

■ **18.24 What is a draw-card and what is its use?**

A draw-card can be obtained when the mechanical indicator is in use on an engine. It is obtained by pulling the indicator cord while the indicator pencil is rising so that the maximum pressure is shown at the mid-length of the card. From the curve drawn, the maximum pressure can be measured from the atmospheric line. A small impulse and a change of slope is also shown; this is used to measure the pressure at which combustion of fuel commences. It must not be confused with the true compression pressure although it is sometimes referred to as the compression pressure. On some engines a separate indicator cam (elliptical shape for 4-stroke cycle engines, eccentric for 2-stroke cycle engines) is used to obtain a draw-card or out-of-phase card.

■ **18.25 List the thermal and physical quantities that must be known by an engineer officer when keeping watch on the main engine of a motor ship. List the instruments used in an engine room where watches are maintained.**

Thermal quantity
Temperature

Physical quantities
(a) Pressures.
(b) Flow.
(c) Viscosity.
(d) Liquid level.

Instruments – thermal quantity
(a) Mercury-filled glass thermometers can be used to measure temperatures up to as high as 400°C (752°F). Alcohol-filled glass thermometers can be used for temperatures up to 65°C (149°F). Alcohol thermometers can also be used for temperatures as low as − 70°C (− 94°F). For exhaust gases, pyrometers of the mercury-in-steel type are used for distant readings.
(b) Nitrogen-filled mercury thermometers can be used for local readings up to 540°C (1004°F). See also Question 18.28.

Instruments – physical quantities

(a) Pressures are often measured with gauges of the Bourdon-tube type. For low pressures, manometers, hydrostatic gauges, or U-tubes using a water or mercury column, are used. Sub-atmospheric pressures or vacuum can be measured with Bourdon-tube type gauges or mercury gauges. Manometers, hydrostatic gauges and similar intruments are often referred to as liquid column gauges.

(b) Flow rates or volumetric quantities are not usually measured except for the fuel supply to the main engines where flow meters *are* sometimes used. The rate of cooling water circulation and lubricating oil circulation is measured indirectly with pressure gauges. It is known that as the velocity of a fluid moving in a pipe or some circuit increases, the frictional resistance to the flow increases; by this means the flow rate can be related to the cooling-water pump and lubricating-oil pump discharge pressures (Bernoulli's Theorem related to total energy and flow). Flow rates related to fuel consumptions for main engines (tonnes/day) are found by volumetric measurements of fuel used within the period of the day and the weight of unit volume of fuel. The product of the volume and the weight of unit volume gives the weight.

(c) Viscosity is measured with a viscometer, fitted in the fuel supply line to the main engine. The viscosity of a fluid is directly related to the pressure drop which occurs when laminar flow takes place in a pipe. By pumping a fixed or constant quantity of fluid through a pipe in unit time to give a constant velocity, and measuring the pressure drop, a value related to the viscosity of the fluid being measured can be obtained. A visual readout is given on a suitably calibrated differential pressure gauge.

(d) Liquid levels can be measured by direct means using a sounding tape in a sounding pipe fitted into the tank. In some cases a sounding rod is used instead of the sounding tape. Other direct means of ascertaining liquid levels is by the use of gauge glasses or sight glasses fitted into the side of a tank. The level in fuel tanks is often measured with an instrument which uses an inverted bell placed on the bottom of the tank. The bell is filled with compressed air, some of which bubbles out through the oil. By connecting a pressure gauge to the air supply pipe the head of oil in the tank can be measured as a pressure when the air supply to the inverted bell is shut off. The head of oil compresses the retained air until balance is reached; the balance pressure gives the head of oil in the tank.

■   **18.26   Pressures may be measured with Bourdon tube-type pressure gauges; how do such instruments operate? What other types of instrument work on a similar principle?**

The Bourdon-tube consists of an elliptical-section tube bent into a circular form, making an arc extending round an angle of approximately 270°. If one end of the tube is sealed and the other end is connected to some source of pressure so that the pressure inside the tube is greater than that outside, the circular-formed tube tends to straighten itself, or increase its radius of curvature. The pressure within the tube is related to the increase in curvature which it undergoes and the increase in curvature is the strain (mechanical) on the tube. Instruments such as the Bourdon-tube operate by virtue of the strain which they undergo when the pressure within the tube is different from that outside it.

Other pressure-measuring or sensing devices which use the strain principle are the bellows, the bellows with a support spring, and elastic membranes which may or may not be supported with springs. The bellows and spring are often used as the sensing device in pressure switches, while elastic membranes are often used in pressure-reducing valves and pressure controllers.

■ 18.27 How can the Bourdon-tube type pressure gauge be used as a temperature measuring device?

It is known that most liquids expand as their temperature rises. If some liquid which has this property is contained in a bulb and the bulb is connected to a pressure gauge, the changes of temperature which occur outside the bulb cause a change in volume of the contained fluid. If the volume of the bulb and the pipe connecting it to the pressure gauge are kept approximately constant or change much less than the change in volume of the liquid, pressure changes will occur when the temperature of the contained liquid changes, and the volume changes are resisted. If the dial of the pressure gauge is calibrated in temperature units it can be used to indicate temperature.

This principle is used in mercury-in-steel type pyrometers used for measuring exhaust gas temperatures locally and also for indicating the temperature at some point remote from the sensing position. In this case the connecting tube between the sensing bulb and the indicator is a steel capillary tube.

■ 18.28 Name methods of measuring temperatures other than by mercury and alcohol glass-bulb thermometers and pressure-recording methods.

Most of us are aware of the compound metallic bar which consists of two flat metal strips joined together. The two metals have different coefficients of expansion. When the temperature of the bar is changed the different change in length of each strip causes curvature of the bar to take place. This principle is used in many dial thermometers. The temperature-sensing element is a bimetallic filament wound into a spiral or helix. One end of the spiral or helix is fastened in the bulb, the other end is connected to a pointer. When the bimetallic filament senses a temperature change the pointer goes through a part revolution. The change in the position of the pointer can then be used to register temperature on a suitably calibrated dial. This type of instrument is used for local temperature measurements of cooling water, lubricating oil and services having similar temperatures. They are also used for local exhaust temperature measurements on diesel generators and similar smaller-size engines.

Other instruments used for exhaust temperature measurement use the thermocouple which consists of two different metallic strips; when connected electrically a small current is caused to flow between the hot junction and cold junction. The current flow is related to temperature which can be measured on a suitably calibrated galvanometer or similar electrical instrument.

■ **18.29 What are the objections to fitting Bourdon-type pressure gauges in control consoles, and control room spaces?**

When Bourdon pressure gauges are fitted remote from the pipe or vessel in which the fluid pressure is being measured, a connection must be made between the pipe and the pressure gauge. If oil or similar inflammable fluids are contained in the pipe or vessel and leakage occurs in the connecting pipe or in the tube of the pressure gauge, dangerous materials may leak out and cause a fire or create other dangers. Should failure occur with a connection or gauge fitted in a control console, parts of the equipment fitted within the console may also be damaged. The height at which the gauge is fitted above or below the point where the pressure is being monitored will affect the accuracy of the gauge reading. This will be equivalent to the amount of static head above or below the pressure gauge and the point from where the pressure measurement is taken.

■ **18.30 How are the magnitudes of the pressures or temperatures at the various parts of an engine transmitted to the control room or control equipment?**

When the pressures and temperatures at various points round an engine are required to be known in the control room they can be transmitted by various means. The common methods use electricity or compressed air as a signal which is easily transmitted to the control room by wiring or small-diameter tubular connections. The essential parts of such instrumentation are first, a sensor which monitors the pressure or temperature changes from some normal value which is commonly zero. If a Bourdon-tube is used as a sensing device the temperature or pressure changes are given out as a small mechanical movement. The second element is a transducer which is used to convert the small mechanical movement from the free end of the Bourdon-tube to an electrical or compressed air signal. The transmitted signal is proportional to the magnitude of the temperature or pressure found by the sensor. The signals are received by the control equipment which will process them to make the desired control changes.

If visual readout of the magnitude of the temperatures or pressures is required in the control room, the signal is fed into an indicator which may give the electrical signal as digital readout of the pressure or temperature on light-emitting diodes (LEDs), or if compressed air is used as the signal, the transmitting line can be connected to a suitably calibrated pressure gauge.

■ **18.31 Briefly describe the parts of a sensor-transducer unit suitable for receiving a pressure measurement and giving out a low-voltage electrical signal.**

The sensor consists of a Bourdon-tube which gives a movement of the free end of the tube that is dependent on the change of pressure being measured. The free end of the Bourdon-tube is connected through links and levers to a small axle which produces rotary movement. This shaft is used with an indicating needle and dial to give a local visual readout of the pressure as it is measured. The opposite end of the shaft has fitted to it an arm which is insulated electrically from the other parts. The end of the arm is fitted with a collector or brush. This connects and makes contact with a wire resistance wound on a drum. The ends of the resistance are connected to an electrical supply with a fixed low voltage. As pressure changes occur and move the end of the Bourdon-tube, the brush or collector moves around the resistance causing a change in the voltage at the brush. The tranducer has two connections to the constant supply voltage and a third connection to a position where the voltage is variable, which is dependent on the pressure being monitored or measured by the Bourdon-tube. The three connections can be connected to a bridge or other type of network and arranged to give a visual readout of a pressure value or used as a signal for processing and connecting into a piece of control equipment such as an electrically actuated valve. The signal can also be connected into a data logger or similar equipment.

■ **18.32   List the component parts in any system used for transmitting the magnitude of main engine exhaust temperatures to the control room or control equipment.**

Electrical means are commonly used to obtain the magnitudes of exhaust gas temperatures at some remote point. The equipment used consists of a temperature-sensitive resistance filament which is fitted in a metal bulb located inside the exhaust-gas outlet branch from the cylinder. The temperature-sensitive resistance and the connecting wiring forms part of a Wheatstone bridge circuit. The other parts of the circuit consist of fixed resistances, an electrical supply with a constant voltage, and a variable resistance with a motorized sliding connection.

In order to measure the resistance of the sensor the Wheatstone bridge is balanced by the variable resistance. The position of the motorized slider working on the variable resistance then indicates the change in resistance of the sensor and can be used to give a visual readout of exhaust temperature.

■ **18.33   What is a differential pressure gauge?**

A differential pressure gauge can be formed with two Bourdon pressure-sensing tubes. They are so arranged that they act in opposition to each other when connected through normal linkage to the shaft which carries the gauge needle. If two different sources of pressure are connected to each Bourdon-tube the higher pressure will overcome the lower pressure and the movement of the needle over the calibrated scale will indicate the difference between the high- and low-pressure inputs. If one of the pressure inputs rises or falls, movement will be indicated on the needle. If both inputs rise or fall the same amount, no change will be indicated as they will cancel each other.

Another form of differential pressure gauge uses a liquid contained in a U-tube. The upper part of each leg of the U-tube is connected to some part of each circuit in which flow takes place.

The difference in the height of the measuring fluid in the U-tube indicates the pressure difference between the points where the connections are made into the flow circuit.

■ **18.34   Where are differential pressure gauges used?**

Differential pressure gauges are used wherever it is required to know the *differences* between two pressures. Common uses for differential pressure gauges are flowmeters working in conjunction with a venturi tube or a restriction orifice. Some types of viscometer also use a differential pressure gauge. Differential pressure gauges are also commonly used to measure the pressure-drop across the inlet air filter on turbo-blowers and also to measure the pressure drop across air coolers in the scavenge or air charging system.

■ **18.35   Describe the means whereby the viscosity of the fuel oil supplied to the main engine is controlled and held at some desired constant value.**

The viscosity of the fuel oil is monitored in a viscometer fitted in the supply line to the engine, downstream of the fuel-oil heater. The viscometer may consist of a small electrically driven gear pump which delivers a constant amount of oil. The pump is supplied with oil taken from the engine fuel supply line. The pump discharges into a pressure chamber and capillary tube. The pressure chamber is connected to one side of a differential pressure gauge while the other side of the gauge is connected to the fuel supply. The flow from the capillary tube goes into the supply line. If the viscosity of the fuel oil departs from the desired value, the pressure-drop along the small bore will change. This pressure-drop variation is shown as a deviation in the differential pressure between the pressure chamber and the main fuel supply line. As the viscosity of the fuel rises the pressure differential will increase in magnitude. A differential pressure gauge can then be used as a sensor for the viscosity of the fuel oil.

A transducer can be incorporated with the differential pressure gauge so that a pneumatic signal is given, the pressure of which will be proportional to the viscosity of the fuel. The signal given out by the transducer can be coupled in

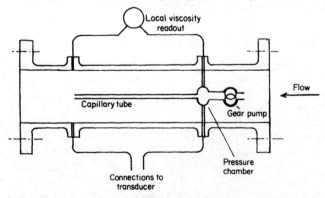

Fig. 18.7 Viscosity meter using a capillary tube and differential pressure gauge calibrated with a viscosity scale.

with an air-controlled pressure valve which is used to control the supply of steam to the engine fuel oil heater. (Fig. 18.7)

■ **18.36 How is it possible to control automatically the temperature of the outlet cooling water from the cylinder jackets and heads?**

Control of the cooling water outlet temperature is often effected by fitting a special type of valve in the jacket cooling water line prior to entry into the jacket water cooler. The valve has one inlet connection or branch and two outlet connections or branches. The inlet connection takes the full flow of cooling water from the engine. One outlet branch is connected to the water cooler while the other is connected to the cooler bypass. A valve within the body controls the flow of jacket cooling water into the cooler and the balance of the flow is diverted round the bypass. The control of the valve is arranged so that any increase of the cooling water temperature from the engine will cause an increase of flow through the cooler (Fig. 18.8).

■ **18.37 How can the temperature of the lubricating or piston cooling oil leaving an engine be controlled?**

The temperature of the lubricating oil leaving the bearings or the cooling oil leaving the pistons can be controlled at the lubricating oil cooler provided that nothing is wrong with any bearing and the rate of flow of oil through the pistons is normal. The temperature of the oil entering the lubricating oil cooler will be related to the temperature of the oil leaving the engine. A difference will exist which will depend on the heat lost by the oil in the drain tank and piping, but these variations are generally negligible in amount over a wide range of ambient sea and air temperatures. The oil temperature can therefore be sensed at the oil inlet to the oil cooler. In order to prevent long-term problems (see Question 15.19) from deposits settling on or within tubes or short-term problems arising with high pour point oil congealing within tubes (sometimes referred to as cold plugging) it is usual with larger-powered engines to keep a full flow of lubricating oil through the cooler and control the flow of cooling water.

This can be done with a valve similar to that described in question 18.36 or with two valves, one being fitted in the cooling-water inlet and one in the bypass. The valves control the amount of water entering the cooler and flowing to the bypass. Naturally the valve actuators work in opposition so that one valve will be opening while the other is closing, or vice versa. In small engines such as a generator or alternator drive, control is sometimes carried out by controlling the oil flow through the cooler with the use of a bypass on the lubricating oil side of the cooler.

■ **18.38 Control valves when fitted on coolers for purposes of temperature control require a sensor and a valve actuator to effect the control. Briefly describe how control is effected.**

The simplest form of flow control valve used for control of temperature of jacket cooling water and lubricating oil is the wax-element control valve. In this type, which has three branches, the medium being cooled, either jacket cooling water or lubricating oil, is passed over a wax-filled sealed bellows which is the temperature sensing element or thermostat. Changes of temperature alter the volume of the semi-molten wax and set up a force which is used to position the valve. The valve is arranged to open the inlet to the cooler and close the inlet to the bypass as the sensed temperature rises.

In other cases the temperature of the medium being cooled is monitored by a sensor and transducer (more information in Questions 18.30 and 18.39). The transducer gives out an air signal which can be passed to a piston-type valve actuator. Linkage from the actuator is connected to the spindle of a butterfly valve. As the signal air pressure rises it forces the piston out against a spring and positions the valve to regulate the flow. Where there is insufficient pressure or flow for follow-up in the air signal the signal can be passed to another valve which is used to control the pressure (from a higher pressure source) to the valve actuator.

Some ships use hydraulically or electrically operated valve actuators, but whichever is used the elements within the system will still consist of a sensor, a transducer, and an actuator.

■   **18.39   Briefly describe the means whereby the movement of the free end of a Bourdon-tube can be made to produce an air pressure signal proportional in magnitude to the amount of movement.**

Compressed air at low pressure is fed into the instrument used, and bled through a small restriction orifice which is fitted into the supply line within the instrument. From the orifice the air passes through a tube to a nozzle. Between the nozzle and the orifice there is a T-branch. The signal air flows into the T-branch and is piped away to where it is required. If air is bled through the restriction orifice, the pressure in the tube can be controlled by controlling the amount of air flowing from the nozzle. If the nozzle flow is reduced by blanking off the nozzle the pressure in the tube rises. If a beam pivoted at one end is fitted over the nozzle so that it is approximately midlength along the beam and the free end of the beam is connected to the Bourdon-tube, movement of the tube will raise or lower the beam. Raising the beam moves it away from the nozzle and the throttling action on the air passing through the nozzle is reduced. This allows more air to pass out of the nozzle and so reduces the pressure in the tube between the restriction and the nozzle. The pressure of the air supply into the T-branch is also reduced. To increase the sensitivity of the instrument, feedback is introduced to the pivoted end of the beam. This is arranged by fitting a spring and bellows with a connection into the T-branch. When movement of the Bourdon-tube moves the beam towards the nozzle a rise in signal air pressure causes outward movement of the bellows against the spring and in turn causes a slight amount of movement of the beam pivot and beam away from the nozzle. The pressure of the air supplied to the instrument is 1.37 bar (20 lb/in²), and the signal pressure will vary from 0.21 bar (3 lb/in²), to 1.03 bar (15 lb/in²).

Note   Other terms used to describe the beam are flapper, force beam, baffle.

■ **18.40** Describe the purpose and action of a relay when fitted to a controller or a signal air transmitter of the type mentioned in 18.39.

Relays can be fitted in pneumatic controllers; they are fitted, where lags are excessive, to give a faster response to changes monitored by the sensor. They consist of a small valve in a branch from the air supply to the instrument, connected to a bellows or diaphragm which is controlled by the air pressure in the small line between the nozzle and the restriction orifice. The action of the relay, which may have a direct or reverse action, is such that a greater air capacity is available in the signal which is not affected by the capacity of the restriction orifice. A relatively large amount of the air flow into the relay valve branch can be bled to atmosphere so that a low-pressure signal is transmitted; as the relay valve reduces the leak-off to atmosphere the output signal pressure rises.

When the output signal pressure is directly proportional to the value of the input signal to the relay bellows, it is said to be a direct-acting relay. If the output signal pressure is inversely proportional to the value of the input signal to the relay bellows, it is said to be a reverse-acting relay.

Note By inversely proportional is meant the action is such that as the input signal to the relay bellows increases, the output signal decreases.

■ **18.41** What do you understand by the term 'proportional band'?

Proportional-type controllers receive input signals from a sensor and give an output signal which is proportional to the input signal. The range or change in value of the input signal, which causes the proportional controller to vary the output signal over its full range, is referred to as the proportional band. Suppose for example we have a temperature controller arranged to give an air output signal of 0.21 bar when the input temperature is 40°C, and of 1.03 bar when the input temperature is 50°C. The proportional band will then be 10°C (50°C to 40°C), when the output signal varies over its pressure range of 0.82 bar (1.03 to 0.21).

Note Other names for proportional band are band width, control band, and throttling band.

Many of the instruments used for proportional control have an adjuster which can be used to control the proportional band; the adjuster is used when a system is being balanced.

■ **18.42** The terms 'open loop control' and 'closed loop control' are used in control engineering. What do you understand by these terms? Give an example of each.

Open loop control refers to manual control systems. In such a system, there must be a sensing device, and a display or indication of a variable value which requires control within certain limits. The value is read on the display unit, its significance is interpreted by the operator, who then takes the necessary action to correct any deviations from the required norm or limits.

Closed loop control refers to automatically controlled systems. In closed loop systems a sensor or sensing device is required to monitor any deviation. The deviations are fed back as a signal from the sensor to a comparator which measures the deviation or change from a required value. The value of the deviation is fed as a signal to a controller which is used to position a regulator. The regulator then makes the necessary changes to correct the deviation. The loop is closed by the connection between the regulator and the sensing device.

An example of an open loop control is the steam pressure gauge on an auxiliary boiler. The pressure gauge is the sensing device and display. Any deviation from the desired steam pressure is regulated by manual control of the oil burner pressure pump or burner supply bypass. An example of a closed loop is the automatic control of the temperature of engine cooling water.

■ **18.43   Where are logic circuits used, what is their purpose and how are they built?**

Logic circuits are used in the control of processes and have a place in both engineering control systems and computers. Their function in control engineering is to effect simple decision-making so that a process can be continued if everything is in order or suspended if some event which should occur does not occur. When an engineer officer serving in an older ship prepares the main engines for sea he makes a whole series of decisions based on certain actions before the engine is started. For example the main switchboard will be checked to ascertain that sufficient electrical generation capacity is available to start the various auxiliaries such as starting-air compressors, pumps, etc. If there is enough reserve electrical capacity, the officer will proceed to the next step. Should the reserve electrical capacity be insufficient he thinks to himself *'not'* enough capacity and does not proceed to check the starting air pressure but instead starts another generator.

If the starting-air pressure is in order he goes to the lubricating oil pumps and will start one *'or'* the other.

The next step is to start the jacket cooling water pump after which the officer will say to himself, 'I have started the lubricating oil pump *'and'* the jacket cooling pump; I will now proceed to the next step.' The steps naturally follow one another in a logical manner.

Logic circuits function and make decisions in a series of steps similar to the engineer officer preparing engines for sea.

Logic circuits are built up of components which are referred to as *'not'*, *'or'*, *'and'*, *'not or'*, *'not and'* circuits. In designing and drawing up logic circuits various symbols are used to simplify the work.

Common uses of logic circuits in marine practice are automatic engine start-up, bridge control of main engines; automatic start-up, synchronizing and switching-in alternators, etc.

Note   The word 'logic' is derived from the Greek words 'logos', which means word or reason, and 'logike', which means the art of reasoning.

■ 18.44 Automatic temperature controls on engine cooling and lubrication systems are sensed at the inlet to the cooler. What happens if the outlet temperature from an individual cylinder jacket, piston or bearing rises to an unsafe value? How are dangerous conditions prevented?

If the coolant outlet temperature from an individual cylinder jacket or piston rises, the rising temperature is masked when the hotter coolant mixes with the normal-temperature coolant from the other cylinders. Further, the degree of masking increases with the number of cylinders on the engine. It therefore follows that if temperature sensing is effected at the cooler inlet there could be a considerable time-lag before the temperature rise at the cooler inlet was sufficient to activate an alarm. In older motor ships flow switches were sometimes fitted in some cooling outlets; these were wired in a fail-safe manner to the engine alarm system.

In more modern motor ships which have been classified for zero manning in the engine room, temperature sensors are connected into each outlet line and a rising temperature will give both a visual and audible alarm at the control station and an audible alarm outside the engine room.

Mini digital computers or micro-processing units are also used for engine alarm systems. When these units are used, the individual outlet temperatures at each of the outlet points are scanned for readout at very short periods. The input is then processed for individual high or low temperatures outside the set limits or reference values. If the reference values are exceeded an alarm is activated. The mean temperature from a series of coolant outlets is also found together with the deviation of any unit from the mean. The average value and deviations can be given on the display unit.

Some engine builders prefer to monitor the piston cooling system temperature on the outlet from the cooler and the jacket cooling system on the inlet to the cooler. The outlet temperature of the piston coolant from the piston cooling system heat exchanger is then kept at the minimum required value. The outlet temperature of the jacket cooling water from the heat exchanger is then kept at its maximum value.

When controlled in this manner the temperature of the piston and hence its diameter is kept at some minimum value. The jacket cooling water and hence the inside diameter of the cylinder bore is kept at a maximum value. This is considered to be a safer way to operate in engines where there is a tendency for the piston or piston skirts to overheat. When controlled in this manner the piston clearance is kept at a maximum possible value.

■ 18.45 Describe a three-way control valve as used on many heat exchangers to control temperatures.

Three-way valves have three branches for the passage of the coolant. The valve operates in such a manner that the flow of coolant passes into the valve through one branch and may leave the valve by one of the other two branches, or the flow may be split so that part of the flow leaves from one branch and the balance leaves by the other branch. One of the outlet branches is connected to

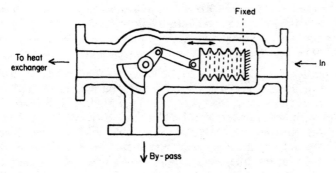

Fig. 18.8 Wax element, three-way control valve.

the inlet side of the heat exchanger while the other is connected to a bypass and prevents the passage of coolant through the heat exchanger.

The inner portion of the valve is designed to be always open to the inlet branch over the operating range of the valve. When maximum cooling is required the valve will close off the bypass branch and open fully to the branch leading to the heat exchanger. When no cooling is required the bypass branch opens fully and the other branch is closed off.

Some three-way valves are operated by a wax element fitted within the valve; other types of valve are pneumatically controlled by air pressure acting on a diaphragm. (Fig. 18.8)

 18.46 The terms 'analogue' and 'digital' are used in control engineering. What do you understand by these terms?

These terms can be understood by reference to a ship's log. Many ships have a log which is passed through the bottom of the ship from the engine room space; it is used to measure the ship's speed in nautical miles per hour (knots) and the distance run by the ship in nautical miles. In the engine room a display unit is provided which gives the speed of the ship indicated by a pointer position on a calibrated dial, and the distance run, from some point in time or place, by a horizontal set of figures.

The numerical value of the speed is represented by the distance that the needle has swung from its zero position. Representation of a numerical value by a stepless physical quantity such as the needle swing is an analogue. The right-hand figure given on the distance-run display changes at the end of each nautical mile covered; it therefore changes a digit or makes a step at the end of each mile run and such a set of figures which change in this manner is referred to as a digital representation.

Note The analogue computer uses physical quantities for its input which may be voltage, current or resistance to represent numerical magnitudes. The digital computer uses input data in the form of digits.

■ 18.47 What are the component parts of a data logger of the type found in the engine room of motor ships? What is their purpose?

In order to write up the log-book in an older ship one has to go around the engine room and observe the values given on the various instruments used for indicating pressures, temperatures, volume and flow, mentally compare the value of the reading to some norm, and then transfer the value noted to the engine log-book. The purpose of the data logger is in effect to take the readings of the various temperatures, pressures, and flows and give a print-out of the readings at set time intervals; it is also arranged so that warnings are given of unacceptable conditions and it continues to give a print-out until the unacceptable condition is corrected or acknowledged.

The component parts of a data logger are first, the sensor-transducers which give an electrical signal to the scanner. The scanner is time-cycled and each of the electrical signals from the sensor-transducers is taken in turn and transmitted to an amplifier. The amplifier takes in the signals from the sensor where they are brought to standardized analogues in the form of d.c. voltage outputs. The analogues are then converted to digital outputs. These digital outputs are passed into logging printers and display units. The logging printer can be timed to give a print-out after any desired time lapse. The display unit can be used to obtain a readout of any of the values of temperature, pressure, or flow (whichever is applicable) at each of the sensor locations. Selector switches are used for obtaining the individual readings.

It is common practice to combine the data logger with an alarm system. The alarm system consists of a comparator unit which receives the digital inputs and compares them with the safe reference values. If any signal is received which falls outside its safe reference value the alarm system is activated and the printer prints-out the alarm condition values. Alarm lamps can also be set to flashing condition on the mimic board to pinpoint the location which is the source of the alarm condition.

■ 18.48 For what purposes can computers be used in conjunction with diesel machinery?

Pre-programmed mini electronic computers are now used in conjunction with diesel machinery. They are connected up with transducers fitted at critical points in the engine. The transducers monitor temperatures, pressures (including pressures within the cylinders and combustion spaces), flow rates and cylinder liner wear. The sensors or transducers feed signals to the computer where they are processed. They are then used in a normal manner for data logging equipment, alarm annunciation and the like. Further uses are to indicate trends in operation changes. For example, it is known that the air inlet filters to the main engine turbo-blowers gradually become fouled and after some time they must be cleaned. In a hard-run ship the problem will sometimes arise as to whether the filters should be cleaned immediately or at the next port which is so many hours' sailing away.

The computer can handle this problem in the following way. The pressure-drop across the filter or pressure differential is accurately measured on two occasions with some reasonable time lapse between measurements. If the increase of fouling, shown by the increase of the pressure difference across the filter is divided by the time lapse in which the fouling increased, we obtain a fouling rate given as pressure-drop per hour. This figure is then divided into the maximum allowable pressure drop, the result giving the number of hours the air filters can remain in operation between successive cleanings. The actual pressure-drop subtracted from the allowable will give the reserve pressure-drop; if this is divided by the pressure-drop rate per hour the number of hours to the next cleaning is obtained.

With information such as this, which would be readily available before the ship's arrival in port, it is possible to effectively plan the maintenance work load so that the most advantageous use is made of the man-hours available after planned maintenance and continuous survey items have been covered, or in an emergency to reschedule planning.

In a similar manner the computer can receive the pressures sensed within the combustion chambers and relate them to a time or stroke base and give a display showing the mean indicated pressure in each cylinder. The computer can sum the values from each cylinder and find the mean value. It can then find the deviation of the m.i.p. of any cylinder from the mean. This information simplifies the balancing of the engine so that the indicated horse-power developed in each cylinder is the same. Within this programme, the engine rev/min which is fed into the computer is used to obtain the indicated power of the engine in both horse-power and kW.

Other programmes are available in the computer to give indications of hull fouling, cylinder liner wear and a whole range of useful information which can effect the economic operation of the ship and its machinery. In the case of accelerated cylinder liner wear arising out of the use of very low sulphur-content fuels with high additive cylinder lubricants, the computer would almost pay for itself in giving early-warning of wear-rate increase. Trouble could then be avoided with little or no lost time (see Question 18.29).

■ **18.49 What do you understand by the term 'semiconductor'? Where are semiconductors likely to be found in the engine or control room in a motor ship?**

Semiconductors, as their name implies, are some form of material which is neither an electrical conductor nor an insulator. Their resistance lies somewhere between that of conductors and insulators. Semiconductors have various properties.

One class changes its resistance when it is subjected to changes of temperature. The semiconducting material can be bonded with other materials to produce sensor-transducers used to sense temperature changes. When this type of semiconductor is used for sensing temperature a current is passed through it. As the semiconductor is subjected to rising temperature its resistance decreases so allowing a greater current flow. This type is referred to as thermo-sensitive. A second classification is referred to as being photo-conductive. Increased exposure to light of one wavelength or another reduces the resistance of the semiconductor. A third classification refers to the ability of some semiconductors to cause a current to flow when the semiconductor is combined with other materials and exposed to light. The cell composed of the semiconductor generates its own e.m.f., thereby causing current flow. This type of semiconductor also has a peculiar effect in that it allows current to flow only in one direction when connected to an electrical supply with alternating voltage.

Semiconductors can be used as transducers for sensing temperature change, as transistors in amplifier circuits, as detectors for boiler oil fuel burner flame indicators, engine and boiler-room fire detectors, static excitation equipment on alternators, and in computer circuits, etc.

■ **18.50 Describe briefly the action of a cathode-ray oscilloscope and state where it is often used.**

The cathode-ray oscilloscope (referred to as an oscilloscope, scope or C.R.O) is used to obtain a record of voltages or currents on a basis of time and bring them to a visual form where they can be easily analysed.

Any electrical variable such as the output voltages in the various parts of an electrical circuit can be used for input. Any mechanical variable capable of obtaining an electrical output from a transducer can also be used. Some examples of these are the displacement, velocity, and acceleration components of a mechanical vibratory movement, air-pressure waves in the form of audible or non-audible frequencies, the pressures of the gas expanding in an engine cylinder, the pressure variations in air inlet and exhaust manifolds, the small movement of engine parts subjected to vibration or when under stress, pressure variations in the blood circulatory system in the body, etc.

An oscilloscope has a conical-shaped glass casing holding some of the components in a high vacuum. The glass casing, commonly called the tube, houses the cathodes which emit a stream of electrons and adjust the intensity of the light spot on the luminescent screen, anodes or grids to accelerate and focus the flow of electrons. These parts are sometimes referred to as the electron gun.

An electron beam can be deflected by a magnetic field or by an electrostatic field.

The other parts fitted in the tube are the horizontal and vertical deflection plates which control the direction of the beam and deflect it vertically or horizontally by electrostatic fields between the plates. Deflection of the beam makes the light spot on the screen move vertically or horizontally or with a combination of both movements at the same time. The amount of movement is proportional to the applied voltage.

The base of the conical tube forms the viewing screen on which the light spot appears. The inside of the screen is coated with a material that is cathodoluminescent; this glows when struck by electrons moving at a very high velocity and forms the luminous point on the screen.

The other components external to the tube are a high-voltage generator to supply current to the cathodes and anodes, time-base generators, amplifiers for amplifying incoming voltages for the deflection plates, synchronizers to hold input to horizontal deflection plates in step with input to the vertical deflection plates, potentiometer controls, change-over switches, connecting wiring and on-off switches. The arrangement and the casing of the equipment will depend on whether it is fitted into a centralized control console for a particular set of duties, or made up into a portable oscilloscope for use as a test instrument.

If the movement of the spot on the screen is considered as a graph of some function ($y = f(x)$) the voltage applied to the vertical deflection plates creates ordinates or 'y' values. The scale of the ordinates is related to the voltage required to give a unit amount of vertical deflection. If a voltage (obtained from the time-base generator) varying from zero to some other value in a uniform manner is applied to the horizontal deflection plates the spot moves from left to right and then 'instantaneously' flies back to its starting position: this sequence is repeated indefinitely and creates the 'x' axis. If a varying voltage is applied to the vertical deflection plates at the same time as the uniformly increasing voltage is applied to the horizontal deflection plates, the spot traces out a graph showing the variation of 'y' with respect to time 'x'.

When the oscilloscope is used to obtain an indicator card the varying voltage corresponding to changes in cylinder pressure given out by a piezoelectric transducer or other form of transducer is amplified and then applied to the vertical deflection plates. The electronic beam and the spot move up and down in unison with the pressures in the cylinder. If the voltage output from the time-base generator is amplified and synchronized with the engine rpm and then applied to the horizontal deflection plates the light spot on the screen traces out a graph of cylinder pressure against time, but without the effect of connecting-rod angularity on the time base.

By using a shaft encoder it is possible to obtain a voltage varying in the same manner as the piston movement throughout the stroke. If a voltage variation in this form is amplified, synchronized and applied to the horizontal deflection plates, a normal indicator diagram is traced out.

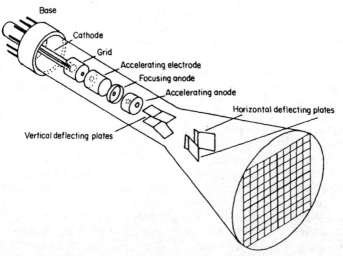

Fig. 18.9 Simplified diagram showing the internal parts of a cathode ray oscilloscope. The internal and external connections are not shown.

This may seem a very complicated method of obtaining cylinder pressure data, but we must remember that this equipment will serve both main propulsion machinery and diesel generators irrespective of speed. It is very convenient to have this equipment immediately available in the machinery console to obtain a quick check on cylinder pressures, combustion behaviour and the like. (Fig. 18.9)

Note Some installations have a cylinder-pressure transducer contained in a portable unit with wiring connections to the console. The unit is connected with each engine cylinder in turn. In other cases a pressure transducer is fitted as a fixture in each cylinder unit. It is desirable to get the pressure transducer fitted as near as possible to the combustion chamber. If the transducer is connected to the combustion chamber through gas passages, extreme care must be taken with their design to ensure that spurious echoes or pressure waves which will be picked up by the transducer, are not generated.

■ **18.51 What are proximity sensors, magnetic pickups, displacement pickups, velocity pickups, and accelerometers? What are they used for?**

Proximity sensors, magnetic pickups, displacement pickups, velocity pickups and accelerometers are the names given to different forms of transducer used to create an output voltage proportional to some mechanical movement. The voltage generated is usually linear in output and normally follows some form of a sinusoid curve that is repeated at equally spaced time intervals (periodic).

These transducers fall into one of three main types:

non-contact type (proximity sensors)
velocity pickups
accelerometers

One form of proximity sensor contains a coil of fine wire which receives a high-frequency current input. A magnetic field is formed at the end of the sensor. When some mass of iron or steel passes the tip of the sensor the field is changed. This change eventually produces a direct current output that is proportional to the distance of the tip from the mass. Beyond a certain distance the sensitivity of the sensor falls away and it ceases to be linear. This type of sensor when located at the end of a shaft may be used in studying axial vibration or used to obtain shaft speeds when located near to some point in the system such as gearing teeth or the teeth cut in a flywheel for engaging the turning or jacking gear. Two sensors located at 90° to each other near a rotating shaft surface can be used to study shaft movement caused by bad balance or shaft coupling misalignment.

A velocity pickup consists of a permanent magnet housed within a coil of wire. Each end of the permanent magnet is held by a spring; axial movement of the magnet can take place between the limit of spring compression. The magnet and the springs are immersed in oil for damping. If the sensor is fastened to a piece of machinery subjected to some force causing vibration, the inertia of the magnet holds it stationary relative to the upward and downward movement of the surrounding coil. The relative movement of the field of the magnet causes a voltage and current flow. The voltage output from the coil is directly proportional to the velocity of the vibratory movement.

Accelerometers are increasingly used in vibration studies because they are smaller and respond to a wider frequency range than the other transducers. The accelerometer consists of a piezoelectric material, part which is held between some mass and the body of the instrument.

If the instrument is fastened to some piece of vibrating machinery the inertia of the mass causes the piezoelectric material to be subjected to a stress and an electrical signal is generated.

As force is equal to the product of mass and acceleration the electrical output is proportionate to the acceleration of the mass acting on the piezoelectric material.

These various transducers are used in many fields of investigation particularly where vibration or noise is concerned. The varying voltage or output signal from the transducer is separated and analysed in various ways using computers and programs, some of which are designed, manufactured and written specifically for this purpose.

The equipment can be used to obtain vibration signatures of rotary machines such as pumps, turbo-chargers and the like, and to make predictions for maintenance programmes. If no change is shown in the vibration signature it can be assumed that there is no deterioration in the moving parts of the equipment that could lead to increased vibration.

■ **18.52   What is a shaft encoder?**

A shaft encoder is used to obtain a series of electrical impulses or a variable output corresponding in amount to some known shaft position during the rotation of a shaft. The output may be impulses occurring at equally spaced intervals of time or angular motion. Some encoders will give an impulse at intervals extending over a fraction of a degree while others may give an impulse only at 90° spacing of shaft rotation. The spacing interval required is related to the equipment and the purpose for which it is used. When used for obtaining data on a diesel engine, a secondary electrical impulse giving the location of the piston position for a reference cylinder can also be given – for example, this might be No. 1 piston when in top dead centre position.

Many encoders work by means of a light source and a photocell. The light source may be projected through a rotating disc with slots. This interrupts the action of the light source on the photocell and causes current flow from the cell in the form of a series of impulses. In some other cases the light source may be projected through a screen with an optical pattern on it to give a continuously varying output that may be used in an analog computer.

Shaft encoders are necessary with some of the electronic equipment used in instrumentation equipment used for monitoring diesel engine performance.

19

SAFETY

■ 19.1 Some ships are fitted with electrical oil heaters. What safety devices are fitted on electrical oil heaters? How do electrical oil heaters differ in their operation compared with steam heaters?

In steam-heated oil heaters the maximum temperature to which the oil can rise will correspond with the saturation temperature of the steam. This maximum temperature will not be exceeded if there is no oil flow through the heater. In electrically operated oil heaters the temperature of the oil can rise to dangerous limits if the oil flow stops and current is left on the heating elements. The important safety device which should come into operation in such a circumstance is the high-temperature cut-out which switches off the electrical supply. The temperature sensor in the heater and the automatic switching devices take various forms; whatever arrangements are used they should be carefully checked at regular intervals to avoid failure of the thermostat controls.

■ 19.2 Oil heaters, heating coils in lubricating-oil and fuel tanks, usually use steam for heating purposes. Where does the steam go to after it is utilized for heating?

After the steam enters the heater or heating coils it gives up its latent heat and condenses to water. The water then passes through the drain trap on the outlet side of the steam system and eventually collects in a drain tank from where it passes through filters and an observation tank.

After passing through the observation tank the water enters into the boiler-feed water system and is pumped back into the boiler.

■ 19.3 What particular attention must be given to the observation tank?

Should any leakage of oil occur in a heater or heating coils, the oil may find its way into the steam space of the heater and eventually contaminate the feed water. If feed water contaminated with oil is pumped into the boiler the oil settles on the heating surfaces. The oil film on the heating surfaces acts as a heat insulator and in restricting the heat flow across the tubes or furnaces it allows the temperature of the material forming the tube or furnace to rise to dangerous limits, which causes its strength to be reduced. The pressure then deforms the overheated material and failure will occur. The observation tank is usually arranged as a cascade-type filter so that if a small oil leak occurs the oil will collect in one space within the tank. This space is often fitted with a set of sight glasses to allow the surface of the water to be illuminated and kept under observation. When steam is being used in heaters and heating coils the return condensate must at all times be kept under close inspection while it is passing through the observation tank.

Note   If any oil is found in the observation tank, a decision must be made as to whether the contamination can be removed by the filter section of the tank. If any doubt exists as to whether the filters can handle the oil and separate it out, the bypass must be opened and the heating condensate return allowed to flow into the bilges. On no account should it be fed into the boiler. The oil should be separated from the bilge water in the oily-bilge-water separator, when bilge water is pumped overboard.

■   **19.4   What do you understand by the term 'water hammer'? Why is it dangerous and how can it be prevented?**

When steam lines are shut down it is possible, from various causes, for them to fill with water. If steam is allowed to enter a line filled with water, the steam starts to move the water down the line. The steam in contact with the water eventually condenses and a vacuum is then formed causing the water to be pulled back at a very high velocity. The water at high velocity returns to the valve which has just been opened and strikes it with a very heavy blow, often fracturing the valve. If the valve fractures it may end disastrously, with risk of loss of life. In order to prevent water hammer it is necessary to open up the drain connections on the steam line which is being brought into use. The water in the line must then be completely drained so that it leaves the line clear. The steam valve may then be very slightly opened (cracked open) so that the line is heated and brought up to near working temperature. Any condensate formed during this period drains out of the line through the drains previously opened. As the line temperature rises and the noise from the drain changes to that of a steam blow, the pressure on the line can be increased by opening the steam valve further. Eventually the drain valves can be closed, after the steam valve is fully opened.

■   **19.5   What are the precautions which must be taken when using a hand-torch to light-up burners in a boiler?**

If a burner shut-off valve leaks or is not shut off correctly, it is possible for fuel oil to find its way into the hot furnace. The oil will vaporize and form gases. When a torch is passed into the furnace and the air supply is opened at the burner front, an explosion or bad blow back may occur. This can injure the person lighting-up the boiler. In lighting-up oil fuel burners with a hand-torch, the first essential is always to stand towards the side of the furnace to manipulate the controls. This brings the boiler operator or engineer out of the line of any blow back should one occur. The next step is to open-up the air supply to the furnace so that any gases present are cleared. After the gases have been cleared, the air can be shut off and the torch entered into the furnace, after which the oil and the air supply can be opened and the torch withdrawn.

Note In carrying out this operation it is essential to watch over oneself and one's staff to ensure that familiarity with a simple operation does not breed contempt for its many dangers.

■ **19.6 Why is it essential that sea valves fitted on the ship's side be opened and closed at regular and frequent intervals? What is a bilge injection valve?**

It is essential that sea-water suction and discharge valves fitted on the ship's side be opened and closed preferably at weekly intervals. This opening and closing prevents the valve spindle seizing in the valve bridge. If the spindle seizes in the bridge it becomes impossible to close or open the valve. If it is impossible to close valves, particularly the large-size valves associated with the main cooling water services, the engine room becomes very vulnerable in the event of pipe failure, and could easily be flooded.

A bilge injection valve is fitted to the end of a branch line connecting with the main sea-water suction line. The bilge injection valve is always of the screw-down non-return type. This valve enables the large main sea-water cooling pump to be used as a bilge pump in an emergency.

■ **19.7 How is a bilge injection valve brought into use during an emergency?**

To bring the bilge injection valve into use during an emergency the bilge injection valve is opened fully and the sea injection or suction valve is fully closed. After it is established that the sea-water pump is capable of lowering the water level in the engine room, the sea-water valve on the ship's side may be opened slowly. This should be done in stages so that the tank top is not pumped dry, as this would cause the sea-water pump to lose its suction.

Note Sea-water cooling pumps are not usually self-priming, if of the centrifugal type.

■ **19.8 Fuel-oil tanks for the supply of fuel to the main engines, boilers or other services, are usually fitted with two outlet valves referred to as high and low suctions. Why are two valves used and what is the purpose of this?**

The high and low valves are actually a safety feature to prevent inadvertent shut-down of engines, boilers or generators, due to water contaminating the fuel. Normally the low suction valve is kept in use. If any water should find its way into the service tanks it will gradually separate towards the bottom of the tank. When it becomes apparent that water is present, either by finding it at drains or by the operation of the engine or boiler burners, it is possible to bring the high suction into use and avoid a shut-down.

Note   Drain valves on fuel tanks should be regularly used to determine whether any water is present in the service tank. They should also be used prior to changing-over fuel service tanks.

■ **19.9   List the various factors which must be present for a scavenge fire to start?**

For any fire to begin there must be present a combustible material, oxygen or air to support combustion, and a source of heat at a temperature high enough to start combustion. In the case of scavenge fires the combustible material is oil. The oil is usually cylinder oil which has drained down from the cylinder spaces; in some cases the cylinder oil residues may also contain fuel oil. The fuel may come from defective injectors, injectors with incorrect pressure setting, fuel particles striking the cylinder, and other similar causes. The oxygen necessary for combustion comes from the scavenge air which is in plentiful supply for the operation of the engines. The heat in the scavenge space, around the cylinder, brings the oil to a condition where it is easily ignited. The high temperature required to start combustion may arise from piston-ring blow-past.

■ **19.10   How would you become aware of a scavenge fire? How would you deal with a scavenge fire?**

The first indications of a scavenge fire may be a slight reduction in the engine speed due to the reduction in power which comes about when a fire starts. Other indications are a higher exhaust temperature at the cylinders where the scavenge fire has started and irregular speed of turbo-blowers. External indications will be given by a smoky exhaust and the discharge of sooty smuts or carbon particles. If the scavenge trunk is oily the fire may spread back from the space around or adjacent to the cylinders where the fire started and will show itself as very hot spots on areas of the scavenge trunk surfaces. In ships where the engine room is periodically unmanned, temperature sensors are fitted at critical points within the scavenge spaces. On uniflow-scavenged engines the sensors are fitted round the cylinder liner just above the scavenge ports. A temperature higher than reference or normal then activates the alarm system.

   If a scavenge fire starts, two immediate objectives arise; they are to contain the fire within the scavenge space of the engine and to prevent or minimize damage to the engine. The engine must be put to dead slow ahead and the fuel must be taken off the cylinders affected by the fire (see Note). The lubrication to these cylinders must be increased to prevent seizure and all scavenge drains must be shut to prevent the discharge of sparks and burning oil from the drains into the engine room. In some cases this allows the fire to burn itself out without damage. Once the fire is out and navigational circumstances allow it, the engine should be stopped and the whole of the scavenge trunk examined and any oil residues found round other cylinders removed. The actual cause of the initiation of the fire should be investigated.

If the scavenge fire is of a more major nature it sometimes becomes necessary to stop the engine and use the steam or extinguishing arrangements fitted to the scavenge trunk. The fire is then extinguished before it can spread to surfaces of the scavenge trunk where it may cause the paint to start burning if special non-inflammable paint has not been used.

Note   The fuel should be taken off the cylinder by lifting the suction valve of the fuel pump or by lifting the fuel pump roller whichever is applicable to the fuel pump concerned (see Question 4.3). The fuel should not be taken off the cylinder by opening the bypass on the fuel valve; fuel may splash out from tundishes or save-alls and create an extra hazard.

If a scavenge fire occurs, care must be taken by engine-room staff to stand clear of the crankcase and scavenge pressure relief devices.

■ **19.11   How can the incidence of scavenge fires be prevented or reduced?**

One of the first things that must receive attention is maintaining the scavenge space in as clean a condition as possible. This can be done by keeping scavenge drain pipes clear and using them regularly to drain off any oil which comes down into the scavenge space drain pockets. The scavenge space and drain pockets should also be cleaned regularly to remove the thicker carbonized oil sludges which do not drain down so easily and which are a common cause of choked drain pipes. The piston rings must be properly maintained and lubricated adequately so that ring blow-by is prevented. At the same time one must guard against excess cylinder-oil usage. With timed cylinder oil injection the timing should be periodically checked. Scavenge ports must be kept clear.

The piston-rod packing rings and scraper rings should also be regularly adjusted so that oil is prevented from entering the scavenge space because of butted ring segments. This may and does occur irrespective of the positive pressure difference between the scavenge trunk and the crankcase space. Crosshead guide adjustment is also important.

The mean indicated pressure in each cylinder must also be carefully balanced so that individual cylinders are not over-loaded.

If cylinder liner wear is up to maximum limits the possibility of scavenge fires will not be materially reduced until the liners are renewed or re-chromed.

■ **19.12   In carrying out an inspection after a scavenge fire where would you direct your attention?**

Heat causes distortion, which sometimes becomes permanent. If the scavenge trunk plating is substantial in thickness as is found on some engines, a bad fire can cause distortion and may upset piston alignment. A check should be made by turning the engine and watching the movement of the piston in the cylinder liner; care must be taken to ascertain that no binding occurs at any part of the stroke. Binding may also be the indication of a bent piston rod. The other points which require attention are springs on scavenge space pressure-relief devices if they were near the seat of the fire. Piston-rod packing should also be examined for garter or other springs which may have become weakened by overheating.

■ **19.13 List the safety devices fitted on fuel-oil settling tanks and daily service tanks. What attention do these devices require?**

The safety devices fitted on fuel-settling tanks and daily service tanks are as follows.

Fuel outlet valves. Remote closing devices with control arranged for operation external to the engine room.

Air pipes. The air pipes are led to above the upper-deck level and external to deck house. The outlets are fitted with metallic gauze screens.

Thermometers. For measuring the temperature of the oil.

Overflow pipes. An overflow pipe is fitted to the top of the tank and led to an overflow tank in the double bottom. An alarm activated by an overflow condition is sometimes fitted to the tank on the overflow pipe.

In ships where the engine room is periodically unmanned, alarms are also fitted to warn of high fuel temperatures and low fuel levels. The remote valve-closing devices should be tested at weekly intervals. The rod or wires leading to the deck controls require lubrication, as do wire pulleys and leads. Care must be exercised when painting to prevent paint from jamming the pulleys. With wire controls, the upper section of the wire is liable to suffer from corrosion before the lower sections; the upper section should be carefully watched and protected. The metal gauzes fitted on the open ends of air pipes also need attention for deterioration. The gauzes should also be examined after the air pipes have been repainted in case paint has got on to the gauze. The alarms must also be subjected to periodic testing. When maintenance has been carried out on safety devices and testing programmes have been run through, appropriate entries must be made in the engine-room log-book.

■ **19.14 What is the reason for fitting thermometers and alarms which give warning of high fuel temperatures in fuel-settling and daily service tanks?**

If fuel oil is heated to temperatures above its flash-point, inflammable vapours are given off. It large quantities of oil vapour are produced it is obvious that dangerous conditions are being created. In order to prevent this danger from arising, thermometer pockets and thermometers are fitted in the settling tanks and daily service tanks on all ships. In ships where the engine room is periodically unmanned, alarms are fitted to warn of dangerous temperatures. The alarm should be set to go off when the temperature of the fuel approaches within 8°C of its flash-point. For example if the flash-point is 80°C, the alarm should give a warning when the temperature of the oil is not more than 72°C. In practice it is wise to set the alarm at some lower figure. It should also be noted that when fuels are heated more than is necessary it may reduce the effective calorific value because of the possibility of releasing volatile constituents.

■ **19.15 Name the factors which must be present for an explosion to occur in the crankcase of a diesel engine.**

Normally when an engine is in operation the lubricating oil used in the bearings is splashed around the crankcase and broken down into moderate-sized particles. When a bearing, guide, piston rod, piston trunk, or skirt becomes overheated, the particles contacting the heated area easily vaporize and form a white vapour which spreads around the crankcase. Some of the vapour condenses to form very small particles which may eventually permeate the whole of the crankcase space. If the mixture of air, very small particles and vapour reaches a certain proportion and the temperature of the hot spot is high enough to initiate combustion, an explosion can occur. In some cases the mixture of oil vapour, particles and air may be too rich to allow combustion to occur. If this condition is present and a crankcase door is opened the ingress of air may then bring the mixture into the explosive range and allow an explosion to occur.

The factors that must be present for an explosion to occur are as follows.
1   A source of heat to cause the lubricating oil to vaporize and permeate the crankcase or part of it with very small oil particles.
2   The correct quantity of air mixed in with the small oil particles to make the oil-air mixture explosive.
3   A source of heat at a temperature high enough to initiate combustion in the oil–air mixture; in most cases the source of heat that causes the crankcase to become permeated with the very small oil particles or mist is also the source of heat which initiates combustion.

Note   The source of heat is sometimes referred to as the hot spot, which is a carry-over of the common name for the flame or ignition tube fitted in the cylinder covers of hot-bulb engines.

■ **19.16 How would it be possible for high pressure to arise in the crankcase of a diesel engine? How is high pressure prevented from building-up if the factors which make it possible are present?**

If some internal part of an engine becomes overheated the creation of lubricating-oil vapour occurs and mist can permeate the atmosphere in the crankcase. When the overheated part initiates combustion of the minute oil particles in the oil mist a pressure rise occurs. The rate of the pressure rise depends initially on the weakness or richness of the oil-particle/air mixture. If the mixture is in the explosive range the pressure-rise occuring after combustion has commenced is so rapid that it is defined as an explosion. In order to combat any pressure rise and cancel out its possible disastrous effects, pressure-relief devices (which must be self-closing) are fitted to the crankcase. They are placed at various points along the crankcase to relieve any pressure wave irrespective of its origin. The pressure-relief devices are often fitted with internal metallic gauzes to prevent flame emerging from the relief device, and external guards to deflect it away from operating localities if the gauze fails.

The necessity for the valve to be self-closing is to prevent a secondary explosion. The self-closing action of the valve is to prevent ingress of air to the crankcase. If there were some oxygen deficiency following the first pressure-rise, ingress of air could trigger off a secondary explosion by supplying the air for combustion.

Note The pressure-relief devices are also called safety doors, explosion doors, or explosion relief valves.

■ **19.17 Describe a pressure-relief device of the type fitted on a diesel engine crankcase. Are these devices fitted on all sizes of engine?**

Crankcase pressure-relief devices are of two main types. One type is hinged with a near-horizontally placed, upward-opening door. The door has a coiled spring fitted around the hinge pin, and is made self-closing both by the action of gravity and the spring. The amount that the door can open is restricted by butts on the door. The other type consists of a circular door or valve disc mounted on a central spindle and spring loaded. The spring holds the valve shut and makes it self-closing. Several thicknesses of wire gauze are sometimes fitted over the opening to the device on the inside of the crankcase. Naturally the free area through the meshed wires is kept well in excess of the area through the valve. Oil tightness on the spring-loaded type door is maintained by having a joint or O-ring of suitable non-stick material fitted into a groove machined around the sealing face of the door. Hinge-type pressure-relief doors are sometimes sealed by a thin plastic sheet held in place by a rubber ring. The thin plastic sheet covers the opening on the inside of the crankcase and prevents oil leakage. Other types of hinged doors have a gutterway on the exterior lower part of the door. Any oil which leaks out is then drained back to the crankcase through a drain tube which is self-sealing by the leakage remaining in a U-bend at the lower end of the drain pipe.

All marine diesel engines with the exception of the very small sizes are fitted with pressure-relief devices on the crankcase. Generally the hinged type are fitted on slow-speed propulsion engines, while the other type is fitted on medium-speed engines used either for propulsion or electrical generation.

■ **19.18 What attention do crankcase pressure-relief devices require?**

Crankcase relief devices are normally examined by the surveyors carrying out machinery surveys. It is also the duty and responsibility of engineer officers to see that they work freely at all times; as they are such simple, trouble-free devices there is a risk that they may go neglected during the period between surveys. If gauzes are fitted they should be examined at the time of a crankcase inspection. The openings in the mesh should be inspected to see that they are clear.

Spring-loaded valves of the type fitted on medium-speed engines should have the valve checked to ascertain that it is free to move on its spindle. This check also ensures that the sealing ring on the valve spindle and the sealing ring on the seat are not causing the valve to stick. The hinged-type of valve used on large engines often becomes stuck following repainting of the engine. Paint enters between the faces of the parts holding the hinge pin and eventually dries out. Paint between the coils of the spring will also reduce its efficiency. Care must be exercised during painting, and any excess paint which could interfere with the working of the door must be cleaned off. Hinged-type doors require periodic lubrication of the hinge pin to prevent rusting and seizing. Most hinge doors are fitted with a handle so that the valve or door can be tested by lifting it and observing that it quickly returns to its seat. If any flame guards or deflectors are removed during examination they must be refitted before the engine is started. After carrying out an examination of the crankcase-relief devices an appropriate entry covering the examination and the conclusions should be entered in the engine-room log-book or port log.

■ **19.19 Describe briefly a device which may be used to detect dangerous conditions within a crankcase. Is the same device used on the main engine when the engine room is periodically unmanned?**

Dangerous conditions come about in a crankcase due to some part over-heating and the creation of oil-vapour mist. We are all aware that mist affects visibility and therefore interferes with the free passage of light. This property of a mist is utilized to detect its presence in an engine crankcase. If mist is detected at its onset dangerous crankcase conditions can be avoided. The devices are referred to as crankcase mist detectors. They consist of a suction fan and rotary valve. In the connecting tube between the suction fan and the rotary valve a photo-electric cell and light source are fitted. Piping is led from the upper part of the crankcase under each cylinder unit to the mist detector. The pipe from each crankcase section is led separately to the rotary valve. When the fan and rotary valve are in operation each section of the crankcase is connected in turn with the fan by the action of the rotary valve. The air from each section of the crankcase is then monitored in turn. If mist exists in any part of the crankcase the fan will suck it through the tube where the photo-electric cell and light source are fitted. The mist will restrict the passage of light which is detected by the photo-electric cell.

There are various ways of arranging the electronic circuits in mist-detection equipment, but in all any reduction in the amount of light reaching the photo-electric cell activates .both visual and audible alarms. Mist detectors are very sensitive and able to detect mist levels in the very-weak mixture range. Time is then available to slow down or stop the engine before the mixture comes within the explosive range.

Mist detectors are fitted on many engines. When a ship is classified for periodic unmanned working in the engine room, the regulations of the governing authorities in some countries and the rules of some of the classification societies are such that a mist detector or similar suitable equipment must be fitted on the main engine.

■ **19.20   Why are bursting devices fitted on starting-air valves or adjacent to the valve on the starting-air manifold?**

If a starting-air valve sticks open during an engine start the starting-air line becomes subject to the maximum pressure in the cylinder which, if the cylinder fires, will be the combustion pressure. Should the inside of the starting-air line be moist with oil it will ignite and the starting-air lines right back to the automatic valve will be subjected to very high pressures. In order to prevent the starting-air line being subject to these high pressures some form of pressure-relieving device is fitted on the starting-air valves or on the branch connecting the starting-air valve to the starting-air manifold. The most commonly used safety device is the bursting cartridge. In external appearance the bursting cartridge looks like a top hat. The wall of the cartridge is machined to a thickness that will ensure that it fractures when the safe pressure is exceeded. The cartridges are usually made of steel which has been tested so that its tensile strength is known accurately. In order to protect the steel cartridges from corrosion they are often copper-plated. Another form of relief device is the lightning full-bore safety valve. This consists of a normal-type valve which is held in place by a piston fitted within a cylinder instead of a spring. Air pressure taken from the starting-air system holds the valve in place.

■ **19.21   What attention do the bursting cartridges fitted on the engine starting-air valves require? If a safety cartridge bursts how can the engine starting system be restored temporarily?**

The starting-air system bursting cartridges require periodic examination to ascertain that they have not corroded and become weakened. If a safety cartridge bursts and the loss of air prevents starting the engine, starting-air services can be quickly restored by fitting a loose piece of pipe over the holes in the cartridge cover. Some ships have pieces of pipe for this purpose placed in a rack on the top platform of the main engine. The emergency pieces should not be used over the cartridge cover until it is seen that the starting-air valve is freely working and closing itself properly.

■ **19.22   How would you conduct the trials on a main engine after major repair work on bearings? Assume that the engine is not fitted with a mist detector.**

In order to carry out a trial safely the vessel should be taken to a trials berth, or moored at a berth which is strong enough to take the thrust set up by the propeller. The trial should be conducted in such a manner that any temperature rise on any bearing or guide slipper is found before the bearing or slipper damages itself or its temperature rises to a dangerous level. The usual way to conduct the trial is to run the engine at its slowest possible speed for a period of up to three minutes, then stop the engine and feel all the bearings for any abnormal temperature rise.

The temperature of the exhaust from the individual cylinders should be noted because at very slow engine speeds some cylinders may not be getting fuel; this makes for bad power balance between the various cylinders. If all the bearings are cool the engine may be started and run at its slowest speed again but for double the previous time, say up to six minutes; it should then be stopped and the feeling of all bearings repeated. If any bearings show a slightly abnormal temperature rise the cause must be ascertained, and rectified, as it will be obvious that a slight temperature rise at these speeds will most likely be a large and dangerous temperature rise at higher speeds. After a six-minute run at the slowest speed, and finding all in order, the engine speed can be increased by ten revolutions per minute and the engine run for two-minute, four-minute and eight-minute periods. The bearings should be felt after each run. Any signs of a temperature rise in one bearing being more than in other similar bearings must be investigated and the cause rectified before the trial proceeds. After this series of runs, the speed may again be put up by ten revolutions per minute and the engine run for two-, four- and eight-minute periods, with stops between each run for feeling bearing temperatures. In this way any temperature rise on a bearing becomes known early and while it is still safe to open the crankcase doors to enter the engine for feeling bearings.

In the early stages of a trial there will sometimes be slight temperature differences among bearings that are similar; small differences may be caused by scraped surfaces bedding themselves in, and by slight differences in the power developed in each cylinder. The power differences are found by carefully noting the exhaust temperature at each cylinder as the trial proceeds.

The power of the engine is gradually increased up to the allowable limits of the berth where the ship is moored. The engine is then run at this power for increasing time periods up to one hour. Full-power trials will then be completed when the vessel gets to sea. Some safe operating speed should be arrived at before the ship leaves the trials berth, and this speed and power should not normally be exceeded during the river or channel passage to the sea.

During the dock trials after major repair work on the main engines the number of people allowed in the engine room during the trial should be kept to a minimum. It often transpires that work on the rest of the ship comes to a halt and the engine room fills with spectators during the trial.

For working the main engines it is usual for the engine room to notify the bridge of the intended movement (such as slow ahead) by ringing the telegraph. When the Master is ready an answer is made on the telegraph which then signifies that the intended movement may be carried out. If it is not safe to carry out the movement the telegraph is rung to stop.

Before running trials, the Master and chief engineer should confer to ensure that there is no possibility of misunderstanding in communication between engine room and bridge.

Note The turning gear *must* be engaged each time the engine is stopped. The crankcase may then be entered for examination and feeling round bearings without fear of injury to the person carrying out the examination.

■ **19.23   What attention must be given to personal safety when pressure-testing fuel valves or fuel injectors?**

The hands and face must be kept well clear of the tip nozzle and piping connections between the test pump and the injector. Fuel under a high pressure and discharging from a leak in the piping or its connections, and also fuel discharged from the nozzle when the needle lifts, leaves with a very high velocity and a large amount of kinetic energy. The fuel has a velocity high enough to puncture and penetrate the eyes or surface of the skin. If penetration occurs it can lead to blindness or to fuel entering the bloodstream and causing poisoning which may be fatal. It is considered a good and safe practice to wear goggles or some form of protective eye-wear.

■ **19.24   Should contact lenses be worn when carrying out welding operations? What dangers are contact lens wearers exposed to when welding?**

On no account should wearers of contact lenses carry out welding operations without removing the lenses and replacing them with corrected safety glasses.

The danger that contact lens wearers are exposed to is one of blindness. This comes about when the wearer is exposed to flashes from checking electrode tongs, moving welding cables before being grounded or earthed, pulling open switches and smiliar activities where an electrical flash can occur.

The contact lenses act in a similar manner to when a magnifiying glass is used to converge the suns rays and cause burning or scorching on the surface exposed to the concentrated rays. In the case of the contact lens wearer, the contact lens in contact with the cornea of the eye converges the light and heat rays from an electrical flash onto the surface of the cornea and causes it to fuse with the lens surface. This results in blindness for the wearer and its tragic consequences.

A welder in the course of his work in a British shipyard has unfortunately suffered blindness from this cause. The opthalmic surgeon attending was not able to save his sight.

INDEX

INDEX